Applied Phycology

Applied Phycology

Edited by

WAI YAN CHEAH

Centre of Research in Development, Social and Environment (SEEDS), Faculty of Social Sciences and Humanities, Universiti Kebangsaan Malaysia, Bangi, Selangor, Malaysia

PAU LOKE SHOW

Center of Excellence for Research, Value Innovation and Entrepreneurship (CERVIE), UCSI University, Cheras, Kuala Lumpur, Malaysia

KIT WAYNE CHEW

Chemical and Petroleum Engineering Department, United Arab Emirates University, Al Ain, Abu Dhabi, United Arab Emirates

KUAN SHIONG KHOO

Algae Bioseparation Research Laboratory, Department of Chemical Engineering and Materials Science, Yuan Ze University, Taoyuan, Taiwan

ELSEVIER

Elsevier
Radarweg 29, PO Box 211, 1000 AE Amsterdam, Netherlands
125 London Wall, London EC2Y 5AS, United Kingdom
50 Hampshire Street, 5th Floor, Cambridge, MA 02139, United States

ISBN: 978-0-443-24758-3

For Information on all Elsevier publications
visit our website at https://www.elsevier.com/books-and-journals

Publisher: Peter Llewellyn
Acquisitions Editor: Pamela Sadhukhan
Editorial Project Manager: Arushi Rawat
Production Project Manager: Gomathi Sugumar
Cover Designer: Arumugam Kothandan

Typeset by MPS Limited, Chennai, India

Contents

List of contributors

Noor Faradilla Abdullah
Department of Diagnostic and Allied Health Science, Faculty of Health and Life Sciences, Management and Science University, Shah Alam, Selangor, Malaysia

Mohd. Azwan Ahmad
MSU Centre for Climate Resilience and Strategy (m-CREST), Management and Science University, Shah Alam, Selangor, Malaysia; Department of Diagnostic and Allied Health Science, Faculty of Health and Life Sciences, Management and Science University, Shah Alam, Selangor, Malaysia

Nazahatul Anis Amaludin
Faculty of Earth Science, Universiti Malaysia Kelantan, Kota Bharu, Kelantan, Malaysia

Khairul Anam
Research Center for Applied Microbiology, National Research and Innovation Agency (BRIN), Cibinong, Indonesia

Mia Yang Ang
Department of Biomedical Sciences, Jeffrey Cheah Sunway Medical School, Faculty of Medical and Life Sciences, Sunway University, Petaling Jaya, Selangor, Malaysia; Sunway Microbiome Centre, Faculty of Medical and Life Sciences, Sunway University, Petaling Jaya, Selangor, Malaysia

Pablo Bolaños-Villegas
School of Agronomy, Faculty of Food and Agricultural Sciences, University of Costa Rica, San Jose, Costa Rica; Fabio Baudrit Agricultural Research Station, University of Costa Rica, Alajuela, Costa Rica; Lankester Botanical Gardens, University of Costa Rica, Cartago, Costa Rica

Zhi Ling Chew
School of Energy and Chemical Engineering, Xiamen University Malaysia, Sepang, Selangor, Malaysia

Devi Bentia Effendi
Research Center for Applied Microbiology, National Research and Innovation Agency (BRIN), Cibinong, West Java, Indonesia

Nur Azalina Suzianti Feisal
MSU Centre for Climate Resilience and Strategy (m-CREST), Management and Science University, Shah Alam, Selangor, Malaysia; Department of Diagnostic and Allied Health Science, Faculty of Health and Life Sciences, Management and Science University, Shah Alam, Selangor, Malaysia

Nadana Raja Vadivu Ganapathy
Department of Biotechnology and Chemical Engineering, Manipal University, Jaipur, Rajasthan, India

Rahul Goswami
Department of Mechanical and Electromechanical Engineering, National Ilan University, Yilan City, Taiwan; Department of Botany and Microbiology, Hemvati Nandan Bahuguna Garhwal University, Srinagar, Uttarakhand, India

Natanamurugaraj Govindan
Department of Botany and Microbiology, Hemvati Nandan Bahuguna Garhwal University, Srinagar, Uttarakhand, India

Nishkarsh Guglani
Diatom Research Laboratory, Amity Institute of Biotechnology, Amity University, Noida, Uttar Pradesh, India

Noor Haziqah Kamaludin
Center for Environmental Health and Safety Study, Faculty of Health Sciences, Universiti Teknologi MARA, Puncak Alam, Selangor, Malaysia; Occupational Health and Safety Risk Management (OHSeRM) Research Initiative Group, Faculty of Health Sciences, Universiti Teknologi MARA, Puncak Alam, Selangor, Malaysia

Muhamad Syahril Khamis
Department of Environmental Health, Faculty of Health Sciences, Universiti Teknologi MARA Cawangan Pulau Pinang, Bertam, Pulau Pinang, Malaysia

Qi Qi Koh
School of Energy and Chemical Engineering, Xiamen University Malaysia, Sepang, Selangor, Malaysia; Centre for Net-Zero Technology (Future Centre), Monash University Malaysia, Subang Jaya, Selangor, Malaysia

Sankaran Krishnamoorthy
Department of Biotechnology, Vignan's Foundation for Science, Technology and Research, Guntur, Andhra Pradesh, India

Vasumathi Krishnan
Department of Mechanical and Electromechanical Engineering, National Ilan University, Yilan City, Taiwan; Department of Biotechnology, Kalasalingam Academy of Research and Education, Virudhunagar, Tamil Nadu, India

Yin Leng Kua
School of Energy and Chemical Engineering, Xiamen University Malaysia, Sepang, Selangor, Malaysia

Fong Fong Liew
Preclinical Sciences Department, Faculty of Dentistry, MAHSA University, Bandar Saujana Putra, Jenjarom, Selangor, Malaysia; Department of Pharmaceutical Life Sciences, Faculty of Pharmacy, Universiti Malaya, Kuala Lumpur, Malaysia

Siti Nurshahida Nazli
Department of Environmental Health, Faculty of Health Sciences, Universiti Teknologi MARA Cawangan Pulau Pinang, Bertam, Pulau Pinang, Malaysia; Child Health Research Centre, The University of Queensland, South Brisbane, QLD, Australia

Der Jiun Ooi
Preclinical Sciences Department, Faculty of Dentistry, MAHSA University, Bandar Saujana Putra, Jenjarom, Selangor, Malaysia

Joko Prayitno
Research Center for Applied Microbiology, National Research and Innovation Agency (BRIN), Cibinong, Indonesia

Laila Dina Amalia Purba
Research Center for Applied Microbiology, National Research and Innovation Agency (BRIN), Cibinong, West Java, Indonesia

K. Rinku
Diatom Research Laboratory, Amity Institute of Biotechnology, Amity University, Noida, Uttar Pradesh, India

Arnaud Ronceret
Instituto de Biotecnología, National Autonomous University of Mexico, Cuernavaca, Mexico

Nurhidayah Sabri
Department of Environmental Health, Faculty of Health Sciences, Universiti Teknologi MARA Cawangan Pulau Pinang, Bertam, Pulau Pinang, Malaysia

Abhishek Saxena
Diatom Research Laboratory, Amity Institute of Biotechnology, Amity University, Noida, Uttar Pradesh, India

Maulin P. Shah
Department of Research Impact and Outcome, Research & Development Cell, Lovely Professional University, Phagwara, Punjab, India

Aida Soraya Shamsuddin
Southeast Asia Disaster Prevention Research Initiative, Institute for Environment and Development (LESTARI), Universiti Kebangsaan Malaysia (UKM), Bangi, Selangor, Malaysia

Pei Fang Sim
Preclinical Sciences Department, Faculty of Dentistry, MAHSA University, Bandar Saujana Putra, Jenjarom, Selangor, Malaysia

Veena Singh
Shri Rawatpura Sarkar College of Pharmacy, Shri Rawatpura Sarkar University, Raipur, Chhattisgarh, India

Vijay Singh
Department of Pharmacy, Shri Rawatpura Sarkar University, Raipur, Chhattisgarh, India

Zoe Tan
Department of Diagnostic and Allied Health Science, Faculty of Health and Life Sciences, Management and Science University, Shah Alam, Selangor, Malaysia

Dhananjay Tandon
Department of Biotechnology, Anjaneya University, Raipur, Chhattisgarh, India

Tengku Nilam Baizura Tengku Ibrahim
Department of Environmental Health, Faculty of Health Sciences, Universiti Teknologi MARA Cawangan Pulau Pinang, Bertam, Pulau Pinang, Malaysia; Occupational Health

studies the downstream processing of bioactive compounds such as proteins, lipids, and carotenoids from microalgae using various extraction technologies (e.g., liquid biphasic system, assisted-liquid biphasic system, and ionic liquid technologies). His current research explores the use of food waste liquor to cultivate microalgae for production of high-value bioactive compounds.

CHAPTER 1

Preservation of microalgae

Abhishek Saxena, Nishkarsh Guglani, Aakash Verma, K. Rinku, and Archana Tiwari
Diatom Research Laboratory, Amity Institute of Biotechnology, Amity University, Noida, Uttar Pradesh, India

1.1 Introduction

Microalgae are microscopic, photosynthetic microorganisms that thrive in aquatic and terrestrial environments (Magnabosco et al., 2025). As primary producers, they contribute significantly to global net primary production and are essential for maintaining ecosystem balance. Their complex cellular structures and pigment systems have attracted substantial research and industrial interest due to their rich biochemical composition, including antioxidants, immunomodulatory compounds, and therapeutic properties.

The unique biological and biochemical properties of microalgae make them valuable in various applications (Sarıtaş et al., 2024). They serve as model organisms for studying photosynthesis, metabolic pathways, and cellular mechanisms, with photosynthetic efficiency surpassing that of terrestrial plants. Recent advances in genetic engineering and omics technologies have improved understanding of microalgal metabolism and gene regulation, enabling the development of strains with optimized photosynthetic capabilities and enhanced biomass productivity. High-value compounds like proteins, polyunsaturated fatty acids (PUFAs), pigments, vitamins, and antioxidants are used in functional foods, dietary supplements, and nutraceutical products. They also serve as sustainable bioresources in the energy sector, pharmaceuticals, and cosmetics due to their antiinflammatory, antioxidant, and antimicrobial activities. These compounds also improve the nutritional quality and health of livestock and aquatic species, promoting sustainable practices and reducing reliance on traditional feedstocks.

However, microalgae preservation is crucial for their long-term viability, genetic integrity, and functional characteristics, ensuring consistent scientific outcomes and sustainable biotechnological applications. It maintains adaptive potential, stabilizes metabolite production, and enables targeted strain improvement through genetic engineering. Preservation techniques ensure reproducibility in scientific research and validate

Applied Phycology
DOI: https://doi.org/10.1016/B978-0-443-24758-3.00004-6

experimental findings. In industrial contexts, large-scale cultivation of microalgae is driven by the need for high-quality, pure, and genetically stable cultures, optimizing biomass productivity and reducing contamination risks (Arguelles et al., 2020). Preservation is also essential for conserving microalgal biodiversity amidst environmental challenges like climate change, habitat degradation, and population declines. Reliable ex situ preservation methods complement in situ conservation by providing repositories for restoration, strain improvement, and bioprospecting, promoting sustainable management of microalgal resources and the circular bioeconomy (Esquivel et al., 1993). This review aims to explore microalgal culture preservation, focusing on its importance for research and industrial purposes, and discussing the biological and technical challenges of preserving diverse species.

1.2 Current laboratory practices for preservation

With developments in microalgae research, its diverse applications across industries have become a major point of attention. As it is used in food, feed, cosmetics, and health products, it has proven to be highly adaptable and biocompatible. This calls for extensive research on various species and strains of microalgae. As microalgae are susceptible to shifts in their environmental conditions, it becomes highly important to ensure the stability of their characteristics and properties (Arguelles et al., 2020). The preservation of an organism's genetic, morphological, and physiological characteristics should be guaranteed by the conservation methodology adopted. Serial subculture is a commonly used preservation technique that requires higher and regular maintenance compared to an alternative such as cryopreservation. Cell immobilization has also emerged as an alternative due to its efficiency in long-term preservation of biochemical characteristics.

1.2.1 Subculturing

Subculturing is a traditional method of microalgal preservation that is often used in laboratories. It is done by simply transferring inoculum from an existing culture to a fresh culture medium that has been sterilized beforehand. This transfer takes place in a sterile environment in order to prevent contamination. It is a continuous process that requires regular maintenance. It is a simple process, hence highly accessible. It is advantageous due to its low requirement for technical skills and the cost of entry. This technique is often used for the preservation of cyanobacteria and eukaryotic microalgae. It is a labour-intensive method that limits the potential of maintaining a larger number of strains (Arguelles et al., 2020). Culture collections like that of the Institute of Plant Physiology of the Russian Academy of Sciences use periodic subculturing for the maintenance of their strains, utilizing both liquid and agar growth media. Storage of plates at a lower temperature increases the viability of each subculture for around seven months.

This technique is highly susceptible to microbial contamination and genetic drifts. Hence, it is used in scenarios where live cells are required for the conduction of a series of experiments in a shorter period, and it is seen as a suboptimal method for long-term preservation. It can potentially lead to the selection of the population that does not properly represent the parent population. Highly impactful changes in morphology and physiology can be observed in extreme cases, like shifts in the pigment concentration in various microalgal species or an observed reduction in diatom cell size.

1.2.2 Cryopreservation

Cryopreservation refers to the storage of living cells at ultralow temperatures below −80°C and around the temperature of liquid nitrogen, which is −196°C. Various cell types respond differently; hence, great diversity in cryobiological responses during the freezing and thawing cycles has been observed. Cryopreservation allows living algae that do not have any normal resting stage to be maintained indefinitely in an arrested state.

1.2.2.1 Components of cryopreservation

Cryoprotective agents (CPAs) refer to chemical substances that help to protect cells during the freezing and thawing processes. These agents have low toxicity and are water-soluble.

Penetrating cryoprotectants, or CPAs, are heavily utilized for microalgal cryopreservation. They lower the temperature of intracellular freezing, which reduces the decrease in cell volume as this phenomenon occurs due to osmotic changes. Cryoprotection is further enhanced by altering membranous properties, for example, the permeability of a variety of solutes through the membrane. They may even decrease membrane potential and ionic permeability. These also act as free radical scavengers.

Commonly used penetrating CPAs include methanol (MeOH), dimethylsulfoxide (DMSO), and glycerol. Culture Collection of Algae and Protozoa (CCAP) and University of Texas Culture Collection of Algae

(UTEX) have preserved a variety of freshwater and terrestrial strains of microalgae. Studies have shown an improved response to MeOH and DMSO. MeOH has emerged as the more preferred CPA. DMSO is used for various marine phytoplankton, and for Tetraselmis, glycerol is used as the major CPA. Other CPAs, such as serum albumin and ethylene glycol, have also been effective (Arguelles et al., 2020). Nonpenetrating cryoprotectants are generally less toxic compared to penetrating CPAs, but they are not widely used for microalgal preservation. Sucrose, trehalose, and exopolysaccharides are some examples. They primarily act externally and are sometimes used in combination with penetrating CPAs.

1.2.2.2 Freezing protocol

A two-step protocol has been developed in order to properly regulate the cooling process. The first step cools the sample to an intermediate temperature (around −30°C) starting from room temperature. Due to this, the cryo-dehydration of the cells starts before it is plunged into liquid nitrogen (−196°C). The sample is now stored in a liquid nitrogen storage system that utilizes either liquid- or vapour-phase nitrogen. To maintain cryopreserved cultures, they need to be stored at ultralow temperatures, which optimally refers to −196°C, the temperature of liquid nitrogen, but there are examples of organisms that can be cryopreserved and stored at a higher range of subzero temperatures (>−70°C) (Arguelles et al., 2020).

1.2.2.3 Thawing process

This step refers to the warming of cryopreserved samples to bring them back to a functional state (*https://pantheracryo.com/what-does-thawing-look-like-in-cryopreservation/*). It can significantly impact cell survival and is done in a manner that prevents cell damage, even referred to as 'freeze slowly and thaw quickly' (*https://www.biocompare.com/Editorial-Articles/597531-Best-Practices-to-Reduce-Cell-Loss-in-Cryopreservation/*). A water bath is the most commonly used method for thawing, which involves maintaining the temperature of the water bath at around 37°C. This process aims for quick thawing with a water bath lasting around 90–120 seconds. Water ensures even heating around and across the sample, yet a water bath can be a source of contamination and needs to be monitored (*https://pantheracryo.com/what-does-thawing-look-like-in-cryopreservation/*). In dry thawing, beads made of aluminium are used to give bead baths at 37°C, similar to water baths, but they use less electricity and do not require refilling. Yet, they may transfer heat less efficiently. Dry automated thawers are devices dedicated to thawing and work in the range of 37°C–50°C, offering a highly sterile environment (*https://pantheracryo.com/what-does-thawing-look-like-in-cryopreservation/*).

1.2.2.4 Cryoinjury

'Cryoinjury' refers to the damage to cells caused by the phase changes that occur during cryopreservation. This includes phase changes of water that occur in extracellular and intracellular environments. This phenomenon is not yet fully established. Velocities during cooling and thawing can greatly impact the physicochemical, biophysical, and cellular processes. Osmotic rupture due to extracellular/intracellular concentrated solutes and intracellular ice formation can be considered an example of cryoinjury. The presence of an intact plasma membrane can be a great indicator of viability.

Determination of cellular viability is an important part of cryopreservation and should be ensured by detecting the metabolic and biochemical activities of the cells. Growth and reproductive studies should be performed in order to further determine the state of cells. A viability level of roughly 30% is taken as the minimum standard postthawing. Significant differences in genotype or phenotype should not be observed compared to the control or parent species.

1.2.3 Cell immobilization

The technique of cell immobilization refers to the encapsulation of a living cell using a capsule, some kind of polymer, or utilizing a support matrix (organic or inorganic) that the cell will bind to (El-Enshasy et al., 1996). These cells remain intact and are metabolically active. These encapsulated cells are implemented in industries as biocatalysts. Immobilized microbial cells have shown higher metabolic activity and resistance to toxic agents. They can maintain microbial populations in a more efficient manner. Microalgal preservation utilizing immobilization strategies has emerged as an alternative that offers greater stability of morphological and biochemical characteristics for a longer period of time.

1.2.3.1 Immobilization techniques (passive versus active)

In order to capture and immobilize algal cells, their natural tendency to attach to a surface is utilized. Various carriers (natural and synthetic) have been developed for this purpose. Loofa sponge and cotton cloth were used for a study on passive immobilization with two microalgal species surviving for up to 18 months on cotton cloth (Sarkarat et al., 2020). Active immobilization is widely used to

Table 1.2 Overview of use of lyophilization preservation for several different algal strains and their limitations.

Method name	Algal strain used	Method used	Limitations	References
Lyophilization	*Chlorella vulgaris*	Frozen at −80°C, protected with 5% trehalose and skim milk; vacuum-dried at 0.005 mbar	Moderate viability loss; oxidative damage if unprotected	Unterlander et al. (2017)
	Nostoc muscorum	Slow freezing with 10% glycerol, followed by gradual drying under vacuum	Requires precise cooling rate; glycerol toxicity possible	McGrath et al. (1978)
	Tetraselmis suecica	Cryoprotected with trehalose and PVP; frozen at −70°C; dried over 36 h	Freezing rate affects cell integrity; protectant concentration crucial	Ashok Kumar et al. (2023)
	Botryococcus braunii	Encapsulated in calcium alginate; frozen; and lyophilized; stored for one year	Lower hydrocarbon recovery; matrix-dependent survival	Morais Junior et al. (2020)
	Dunaliella salina	The freeze-drying process was carried out under -50°C and 1.63 hectopascals using the freeze dryer of type Heto Power Dry PL 3000	Osmotic stress: recovery varies with strain	Emeish (2012)
	Spirulina platensis	Pretreated with mannitol and ascorbic acid, frozen at −60°C, freeze-dried	Loss of phycobiliproteins; structural collapse without protectants	Takano et al. (1973)

architecture at ultralow moisture levels and controlled thermal conditions ensures that the algal cells can be stored for extended periods without undergoing substantial metabolic shifts or structural compromise. As a result, lyophilization holds immense potential for enhancing the reliability and scalability of algal preservation practices in both research and industrial settings (Bhambere et al., 2015).

The lyophilization process is characterized by a sequence of three distinct yet interdependent stages—namely, freezing, primary drying, and secondary drying—each of which plays a critical role in ensuring the effectiveness and stability of the preservation outcome. The steps of lyophilization are explained as follows:

1. The first stage, known as the freezing phase, involves the rapid cooling of the algal biomass to extremely low temperatures, typically ranging from −40°C to −80°C. This rapid thermal descent induces the crystallization of both intracellular and extracellular water, thereby converting it into ice. The goal during this phase is to solidify the aqueous components without causing excessive damage to cellular structures, which can occur due to ice crystal formation if the cooling rate is not properly optimized.
2. Following freezing, the process enters the primary drying phase, which constitutes the most significant portion of water removal. During this stage, the frozen algal sample is exposed to a high-vacuum environment. Under these reduced pressure conditions, the solid ice sublimates directly into water vapour, thereby bypassing the intermediate liquid phase entirely. This sublimation process not only facilitates efficient dehydration but also preserves the overall structural morphology of the cells. It is estimated that this primary drying stage is responsible for eliminating more than 90% of the total water content present in the sample.
3. The final step, referred to as secondary drying, is designed to eliminate residual moisture that remains tightly bound to cellular components. During this phase, the temperature is carefully and gradually increase-d—usually reaching a range of approximately 20°C–30°C. This controlled heating promotes the desorption of unfrozen, bound water molecules, further reducing the sample's overall moisture content. The end product is a dry, porous matrix that retains the essential physical architecture of the algal cells while lowering the water content to below 2%. This ultralow moisture level is crucial, as it allows the preserved material to be stored safely under ambient conditions for extended durations without significant degradation.

1.3.3 Algal variation—a challenge for preservation

Despite the considerable potential of lyophilization as a preservation technique, its application to microalgal species presents several significant technical and biological challenges. Unlike the relatively simpler cellular structures of many bacterial or yeast cells, microalgal cells often possess complex and diverse cell wall compositions. These intricate cell walls may be composed of an array of biomolecules, including cellulose, pectins, glycoproteins, and, in certain species, even silica-based elements. The composition and architecture of these walls are highly species-specific and contribute to the structural rigidity and functional resilience of the cells. However, they also render algal cells particularly vulnerable to the mechanical and osmotic stresses that arise during the freezing and dehydration phases of the lyophilization process (Fuertes-Rabanal et al., 2024).

During freezing, the formation of ice crystals within and around the cells can cause significant physical disruption, especially in cell walls that are less flexible or highly mineralized. Similarly, the rapid removal of water during dehydration can create osmotic imbalances that further compromise membrane integrity and cellular viability. These stressors collectively reduce the survival rate of algal cells postpreservation and can impair their physiological functionality upon rehydration (Meryman 1971).

To counteract these deleterious effects, the inclusion of protective agents has become a standard practice in lyophilization protocols for microalgae. These substances, broadly categorized as cryoprotectants and lyoprotectants, are introduced into the microalgal suspension before initiating the freeze-drying process. Cryoprotectants help minimize ice crystal formation and buffer the cells against freezing damage, while lyoprotectants stabilize membrane structures and intracellular macromolecules during the dehydration and storage phases. The strategic use of these additives is critical for enhancing cell survival, maintaining post-preservation viability, and ensuring the long-term utility of lyophilized algal cultures (McGrath et al., 1978).

1.3.3.1 Encapsulation-drying-vitrification (Encasulation-Drying-Vitrification) method

Encasulation-Drying-Vitrification method has emerged as a novel and highly effective preservation technique for microalgal cultures, offering an alternative to conventional cryopreservation by integrating the principles of controlled dehydration and glass-transition stabilization. Originally adapted from plant cryopreservation protocols, the Encasulation-Drying-Vitrification method has gained traction as a novel approach to stabilize delicate photosynthetic cells that are otherwise poorly suited for traditional freezing or drying. Its strength lies in a multistage process that allows cells to transition into a glass-like state—vitrification—while protected within a hydrated yet immobilized matrix. This approach minimizes structural damage, inhibits ice crystal formation, and helps retain cellular function post-thaw or rehydration.

The conceptual framework of Encasulation-Drying-Vitrification draws inspiration from the foundational principles established through cryobiological research. For instance, the two-step cooling approach initially proposed by Holm-Hansen in 1963, which emphasized that a gradual cooling rate enhances post-thaw viability in microalgae, shares conceptual parallels with the preconditioning and staged dehydration phases in Encasulation-Drying-Vitrification. It has been established that holding microalgal culture vials at an intermediate subzero temperature of −30°C for 15 minutes before immersion in liquid nitrogen significantly improved survival outcomes. In that particular study, 252 out of 284 strains exhibited recovery rates above 50% within 24 hours, and notably, eight strains maintained post-preservation viability exceeding 70% after one year.

The Encasulation-Drying-Vitrification process similarly emphasises the importance of controlled pretreatment steps, such as temperature acclimation and osmotic balancing, before the vitrification phase, which parallels the use of a −80°C hold for up to 10 hours in modern cryopreservation protocols to facilitate cold acclimation and minimize cryoinjury. Furthermore, Encasulation-Drying-Vitrification protocols also recognize the importance of biological factors, such as the physiological age of cultures. As shown in Morris's experiment, of the 17 *Chlorococcales* strains tested, 11 achieved high post-thaw viability when harvested during the stationary phase—underscoring the influence of growth phase on preservation success. A schematic diagram illustrating the Encasulation-Drying-Vitrification method for microalgae preservation, showing algal encapsulation in alginate beads, dehydration, air drying, and cryogenic storage in liquid nitrogen is presented in Fig. 1.1.

It needs to be established that Encasulation-Drying-Vitrification method relies heavily on the use of hydrogel-based encapsulation, typically employing calcium alginate, to form semipermeable beads or droplets that physically entrap the algal cells. This initial step not only localizes the cells but also acts as a mechanical buffer, shielding them from abrupt osmotic

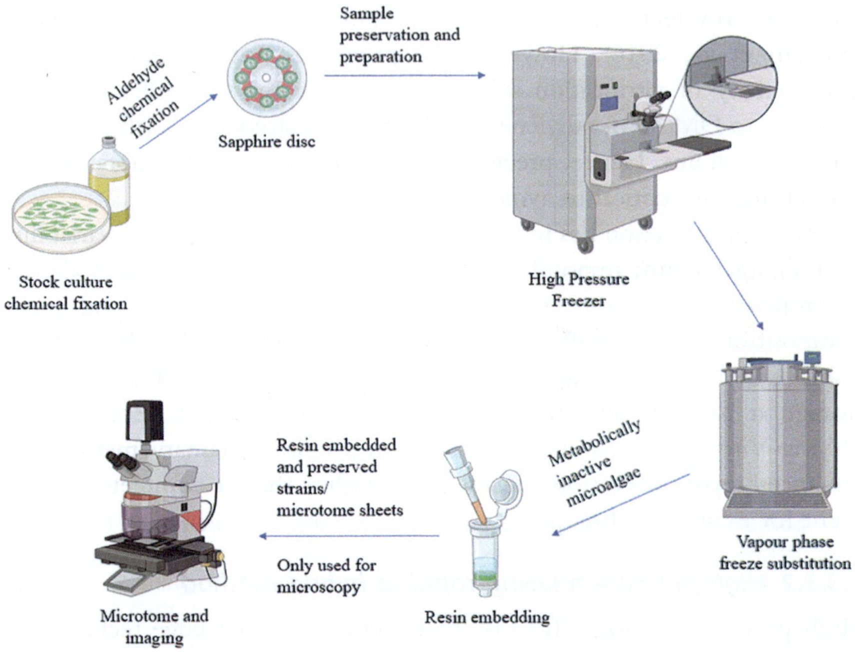

Figure 1.2 ***Adapted high-pressure freezing* (HPF) *method for microalgae preservation.*** The figure illustrates the adaptation of the HPF technique, conventionally applied for transmission electron microscopy, to improve the preservation of microalgae through vapor-phase freezing and resin embedding. The workflow is based on established protocols for cryo-fixation and resin embedding of biological samples used in electron microscopy and chemical imaging (Sachse et al., 2024; Gallet et al., 2024). M. Sachse, S.Y. Fernández-Sánchez, R. Tenorio, I.F. de Castro, C. Risco, Imaging Bunyavirus infections by transmission electron microscopy: Conventional sample preparation vs high-pressure freezing and freeze substitution. In: Methods in Molecular Biology, 2024, 241-258 https://doi.org/10.1007/978-1-0716-3926 9_16.

membrane-permeating cryoprotectants, such as DMSO. This makes the technique particularly appealing for the preservation of sensitive or fastidious microalgal strains that are easily damaged by chemical agents.

The effectiveness of HPF in preserving cellular fine structure has been well demonstrated in model algal organisms. For instance, in *Chlamydomonas reinhardtii*, a widely studied green alga, HPF has been successfully employed to preserve critical ultrastructural elements such as chloroplasts, thylakoid membranes, and pyrenoids. Comparative imaging studies using electron micrographs and cryo-electron tomography revealed that HPF preserves internal chloroplast architecture with far greater fidelity than either plunge freezing or chemical fixation techniques. In applied contexts, HPF has also shown merit in preserving structural integrity in biotechnologically relevant strains such as *Nannochloropsis oceanica*, which is known for its lipid-rich cytoplasm. HPF-treated samples retained the morphology and integrity of intracellular lipid bodies—an essential feature for downstream applications in biofuel production and nutraceutical extraction. A diagram illustrating a modified HPF workflow adapted for the preservation of algal samples is presented in Fig. 1.3.

Despite its remarkable potential, HPF remains underutilized in the routine long-term preservation of microalgae, primarily due to several

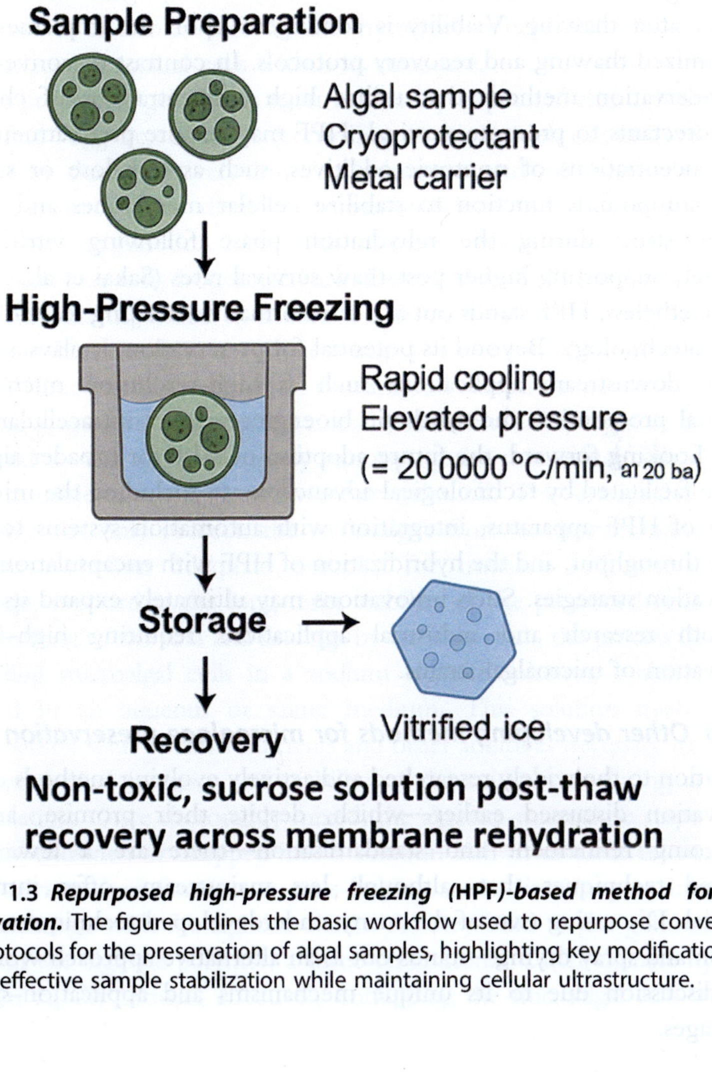

Figure 1.3 ***Repurposed high-pressure freezing*** **(HPF)*-based method for algal preservation.*** The figure outlines the basic workflow used to repurpose conventional HPF protocols for the preservation of algal samples, highlighting key modifications that enable effective sample stabilization while maintaining cellular ultrastructure.

Table 1.3 The use and employment of the ionic gelation preservation method in various studies and their respective purposes/applications.

Microalgal strain	Ionic gelation matrix	Purpose/application	Key outcomes	References
Spirulina sp.	Ca(II)-alginate	Protein delivery/ nutraceutical encapsulation	Uniform bead formation suitable for food use	Rajmohan and Bellmer (2019)
Chlorella spp. and *Trichosporonoides spathulata*	Alginate beads (coculture)	Lipid production enhancement.	Mutualistic interaction; increased lipid yield	Kitcha and Cheirsilp (2014)
Synechococcus sp. PCC 7002	Alginate encapsulation	Coculture with *Chlamydomonas* for acetate production	Enabled acetate-free growth	Therien et al. (2014)
Dunaliella salina	Alginate/SiO_2 core-shell with PDADMAC	Photobioreactor deployment	Nutrient selectivity, predator protection	Desmet et al. (2014)
Euglena gracilis	Ca(II)-alginate	Cryopreservation (12 months)	40% postthaw viability	Day et al. (2000)
Chlamydomonas reinhardtii	Alginate-PLL/Alginate-PDADMAC/ hybrid gels	Biosensors, bioreactors, microdevices	High encapsulation compatibility	Perullini et al. (2014)

semisolid gel structure. This process results in instant gelation, providing both structural stability and a protective microenvironment for the encapsulated cells.

Ionic gelation presents numerous benefits when applied to the encapsulation of microalgae. Its key advantages include operation under mild and nondenaturing conditions, excellent biocompatibility, and the capacity to entrap living, sensitive biological material without compromising cell viability. The hydrogel matrix formed offers not only protection from physical and chemical stress but also facilitates the retention of metabolic activity by allowing selective permeability for nutrients and gases. Encapsulated microalgae produced via this technique can be employed across a wide spectrum of industrial and research applications. These include bioremediation of contaminated environments, development of pharmaceutical agents, incorporation into cosmetic formulations, and use in nutraceuticals or dietary supplements. Furthermore, the matrix's inherent ability to allow sustained and controlled release of entrapped materials enhances its potential.

1.3.5 Spray drying: a rapid and scalable approach

Spray drying is a well-established, industrial-scale dehydration technique that has recently gained attention for its applicability in the preservation of microalgal biomass (Table 1.4). Initially developed for use in the food, pharmaceutical, and probiotic industries, spray drying is now being recognized as a cost-efficient and scalable alternative to lyophilization, particularly in scenarios where the preservation of long-term cell viability is not a primary requirement. Instead, the method is favoured for its ability to maintain the biochemical stability of key microalgal constituents, such as pigments, proteins, and lipids.

The spray drying process involves the atomization of a microalgal suspension into fine droplets, which are then rapidly introduced into a stream of hot air or gas—typically at inlet temperatures ranging from 150°C to 200°C. Under these conditions, water evaporates almost instantaneously from the droplets, resulting in the formation of dry, free-flowing powder particles. These particles retain a substantial portion of their bioactive compounds, owing to the brief exposure time (usually in the range of milliseconds to a few seconds), which minimizes thermal degradation.

To further enhance the thermal stability and protect sensitive cellular components during drying, various excipients and protectants can be

Table 1.4 Summary of recent applications of spray drying in microalgal preservation and functional product development.

Microalgal strain	Wall materials/protectants	Application	Key outcome	References
Dunaliella salina, Porphyridium cruentum, Chlorella vulgaris	Maltodextrin	Sustainable food and feed	Improved stability; enhanced nutritional profile	Zhang et al. (2023)
Nannochloropsis oculata	Sugar, hydrogenated palm kernel oil, skimmed milk powder, whey, lecithin, PGPR, vanillin	Functional white chocolate spread	Improved pigment retention; formulation stability	Acién Fernández et al. (2018)
Spirulina maxima	Octenyl succinic anhydride-modified starch	Vegan biscuit enrichment	Improved nutritional value with no sensory loss	Paula da Silva et al. (2021)
Spirulina sp. LEB-18	Maltodextrin and soy lecithin	Functional chocolate milk	High encapsulation efficiency; functional enrichment	Batista de Oliveira et al. (2021)
Chlorella vulgaris	Deactivated baker's yeast	Colorant/protein additive in food	Optimal encapsulation yield and pigment retention	Tamtürk et al. (2023)
Spirulina extract	Chitosan and mannitol	Oral administration; functional foods	100% encapsulation efficiency; improved intestinal permeability	Aquino et al. (2020)
Chlorella vulgaris and *Arthrospira platensis*	Silk sericin	Wound healing	Enhanced tissue regeneration properties	Bari et al. (2017)
C-phycocyanin	Aqueous two-phase systems	Shelf-life extension	Double encapsulation achieved with purity enhancement	Chandralekha et al. (2021)

incorporated into the algal suspension prior to atomization. Commonly used additives include trehalose, mannitol, maltodextrin, and whey protein isolates, all of which function as stabilizers by forming a protective matrix around the cells and bioactives. The inclusion of such compounds significantly improves the retention of heat-sensitive molecules and supports the production of a dry powder that is not only stable over extended storage periods but also suitable for downstream formulation in food, feed, cosmetics, or pharmaceutical applications.

1.4 Challenges, limitations, and future perspective

Microalgal preservation is important for safeguarding the diversity of strains and maintaining their genetic stability while ensuring a continuous supply of viable cultures for ecological and biotechnological applications. However, challenges posed in designing a universal, effective preservation protocol include the immense diversity in taxonomy, structural complexity, and physiological variability of microalgae. Although there have been advancements in cryopreservation and alternative strategies for microalgal preservation, limitations like species-specific responses, loss of viability postpreservation, and scalability constraints persist, which highlights the need for further innovation and refinement of these strategies (Table 1.5).

1.4.1 Postpreservation vitality (thawing, re-cultivation, and viability assay [loss of viability])

For effective utilization of microalgal cultures in research and industry, ensuring their high post-preservation viability and functionality is critical. During preservation, microalgae encounter osmotic imbalance, ice crystal formation, and reactive oxygen species (ROS) generation, which can damage cellular membranes, organelles, and metabolic pathways, particularly under cryogenic conditions. These stressors frequently lead to delayed recovery of photosynthetic efficiency, reduced chlorophyll fluorescence, and prolonged lag phases upon re-cultivation, thus compromising the functionality of preserved cultures. Thawing, which is often performed rapidly in water baths to minimize ice recrystallization, must be carefully controlled to minimize osmotic shock during rehydration. Typically, post-thaw viability assessments utilize propidium iodide exclusion for membrane integrity, fluorescein diacetate staining to evaluate esterase activity, and chlorophyll fluorescence for photosystem II efficiency, along with regrowth monitoring under optimum culture conditions. These methods can, however, sometimes

Table 1.5 Key challenges in microalgae preservation and proposed solutions.

Challenge	Cause	Impact	Mitigation strategy	References
Posthaw viability loss	Intracellular ice formation and reactive oxygen species generation during freezing and thawing	Reduced membrane integrity, chlorophyll degradation	Optimization of cryoprotectant type and concentration	Foo et al. (2023)
Cryoprotectant cytotoxicity	High concentrations and prolonged exposure to DMSO or glycerol	Cellular toxicity, reduced viability	Use of lower concentrations, combination cryoprotectants	Kapoore et al. (2019)
Inefficient cryoprotectant penetration in diatoms	Presence of rigid silica frustules	Uneven cryoprotectant distribution, ice damage in cytoplasm	Extended equilibration times, use of penetrating and nonpenetrating cryoprotectant mixtures	Stock et al. (2018)
Lipid peroxidation in PUFA-rich strains	High unsaturated lipid content prone to oxidation	Membrane damage, loss of viability	Antioxidant supplementation during preservation	Molina Grima et al. (1994)
Species-specific preservation response	Variability in cell wall structure,	Inconsistent postpreservation viability across taxa	Strain-specific protocol development with optimized cooling/thawing rates	Day (2007)
Structural damage during thawing	Rapid osmotic changes, ice recrystallization	Cell lysis, loss of structural integrity	Controlled thawing rates, stepwise rehydration	Demirel et al. (2018)
Long-term genetic and phenotypic instability	Accumulated stress during storage	Loss of strain authenticity, functional impairment	Long-term viability assessments and stability monitoring	Foo et al. (2023)

overestimate or underestimate viability, as sublethal injuries may not immediately manifest but may affect subsequent growth and productivity. Cryoprotectants such as glycerol and DMSO are widely used to stabilize cellular structures and reduce intracellular ice formation, but their cytotoxicity at higher concentrations remains a challenge (Taylor and Fletcher, 1999). Strategies including pretreatment with antioxidants like glutathione and ascorbic acid have been explored to mitigate ROS-mediated damage and enhance postthaw viability. Despite these measures, many microalgal strains exhibit reduced viability post-preservation, which indicates the necessity for further protocol refinement and strain-specific optimization.

1.4.2 Diverse classes of microalgae (taxonomic diversity, species-specific response)

The taxonomic diversity of microalgae, which encompasses various groups such as chlorophyta, bacillariophyta, cyanobacteria, haptophyta, and dinoflagellates, plays a significant role in determining their responses to preservation strategies. Diatoms often face challenges in cryoprotectant permeability due to their rigid siliceous frustules and are prone to mechanical stress during freezing and thawing, whereas many green algae possess flexible cell walls that facilitate cryoprotectant penetration, compromising viability. Additional complexity in cyanobacteria occurs due to mucilaginous sheaths and gas vesicles, which influence dehydration and rehydration dynamics, whereas haptophytes and dinoflagellates can form resistant cysts under environmental stress. Species-specific optimization of preservation protocols is essential, as factors such as cryoprotectant concentrations, cooling rates, and thawing procedures yield varying outcomes among various strains. Acclimation to sublethal cold temperatures before preservation has been shown to improve cryotolerance in specific strains by inducing cold shock protein expression; however, such strategies require validation on a strain-by-strain basis.

1.4.3 Futuristic innovation in preservation techniques

While improving postthaw viability and enabling scalability, emerging innovations in microalgal preservation aim to overcome current limitations (Table 1.5). One such method, vitrification, involves transitioning cellular water into an amorphous glassy state without ice formation; it has demonstrated potential for preservation of sensitive strains, yet challenges related to osmotic stress and cryoprotectant toxicity persist. Another method, encapsulation-dehydration techniques, where cells are immobilized in alginate or silica matrices followed by controlled dehydration and storage, has demonstrated the ability to maintain the viability of strains sensitive to conventional freezing methods. Nanotechnology-assisted preservation is gaining interest, with magnetic and gold nanoparticles being investigated for their ability to facilitate uniform heat distribution and reduce ice nucleation during freezing, thereby enhancing cell survival rates. Additionally, there is growing interest in developing low-toxicity cryoprotectants, which include natural compounds such as trehalose, proline, and ectoine, that can stabilize cellular structures without inducing significant cytotoxicity. The integration of omics approaches, including genomics, transcriptomics, and metabolomics, has improved the understanding of cryotolerance mechanisms, enabling the identification of biomarkers and pathways associated with successful preservation. Furthermore, the use of artificial intelligence and machine learning to predict optimal preservation conditions based on strain-specific physiological and molecular data has the potential to streamline protocol development and reduce empirical testing.

1.5 Conclusion

Microalgae preservation is crucial for research, biotechnology applications, and commercial exploitation of microalgal resources. Traditional methods like continuous subculturing and cryopreservation are widely used, but advanced preservation approaches offer improved long-term viability and genetic stability. Challenges include species-specific responses, loss of viability post-thaw, and scalability. Cross-disciplinary innovations, such as adapting bacterial preservation strategies for microalgae, hold promise for overcoming these limitations. Research should focus on tailoring preservation protocols to specific taxa, developing cost-effective large-scale storage techniques, and ensuring postpreservation physiological integrity. With ongoing advancements, microalgae preservation is poised to support laboratory sustainability and large-scale biotechnological applications across sectors.

Acknowledgment

This work was supported by ASEAN-India Science, Technology, and Innovation Cooperation; AISTDF Secretariat; Department of Science and Technology; Government of India, India (Grant number CRD/2022/000595).

Declaration of competing interest

The authors declare no competing interests.

References

Acién Fernández, F.G., Fernández Sevilla, J.M., Molina Grima, E., 2018. Costs analysis of microalgae production. Biomass, Biofuels, Biochemicals: Biofuels from Algae, 2nd ed. Elsevier, pp. 551–566. Available at https://doi.org/10.1016/B978-0-444-64192-2.00021-4.

Aljabri, H., Cherif, M., Siddiqui, S.A., Bounnit, T., Saadaoui, I., 2023. Evidence of the drying technique's impact on the biomass quality of *Tetraselmis subcordiformis* (Chlorophyceae). Biotechnology for Biofuels and Bioproducts 16 (1), 1–11. Available at https://doi.org/10.1186/s13068-023-02335-x.

Almeida, A.C.A.de, Vaz, B.da S., Costa, J.A.V., Morais, M.G.de, 2024. Innovative technology for microalgal cell preservation through immobilization in polylactic acid nanofibers. Algal Research 84, 103781. Available at https://doi.org/10.1016/j.algal.2024.103781.

Aquino, R.P., Auriemma, G., Conte, G.M., Esposito, T., Sommella, E., Campiglia, P., et al., 2020. Development of chitosan/mannitol microparticles as delivery system for the oral administration of a *Spirulina* bioactive peptide extract. Molecules (Basel, Switzerland) 25 (9), 2086. Available at https://doi.org/10.3390/molecules25092086.

Arguelles, E.D., NH, T.G., Monsalud, R.G., 2020. Maintenance and preservation of microalgal cultures. Methods in Microalgal Studies 1, 53–60.

Ashok Kumar, N., Sridhar, S., Jayappriyan, K.R., Raja, R., 2023. Applications of microalgae in aquaculture feed. Handbook of Food and Feed from Microalgae: Production, Application, Regulation, and Sustainability. Elsevier, pp. 421–433. Available at https://doi.org/10.1016/B978-0-323-99196-4.00011-5.

Bari, E., Arciola, C.R., Vigani, B., Crivelli, B., Moro, P., Marrubini, G., et al., 2017. In vitro effectiveness of microspheres based on silk sericin and *Chlorella vulgaris* or *Arthrospira platensis* for wound healing applications. Materials 10 (9), 983. Available at https://doi.org/10.3390/ma10090983.

Batista de Oliveira, T.T., Miranda dos Reis, I., Bastos de Souza, M., da Silva Bispo, E., Fonseca Maciel, L., Druzian, J.I., et al., 2021. Microencapsulation of *Spirulina sp.* LEB-18 and its incorporation in chocolate milk: properties and functional potential. LWT 148, 111674. Available at https://doi.org/10.1016/j.lwt.2021.111674.

Chandralekha, A., Prashanth, H.S., Tavanandi, H., Raghavarao, K.S.M.S., 2021. A novel method for double encapsulation of C-phycocyanin using aqueous two phase systems for extension of shelf life. Journal of Food Science and Technology 58 (5), 1750–1763. Available at https://doi.org/10.1007/s13197-020-04684-z.

Day, J.G., 2007. Cryopreservation of microalgae and cyanobacteria. In: Day, J.G., Stacey, G.N. (Eds.), Cryopreservation and Freeze-Drying Protocols. Humana Press, pp. 141–151. Available at https://doi.org/10.1007/978-1-59745-362-2_10.

Day, J.G., Fleck, R.A., Benson, E.E., 2000. Cryopreservation-recalcitrance in microalgae: novel approaches to identify and avoid cryo-injury. Journal of Applied Phycology 12 (3–5), 369–377. Available at https://doi.org/10.1023/A:1008107229005.

de-Bashan, L.E., Bashan, Y., 2010. Immobilized microalgae for removing pollutants: review of practical aspects. Bioresource Technology 101 (6), 1611–1627. Available at https://doi.org/10.1016/j.biortech.2009.09.043.

Day, J.G., Slocombe, S.P., Stanley, M.S., 2012. Overcoming biological constraints to enable the exploitation of microalgae for biofuels. Bioresource Technology 109, 245–251. Available at https://doi.org/10.1016/j.biortech.2011.05.033.

Demirel, Z., Imamoglu, E., Deniz, İ., Dalay, M.C., 2018. Optimization of cryopreservation process using response surface methodology for *Chlorella saccharophila* and *Chlorella zofingiensis*. Celal Bayar University Journal of Science 14 (4), 405–412. Available at https://doi.org/10.18466/cbayarfbe.426444.

Desmet, J., Meunier, C.F., Danloy, E.P., Duprez, M.E., Hantson, A.L., Thomas, D., et al., 2014. Green and sustainable production of high value compounds via a microalgae encapsulation technology that relies on CO_2 as a principal reactant. Journal of Materials Chemistry A 2 (48), 20560–20569. Available at https://doi.org/10.1039/C4TA04659E.

El-Enshasy, H.A., Farid, M.A., El-Diwany, A.I., 1996. Oxytetracycline production by free and immobilized cells of *Streptomyces rimosus* in batch and repeated batch cultures. Progress in Biotechnology 11, 437–443. Available at https://doi.org/10.1016/S0921-0423(96)80060-8.

Emeish, S., 2012. Production of natural β-carotene from *Dunaliella* living in the Dead Sea. Jordan Journal of Earth and Environmental Sciences 4 (2). Available at https://www.researchgate.net/publication/287210526.

Esquivel, B.C., Lobina, D.V., Sandoval, F.C., 1993. The biochemical composition of two diatoms after different preservation techniques. Comparative Biochemistry and Physiology Part B: Comparative Biochemistry 105 (2), 369–373.

Fernandez-Valenzuela, S., Chávez-Ruvalcaba, F., Beltran-Rocha, J.C., San Claudio, P.M., Reyna-Martínez, R., 2021. Isolation and culturing axenic microalgae: mini–review. The Open Microbiology Journal 15 (1), 111–119. Available at https://doi.org/10.2174/1874285802115010111.

Foo, S.C., Mok, C.Y., Ho, S.Y., Khong, N.M.H., 2023. Microalgal culture preservation: progress, trends and future developments. Algal Research 71, 103007. Available at https://doi.org/10.1016/j.algal.2023.103007.

Fuertes-Rabanal, M., Rebaque, D., Largo-Gosens, A., Encina, A., Mélida, H., 2024. Cell walls: a comparative view of the composition of cell surfaces of plants, algae, and microorganisms. Journal of Experimental Botany 76 (10), 2614. Available at https://doi.org/10.1093/jxb/erae512.

Hosseinizand, H., Sokhansanj, S., Lim, C.J., 2018. Studying the drying mechanism of microalgae *Chlorella vulgaris* and the optimum drying temperature to preserve quality characteristics. Drying Technology 36 (9), 1049–1060. Available at https://doi.org/10.1080/07373937.2017.1369986.

Jang, T.H., Park, S.C., Yang, J.H., Kim, J.Y., Seok, J.H., Park, U.S., et al., 2017. Cryopreservation and its clinical applications. Integrative Medicine Research 6 (1), 12–18. Available at https://doi.org/10.1016/j.imr.2016.12.001.

Bhambere, D. B., Gaidhani, K., Harwalkar, M., & Nirgude, P. (2015). Lyophilization/freeze drying – A review. World Journal of Pharmaceutical Research, 4, 516–543

Khandagale, P.M., Bhairav, B., Saudagar, R.B., 2016. Lyophilization technique: A review. Asian Journal of Research in Pharmaceutical Science 6 (4), 269. Available at 10.5958/2231-5659.2016.00038.2.

Kapoore, R.V., Huete-Ortega, M., Day, J.G., Okurowska, K., Slocombe, S.P., Stanley, M.S., et al., 2019. Effects of cryopreservation on viability and functional stability of an industrially relevant alga. Scientific Reports 9, 3849. Available at https://doi.org/10.1038/s41598-019-38588-6.

Karnieli, O., 2016. Bioreactors and downstream processing for stem cell manufacturing. Stem Cell Manufacturing. Elsevier, pp. 141–160. Available at https://doi.org/10.1016/B978-0-444-63265-4.00006-6.

Khan, M.I., Shin, J.H., Kim, J.D., 2018. The promising future of microalgae: current status, challenges, and optimization of a sustainable and renewable industry for biofuels, feed, and other products. Microbial Cell Factories 17 (1), 36. Available at https://doi.org/10.1186/s12934-018-0879-x.

Kitcha, S., Cheirsilp, B., 2014. Enhanced lipid production by co-cultivation and co-encapsulation of oleaginous yeast *Trichosporonoides spathulata* with microalgae in alginate gel beads. Applied Biochemistry and Biotechnology 173 (2), 522–534. Available at https://doi.org/10.1007/s12010-014-0859-5.

Kumari, N., Gupta, M.K., Singh, R.K., 2016. Open encapsulation-vitrification for cryopreservation of algae. Cryobiology 73 (2), 232–239. Available at https://doi.org/10.1016/j.cryobiol.2016.07.005.

Meryman, H.T., 1971. Osmotic stress as a mechanism of freezing injury. Cryobiology 8 (5), 489–500. Available at 10.1016/0011-2240(71)90040-x.

Lukavský, J., 1988. Long-term preservation of algal strains by immobilization. Archiv für Protistenkunde 135 (1-4), 65–68. Available at 10.1016/s0003-9365(88)80054-x.

Magnabosco, C., Santaniello, G., Romano, G., 2025. Microalgae: a promising source of bioactive polysaccharides for biotechnological applications. Molecules (Basel, Switzerland) 30 (9), 2055.

McGrath, M.S., Daggett, P.M., Dilworth, S., 1978. Freeze-drying of algae: Chlorophyta and Chrysophyta. Journal of Phycology 14 (4), 521–525. Available at https://doi.org/10.1111/j.1529-8817.1978.tb02480.x.

Molina Grima, E., Sánchez Pérez, J.A., García Camacho, F., Acién Fernández, F.G., López Alonso, D., Segura del Castillo, C.I., 1994. Preservation of the marine microalga *Isochrysis galbana*: influence on the fatty acid profile. Aquaculture (Amsterdam, Netherlands) 123 (3–4), 377–385. Available at https://doi.org/10.1016/0044-8486(94)90072-8.

Morais Junior, W.G., Gorgich, M., Corrêa, P.S., Martins, A.A., Mata, T.M., Caetano, N.S., 2020. Microalgae for biotechnological applications: cultivation, harvesting and biomass processing. Aquaculture (Amsterdam, Netherlands) 528, 735562. Available at https://doi.org/10.1016/j.aquaculture.2020.735562.

Paula da Silva, S., Ferreira do Valle, A., Perrone, D., 2021. Microencapsulated *Spirulina maxima* biomass as an ingredient for the production of nutritionally enriched and sensorially well-accepted vegan biscuits. LWT 142, 110997. Available at https://doi.org/10.1016/j.lwt.2021.110997.

Perullini, M., Ferro, Y., Durrieu, C., Jobbágy, M., Bilmes, S.A., 2014. Sol-gel silica platforms for microalgae-based optical biosensors. Journal of Biotechnology 179 (1), 65–70. Available at https://doi.org/10.1016/j.jbiotec.2014.02.007.

Prasad, R., Shabnam, N., Pardha-Saradhi, P., 2016. Immobilization on cotton cloth pieces is ideal for storage and conservation of microalgae. Algal Research 20, 172–179. Available at https://doi.org/10.1016/j.algal.2016.10.005.

Rajmohan, D., Bellmer, D., 2019. Characterization of spirulina-alginate beads formed using ionic gelation. International Journal of Food Science 2019, 7101279. Available at https://doi.org/10.1155/2019/7101279.

Rohman, A., Dijkstra, B.W., 2021. Application of microbial 3-ketosteroid Δ1-dehydrogenases in biotechnology. Biotechnology Advances 49, 107751. Available at https://doi.org/10.1016/j.biotechadv.2021.107751.

Sachse, M., Fernández-Sánchez, S.Y., Tenorio, R., de Castro, I.F., Risco, C., 2024. Imaging Bunyavirus infections by transmission electron microscopy: conventional sample preparation vs high-pressure freezing and freeze-substitution. Methods in Molecular Biology 241–258. Available at https://doi.org/10.1007/978-1-0716-3926-9_16.

Sakai, A., Hirai, D., Niino, T., 2008. Development of PVS-based vitrification and encapsulation-vitrification protocols. In: Reed, B.M. (Ed.), Plant Cryopreservation: A Practical Guide. Springer, New York, pp. 33–57. Available at https://doi.org/10.1007/978-0-387-72276-4_3.

Sarıtaş, S., Kalkan, A.E., Yılmaz, K., Gurdal, S., Göksan, T., Witkowska, A.M., et al., 2024. Biological and nutritional applications of microalgae. Nutrients 17 (1), 93.

Sarkarat, R., Tavakoli, O. and Ajayebi, S., 2020. Applying immobilization and gel entrapment techniques for microalgal cell preservation and evaluation of their efficiencies. SSRN. https://doi.org/10.2139/ssrn.4924054.

Stock, W., Pinseel, E., De Decker, S., Sefbom, J., Blommaert, L., Chepurnova, O., et al., 2018. Expanding the toolbox for cryopreservation of marine and freshwater diatoms. Scientific Reports 8, 4279. Available at https://doi.org/10.1038/s41598-018-22460-0.

Studer, D., Humbel, B.M., Chiquet, M., 2008. Electron microscopy of high-pressure frozen samples: bridging the gap between cellular ultrastructure and atomic resolution. Histochemistry and Cell Biology 130 (5), 877–889. Available at https://doi.org/10.1007/s00418-008-0500-1.

Gallet, B., Moriscot, C., Schoehn, G., & Decelle, J. (2024). Cryo-fixation and resin embedding of biological samples for electron microscopy and chemical imaging v1. Available at https://doi.org/10.17504/PROTOCOLS.IO.BP2L62KNDGQE/V1.

Takano, M., Sado, J.I., Ogawa, T., Terui, G., 1973. Freezing and freeze-drying of *Spirulina platensis*. Cryobiology 10 (5), 440–444. Available at https://doi.org/10.1016/0011-2240(73)90073-4.

Tamtürk, F., Gürbüz, B., Toker, Ö.S., Dalabasmaz, S., Malakjani, N., Durmaz, Y., et al., 2023. Optimization of *Chlorella vulgaris* spray drying using various innovative wall materials. Algal Research 72, 103115. Available at https://doi.org/10.1016/j.algal.2023.103115.

Taylor, R., Fletcher, R.L., 1999. Cryopreservation of eukaryotic algae—a review of methodologies. Journal of Applied Phycology 10 (5), 481–501. Available at https://doi.org/10.1023/A:1008094622412.

Therien, J.B., Zadvornyy, O.A., Posewitz, M.C., Bryant, D.A., Peters, J.W., 2014. Growth of *Chlamydomonas reinhardtii* in acetate-free medium when co-cultured with alginate-encapsulated, acetate-producing strains of *Synechococcus sp.* PCC 7002. Biotechnology for Biofuels 7 (1). Available at https://doi.org/10.1186/s13068-014-0154-2.

Unterlander, N., Champagne, P., Plaxton, W.C., 2017. Lyophilization pretreatment facilitates extraction of soluble proteins and active enzymes from the oil-accumulating microalga *Chlorella vulgaris*. Algal Research 25, 439–444. Available at https://doi.org/10.1016/j.algal.2017.06.010.

Zhang, Z., Chen, Y., Klausen, L.H., Skaanvik, S.A., Wang, D., Chen, J., et al., 2023. The rational design and development of microalgae-based biohybrid materials for biomedical applications. Engineering 24, 102–113. Available at https://doi.org/10.1016/j.eng.2022.09.016.

CHAPTER 2

Techniques for isolation and characterization of bioactive compounds from algae

Madhavi Tiwari[1], Dhananjay Tandon[2], Veena Singh[3], and Vijay Singh[4]
[1]Department of Applied Science, Shri Rawatpura Sarkar University, Raipur, Chhattisgarh, India
[2]Department of Biotechnology, Anjaneya University, Raipur, Chhattisgarh, India
[3]Shri Rawatpura Sarkar College of Pharmacy, Shri Rawatpura Sarkar University, Raipur, Chhattisgarh, India
[4]Department of Pharmacy, Shri Rawatpura Sarkar University, Raipur, Chhattisgarh, India

2.1 Introduction

Algae, encompassing both microalgae and macroalgae, are photosynthetic organisms distributed across a wide range of submarine and, to a lower extent, terrestrial ecosystems. These organisms have gained considerable scientific and artificial attention over recent decades as rich sources of biologically active composites with different operations in medicinals, nutraceuticals, husbandry, cosmetics, and environmental remediation (Borowitzka, 2013). The rising global demand for sustainable, renewable, and naturally deduced bioresources has boosted exploration efforts aimed at exploring the biochemical eventuality of algal species as druthers to conventional factory- and synthetic-deduced composites.

As primary directors at the base of submarine food webs, algae exhibit exceptional metabolic inflexibility and biosynthetic effectiveness, enabling them to synthesize a broad range of structurally different primary and secondary metabolites. These composites include colors, polysaccharides, polyphenols, adipose acids, sterols, peptides, and other low-molecular-weight bioactives that play critical ecological roles in stress forbearance, defense mechanisms, and cellular regulation. Numerous of these metabolites have demonstrated significant health-promoting parcels, including antioxidant, antimicrobial, antiinflammatory, anticancer, and immunomodulatory conditioning, thereby situating algae as promising sources of new functional constituents and remedial agents (Plaza et al., 2009).

Likewise, the rapid-fire growth rates, high photosynthetic effectiveness, and capability of algae to thrive under controlled civilization conditions

Applied Phycology
DOI: https://doi.org/10.1016/B978-0-443-24758-3.00015-0

make them attractive campaigners for large-scale biotechnological exploitation. These attributes, coupled with their capacity to use nonarable land and wastewater coffers, further enhance the applicability of algae in the environment of sustainable bioactive emulsion product. Accordingly, methodical investigation into the insulation and characterization of algal bioactive composites has become a pivotal area of applied phycological exploration, bolstering the development of effective birth technologies and advanced logical methodologies.

2.1.1 Significance of algae as a source of bioactive composites

Algae constitute a phylogenetically different group of photosynthetic organisms that inhabit marine, brackish, and terrestrial ecosystems, ranging from polar to tropical surroundings. Due to their long evolutionary history and adaptation to extreme ecological conditions—similar to high saltness, violent ultraviolet radiation, nutrient limitation, and oxidative stress—algae have developed the capacity to synthesize a wide array of structurally different bioactive composites with significant natural functions (Borowitzka, 2013; Plaza et al., 2008). In recent times, algae have surfaced as a strategic bioresource for natural product discovery owing to several essential advantages over terrestrial shops. These include rapid-fire growth rates, high biomass productivity, minimum land demand, time-round civilization eventuality, and the capability to use nonpotable water sources and artificial backwaters (Markou and Nerantzis, 2013; Pulz and Gross, 2004). Similar attributes make algae particularly attractive in the environment of sustainable bioprospecting and indirect bioeconomy fabrics. From a biochemical perspective, algal bioactive composites play critical ecological roles, including photoprotection, antioxidative defense, chemical signaling, and protection against pathogens and scrape. These naturally being functions restate into a wide range of pharmacological and nutraceutical conditioning, similar to antioxidant, antimicrobial, antiviral, anticancer, antiinflammatory, antidiabetic, antihypertensive, and immunomodulatory goods (Gupta and Abu-Ghannam, 2011; Lordan et al., 2011). Growing public concern regarding the toxicological pitfalls and environmental continuity of synthetic complements has further boosted interest in algae-deduced natural composites. Regulatory pressure and consumer demand for "clean-marker" products have encouraged diligence to explore algal metabolites as safer and eco-friendly druthers for medicinals, foods, cosmetics, and functional accoutrements (Wells et al., 2017). Accordingly,

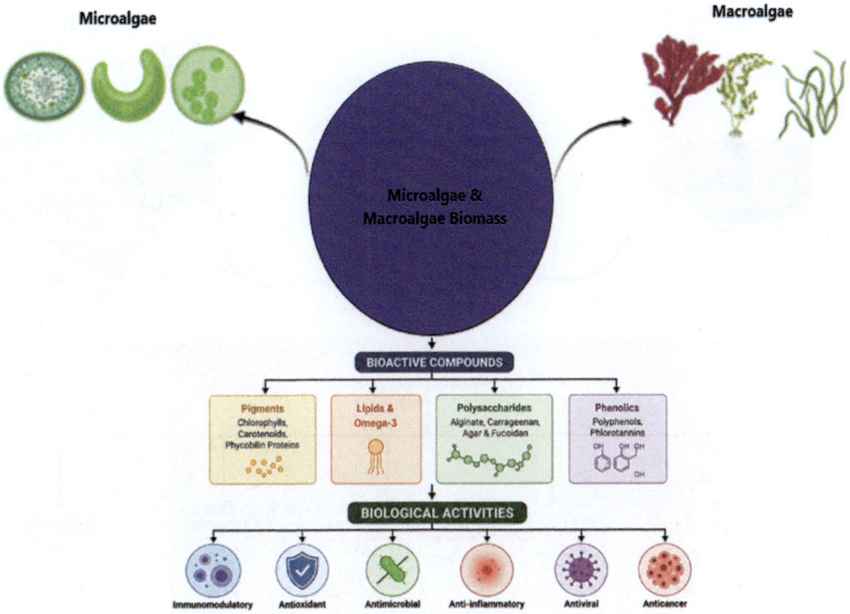

Figure 2.1 ***Overview of major bioactive compounds derived from microalgae and macroalgae and their associated biological activities.*** The figure highlights key classes of algal bioactive compounds, including pigments, polysaccharides, lipids and omega-3 fatty acids, phenolic compounds, and proteins, and their major biological activities such as antioxidant, antimicrobial, anti-inflammatory, antiviral, anticancer, and immunomodulatory properties.

algae are no longer viewed simply as primary directors in submarine ecosystems but as artificial cell manufactories able to deliver high-value bioactives. Algae produce a diverse range of bioactive compounds, including pigments, polysaccharides, lipids, phenolics, and bioactive peptides, which collectively contribute to their antioxidant, antimicrobial, antiinflammatory, antiviral, anticancer, and cardioprotective activities (Fig. 2.1).

2.1.2 Bracket and diversity of bioactive substances in algae

The diversity of algal bioactive composites is nearly linked to the taxonomic and physiological diversity of algae. Astronomically, algae are distributed into microalgae (including cyanobacteria and unicellular eukaryotic algae) and macroalgae (seaweeds), which are further classified into green (Chlorophyta), red (Rhodophyta), and brown (Phaeophyta) algae.

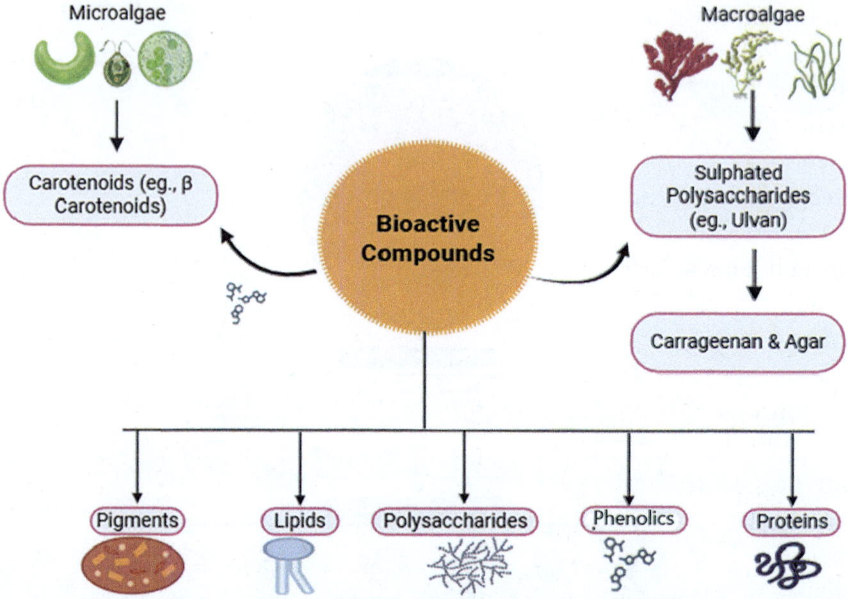

Figure 2.2 ***Bioactive compounds associated with different algal groups.*** The figure illustrates microalgae and macroalgae, which produce carotenoids (β-carotenoids) and specialized sulphated polysaccharides (ulvan, carrageenan, and agar), respectively. Both groups serve as primary sources for pigments, lipids, polysaccharides, phenolics, and proteins.

Each group exhibits distinctmetabolic autographs and biosynthetic pathways, performing in the product of chemically different metabolites (Delattre et al., 2016). The diversity of algal bioactive compounds is closely linked to algal taxonomy, with distinct metabolite profiles observed among microalgae and different macroalgal groups (Fig. 2.2). The taxonomic classification of algae and the associated diversity of bioactive compounds produced by different algal groups are schematically illustrated in Fig. 2.2.

Algal bioactive composites are astronomically divided into primary metabolites, essential for growth and metabolism, and secondary metabolites, which confer adaptive advantages. Among these, secondary metabolites are of particular interest due to their potent natural conditioning and artificial applicability. Major classes of algal bioactive composites include colors, similar to chlorophylls, carotenoids (β- carotene, astaxanthin, and fucoxanthin), and phycobiliproteins, which parade strong antioxidant, photoprotective, antiobesity, and antiinflammatory parcels (Guedes et al., 2011). Polysaccharides, including agar, carrageenan, alginate, fucoidan, and ulvan, are known for their antiviral, anticoagulant, immunomodulatory, prebiotic, and crack- mending conditioning (Wijesekara et al., 2011). Phenolic composites, particularly phlorotannins from brown algae, display exceptional antioxidant, antimicrobial, and antidiabetic properties (Li et al., 2011). Proteins and bioactive peptides, generated either naturally or through enzymatic hydrolysis, flaunt antihypertensive, antioxidant, and antimicrobial conditioning (Harnedy and FitzGerald, 2012). Lipids and polyunsaturated adipose acids (polyunsaturated fatty acids [PUFAs]), especially omega- 3 adipose acids similar to EPA and DHA, are critical for cardiovascular, neural, and antiinflammatory health (Adarme-Vega et al., 2012). Terpenoids, sterols, and alkaloids contribute to anticancer, antiviral, neuroprotective, and anticholesterol conditioning (Blunt et al., 2018). The qualitative and quantitative composition of these bioactives is explosively determined by species, environmental stressors, nutrient vacuity, light intensity, and growth phase, emphasizing the significance of acclimatized birth and characterization ways. The major classes of algal bioactive compounds, along with their representative examples and biological functions, are summarized in Table 2.1

Table 2.1 Major classes of algal bioactive compounds and their biological significance.

Class of compound	Representative examples	Major biological activities	Key application areas
Pigments	Astaxanthin, fucoxanthin, phycocyanin	Antioxidant, antiinflammatory	Food colorants, nutraceuticals, cosmetics
Polysaccharides	Alginate, carrageenan, fucoidan	Antiviral, anticoagulant, immunomodulatory	Pharmaceuticals, food hydrocolloids
Phenolics	Phlorotannins	Antioxidant, antimicrobial	Functional foods, cosmeceuticals
Proteins and peptides	Bioactive peptides	Antihypertensive, antioxidant	Nutraceuticals, therapeutic agents
Lipids	EPA, DHA	Cardioprotective, neuroprotective	Infant nutrition, supplements

2.1.3 Compass and industrial relevance

The artificial applicability of algal bioactive composites has expanded vastly due to their multifunctionality, renewability, and comity with sustainable product systems. In the pharmaceutical sector, algae serve as promising sources of lead motes for medicine development, particularly for habitual and life-related conditions where oxidative stress and inflammation play central roles (Newman and Cragg, 2020).

In the food and nutraceutical diligence, algal bioactives are extensively used as functional constituents, natural antioxidants, colorings, stabilizers, and omega-3 supplements. Microalgae similar to Spirulina and Chlorella have formerly achieved marketable success due to their high protein content and health-promoting parcels (Becker, 2007). Also, seaweed-deduced polysaccharides dominate the global request for food hydrocolloids. The ornamental and cosmeceutical diligence considerably exploit algal excerpts for antiaging, skin-whitening, ultraviolet (UV)-protection, and moisturizing phrasings, owing to their rich antioxidant and bioprotective biographies (Kim and Pangestuti, 2018). In husbandry, algal bioactives serve as biostimulants, enhancing plant growth, nutrient uptake, and stress forbearance while reducing dependence on synthetic agrochemicals. Importantly, the integration of algal bioactive product within biorefinery and wastewater-grounded civilization systems enhances profitable feasibility and environmental sustainability. Similar intertwined approaches enable contemporaneous biomass generation, contaminant junking, and high-value emulsion birth, making algae a foundation of unborn bio-based diligence (Chisti, 2007). Given this expanding compass, the development and optimization of robust ways for insulation and characterization are critical for maximizing yield, conserving bioactivity, icing product safety, and enabling nonsupervisory blessing. Thus a comprehensive understanding of these ways forms the foundation for rephrasing algal bioactive exploration from laboratory examinations to artificial perpetration. The wide range of industrial applications of algal bioactive compounds across pharmaceutical, nutraceutical, cosmetic, agricultural, and animal feed sectors is schematically illustrated in Fig. 2.3.

2.2 Sample collection and preparation

Sample collection and medication represent a critical preanalytical phase in the insulation and characterization of algal bioactive composites. Any variation at this stage can profoundly impact the yield, chemical composition, stability, and

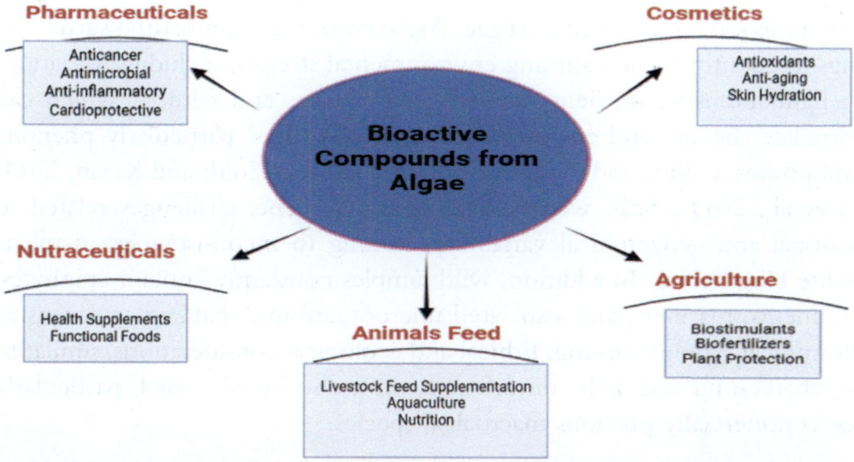

Figure 2.3 ***Applications and industrial relevance of bioactive compounds derived from algae.*** The figure illustrates the diverse applications of algal bioactive compounds in pharmaceuticals, cosmetics, nutraceuticals, agriculture, and animal feed. These compounds exhibit antioxidant, antimicrobial, anti-inflammatory, anticancer, cardioprotective, anti-aging, and skin-hydrating properties, while also supporting functional foods, dietary supplements, biostimulants, biofertilizers, plant protection, livestock feed supplementation, and aquaculture nutrition.

natural exertion of target metabolites. Improper harvesting or inadequate preprocessing may lead to enzymatic degradation, oxidation, or loss of thermolabile compounds, thereby compromising downstream extraction and analytical outcomes (Guedes et al., 2011; Plaza et al., 2008).

Therefore standardized and precisely optimized procedures are essential to ensure the reproducibility and reliability of algal bioactive research.

2.2.1 Algal harvesting styles

Algal biomass used for bioactive emulsion birth is generally attained through wild collection or controlled civilization. The selection of harvesting strategy depends on algal type (microalgae or macroalgae), target composites, scale of operation, and intended operation.

2.2.1.1 Wild collection

Wild harvesting involves the direct collection of algae from natural territories similar to marine littoral zones, arms, brackish lakes, gutters, and ponds. This system is generally used for macroalgae (seaweeds) and certain bloom-forming microalgae. Algae collected from natural surroundings frequently witness shifting environmental stresses, including variations in light intensity, nutrient vacuity, temperature, and saltness, which can stimulate the biosynthesis of secondary metabolites, particularly phenolic composites, colors, and defensive polysaccharides (Holdt and Kraan, 2011; Li et al., 2011). Still, wild collection also presents challenges related to seasonal and geographical variability, leading to inconsistencies in metabolite biographies. In addition, wild samples constantly contain epiphytes, sediments, mariners, and associated microorganisms, challenging extensive cleaning and preprocessing. Ethical and ecological considerations, similar to overharvesting and niche disturbance, must also be addressed, particularly for commercially precious macroalgal species.

2.2.1.2 Civilization-grounded harvesting

Controlled civilization is the favored system for producing microalgal and cyanobacterial biomass and is decreasingly espoused for named macroalgal species. Civilization systems include open ponds, waterway ponds, and closed photobioreactors, which allow precise regulation of environmental parameters similar to light intensity, temperature, nutrient attention, pH, and carbon dioxide (CO_2) force (Markou and Nerantzis, 2013; Pulz and Gross, 2004). The major advantage of civilization-grounded harvesting lies in its reproducibility, scalability, and thickness of metabolite biographies, making it suitable for medicinal, nutraceutical, and artificial operations. Also, civilization conditions can be designedly manipulated to enhance specific bioactive composites; for illustration, nitrogen limitation to stimulate lipid and carotenoid accumulation or highlight stress to enhance antioxidant color conflation (Adarme-Vega et al., 2012). Harvesting of microalgal biomass from liquid societies involves downstream separation processes similar to centrifugation, filtration, sedimentation, or chemical flocculation. Each system differs in effectiveness, energy demand, and felicity depending on cell size and viscosity.

2.2.2 Drying and preprocessing ways

Following harvesting, lately collected algal biomass contains high humidity content and active endogenous enzymes that can quickly degrade sensitive metabolites. Thus immediate stabilization through applicable drying and preprocessing is essential to save bioactive emulsion integrity (Wells et al., 2017).

2.2.2.1 Washing and drawing

Gathered algae are completely washed using distilled water or filtered seawater to remove stuck mariners, beach patches, epiphytes, and microbial pollutants. For logical-grade studies, repeated washing ways are recommended to minimize matrix hindrance during chromatographic and spectroscopic analyses. Care is taken to avoid inordinate mechanical stress that may rupture cells and promote unseasonable metabolite loss.

2.2.2.2 Drying ways

Drying is a decisive step that significantly affects the stability, passage, and bioactivity of algal metabolites. Several drying styles are employed depending on the target composites and available structure. Sun drying is a traditional and low-cost system generally applied to macroalgae, particularly in littoral regions. Still, dragged exposure to sun and oxygen may degrade photosensitive colors and phenolic composites. Roaster drying at controlled low temperatures (40°C–60°C) offers better reproducibility and reduced drying time but may still compromise thermolabile metabolites. Snap-drying (lyophilization) is extensively regarded as the most effective system for conserving bioactive composites, as it minimizes thermal and oxidative declination by removing water under vacuum at low temperatures (Guedes et al., 2011). Spray drying is constantly used in artificial-scale microalgal processing, particularly for nutraceutical maquillages, owing to its effectiveness and scalability, although careful optimization is needed to help heat-convinced declination.

2.2.2.3 Grinding and storehouse

Dried algal biomass is mechanically based into a fine greasepaint to increase face area and ameliorate solvent penetration during birth. The powdered samples are stored in airtight, light-defended holders at low temperatures to minimize oxidative and photodegradative losses previous to birth and characterization.

A comparative overview of algal harvesting and preprocessing strategies and their influence on bioactive compound preservation is presented in Table 2.2.

2.3 Extraction of bioactive composites from algae

Extraction is a vital step in the recovery of algal bioactive composites, as it directly influences birth yield, chemical composition, functional integrity,

Table 2.2 Algal harvesting and preprocessing styles and their influence on bioactive emulsion preservation.

Process stage	Method	Advantages	Limitations
Harvesting	Wild collection	High chemical diversity	Seasonal variability, contamination
Harvesting	Cultivation	Reproducibility, scalability	Higher operational cost
Drying	Sun drying	Low cost	Loss of sensitive metabolites
Drying	Oven drying	Controlled conditions	Thermal degradation risk
Drying	Freeze-drying	Excellent metabolite preservation	High cost
Preprocessing	Grinding	Improved extraction efficiency	Requires controlled storage

and natural efficacity of the insulated metabolites. Algal biomass contains a chemically different array of composites, including polar polysaccharides, relatively polar phenolics and proteins, and nonpolar lipids and colors. Accordingly, no single birth system is widely applicable, and the selection of an applicable birth strategy must be guided by the chemical nature of the target emulsion, intended operation, and scale of operation (Herrero et al., 2010; Plaza et al., 2008). In addition, algal cell walls are frequently complex and resistant to solvent penetration due to the presence of rigid polysaccharides, glycoproteins, and, in some species, mineralized factors. Effective birth thus requires not only suitable detergents but also acceptable cell dislocation mechanisms to grease metabolite release.

2.3.1 Abecedarian principles of algal extraction

The birth of bioactive composites from algae is governed by abecedarian physicochemical principles, including solute–detergent relations, prolixity kinetics, and mass transfer effectiveness. The primary end is to widely solubilize target composites while minimizing coextraction of uninvited factors and precluding declination. Crucial factors impacting algal birth include solvent opposition and selectivity, which determine solubility of specific metabolite classes; birth temperature, affecting not only solubility and prolixity but also thermal stability; extraction time, impacting yield and declination threat; flyspeck size of biomass, which controls face area and solvent penetration; solid-to-detergent rate, impacting birth effectiveness and attention slants; and inappropriate control of these parameters may lead to oxidation, hydrolysis, or denaturation of bioactive composites, particularly colors, phenolics, and proteins (Chemat et al., 2012). Thus birth protocols must be precisely designed and formalized to insure reproducibility and bioactivity retention.

2.3.2 Conventional extraction ways

2.3.2.1 Maceration and solvent extraction

Maceration is one of the simplest and most extensively used extraction ways, involving soaking dried and pulverized algal biomass in suitable organic detergents similar to methanol, ethanol, acetone, hexane, or their fusions. Solvent choice depends on the composition of the target composites—polar detergents for phenolics and polysaccharides and nonpolar detergents for lipids and colors (Gupta and Abu-Ghannam, 2011). Although maceration is easy to perform and requires minimum outfit, it is frequently time-consuming and may affect in low-extraction effectiveness. Dragged exposure to detergents may also lead to partial declination of sensitive metabolites. The general workflow of maceration and solvent-based extraction of algal bioactive compounds, including solvent addition, agitation, filtration, and recovery of crude extract, is illustrated in Fig. 2.4.

2.3.2.2 Soxhlet extraction

Soxhlet extraction is a classical hot birth fashion that enables nonstop solvent rotation through algal biomass under influx conditions. This system is particularly effective for rooting nonpolar composites, similar to lipids, carotenoids, and sterols, from both microalgae and macroalgae (Adarme-Vega et al., 2012). While Soxhlet extraction ensures total recovery of target composites, it suffers from several downsides, including prolonged extraction time, high energy demand, inordinate solvent operation, and implicit thermal declination of heat-sensitive metabolites. Accordingly, its operation is decreasingly confined to logical purposes rather than large-scale products.

2.3.2.3 Hot water extraction

Hot water extraction is generally employed for the recovery of high-molecular- weight polysaccharides, similar to agar, carrageenan, alginate, and ulvan from macroalgae. Elevated temperatures promote the

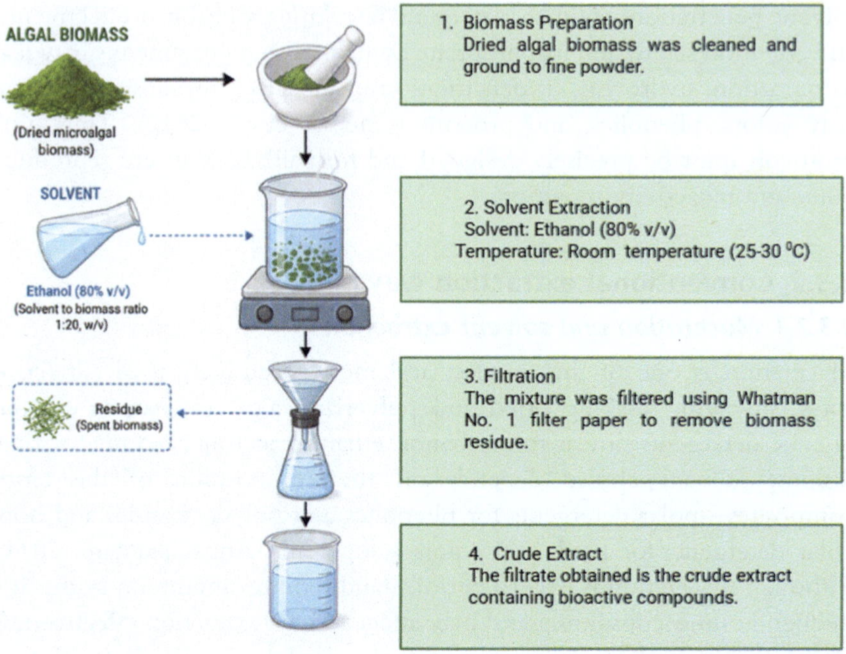

Figure 2.4 ***Schematic representation of maceration and solvent extraction of bioactive compounds from algal biomass.*** The figure illustrates the sequential steps of biomass preparation, solvent addition, agitation, filtration, and collection of the crude algal extract containing soluble bioactive compounds for subsequent concentration, purification, and downstream analysis.

solubilization of cell wall polysaccharides, enabling their release into the waterless phase (Wijesekara et al., 2011). Although this system is environmentally benign and suitable for food-grade operations, dragged heating may affect in partial depolymerization and loss of functional parcels, similar to density and bioactivity. Thus temperature and extraction time must be precisely optimized.

2.3.3 Advanced and green extraction technologies

To overcome the limitations of conventional ways, advanced extraction styles have been developed with a focus on effectiveness, selectivity, reduced solvent operation, and environmental sustainability. These approaches are decreasingly favored in artificial and pharmaceutical surroundings.

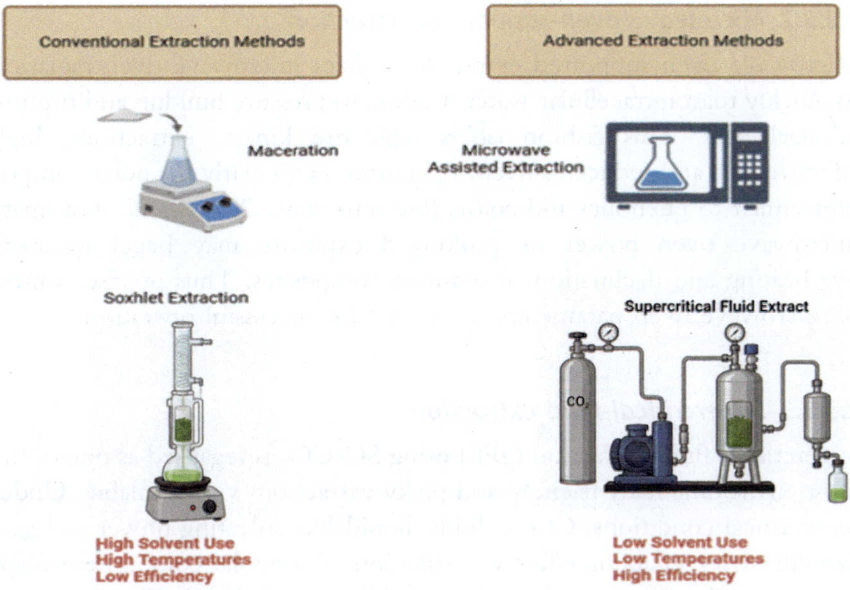

Figure 2.5 ***Comparison of conventional and advanced extraction techniques for algal bioactive compounds.*** The figure compares conventional methods, including maceration and Soxhlet extraction, with advanced techniques such as microwave-assisted extraction and supercritical fluid extraction, highlighting differences in solvent consumption, extraction temperature, efficiency, and suitability for sustainable recovery of algal bioactive compounds.

A comparative overview highlighting the key differences between conventional and advanced extraction techniques in terms of solvent consumption, operating temperature, and extraction efficiency is presented in Fig. 2.5.

2.3.3.1 Ultrasound-supported extraction

Ultrasound-supported extraction (UAE) employs high-frequency sound swells to induce cavitation bubbles within the detergent. The collapse of these bubbles produces microjets and shock swells that disrupt algal cell walls, thereby enhancing solvent penetration and mass transfer (Chemat et al., 2017). UAE significantly reduces extraction time and detergent consumption while perfecting yields of phenolics, colors, and antioxidant composites. Also, UAE can be conducted at fairly low temperatures, minimizing thermal declination and making it suitable for thermolabile metabolites.

2.3.3.2 *Microwave oven-supported extraction*

Microwave oven-supported extraction utilizes microwave oven radiation to quickly toast intracellular water, leading to pressure buildup and rupture of algal cells. This fashion offers rapid-fire kinetic extractions, high effectiveness, and reduced solvent operation, particularly for polar composites similar to phenolics and colors (Herrero et al., 2010). Still, inordinate microwave oven power or prolonged exposure may beget localized overheating and declination of sensitive composites. Thus precise control of microwave oven parameters is essential for successful operation.

2.3.3.3 *Supercritical fluid extraction*

Supercritical fluid extraction (SFE) using SC-CO_2 is regarded as one of the most environmentally friendly and picky extraction ways available. Under supercritical conditions, CO_2 exhibits liquid-like solvating power and gas-like diffusivity, enabling effective extraction of nonpolar composites similar to carotenoids, adipose acids, and sterols (Plaza et al., 2010). The absence of poisonous organic detergents and low temperatures makes SFE particularly attractive for food, nutraceutical, and medicinal diligence. The addition of small quantities of cosolvents (e.g., ethanol) can further enhance the birth of relatively polar composites.

2.3.4 Selection of birth strategy grounded on target bioactive composites

Given the chemical diversity of algal metabolites, birth strategies must be composite-specific. Polar polysaccharides bear waterless systems, whereas lipophilic composites bear nonpolar or supercritical detergents. Advanced ways are frequently preferred when high chastity and bioactivity preservation are needed. The selection of applicable birth ways grounded on the chemical nature of algal bioactive composites is presented in Table 2.3.

2.3.5 Optimization of extraction parameters

Optimization of extraction parameters is a pivotal step in maximizing the recovery and conserving the natural exertion of algal-deduced bioactive composites. Indeed, when an applicable birth fashion is named, suboptimal functional conditions can affect low yields, coextraction of undesirable composites, or declination of sensitive metabolites. Thus methodical

Table 2.3 Extraction techniques for major classes of algal bioactive compounds.

Compound class	Preferred extraction method	Key solvents/conditions
Polysaccharides	Hot water extraction	Water, mild heating
Phenolics	UAE, MAE	Ethanol, methanol
Pigments	SFE, Soxhlet	CO_2, acetone
Lipids and PUFAs	SFE, Soxhlet	CO_2, hexane
Proteins and peptides	Aqueous extraction	Buffer systems

MAE, Microwave oven-supported extraction; *PUF*, polyunsaturated fatty acids; *SFE*, supercritical fluid extraction; *UAE*, ultrasound-supported extraction.

optimization is essential to ensure effectiveness, reproducibility, and scalability of birth processes (Bezerra et al., 2008; Chemat et al., 2012).

2.3.5.1 Influence of solvent type and concentration

Solvent selection plays a decisive role in determining birth effectiveness, as it governs the solubility of target composites and the selectivity of the process. The opposition of the detergent must be compatible with the chemical nature of the said bioactive composites. Polar detergents similar to water, methanol, and ethanol are effective for rooting polysaccharides, phenolics, and proteins, whereas nonpolar detergents similar to hexane and supercritical CO_2 are preferred for lipids and carotenoids (Plaza et al., 2008). In numerous cases, double detergent systems (e.g., ethanol–water fusions) give superior birth effectiveness compared to single detergents by enhancing solubility and mass transfer. Still inordinate detergent attention may also increase the birth of snooping substances, challenging careful optimization to balance yield and chastity.

2.3.5.2 Effect of temperature and birth time

Extraction temperature and time significantly impact prolixity kinetics, solvent penetration, and emulsion stability. Adding temperature generally enhances solubility and extraction rate; still, elevated temperatures can lead to thermal declination of heat-sensitive composites, similar to colors, phenolics, and certain bioactive peptides (Herrero et al., 2010). Also, prolonged extraction time may ameliorate yield up to an optimum point, beyond which oxidative declination and structural revision of metabolites may do. Thus relating the optimal temperature–time combination is critical to achieving high-birth effectiveness without compromising bioactivity.

2.3.5.3 Solid- to-detergent rate and flyspeck size

The solid-to-detergent rate directly affects attention slants and mass transfer during extraction. A advanced detergent volume generally improves birth effectiveness by enhancing solute prolixity; still, inordinate solvent use increases recycling cost and downstream solvent recovery conditions. Optimizing this rate is particularly important for artificial-scale operations.

Flyspeck size of algal biomass also plays an important role, as lower patches give a larger face area for solvent commerce. Nonetheless, exorbitantly fine patches may beget filtration difficulties and hamper solvent rotation. Thus an optimal flyspeck size must be named to balance birth effectiveness and process practicality (Chemat et al., 2017).

2.3.5.4 Statistical and experimental optimization approaches

Traditional one-factor-at-a-time optimization styles are time-consuming and frequently fail to capture relations between variables. In discrepancy, statistical optimization tools, similar to response face methodology, central compound design, and Box–Behnken design, enable contemporaneous evaluation of multiple parameters and their relations with a reduced number of trials (Bezerra et al., 2008). These approaches not only ameliorate birth yield but also enhance process reproducibility and robustness. Optimization models generated through statistical styles are particularly precious for spanning up laboratory protocols to air man- and artificial-scale operations.

2.3.5.5 Significance of optimization for industrial and sustainable operations

From an artificial perspective, optimized birth conditions reduce solvent consumption, energy input, and processing time, thereby perfecting profitable feasibility and environmental sustainability. Optimized protocols also ensure batch-to-batch thickness, which is essential for nonsupervisory compliance in food, nutraceutical, and medicinal diligence (Wells et al., 2017). Also, optimization contributes to the development of green extraction processes by minimizing waste generation and enabling the integration of environmentally friendly detergents and energy-effective technologies.

2.4 Purification and fractionation approaches

The algal or plant extract comprises a diverse array of secondary metabolites. It is hard to separate bioactive chemicals from a group of compounds since it takes a lot of biological screening tests. When we look at the ancient ways of doing things, we can see that many new ways have been made. We have discussed bioactive molecule separation methods in this chapter, including column chromatography, solid-phase extraction (SPE), size-exclusion, and ion-exchange approaches.

2.4.1 Column chromatography

Column chromatography is a simple and common way to separate and clean things. It is a good way to separate and cleanup both solid and liquid samples. Column chromatography uses a solid phase that stays still and a liquid phase that moves through it to adsorb and separate chemicals. The chemical makeup of a substance determines how well it sticks to the adsorbent. Elution happens when the adsorbent sticks to a substance in different amounts. Column chromatography uses a variety of stationary phases, such as silica, alumina, calcium phosphate, calcium carbonate, starch, and magnesia, as well as different solvent compositions that are made to fit the unique properties of the compounds that need to be separated and isolated. Improving the procedure is an important step in separating different groups of chemicals in extracts. Column chromatography uses a cylindrical glass tube that is blocked at the bottom by a piece of glass wool or a porous disc. The tube is filled with slurry (adsorbent) and a solvent that works well with the slurry. To separate samples, they are mixed with silica and put at the top of the column, where they can move with the solvent. Because of differences in polarity, substances stick to different places and can be removed by solvents that have the right polarity. The compound that can hold more will be on top, while the one that can hold less will be on the bottom. Elution is the process of adding a solvent at the top of a column to help chemicals leave the column. This lets them move through the column (Coskun, 2016; Janson and Jönsson, 2011; Kondeti et al., 2014). The basic setup and working principle of column chromatography, including stationary phase packing, sample loading, solvent elution, and fraction collection, are illustrated in Fig. 2.6.

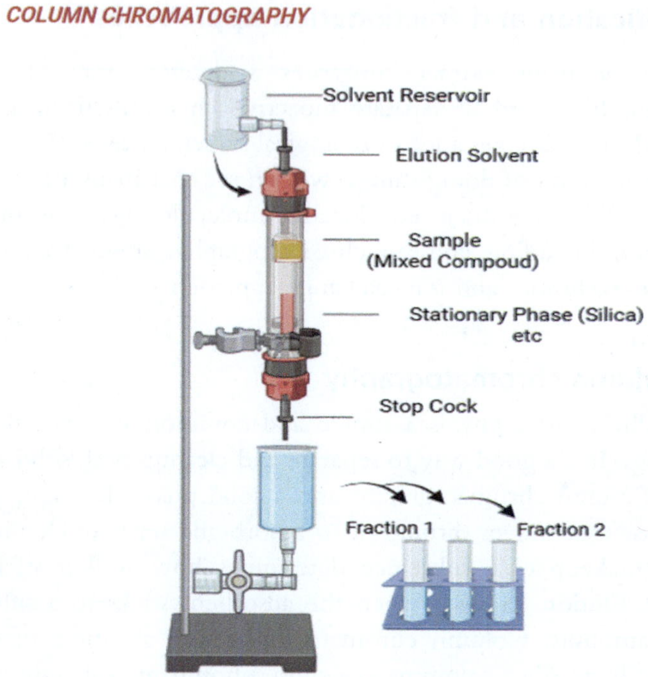

Figure 2.6 ***Schematic illustration of column chromatography for the separation of bioactive compounds.*** The figure depicts the application of a sample containing mixed compounds onto a stationary phase, followed by controlled elution with a suitable solvent. Differential interactions between the compounds and stationary phase facilitate their separation, resulting in the collection of individual fractions for further purification and analysis.

2.4.2 Solid-phase extraction

Analytical workflows that include preparing samples, separating analytes, and measuring them in complicated matrices are very important for modern chemical analysis. Each step is very important for making sure that the results are correct and useful. Sample preparation is one of these things that has a big impact on both the accuracy and speed of the analytical procedure. In the past few years, new and environmentally friendly ways to prepare samples have come out that work better. One such technology, SPE, has become very popular in many fields since it has many advantages over traditional methods. SPE is a better choice than liquid–liquid extraction (LLE) because it solves LLE's problems, like using too many organic solvents, taking too long to process, having too many processes, making mistakes more likely, and costing more.

SPE is versatile because it uses innovative sorbent technologies that are designed for specific extraction methods, such as reversed-phase, normal-phase, cation exchange, anion exchange, and mixed-mode interactions. Most commercial SPE cartridges are meant to be thrown away after use. However, new technologies have made it possible to make reusable cartridges using restricted access materials (RAM) and molecularly imprinted polymers. This chapter examines the theoretical principles and practical implementations of SPE, focusing specifically on sorbent packing materials, cartridge configurations, and performance metrics. It also talks about new developments in SPE methods and tools, such as solid-phase microextraction (SPME), in-tube SPME (IT-SPME), and magnetic SPE. The combination of SPE with chromatographic methods like liquid chromatography (LC) and gas chromatography (GC) is discussed, as well as its many uses in testing drugs, biological samples, natural products, pesticides, environmental pollutants, and food and drink samples (Badawy and Rabea, 2022). The sequential steps involved in SPE, including sample loading, washing, elution, and final collection of target analytes, are schematically illustrated in Fig. 2.7.

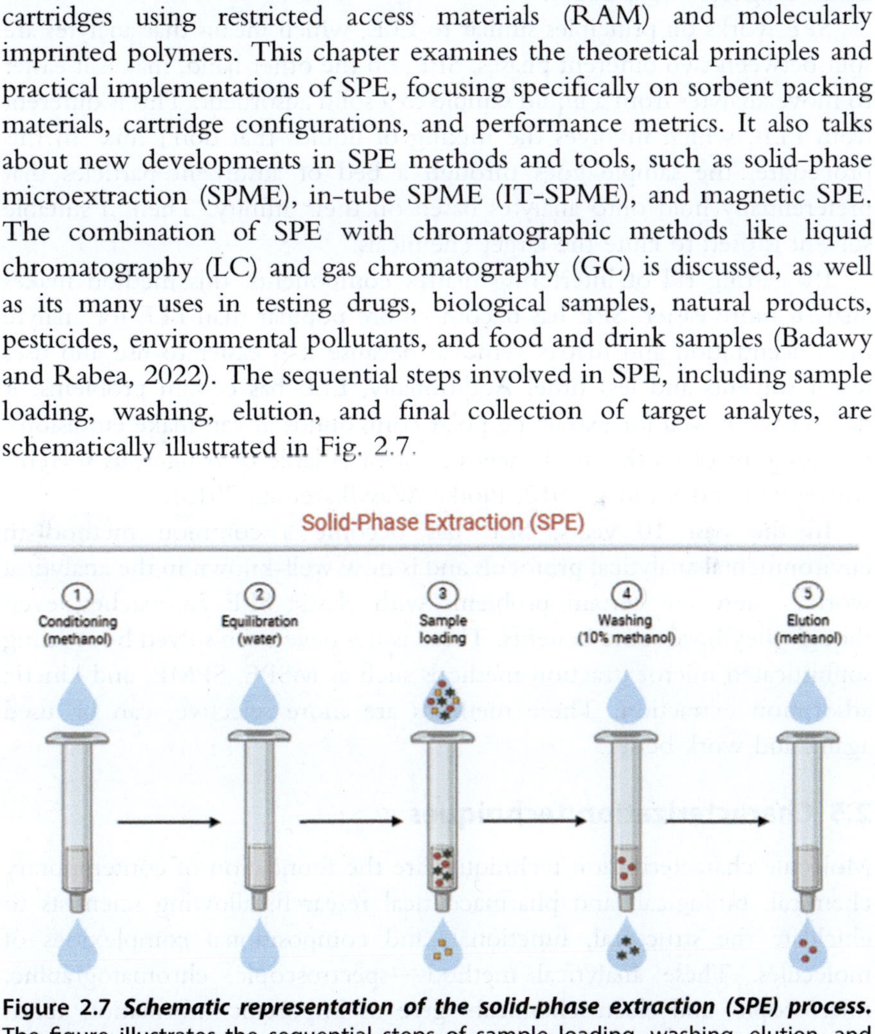

Figure 2.7 ***Schematic representation of the solid-phase extraction (SPE) process.*** The figure illustrates the sequential steps of sample loading, washing, elution, and collection, demonstrating the selective retention of target analytes on a solid sorbent and their subsequent recovery using an appropriate elution solvent.

2.4.3 Theoretical framework and fundamentals of solid-phase extraction

SPE is a commonly used method for preparing chemical samples. It fills the gap between collecting samples and analyzing them. SPE is usually not done alone; it is generally preceded with measures to improve extraction efficiency, such as dilution or pH adjustment (Simpson, 2000; Ozturk, 2022; Dugheri et al., 2021).

SPE works on principles similar to LLE, which means that analytes are split between two different phases. SPE, on the other hand, makes it easier to move analytes from a liquid sample to a solid adsorbent. This is different from LLE, which involves the mixing of liquids that don't mix. In this procedure, the sample goes through a bed of adsorbent particles that preferentially hold onto analytes based on their affinity. Then, a suitable solvent is used to elute the target chemicals.

By getting rid of interfering matrix components, this method makes analysis more easier. SPE has become more popular than LLE for analyte preconcentration and matrix removal because it is easier to use and uses fewer solvents and less time. Additionally, LLE has certain problems: it doesn't work well for extracting polar compounds, it can make emulsions, it takes a lot of work, and it needs a lot of volatile or dangerous solvents (Buszewski and Szultka, 2012; Plotka-Wasylka et al., 2015).

In the past 10 years, SPE has become a common method in environmental analytical protocols and is now well-known in the analytical world. There are certain problems with classic SPE approaches, even though they have some benefits. These issues have been solved by creating sophisticated microextraction methods such as MSPE, SPME, and kinetic adsorption extraction. These methods are more selective, can be used again, and work better.

2.5 Characterization techniques

Molecule characterization techniques are the foundation of contemporary chemical, biological, and pharmaceutical research, allowing scientists to elucidate the structural, functional, and compositional complexities of molecules. These analytical methods—spectroscopic, chromatographic, microscopic, and electrochemical—give us important information about a molecule's identification, purity, shape, and how it interacts with other molecules. Researchers can use technologies like nuclear magnetic resonance (NMR) spectroscopy, mass spectrometry (MS), infrared spectroscopy, and X-ray crystallography to figure out how complex chemicals are built, check synthetic routes, and learn more about how biomolecules behave. This basic information not only helps scientists learn more, but it also leads to new ideas in medicine research, materials science, and environmental monitoring.

2.5.1 Chromatographic techniques

2.5.1.1 Thin layer chromatography

Thin layer chromatography (TLC) is a quick, cheap, and flexible way to analyze mixtures of substances. It is often used to separate, identify, and qualitatively analyze molecules. TLC works on the idea of differential migration of analytes under capillary action in an appropriate solvent system. It uses a thin layer of adsorbent material, usually silica gel or alumina, that is deposited onto a glass, plastic, or aluminum plate. This method is very useful for organic synthesis, pharmaceutical analysis, and natural product screening since it gives immediate information about the purity of a molecule, the progress of a reaction, and the profile of its components. TLC is an essential tool in both research and everyday lab work because it is simple, doesn't require a lot of samples, and can be used in many different ways.

2.5.1.2 High-performance liquid chromatography

High-performance liquid chromatography (HPLC) is a strong and frequently used method for breaking down, identifying, and measuring the parts of complicated mixtures. It works on the idea of differential partitioning between a mobile phase (a liquid solvent) and a stationary phase (usually packed columns with silica or polymer-based materials). The mobile phase moves the sample through the column under high pressure. Each analyte interacts with the stationary phase in a unique way, which leads to various retention durations. HPLC is very important for pharmaceutical analysis, environmental monitoring, food safety, and biochemical research because it has great resolution, sensitivity, and reproducibility. Different modes, like reverse-phase, normal-phase, ion-exchange, and size-exclusion, are used to separate molecules based on their properties. Detection is usually done with UV–visible (UV–vis), fluorescence, or MS for both qualitative and quantitative analysis. HPLC is a key part of modern analytical chemistry since it is aut–omated, accurate, and can work with many different types of samples.

2.5.1.3 Gas chromatography

GC is a very sensitive and useful way to separate and study chemicals that can be turned into gas without breaking down. It works by separating a mobile phase, which is an inert carrier gas like helium or nitrogen, from a stationary phase, which is usually a liquid or polymer that is deposited onto a solid support in a column. As the sample is injected and moved along the column by the gas, each part interacts with the stationary phase in its own way, which causes the retention durations to be varied.

GC is great for testing volatile organic substances, which is why it is so important in environmental analysis, forensic research, petrochemical testing, and food safety. Flame ionization detectors, thermal conductivity detectors, or mass spectrometry (gas chromatography–mass spectrometry [GC–MS]) are often used to find things. These methods can be used for both qualitative and quantitative investigation. GC is a key method in analytical labs because it has excellent resolution, can be repeated, and works quickly.

2.5.2 Spectroscopic and spectrometric tools

2.5.2.1 UV–visible spectrophotometer

UV–visible spectroscopy is a common way to analyze molecules by measuring how much ultraviolet and visible light they absorb. It uses the theory of electronic transitions to find out how molecules absorb light in the 200–800 nm range, which moves electrons from ground to excited states, usually $\pi \rightarrow \pi^*$ or $n \rightarrow \pi^*$ transitions. The resulting absorbance spectrum gives us important information about the structure, concentration, and functional groups of the molecules. UV–vis is very useful for measuring the amount of analytes in a solution, keeping track of how fast reactions happen, and describing conjugated systems, dyes, and biomolecules like proteins and nucleic acids. Because it is simple, quick, and doesn't damage things, it is a common tool in pharmaceutical, environmental, and biochemical labs. The Beer–Lambert law tells us how to use it in a quantitative way by connecting absorbance to concentration and path length.

2.5.2.2 Fourier transform infrared spectroscopy

Fourier transform infrared spectroscopy (FTIR) is a strong way to figure out what materials are by looking at their infrared absorption spectra. It can do this for organic, polymeric, and even inorganic materials. It works on the idea that chemical vibrations absorb certain infrared light frequencies, which gives each compound a unique spectral fingerprint. FTIR employs an interferometer to gather all wavelengths at once. It then uses a mathematical Fourier transform to turn the raw data into a spectrum that can be understood.

FTIR is often used to identify functional groups, check the quality of materials, and do forensic analysis. It doesn't hurt the sample, doesn't need any preparation, and can look at solids, liquids, and gasses. Some of the main benefits are quick data collection, great sensitivity, and the capacity to find small changes in structure. Attenuated total reflectance, transmission, and diffuse reflectance are all common ways to sample.

2.5.2.3 Nuclear magnetic resonance

NMR spectroscopy is a very useful analytical tool for figuring out the structure, dynamics, and surroundings of molecules, especially organic compounds. It works by having nuclei (usually ^1 H and ^13 C) that absorb radiofrequency radiation in a magnetic field, which causes changes in the nuclear spin states. The chemical shifts, coupling constants, and signal intensities that come from this give us a lot of information on molecular connectivity, functional groups, and stereochemistry.

NMR is essential for determining structure, checking purity, and looking at conformations. 1D (^1 H, ^13 C) and 2D (COSY, HSQC, HMBC) NMR are two examples of techniques that can be used to map molecular frameworks in great detail. It is a key part of organic synthesis, medicines, metabolomics, and polymer science because it doesn't destroy samples and can examine them in solution.

2.5.3 Mass spectrometry

MS is a very sensitive way to find out the molecular weight, structure, and makeup of chemical compounds. It operates by ionizing molecules to generate charged particles. These particles are then sorted using electric and magnetic fields based on their mass-to-charge ratio (m/z). The mass spectrum that comes out of this is a unique fingerprint that shows chemical identification, isotopic patterns, and fragmentation processes. MS is frequently used in proteomics, metabolomics, drug testing, environmental monitoring, and forensic science. It is typically used with chromatographic methods like GC–MS or LC–MS to improve separation and detection. MS allows for both qualitative and quantitative examination of complex mixtures, even at trace levels, because it has excellent resolution, precision, and sensitivity.

2.5.3.1 Liquid chromatography–mass spectrometry

LC–MS is a strong hybrid analytical method that combines the separation powers of LC with the detection and identification powers of MS. LC separates complicated mixtures depending on their chemical properties, like size or polarity. MS then examines the eluted chemicals by detecting their m/z ratios following ionization. People utilize LC–MS a lot in drug development, proteomics, metabolomics, environmental investigation, and clinical diagnostics. It has a high level of sensitivity and specificity, and it can find trace-level chemicals in complicated matrices. The method can be used for both qualitative and quantitative investigation, and it is especially useful for finding new substances, keeping an eye on drug metabolites, and profiling biomolecules. LC–MS is an essential instrument in current analytical and bioanalytical chemistry because it is versatile, accurate, and works with many different types of samples.

2.5.3.2 Gas chromatography–mass spectrometry

GC–MS is a complex method of analysis that combines the ability of GC to separate substances with the ability of MS to identify molecules. In this procedure, the gas chromatograph initially separates volatile and thermally stable chemicals based on their boiling points and how they interact with the stationary phase. After being separated, these chemicals are put into a mass spectrometer, where they are ionized and broken up into smaller pieces to make a mass spectrum. This is like a fingerprint that shows the molecular structure and makeup of each analyte. GC–MS is well-known for being very sensitive, specific, and able to find very small amounts of chemicals in complicated combinations. It is widely utilized in areas like food safety, forensic toxicology, environmental monitoring, and pharmaceutical analysis. The method works for both qualitative and quantitative examination, and it is notably useful for finding novel compounds and verifying the presence of known substances with a high level of confidence. GC–MS is the best tool for analytical labs throughout the world since it is strong and accurate.

2.6 Challenges and future perspectives

Algies such as microalgae and macroalgae characterize one of the richest sources of structurally diverse bioactive compounds with enormous potential in nutraceuticals, pharmaceuticals, functional foods, cosmetics, and so on. These organisms produce number of biomolecules, such as protein, polysaccharides, sterols, lipids, phenolics, pigments, and PUFAs, which are having various therapeutic properties including antioxidant, anticancer antimicrobial, antiinflammatory, and antidiabetic activities (Silva et al., 2025). However, the isolation, purification, and characterization of these algal bioactive compounds possess significant technical, economic, and regulatory challenges. Furthermore, despite advancement in extraction and analytical methods, the exploitation of algal bioactives remains controlled by issues of scalability, reproducibility, safety, and standardization. However, the algal composition may be altered due to various factors such as seasonal changes, habitat differences, and cultivation conditions. These factors lead to varying the extraction value, posing challenges for reproducible bioactive potential and quality control. Moreover, the presence of complex polysaccharide compounds and pigments frequently affects with downstream purification and quantification (Bouafir et al., 2025). Therefore advancement of hybrid extraction systems by combining physical, enzymatic, and solvent-based methods may help to overcome these challenges by improving selectivity and sustainability. This chapter highlights the emerging challenges and future perspectives in advancing techniques for the isolation and characterization of bioactive compounds from algae. It predominantly underlines three major areas: (1) technical and regulatory challenges hindering standardization and commercialization, (2) the integration of omics-driven technologies for comprehensive biochemical profiling, and (3) future prospects in synthetic biology and metabolomic engineering to improve yield, precision, and sustainability (Ali et al., 2025).

2.6.1 Technical and regulatory challenges

The extraction and characterization of algal bioactive compounds require not only advanced scientific tools but also compliance with stringent safety and regulatory frameworks. The complexity of algal species diversity and their biochemical variability pose a unique challenge for developing standardized extraction protocols and analytical methods (Cadar et al., 2025; Sathasivam et al., 2019). Many microalgal species, for instance, produce secondary metabolites that can act as potent toxins, such as microcysteines, saxitoxins, anatoxins, and autotoxins, which may contaminate extracts during isolation (Caruana and Amzil, 2018). Ensuring the exclusion of such toxin-producing strains through molecular screening, genomic authentication, and cultivation under controlled conditions is essential to maintain safety in downstream processing.

Traditionally, extraction methods using organic solvents, acids, and alkalies often damage the natural structure or activity of delicate compounds. The industry is moving toward eco-friendly green extraction techniques, such as supercritical critical fluid extraction, pressurized liquid extraction, microwave-assisted extraction, and enzymatic hydrolysis. The main hurdle remains scaling these newer, gentler technologies for industrial use while preserving compound integrity remains a challenge.

Overcoming regulatory requirements is a major difficulty for the commercial success of algal products. Currently, there is no globally harmonized framework for approving algal-derived bioactives in food or pharmaceutical formulation (Yu et al., 2024). While Spirulina and Chlorella are officially recognized as safe for consumption (GRAS) in some countries, most other algae and their compounds have not been thoroughly tested for safety. Without this approval, their use in foods or supplements remains limited and requires further safety evaluation (Torres-Tiji et al., 2020). These regulatory challenges hinder industrial investment and slow down the development of new products. Moreover, the absence of standardized quality control markers and reference compounds makes it difficult to ensure batch-to-batch consistency, purity, and potency (Milana et al., 2025).

Allergenic potential and nucleic acid content also present safety considerations. Some algal proteins and polysaccharides may cause allergic responses, as observed in clinical trial studies on *Coccomyxa* sp. (Milana et al., 2025). Similarly, the high nucleic acid content in microalgal biomass can increase uric acid levels and cause metabolic disturbances such as gout. These issues necessitate toxicological profiling, genetic selection, and postharvest processing techniques (e.g., enzymatic degradation or mild thermal treatment) to reduce potential risks (Ampofo and Abbey, 2022).

Finally, economic feasibility and environmental sustainability remain significant challenges. Cultivation systems for microalgae demand high energy inputs for aeration, lighting, and temperature regulation, and inefficient extraction systems inflate operational costs (Ampofo and Abbey, 2022; Martínez-Ruiz et al., 2025). Adoption of integrated biorefinery models that utilize every biomass component, including pigments, proteins, carbohydrates, and lipids, can enhance profitability and reduce environmental burden. Life cycle assessment approaches are increasingly used to assess environmental sustainability across extraction, purification, and waste management stages (Goswami et al., 2022).

2.6.2 Integration with omics technologies

Recent advances in omics technologies such as proteomics, metabolomics, genomics, transcriptomics, and lipidomics have changed our understanding of algal biology and significantly enhanced the precision of bioactive compound discovery and characterization (Mishra et al., 2019).

Genomics and transcriptomics allow for the identification of genes and regulatory elements that are responsible for the synthesis of rich algal bioactive compounds such as carotenoids, phycobiliproteins, or sulfated polysaccharides (Rai et al., 2016). Comparatively, genomic studies have revealed vital metabolic pathways and regulatory genes that can be targeted for enhancing bioactive yield (Wei et al., 2023). Transcriptomic profiling under different environmental or stress conditions such as light intensity, salinity, or nutrients provides insights into transcriptional responses that modulate metabolite accumulation (Wei et al., 2023).

Proteomics complements these findings by identifying enzymes involved in critical biosynthetic steps. Integrated with MS-based metabolomics, researchers can identify enzyme–substrate interactions, assess metabolic fluxes, and discover new secondary metabolites of algae with bioactive potential (González-López et al., 2021). The utilization of metabolomics in algal studies enables the comprehensive quantification of small molecules, facilitating the identification of biomarkers for quality control and authenticity testing of algal extracts (Gupta et al., 2025; Yang et al., 2023).

Furthermore, integrating multiomics data with bioinformatics, artificial intelligence, and machine learning algorithms allows predictive modeling of metabolic pathways (Guo et al., 2025; Helmy et al., 2023). For instance, omics-guided studies in Chlorella and Nannochloropsis have identified critical transcriptional regulators for lipid and carotenoid biosynthesis (Chen et al., 2017; Udayan et al., 2018). These insights have accelerated the development of metabolic engineering strategies to improve yield and optimize culture conditions. Thus omics-driven analytics form the backbone of next-generation precision isolation and characterization techniques, enabling researchers to target specific metabolites based on molecular signatures rather than random screening (Lauritano et al., 2019).

2.6.3 Future prospects in synthetic biology and metabolomic

The future of algal bioactive research is becoming closely linked with synthetic biology and metabolomic engineering. These advanced fields make it possible to create improved algal strains that can produce more valuable compounds and make the extraction process easier and more efficient. Synthetic biology acts as advanced tools such as CRISPR-Cas9, TALENs, and recombinational genetic engineering methods to precisely control algal genomes, introduce heterologous genes, or silence unwanted pathways. These interventions can increase the yield of metabolites, reduce nucleic acid content, or eliminate toxin-producing genes, thus enhancing both productivity and safety (Bouafir et al., 2025).

In this context, using synthetic biology, scientists can modify the metabolic pathways as well as gene promoters to control activity of cell toward producing specific beneficial compounds like phenolic antioxidants, omega-3 fatty acids, and fucoxanthin. This method also helps reduce energy loss by controlling and minimizing unnecessary or competing processes within the cell (Tanaka et al., 2025). Furthermore, cocultivation or synthetic strain of algae with valuable bacteria can help to promote mutual exchange of metabolites and also improve growth and productivity.

Metabolomic engineering, integrated with hyphenated analytical techniques such as LC–MS/MS, GC–MS, and NMR, provides comprehensive insights into metabolite dynamics and flux distribution. When combining the metabolomics with metabolic flux analysis, scientists can recognize rate-limiting enzymatic steps and strategically upregulate or downregulate specific genes to improve yield. This type of precision-driven metabolic transformation is crucial for identifying algal bioactives to commercial production, though sustaining reliability and bioefficacy (Schmitz and Nunes, 2022).

In addition to these biological advancements, eco-friendly and sustainable extraction methods are gaining more interest. Modern extraction techniques such as ultrasound-assisted extraction, enzyme-assisted extraction, and SFE support to decrease the use of hazardous solvents and energy though maintaining the natural properties of the extracted compounds (Sun et al., 2025). In the field of algal bioactives, advanced approaches such as nanoencapsulation and biopolymeric delivery systems are gaining importance for improving the stability, solubility, and absorption of some algal bioactives. These include carotenoids (e.g., astaxanthin and fucoxanthin), polyphenols (e.g., phlorotannins), and omega-3 polyunsaturated fatty acids (e.g., EPA and DHA). By protecting these sensitive bioactives from degradation, which is caused by light, heat, or oxidation, such delivery systems sustained their stability. Moreover, they allow for targeted and controlled release, enhancing the effectiveness of algal-derived ingredients in functional foods, supplements, and pharmaceutical formulations (Chadorshabi et al., 2025).

2.7 Conclusion

Marine algae serve as a sustainable and renewable source of bioactive compounds with immense potential for applications in the food and agriculture industries. Rich in PUFAs, proteins, polysaccharides, and polyphenols, these marine organisms offer diverse functional and therapeutic benefits, enhancing nutritional value and promoting human health. Apart from food applications, algal bioactives play vital roles in agriculture as biostimulants, biofertilizers, and biopesticides, contributing to improved crop productivity and environmental sustainability.

Despite their various therapeutic potential, the large-scale adoption of algae-derived bioactive compounds faces challenges related to safety regulations, quality assurance, and consumer acceptance. Future progress in this field will depend on the integration of omics technologies, synthetic biology, and advanced extraction techniques. Omics approaches will facilitate the precise identification of biosynthetic pathways, while metabolomic engineering and synthetic biology will enable the development of optimized algal strains with enhanced bioactive yields and reduced ecological impact.

Furthermore, integrating nanotechnology enhances the bioavailability, stability, and efficacy of algal compounds, increasing their commercial and therapeutic potential. However, to ensure global acceptance, robust regulatory frameworks, standardized safety assessments, and international quality control guidelines are essential. Collaborative efforts among academia, industry, and policymakers are equally important to bridge laboratory innovations with industrial-scale applications.

As green and high-throughput extraction methods continue to evolve, algae are poised to become the cornerstone of sustainable biotechnology, offering novel solutions for human health, food security, and a circular bioeconomy, truly positioning them as the "green factories" of the future.

Acknowledgment

The authors gratefully acknowledge the use of BioRender for the creation of schematic illustrations and graphical representations included in this chapter. BioRender was used solely as a visualization tool to enhance the clarity and presentation of scientific concepts, and all figures were designed and interpreted by the authors. The authors also acknowledge the support of available scientific literature and standard laboratory methodologies that contributed to the development of this work. No external funding was received for the preparation of this chapter.

AI disclosure

During the preparation of this manuscript, the authors used AI and AI-assisted technologies in image creation. The authors take full responsibility for the content, scientific accuracy, and integrity of all figures presented in this work.

References

Adarme-Vega, T.C., Lim, D.K.Y., Timmins, M., Vernen, F., Li, Y., Schenk, P.M., 2012. Microalgal biofactories: A promising approach towards sustainable omega-3 fatty acid production. Microbial Cell Factories 11, 96. Available at https://doi.org/10.1186/1475-2859-11-96.

Ali, S.S., Al-Tohamy, R., Al-Zahrani, M., Schagerl, M., Kornaros, M., Sun, J., 2025. Advancements and challenges in microalgal protein production: A sustainable alternative to conventional protein sources. Microbial Cell Factories 24 (1), 61. Available at https://doi.org/10.1186/s12934-025-02685-1.

Ampofo, J., Abbey, L., 2022. Microalgae: Bioactive composition, health benefits, safety and prospects as potential high-value ingredients for the functional food industry. Foods 11 (12), 1744. Available at https://doi.org/10.3390/foods11121744.

Badawy, M.E.I., Rabea, E.I., 2022. Solid-phase extraction techniques: Principles, developments and applications in analytical chemistry. TrAC Trends in Analytical Chemistry 147, 116531.

Becker, E.W., 2007. Microalgae as a source of protein. Biotechnology Advances 25 (2), 207–210. Available at https://doi.org/10.1016/j.biotechadv.2006.11.002.

Bezerra, M.A., Santelli, R.E., Oliveira, E.P., Villar, L.S., Escaleira, L.A., 2008. Response surface methodology (RSM) as a tool for optimization in analytical chemistry. Talanta 76 (5), 965–977. Available at https://doi.org/10.1016/j.talanta.2008.05.019.

Blunt, J.W., Carroll, A.R., Copp, B.R., Davis, R.A., Keyzers, R.A., Prinsep, M.R., 2018. Marine natural products. Natural Product Reports 35 (1), 8–53. Available at https://doi.org/10.1039/C7NP00052A.

Borowitzka, M.A., 2013. High-value products from microalgae—Their development and commercialisation. Journal of Applied Phycology 25, 743–756. Available at https://doi.org/10.1007/s10811-013-9983-9.

Bouafir, Y., Bouhenna, M.M., Nebbak, A., Belfarhi, L., Aouzal, B., Boufahja, F., et al., 2025. Algal bioactive compounds: A review on their characteristics and medicinal properties. Fitoterapia 183, 106591. Available at https://doi.org/10.1016/j.fitote.2025.106591.

Buszewski, B., Szultka, M., 2012. Past, present, and future of solid phase extraction: A review. Critical Reviews in Analytical Chemistry 42 (3), 198–213. Available at https://doi.org/10.1080/10408347.2011.644528.

Cadar, E., Popescu, A., Dragan, A.M., Pesterau, A.M., Pascale, C., Anuta, V., et al., 2025. Bioactive compounds of marine algae and their potential health and nutraceutical applications: A review. Marine Drugs 23 (4), 152. Available at https://doi.org/10.3390/md23040152.

Caruana, A., Amzil, Z., 2018. Microalgae and toxins. In: Levine, I., Fleurence, J. (Eds.), Microalgae in Health and Disease Prevention. Academic Press, London, pp. 263–305. Available at https://doi.org/10.1016/B978-0-12-811405-6.00013-X.

Chadorshabi, S., Mohammadi, M., Shamloo, E., Mahmoudzadeh, M., 2025. Marine algae-derived bioactives: A sustainable resource for the food and agriculture industries. Food Frontiers 6 (6), 2514–2541. Available at https://doi.org/10.1002/fft2.70082.

Chemat, F., Rombaut, N., Sicaire, A.G., Meullemiestre, A., Fabiano-Tixier, A.S., Abert-Vian, M., 2017. Ultrasound assisted extraction of food and natural products: Mechanisms, techniques, combinations, protocols and applications. A review.

Ultrasonics Sonochemistry 34, 540–560. Available at https://doi.org/10.1016/j.ultsonch.2016.06.035.
Chemat, F., Vian, M.A., Cravotto, G., 2012. Green extraction of natural products: Concept and principles. International Journal of Molecular Sciences 13 (7), 8615–8627. Available at https://doi.org/10.3390/ijms13078615.
Chen, H., Zheng, Y., Zhan, J., He, C., Wang, Q., 2017. Comparative metabolic profiling of the lipid-production green microalga Chlorella reveals that nitrogen and carbon metabolic pathways contribute to lipid metabolism. Biotechnology for Biofuels 10, 153. Available at https://doi.org/10.1186/s13068-017-0839-4.
Chisti, Y., 2007. Biodiesel from microalgae. Biotechnology Advances 25 (3), 294–306. Available at https://doi.org/10.1016/j.biotechadv.2007.02.001.
Coskun, O., 2016. Separation techniques: Chromatography. Northern Clinics of Istanbul 3 (2), 156–160. Available at https://doi.org/10.14744/nci.2016.32757.
Delattre, C., Fenoradosoa, T.A., Michaud, P., 2016. Galactans: An overview of their extraction, purification, and biological activities. Marine Drugs 14 (1), 5. Available at https://doi.org/10.3390/md14010005.
Dugheri, S., Bonari, A., Pompilio, I., Boccalon, P., Mucci, N., Arcangeli, G., 2021. Solid-phase extraction techniques in occupational and environmental exposure assessment. Separations 8 (3), 37. Available at https://doi.org/10.3390/separations8030037.
González-López, N.M., Huertas-Ortiz, K.A., Leguizamon-Guerrero, J.E., Arias-Cortés, M.M., Tere-Peña, C.P., García-Castañeda, J.E., et al., 2021. Omics in the detection and identification of biosynthetic pathways related to mycotoxin synthesis. Analytical Methods 13 (36), 4038–4054. Available at https://doi.org/10.1039/D1AY01017D.
Goswami, R.K., Agrawal, K., Upadhyaya, H.M., Gupta, V.K., 2022. Microalgae conversion to alternative energy, operating environment and economic footprint: An influential approach towards energy conversion and management. Energy Conversion and Management 269, 116118. Available at https://doi.org/10.1016/j.enconman.2022.116118.
Guedes, A.C., Amaro, H.M., Malcata, F.X., 2011. Microalgae as sources of carotenoids. Marine Drugs 9 (4), 625–644. Available at https://doi.org/10.3390/md9040625.
Guo, B., Lu, X., Jiang, X., Shen, X.L., Wei, Z., Zhang, Y., 2025. Artificial intelligence in advancing algal bioactive ingredients: Production, characterization, and application. Foods 14 (10), 1783. Available at https://doi.org/10.3390/foods14101783.
Gupta, S., Abu-Ghannam, N., 2011. Bioactive potential and possible health effects of edible brown seaweeds. Trends in Food Science & Technology 22 (6), 315–326. Available at https://doi.org/10.1016/j.tifs.2011.03.011.
Gupta, S., Gaur, P.K., Lal, N., 2025. Recent progress in marine algae research: Bridging biodiversity with pharmaceutical potential. Bulletin of the National Research Centre 49, 77. Available at https://doi.org/10.1186/s42269-025-01372-w.
Harnedy, P.A., FitzGerald, R.J., 2012. Bioactive peptides from marine processing waste and shellfish: A review. Journal of Functional Foods 4 (1), 6–24. Available at https://doi.org/10.1016/j.jff.2011.09.001.
Helmy, M., Elhalis, H., Liu, Y., Chow, Y., Selvarajoo, K., 2023. Perspective: Multiomics and machine learning help unleash the alternative food potential of microalgae. Advances in Nutrition 14 (1), 1–11. Available at https://doi.org/10.1016/j.advnut.2022.11.002.
Herrero, M., Sánchez-Camargo, A.P., Cifuentes, A., Ibáñez, E., 2010. Plants, seaweeds, microalgae and fungi as sources of functional ingredients obtained using supercritical fluid extraction. Journal of Chromatography. A 1217 (16), 2495–2511. Available at https://doi.org/10.1016/j.chroma.2009.06.019.
Holdt, S.L., Kraan, S., 2011. Bioactive compounds in seaweed: Functional food applications and legislation. Journal of Applied Phycology 23 (3), 543–597. Available at https://doi.org/10.1007/s10811-010-9632-5](10.1007/s10811-010-9632-5).

Janson, J.C., Jönsson, J.A., 2011. Protein Purification: Principles, High Resolution Methods, and Applications, Second ed. Wiley-VCH, Weinheim.

Kim, S.K., Pangestuti, R., 2018. Cosmetic applications of marine algae. Journal of Cosmetic Dermatology 17 (4), 528–536. Available at https://doi.org/10.1111/jocd.12320.

Kondeti, V.S., Ramesh, B., Reddy, N.V., 2014. Isolation and purification of bioactive compounds from natural sources by column chromatography. Journal of Chemical and Pharmaceutical Research 6 (7), 220–226.

Lauritano, C., Ferrante, M.I., Rogato, A., 2019. Marine natural products from microalgae: An omics overview. Marine Drugs 17 (5), 269. Available at https://doi.org/10.3390/md17050269.

Li, Y.X., Wijesekara, I., Li, Y., Kim, S.K., 2011. Phlorotannins as bioactive agents from brown algae. Process Biochemistry 46 (12), 2219–2224. Available at https://doi.org/10.1016/j.procbio.2011.09.015.

Lordan, S., Ross, R.P., Stanton, C., 2011. Marine bioactives as functional food ingredients: Potential to reduce the incidence of chronic diseases. Marine Drugs 9 (6), 1056–1100. Available at https://doi.org/10.3390/md9061056.

Markou, G., Nerantzis, E., 2013. Microalgae for high-value compounds and biofuels production: A review with focus on cultivation under stress conditions. Biotechnology Advances 31 (8), 1532–1542. Available at https://doi.org/10.1016/j.biotechadv.2013.07.011.

Martínez-Ruiz, F.E., Andrade-Bustamante, G., Holguín-Peña, R.J., Renganathan, P., Gaysina, L.A., Sukhanova, N.V., et al., 2025. Microalgae as functional food ingredients: Nutritional benefits, challenges, and regulatory considerations for safe consumption. Biomass 5 (2), 25. Available at https://doi.org/10.3390/biomass5020025.

Milana, M., van Asselt, E., van der Fels-Klerx, I.H.J., 2025. A review of the toxicological effects and allergenic potential of emerging alternative protein sources. Comprehensive Reviews in Food Science and Food Safety 24 (1), e70123. Available at https://doi.org/10.1111/1541-4337.70123.

Mishra, A., Medhi, K., Malaviya, P., Thakur, I.S., 2019. Omics approaches for microalgal applications: Prospects and challenges. Bioresource Technology 291, 121890. Available at https://doi.org/10.1016/j.biortech.2019.121890.

Newman, D.J., Cragg, G.M., 2020. Natural products as sources of new drugs over the nearly four decades from 01/1981 to 09/2019. Journal of Natural Products 83 (3), 770–803. Available at https://doi.org/10.1021/acs.jnatprod.9b01285.

Ozturk, E., 2022. Solid-phase extraction techniques in analytical chemistry: Recent advances and applications. Journal of Chromatographic Science 60 (3), 245–258. Available at https://doi.org/10.1093/chromsci/bmab099.

Plaza, M., Cifuentes, A., Ibáñez, E., 2009. In the search of new functional food ingredients from algae. Trends in Food Science & Technology 20 (1), 31–39. Available at https://doi.org/10.1016/j.tifs.2008.07.012.

Plaza, M., Herrero, M., Cifuentes, A., Ibáñez, E., 2008. Innovative natural functional ingredients from microalgae. Journal of Agricultural and Food Chemistry 56 (16), 7221–7230. Available at https://doi.org/10.1021/jf800775s.

Plaza, M., Turner, C., Ibáñez, E., Herrero, M., 2010. Pressurized hot water extraction of bioactives. Trends in Analytical Chemistry 29 (11), 1165–1175. Available at https://doi.org/10.1016/j.trac.2010.07.001.

Plotka-Wasylka, J., Szczepańska, N., de la Guardia, M., Namieśnik, J., 2015. Modern trends in solid-phase extraction: New sorbents, formats and applications. TrAC Trends in Analytical Chemistry 77, 23–43. Available at https://doi.org/10.1016/j.trac.2015.10.010.

Pulz, O., Gross, W., 2004. Valuable products from biotechnology of microalgae. Applied Microbiology and Biotechnology 65 (6), 635–648. Available at https://doi.org/10.1007/s00253-004-1647-x.

Rai, V., Karthikaichamy, A., Das, D., Noronha, S., Wangikar, P.P., Srivastava, S., 2016. Multi-omics frontiers in algal research: Techniques and progress to explore biofuels in the postgenomics world. OMICS: A Journal of Integrative Biology 20 (7), 387–399. Available at https://doi.org/10.1089/omi.2016.0061.

Sathasivam, R., Radhakrishnan, R., Hashem, A., Abd Allah, E.F., 2019. Microalgae metabolites: A rich source for food and medicine. Saudi Journal of Biological Sciences 26 (4), 709–722. Available at https://doi.org/10.1016/j.sjbs.2017.11.003.

Schmitz, C., Nunes, A., Munaro, D., Gerber, T., Maraschin, M., 2022. Metabolomics applied to the discovery of new bioactive pharmaceuticals in complex matrices. In: In: Rahman, A. (Ed.), Studies in Natural Products Chemistry Vol. 74. Elsevier, Amsterdam, pp. 165–195. Available at https://doi.org/10.1016/B978-0-323-91099-6.00006-2.

Silva, M., Geada, P., Pereira, R.N., Teixeira, J.A., 2025. Microalgae biomass – A source of sustainable dietary bioactive compounds towards improved health and well-being. Food Chemistry Advances 6, 100926. Available at https://doi.org/10.1016/j.focha.2025.100926.

Simpson, N.J.K., 2000. Solid-Phase Extraction: Principles, Techniques, and Applications. Marcel Dekker, New York.

Sun, S., Yu, Y., Jo, Y., Han, J.H., Xue, Y., Cho, M., et al., 2025. Impact of extraction techniques on phytochemical composition and bioactivity of natural product mixtures. Frontiers in Pharmacology 16, 1615338. Available at https://doi.org/10.3389/fphar.2025.1615338.

Tanaka, K., Lan, J.C., Kondo, A., Hasunuma, T., 2025. Metabolic engineering and cultivation strategies for efficient production of fucoxanthin and related carotenoids. Applied Microbiology and Biotechnology 109 (1), 57. Available at https://doi.org/10.1007/s00253-025-13441-1.

Torres-Tiji, Y., Fields, F.J., Mayfield, S.P., 2020. Microalgae as a future food source. Biotechnology Advances 41, 107536. Available at https://doi.org/10.1016/j.biotechadv.2020.107536.

Udayan, A., Kathiresan, S., Arumugam, M., 2018. Kinetin and gibberellic acid act synergistically to produce high-value polyunsaturated fatty acids in *Nannochloropsis oceanica*. Algal Research 32, 182–192. Available at https://doi.org/10.1016/j.algal.2018.03.007.

Wei, F., Wan, R., Shi, Z., Ma, W., Wang, H., Chen, Y., et al., 2023. Transcriptomics and metabolomics reveal the critical genes of carotenoid biosynthesis and color formation. Plants 12 (15), 2791. Available at https://doi.org/10.3390/plants12152791.

Wells, M.L., Potin, P., Craigie, J.S., Raven, J.A., Merchant, S.S., Helliwell, K.E., et al., 2017. Algae as nutritional and functional food sources: Revisiting our understanding. Journal of Applied Phycology 29 (2), 949–982. Available at https://doi.org/10.1007/s10811-016-0974-5.

Wijesekara, I., Pangestuti, R., Kim, S.K., 2011. Biological activities and potential health benefits of sulfated polysaccharides derived from marine algae. Carbohydrate Polymers 84 (1), 14–21. Available at https://doi.org/10.1016/j.carbpol.2010.10.062.

Yang, Y., Hassan, S.H.A., Awasthi, M.K., Gajendran, B., Sharma, M., Ji, M.K., et al., 2023. Recent progress on the bioactive compounds from algal biomass for human health applications. Food Bioscience 51, 102267. Available at https://doi.org/10.1016/j.fbio.2022.102267.

Yu, B.S., Pyo, S., Lee, J., Han, K., 2024. Microalgae: A multifaceted catalyst for sustainable solutions in renewable energy, food security, and environmental management. Microbial Cell Factories 23 (1), 308. Available at https://doi.org/10.1186/s12934-024-02588-7.

CHAPTER 3

Deep eutectic solvents for extracting bioproducts from algae

Qi Qi Koh[1,2], Yin Leng Kua[1], and Zhi Ling Chew[1]
[1]School of Energy and Chemical Engineering, Xiamen University Malaysia, Sepang, Selangor, Malaysia
[2]Centre for Net-Zero Technology (Future Centre), Monash University Malaysia, Subang Jaya, Selangor, Malaysia

3.1 Deep eutectic solvents

The term “eutectic," originating from the Greek for "easy to melt" or "low melting," was introduced by Frederick Guthrie in 1884 to describe materials that liquefy at a lower temperature than other combinations (Martins and Pinho, 2019). In 2003, Abbott coined the term deep eutectic solvents (DES) for eutectic mixtures of amides and quaternary ammonium salts with low melting points, highlighting their potential as alternatives to ionic liquids. This definition was broadened in 2004 to include mixtures of Lewis or Brønsted acids and bases with various ionic species (Kovács and Neyts, 2020). DES applications expanded to electrochemistry, such as cyclic voltammetry, in 2007 (Paiva and Craveiro, 2014). In 2011, Choi et al. (2011) introduced "natural deep eutectic solvents" (NADES), derived from plant metabolites, gaining attention for their biodegradability, sustainability, and nontoxicity.

3.1.1 Preparation method

DES is typically formed by mixing a hydrogen bond donor (HBD) and a hydrogen bond acceptor (HBA) at specific molar ratios and temperatures, usually between 50°C and 100°C. High temperatures may cause degradation through esterification reactions (El Achkar et al., 2021). Various preparation methods for DES are summarized in Table 3.1.

3.1.2 Phase behavior

DES is formed through the complexation of HBA and HBD at a certain molar ratio. Upon mixing, the overall melting point is lowered as

Applied Phycology
DOI: https://doi.org/10.1016/B978-0-443-24758-3.00014-9

Table 3.1 DES preparation methods.

Method of preparation	Descriptions	References
Heating and stirring	The HBA and HBD are heated at 80°C with stirring until a clear, homogeneous liquid forms.	Dai et al. (2013); Doldolova and Bener (2021); Gutiérrez and Ferrer (2009); Meneses and Santos (2019); Popovic et al. (2022)
Ultrasonic assisted	The HBA and HBD are ultrasonicated at 50°C, 30 W, and 37 kHz until a clear, homogeneous liquid forms.	Bajkacz and Adamek (2018); Popovic et al. (2022)
Microwave assisted	The HBA and HBD are microwaved for 3 × 7 s with 5-s pauses for cooling and vortexing to prevent overheating.	Popovic et al. (2022)
Vacuum evaporation	The HBA and HBD are dissolved in deionized water, mixed, and evaporated using a rotary evaporator until a clear, homogeneous liquid forms.	Dai et al. (2013); Meneses and Santos (2019)
Freeze drying	The HBA and HBD are dissolved in deionized water, mixed, frozen with liquid nitrogen, and freeze-dried for 48 h to ensure complete dryness.	Gutiérrez and Ferrer (2009); Meneses and Santos (2019)

DES, Deep eutectic solvents; *HBA*, hydrogen bond acceptor; *HBD*, hydrogen bond donor.

compared to the raw materials, which can be represented through a solid–liquid phase diagram (El Achkar et al., 2021). Fig. 3.1 shows the solid–liquid phase diagram, whereby the eutectic point represents the minimum temperature required for the formation of the DES. The degree of the interaction between A and B is linked to the difference in the freezing point at the eutectic composition of a binary mixture of A + B compared to that of a theoretical ideal mixture. The freezing point of the DES lies between –69°C and 150°C. According to El Achkar et al. (2021),

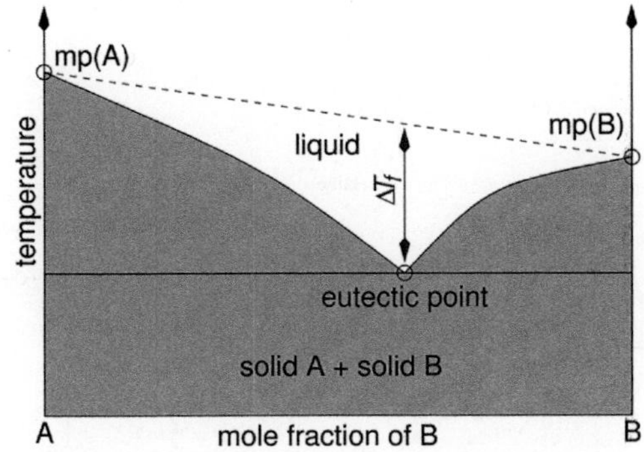

Figure 3.1 Solid–liquid phase diagram of deep eutectic solvents (Smith et al., 2014).

the larger the interaction, the greater the difference in freezing point. Based on Smith et al. (2014), DES has a melting point lower than the ideal eutectic temperature and presents as a liquid state at operating temperature. The melting point depression in DES depends on its composition (Kovács and Neyts, 2020). Adding a third component disrupts hydrogen bonding, lowering the melting point, as observed by Maugeri et al. (2012) with glycerol (Gly), reducing viscosity, density, and melting point. Factors such as cation asymmetry, lower HBD molecular weight, and high anion electron affinity further reduce the melting point by weakening interactions and lattice energy.

3.1.3 Structure

DES are eutectic solvents formed through the complexation of an HBA and an HBD, where the interaction significantly lowers the freezing point. Fig. 3.2 highlights common HBA and HBD used in DES synthesis. Choline chloride (ChCl) is the most common HBA in DES due to its affordability, nontoxicity, and environmental friendliness. It is also approved as a natural additive for various animal species (El Achkar et al., 2021). Upon DES formation, hydrogen bonds between the HBA and HBD can be detected using spectroscopic techniques such as fourier transform infrared spectroscopy (FT-IR), nuclear magnetic resonance (NMR), and Overhauser spectroscopy. Studies confirm extensive hydrogen bonding within DES components, forming a supramolecular structure with a robust hydrogen-bond network (Choi et al., 2011; Dai and Witkamp, 2013, 2015). For instance, Delgado et al. (2018) observed

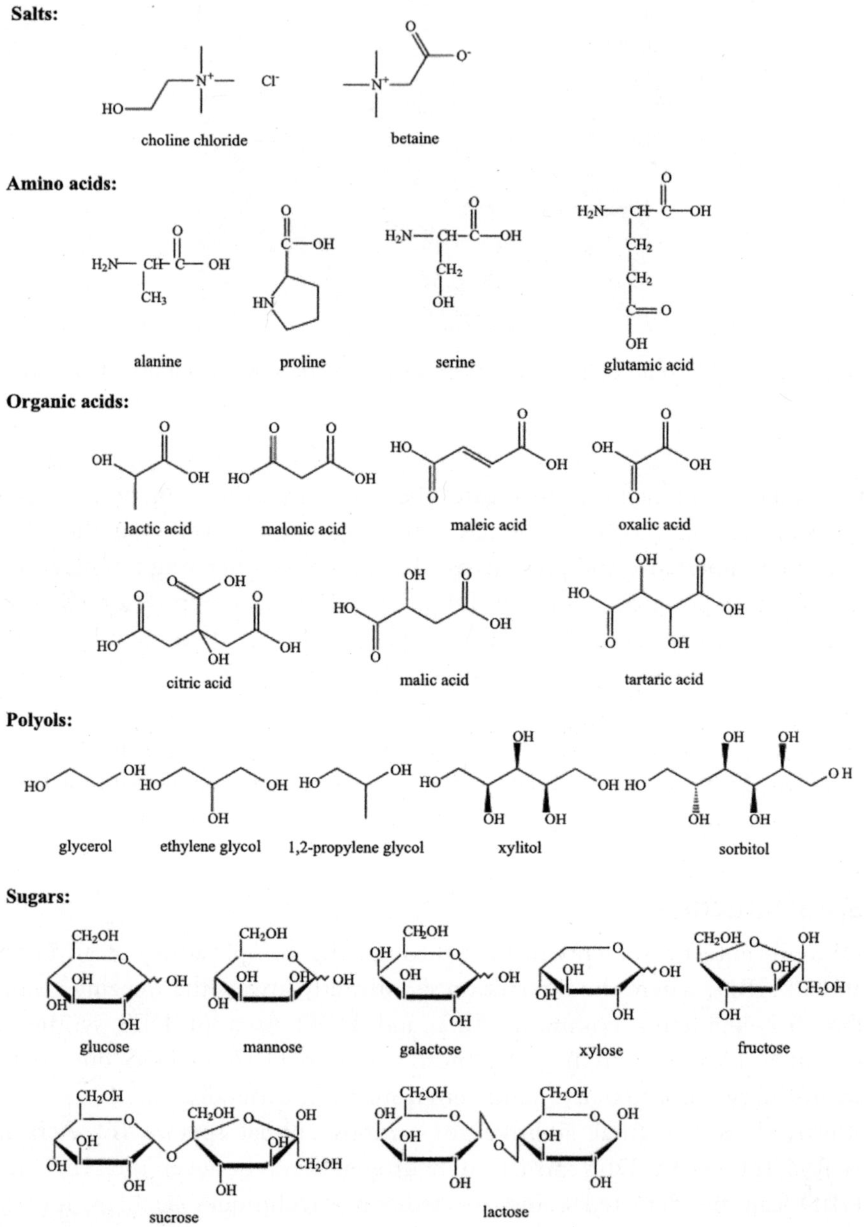

Figure 3.2 Common natural compound used in natural deep eutectic solvent preparation (Yang, 2018).

hydroxyl stretching and bending at 2920/cm and 1414/cm, while Dai and Witkamp (2015) reported a stretching vibration shift at 1717/cm and 2450/cm, indicating hydrogen-bond formation. However, excessive water disrupts the hydrogen bonds in DES, rupturing its supramolecular structure. Dai and Witkamp (2015) found that adding over 50% water destroys the hydrogen bonds, while lower amounts (less than 50%) preserve the supramolecular characteristics. For DES formed from 1,2-propanediol (PDO), ChCl, and water, a downshift in the hydroxyl group peak in the NMR spectrum indicated progressive hydrogen-bond rupture.

3.2 Application of deep eutectic solvents

3.2.1 Textile industry

Cellulose nanofiber is widely used in textiles due to its mechanical strength, colorfastness, and comfort (Mehdi et al., 2022). Materials like viscose, rayon, Tencel, and modal are derived from cellulose fibers. Traditional dyeing methods rely on toxic sodium salts and sulfates (Lellis et al., 2019; Mehdi et al., 2022), prompting interest in DES as a safer alternative. Mehdi et al. (2022) demonstrated that ChCl:ethylene glycol (EG) DES outperformed conventional media in dyeing cellulose nanofiber, achieving higher color strength (4.87 vs 3.1). Similarly, Pawar and Athalye (2021) found that ChCl:urea DES-assisted dyeing enhanced color strength on silk in less time (37 minutes) compared to conventional methods. DES improves hydrophilicity and swelling, enhancing dye diffusivity and color strength. DES-treated fibers exhibited improved wash fastness, tensile strength, and crystallinity, attributed to the ChCl cation's dye substantivity and the OH group's effect on the fiber structure (Mehdi et al., 2022; Pawar and Athalye, 2021).

Keratin, a fibrous protein from hair, feathers, and nails, is valued for its biodegradability, nontoxicity, and mechanical properties like tensile strength and elasticity, making it ideal for textiles. DES-based methods, such as ChCl:oxalic acid, have been used to extract keratin from rabbit hair, yielding flakes with a molecular weight of 3.8–5.8 kDa (Asla et al., 2023). The extracted keratin has lower thermal stability due to a disordered structure and reduced crystallinity, but is suitable for low molecular weight textile applications. Incorporating keratin enhances fabric properties, such as dye uptake, making it a promising material for sustainable textiles. Bamboo fibers are valued in textiles for their moisture absorption, tensile strength, and anti-UV properties, but are limited by their rough texture. DES has been used to improve bamboo fiber texture and softness, enhancing their suitability for textiles (Ye and Zhu, 2022). This modification broadens bamboo fiber applications, enabling the production of moisture-wicking fabrics, comfortable clothing, and high-performance sportswear while meeting the growing demand for eco-friendly materials.

Lastly, the textile industry generates significant wastewater, with dyeing producing harmful effluents. Traditional treatment methods, such as filtration, coagulation, and oxidation, often rely on volatile organic compounds, leading to dye degradation or sludge disposal issues. DES offers a greener alternative for wastewater treatment due to its well-established extraction technologies. A DES composed of tetrabutylammonium chloride, thymol, and menthol achieved over 95% dye removal across all concentrations, with minimal DES usage (DES:aqueous phase ratio of 1:32) and reduced residue generation (Asla et al., 2023). DES requires less processing time and no additional extraction stages, making it an efficient, eco-friendly solution for dyeing wastewater. DES presents a sustainable and versatile solution for the textile industry by enhancing the properties of cellulose nanofibers, keratin, and bamboo fibers while offering eco-friendly alternatives for dyeing processes and wastewater treatment.

3.2.2 Food industry

Land activities like mining and fertilization contribute to cadmium accumulation in soil, with rice being particularly prone to absorbing it (Huang and Feng, 2018). Despite the 0.4 mg/kg cadmium limit, contaminated rice is still produced, posing health risks through kidney accumulation (Huang and Feng, 2018; Mišan et al., 2020). To address this, Huang and Feng (2018) explored DES as eco-friendly washing agents for cadmium removal from rice flour. Using a heated solution and ultrasound-assisted extraction, ChCl:xylitol DES achieved 90.4% cadmium removal without altering the rice flour's composition. DES are water-soluble, easy to separate, and offer sustainable applications in the cereal industry. Additionally, DES can be easily dissolved in water and separated from the rice flour, making it a convenient washing agent with broad applications in the cereal-based industry (Huang and Feng, 2018; Mišan et al., 2020).

DES, such as ChCl:Gly and ChCl:urea, also effectively removed ochratoxin A from wheat, bread crumbs, and biscuits with an 89%

recovery rate, meeting European Commission standards (Piemontese and Perna, 2017). Additionally, DES-based microextraction using ChCl:urea successfully removed lead and cadmium from edible oils, achieving recovery rates of 95%–104% with excellent accuracy and precision (Karimi and Dadfarnia, 2015; Mišan et al., 2020). These findings highlight DES as a safer, eco-friendly alternative to volatile organic solvents for heavy metal removal in the food industry. On top of that, gluten, a protein found in grains, can cause adverse effects in individuals with celiac disease or gluten intolerance. Celiac disease, linked to HLA-DQ2 and HLA-DQ8 genotypes, triggers an autoimmune response to gluten, leading to inflammation, nutrient deficiencies, and other health issues. Traditional gluten extraction methods using ethanolic solutions pose environmental and health risks, prompting interest in safer alternatives like DES. Lores et al. (2017) demonstrated that NADES made from fructose and citric acid achieved comparable results to conventional solvents, with extractability enhanced by adding 50% water. Citric acid also stabilized the protein, enabling its reuse as a food additive. Thus DES provides a sustainable and safe alternative for gluten extraction.

The demand for healthier foods with reduced salt, sugar, and monosodium glutamate (MSG) has highlighted the need for innovative, food-grade taste enhancers. DES has gained attention for its nontoxic, affordable, and safe properties. Kranz et al. (2018) demonstrated that DES significantly accelerated the Maillard reaction, producing various taste enhancers. Sugar-based DES (glucose and sucrose) yielded the highest flavor enhancer production (489.0 μmol/mmol). Key molecules, such as 1-deoxy-D-fructosyl-*N*-*β*-alanyl-L-histidine (49%), *N*-(1-methyl-4-oxoimidazolidin-2-ylidene) aminopropionic acid (54%), and *N*2-(1-carboxyethyl) guanosine 50-monophosphate (22%), were produced at 80°C–100°C within 2 hours, imparting distinct flavors like meaty, sour, and umami to food. In conclusion, DES has demonstrated immense potential as a sustainable, nontoxic alternative for various food industry applications, including the removal of heavy metals and gluten, as well as enhancing food flavor through accelerated Maillard reactions, aligning with the demand for healthier and eco-friendly solutions.

3.2.3 Biomass conversion

High lignin content in the biomass requires severe pretreatment conditions, which in turn increase the cost of the biomass supply chain. Traditional extraction methods are costly and may damage target components, prompting interest in eco-friendly alternatives such as DES, which effectively dissolve lignin and enable milder pretreatment conditions (Kalhor and Ghandi, 2019).

To reduce costs, pretreatment temperatures are kept below 150°C, as higher temperatures lower lignin content but increase biomass loss and operating expenses. For instance, potato peel biomass recovery using ChCl:Gly DES fell from 85% at 60°C to 53% at 150°C. Increasing solvent content offers an alternative approach. Doubling the solvent ratio from 8 to 16 g/g biomass reduced lignin content from 26.3% to 24.5%, with a slight decrease to 23.2% at 32 g/g biomass. Additionally, DES can be efficiently recovered by distillation, minimizing solvent waste.

Numerous investigations have been carried out to explore the potential of DES and NADES for the pretreatment of various types of lignocellulosic biomass. For example, Xu et al. (2016) examined the performance of seven DES for pretreating corn stover to generate biobutanol. They found that ChCl:formic acid DES, which is an acidic DES, was the most effective at eliminating hemicellulose and lignin. In a similar study, Chen et al. (2018) utilized ChCl:EG DES in acidic conditions to pretreat switchgrass, resulting in significant removal of lignin and hemicellulose, with cellulose enrichment up to 72.6%. The resulting fermentable sugars were transformed into 2,3-butanediol with a concentration of 90.2 g/L. Mamilla et al. (2019) investigated several ChCl-based DES for the fractionation of lignocellulosic biomass, discovering that ChCl:oxalic acid and ChCl:potassium hydroxide DES were successful in dissolving beech wood polymers. Procentese et al. (2018b) employed ChCl:Gly and ChCl:EG DES to pretreat various biomass residues, including apple, potato, coffee, and spent brewer's grains. The authors discovered that the optimal pretreatment conditions were a 3-hour treatment with ChCl:Gly DES at a solid:solvent ratio of 1:16 and a temperature of 115°C. These studies highlighted the potential of DES and NADES as effective and sustainable pretreatment agents for biomass conversion.

Adding water to eutectic solvents can enhance their viscosity and lignin removal capacity. New et al. (2019) demonstrated that using ChCl:urea DES with added water improved delignification of oil palm fronds at 120°C for 4 hours compared to pure DES. Similarly, Procentese et al. (2017) found that ChCl:Gly DES required less energy for lettuce leaf pretreatment in biobutanol production than NaOH or steam explosion methods. These studies highlight the importance of optimizing eutectic solvent composition and treatment conditions for efficient and sustainable biomass conversion.

3.2.4 Energy storage industry

Lithium-ion batteries are widely used due to their high energy density, long lifespan, and low self-discharge (Jaumaux et al., 2020; Wu et al., 2021). However, the use of carbonate and ether-based liquid electrolytes poses safety risks, such as short circuits and thermal runaway. Safer alternatives are needed to enhance electrochemical stability and ionic conductivity while addressing these concerns.

Jaumaux et al. (2020) developed a DES electrolyte composed of 2-(3-(6-methyl-4-oxo-1,4-dihydropyrimidin-2-yl)ureido)ethylmethacrylate (UPyMA), pentaerythritol tetraacrylate (PETEA) monomers, lithium bis (trifluoromethanesulfonyl)imide (LiTFSI), *N*-methylacetamide (NMA), and fluoroethylene carbonate (FEC) to address issues with lithium (Li) metal anodes, such as dendrite growth and unstable solid electrolyte interphase (Wu et al., 2021). The addition of FEC improved ionic conductivity, oxidative stability, and facilitated the formation of protective interfaces on both the Li anode and the $LiMn_2O_4$ cathode, preventing structural degradation and enhancing safety. The UPyMA-PETEA polymeric network maintained the DES in a quasisolid state, preventing electrolyte leakage, improving electrode–electrolyte integration, and suppressing manganese dissolution from the cathode. The DES electrolyte exhibited nonflammability, high ionic conductivity (1.79×10^{-3} S/cm at 25°C), and superior electrochemical stability (Jaumaux et al., 2020). Jhong et al. (2009) explored the use of DES in dye-sensitized solar cells as a substitute for volatile acetonitrile, which, despite achieving over 11% light-to-electricity conversion efficiency, poses leakage issues. Replacing acetonitrile with a DES composed of ChCl and Gly, the solar cell achieved a short-circuit photocurrent density (Jsc) of 12.0 mA cm^2, an open-circuit photovoltage (Voc) of 553 mV, a fill factor (FF) of 0.582, and a photovoltaic conversion efficiency of 3.88%. In comparison, the acetonitrile-based cell achieved 4.89% efficiency, with DES delivering 79% of its performance.

Zinc-ion batteries (ZIBs) have gained attention due to their unique composition, including a zinc metal anode, zinc-containing electrolyte, and Zn-ion accommodating cathode. ZIBs offer advantages like low toxicity, high volumetric capacity, and resistance to oxygen and humidity in the zinc anode, which lowers processing costs (Ming et al., 2019). However, challenges such as uncontrolled dendrite growth and side reactions at the Zn-electrolyte interface, especially in aqueous electrolytes, hinder the development of rechargeable Zn batteries (Kao-ian et al., 2019; Ming et al., 2019). A study by Kao-ian et al. (2019) explored a DES composed of ChCl and urea as an electrolyte for rechargeable ZIBs, demonstrating excellent electrochemical performance with high specific capacity (170 mAh/g) and dendrite-free cycling. Additionally, Zhao et al. (2019) developed a stable aqueous electrolyte by incorporating 6 wt.% water into a urea-based DES, improving zinc-electrolyte compatibility. The resulting water-in-DES electrolyte exhibited high energy density (52 Wh/kg) and enhanced stability, offering superior rechargeability and a long lifespan for ZIBs. In conclusion, DES electrolytes offer safer, more efficient alternatives, enhancing the performance and safety of lithium-ion batteries and ZIBs and improving dye-sensitized solar cells.

3.3 Usage of deep eutectic solvents for algae bioproducts

Algae have long been valued for their beneficial properties across various industries. Recent advancements in extraction methods, particularly DES, have enhanced the isolation of bioactive compounds from algae, with growing interest from fields such as pharmaceuticals, nutraceuticals, and cosmetics. As shown in Fig. 3.3, research on DES-based extraction has steadily increased from 2012 to 2022, highlighting its potential. The cosmetic industry, in particular, has shown significant interest in algae-derived bioactive compounds for their skin health benefits. However, the use of DES remains limited due to the complex extraction process and a lack of understanding of optimal conditions. With continued research, DES-based extraction is expected to become a more widely used method.

3.3.1 Carotenoids

Astaxanthin, a carotenoid with remarkable antioxidant, antiinflammatory, and antitumor properties, surpasses α-tocopherol in antioxidant activity due to its molecular structure, which includes hydroxyl and carbonyl groups (Couteau and Coiffard, 2018; Lee and Row, 2016). *Haematococcus pluvialis*, a major source of astaxanthin, can produce over 3 g of the pigment per kg of dry biomass, making it an important industrial crop (Couteau and Coiffard, 2018). However, traditional extraction methods are inefficient,

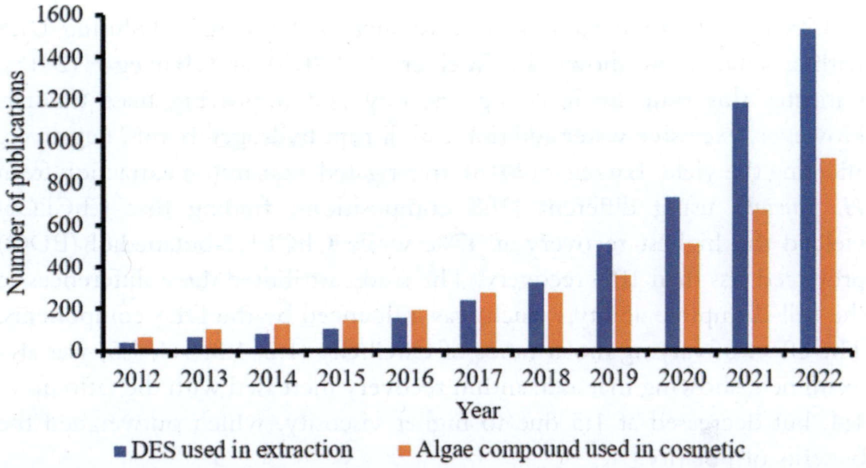

Figure 3.3 Number of publications in 2012–22 related to the application of deep eutectic solvent (DES) as an extraction solvent and the use of algae bioactive compounds in cosmetic products, based on the ScienceDirect database (searched keywords: "deep eutectic solvent," "extraction," "bioactive compound," "algae," "cosmetics").

leading researchers to explore DES as an additive to enhance yield. DES offers an environmentally friendly and cost-effective alternative to conventional solvents, significantly improving astaxanthin extraction (Lee and Row, 2016). Astaxanthin was extracted from various marine organisms, including *Fenneropenaus chinensis*, *Portunus trituberculatus* waste, *Laminaria japonica*, and *Undaria pinnatifida*, with the highest concentrations found in *P. trituberculatus* waste (25.17 μg/g) (Couteau and Coiffard, 2018). A preliminary study using dichloromethane/methanol (25:75, v/v) identified the most astaxanthin-rich organisms. To optimize extraction, six organic solvents were tested, with acetone yielding the highest amount (5.47 μg/g), while hexane produced no extractable astaxanthin. These findings highlight the importance of solvent choice, as astaxanthin's solubility in polar solvents led to higher yields with acetone (Couteau and Coiffard, 2018). The addition of DES (phosphonium: 1,2-butanediol) significantly improved astaxanthin extraction from *P. trituberculatus* waste, with a 73.59% increase in yield (39.37 μg/g) compared to organic solvents under optimal conditions (1 g powder, 40 mL acetone, 1.25 mg/mL DES, 65 W sonication for 90 minutes) (Couteau and Coiffard, 2018). The DES's supermolecular structure, with extensive hydrogen bonding, enhanced the extraction efficiency. However, a higher DES concentration

(>0.25 mg/mL) reduced yield due to increased viscosity. Diluting DES with a solvent, as shown by Wei et al. (2015) and Barzegar (2018), mitigates this issue by lowering viscosity and improving mass transfer. However, excessive water addition can disrupt hydrogen bonds, negatively affecting the yield. Barzegar (2018) investigated astaxanthin extraction from *H. pluvialis* using different DES compositions, finding that ChCl:Gly yielded the highest recovery at 47%, while ChCl:1,2-butanediol (BDO) produced less than 10% recovery. The study attributed these differences to the cell-disruptive ability, which was influenced by the DES components. The effect of varying molar ratios of ChCl:Gly (1:2, 1:3, 1:4, 1:5) was also examined, showing that astaxanthin recovery increased with the ratio up to 1:4, but decreased at 1:5 due to higher viscosity, which outweighed the benefits of polarity.

Fucoxanthin, a carotenoid from brown algae, offers antioxidant, antiobesity, anticancer, and antidiabetic properties, helping regulate fat metabolism and reduce body weight, fat, and triglycerides (Gille et al., 2019; Abidov et al., 2010). It also inhibits cancer cell growth and improves glucose (Glu) and lipid metabolism, making it a promising natural therapeutic agent for various diseases. Xu et al. (2022) screened DES for fucoxanthin extraction using COSMO-RS, identifying thymol:dodecanoic acid as the most effective solvent, yielding 7.69 mg/g DW, 22.1% higher than methanol and 14.8% higher than ethanol. The study found that increasing extraction temperature (up to 36 °C), time (up to 3 hours), and biomass percentage (11.36%) enhanced the yield, with a maximum of 22.03 mg/g DW. The results were attributed to thymol:dodecanoic acid's low polarity and the ability to interact with both polar and nonpolar compounds, making it an ideal solvent for fucoxanthin extraction. Obluchinskaya and Pozharitskaya (2021) compared hydrophilic DES (ChCl:lactic acid (LA) and LA:Glu:water) with ethanol for fucoxanthin extraction from *Fucus vesiculosus*, finding ethanol yielded the highest amount (1.10 mg/g DW) followed by ChCl:lLA (0.24 mg/g DW) and LA:Glu:water (0.17 mg/g DW). The fucoxanthin yield was significantly lower in all three methods compared to the findings in Xu et al. (2022), which was attributed to the high viscosity and polarity of the solvents, unsuitable for fucoxanthin's hydrophobic nature. The viscosities of LA:Glu:water (185.14 mPa.s) and ChCl:LA (12.20 mPa.s) were reported by Rebocho et al. (2022) and Che Zain et al. (2021). Despite both being hydrophilic, ChCl:LA's lower viscosity results in better extractability. High viscosity impedes mass transfer, reducing fucoxanthin yield. Adding water to reduce viscosity isn't effective for fucoxanthin, as it is hydrophobic and insoluble in water, as shown by Obluchinskaya and Pozharitskaya (2021). Hydrophobic DES, with a polarity closer to fucoxanthin, extracts higher yields than hydrophilic ones, as confirmed by Xu et al. (2022) and Obluchinskaya and Pozharitskaya (2021). This highlights the advantage of hydrophobic DES for extracting nonpolar compounds.

Lutein, a yellow-orange carotenoid in algae, is a potent antioxidant and hydrophobic compound. Fan et al. (2022) used hydrophobic DES, specifically Fen:Thy, to extract lutein from *Scenedesmus sp.* and achieved a yield of 6.26 mg/g, higher than the 3.91 mg/g obtained with ethyl acetate. The highest lutein yield was due to Fen:Thy's strong Van der Waals and hydrogen bond interactions with lutein. Meanwhile, Muhammad et al. (2022) used polyol-based DES (ChCl:Gly) to extract lutein from *Chlorella pyrenoidosa*, obtaining 0.79 mg/g, which was 119.44% higher than conventional potassium hydroxide extraction. Both studies highlighted the importance of optimal temperature, time, and biomass-to-solvent ratio for maximizing lutein yield. Hydrophobic DES offers a sustainable, efficient method for lutein extraction from microalgae.

3.3.2 Polysaccharides

Shang et al. (2021) studied polysaccharide extraction from seaweed using various DES, finding alcohol-based DES yielded the highest (108.89 mg/g), outperforming organic acid- and urea-based DES. Alcohol-based DES also showed better extraction efficiency than methanol and ethanol. Dai et al. (2013) confirmed that alcohol-based DES had lower viscosity (51.3 mm^2/s) than organic acid-based DES (445.9 mm^2/s), enhancing its effectiveness. Nie et al. (2020) compared DES extraction from *Sargassum horneri* with hot water extraction, achieving a 75% time savings with DES, though the conventional method yielded slightly more polysaccharides (13.52% vs 11.31%). Optimized DES conditions (1:2 molar ratio, 1:30 solid–liquid ratio, 30% water, 70°C, 30 minutes) preserved polysaccharide structure. Both studies confirmed the biological activity of extracted polysaccharides, highlighting DES as a promising alternative for efficient, eco-friendly polysaccharide extraction with potential food and pharmaceutical applications.

Carrageenan, a hydrophilic, water-soluble sulfated polysaccharide found in red algae such as *Kappaphycus alverezii*, is widely used in the cosmetics industry due to its stabilizing and gelling properties (Lochhead, 2017; Shafie et al., 2022). It has antimicrobial, antitumor, and antiaging benefits, and enhances product stability without forming emulsions (Morais et al., 2021; Shafie et al., 2022). Traditional extraction methods, involving strong alkalis, suffer from low yield, high water consumption, and costly waste treatment (Liu et al., 2022). A study by Das et al. (2016) compared DES (ChCl-based) extraction with conventional methods and found that DES extraction (up to 60.25% yield) outperformed traditional methods (36.58% yield). Adding water to DES improved yield by reducing viscosity, with optimal extraction occurring at viscosities between 34 and 52 cP. The choline ion in DES enhanced carrageenan's affinity, making DES a more efficient solvent than water or conventional methods. This highlights DES as a promising, eco-friendly alternative for carrageenan extraction, offering potential for cost and environmental savings.

Fucoidan, a bioactive polysaccharide found in brown seaweeds, is known for its antioxidant, anticoagulant, antiinflammatory, and anticancer properties but is difficult to extract due to its high molecular weight, complexity, and structural variability. Several studies have explored the use of DES for its extraction. Li and Dai (2019) demonstrated that ChCl-Urea DES yielded the highest fucoidan, attributed to electrostatic and ion-exchange interactions that improved the extraction process. The urea-based DES enhanced the polymerization process, increasing extraction efficiency. Saravana and Cho (2018) also studied fucoidan extraction from *Saccharina japonica* using DES, with ChCl-Gly achieving the highest yield due to the steric hindrance of Gly's three hydroxyl groups, reducing interactions between polysaccharides and chloride ions. The study showed that sonication time and solid–liquid ratio significantly impacted yield, with the highest fucoidan yield (90 mg/g) achieved at 25 minutes of sonication. Beyond this, the yield plateaued. An optimal solid–liquid ratio of 30–50 mg/mL was found, as increasing the ratio further did not enhance yield. Temperature played a crucial role, with 150°C being optimal for fucoidan extraction, as it improved mass transfer and solvent diffusion while reducing DES viscosity and surface tension. However, excessive heat could degrade fucoidan. The study also found that the optimal DES volume was 2 mL; higher volumes reduced yield due to mutual electronic effects that hindered bonding between DES and fucoidan. Thus optimizing factors like sonication time, solid–liquid ratio, temperature, and DES volume is crucial for efficient fucoidan extraction (Li and Dai, 2019; Saravana and Cho, 2018).

3.3.3 Other natural pigments

Pigments from algae, such as chlorophyll, polyphenol, and phycocyanin, are valued for their antipollutant, antiinflammatory, and antiaging properties, making them sought-after in cosmetics. Traditional organic solvents like chloroform and methanol, though effective, are toxic and environmentally unsustainable. Researchers have explored using DES, which are nontoxic, biodegradable, and eco-friendly, as an alternative. A study by Wils et al. (2021) found that hydrophilic DES, particularly Glu and Gly, efficiently extracted phycocyanin from *Arthrospira platensis*, achieving 21.9 mg/g. However, the extraction of nonpolar bioactive compounds like carotene was low in this DES. A similar study by Martins et al. (2023) reported Glu:Gly as more efficient in extracting phycocyanin and chlorophyll compared to water. Water was better for carotene extraction, with a yield of 0.20 mg/g compared to 0.13 mg/g from Glu:Gly. Furthermore, adding water to DES improved extraction efficiency, particularly for phycocyanin, which increased from 21.9 to 70.8 mg/g with a 10%–20% water addition (Wils et al., 2021). Water reduced DES viscosity and surface tension, enhancing solubility and yield. Thus DES, especially with added water, proves to be a promising method for extracting multiple pigments from algae. In hydrophobic DES, no phycocyanin was detected due to polarity differences, while carotene and chlorophyll yields were similar to ethanol extraction (Wils et al., 2021). Both DES types showed selectivity for chlorophyll; hydrophilic DES preserved chlorophyll, preventing its degradation to pheophytin, whereas hydrophobic DES did not. These results highlight how DES polarity and composition influence selective extraction.

Polyphenols, found in plants and algae, are typically extracted using organic solvents like ethyl acetate, methanol, and acetone, but these can leave harmful residues (Wan Mahmood and Lorwirachsutee, 2019). Green DES extraction has been developed as a safer alternative. Research on *Chlorella vulgaris* showed that DES, such as ChCl:Gly 1:2, ChCl:EG 1:4, ChCl:PDO 1:4, and ChCl:BDO 1:4, outperformed ethyl acetate in polyphenol extraction, yielding up to twice the amount. The higher efficiency of DES is attributed to its matching polarity and viscosity with polyphenols. Similar results were found in other studies, with DES performing better than water and ethanol for extracting phlorotannin (Obluchinskaya et al., 2019). However, methanol showed higher efficiency than DES for extracting phlorotannin due to its polarity.

Extraction efficiency can be improved by adjusting viscosity, temperature, extraction time, and biomass-to-solvent ratio. Higher polyol concentrations reduce the viscosity of DES, increasing the mass transfer rate and boosting polyphenol yield. However, high Gly concentrations increase viscosity, reducing yield (Wan Mahmood and Lorwirachsutee, 2019). Adding water to DES also reduces viscosity and enhances polyphenol bonding, significantly improving yields, with up to 40% water increasing phlorotannin yield (Obluchinskaya and Pozharitskaya, 2021). Temperatures around 60°C are optimal, as higher temperatures may degrade polyphenols. Longer extraction times increase yield, but oxidation occurs beyond 100 minutes. The optimum liquid-to-solvent ratio was found to be 20:1, balancing yield without wastage (Wan Mahmood and Lorwirachsutee, 2019).

3.4 Challenges and future prospects of deep eutectic solvents

Research on the recovery, recycling, and scalability of DES in extraction processes is limited (Vigier and Chatel, 2015). A study on DES reuse in switchgrass pretreatment showed that 95% of DES could be recovered per cycle without loss of effectiveness (Satlewal and Agrawal, 2018). However, challenges remain in scaling up DES-based processes for commercial use, requiring careful evaluation of feasibility, cost-effectiveness, and regulatory factors for broader industrial adoption (Vigier and Chatel, 2015). The low conductivity of DES limits their use in energy materials (Vigier and Chatel, 2015), while their high hygroscopicity leads to poor weather resistance and reduced voltage window. Achieving anhydrous DES is crucial, as current research on hydrophobic DES for this purpose is limited, and addressing moisture sensitivity is key to expanding their applications. Current DES research focuses on common HBA and HBD combinations, limiting their potential applications. To expand DES uses, computational studies are needed to better understand DES properties across various HBA and HBD combinations, providing valuable insights into their behavior and opening new opportunities for utilization. Lastly, there is a lack of comprehensive technoeconomic and life cycle analyses for DES-based technologies. Technoeconomic analysis evaluates the economic feasibility, while life cycle analysis assesses environmental impacts from production to disposal. The absence of these analyses highlights the need to demonstrate the sustainability and economic viability of DES-based technologies.

Additionally, while DES are considered more environmentally friendly than traditional solvents, challenges remain in their production, disposal, and potential environmental accumulation. Further research is needed to address these sustainability concerns and fully harness DES as a greener alternative.

3.5 Conclusion

In conclusion, the integration of DES with algae offers significant potential in cosmeceuticals and other industries, including textiles, food, and energy. DES extraction methods have proven effective for extracting bioactive compounds from algae, highlighting the versatility of DES. Despite challenges such as viscosity control, recycling, and low conductivity, further research is needed to enhance the practicality and scalability of DES-based systems. Additionally, exploring new combinations of DES and conducting technoeconomic and life-cycle analyses will help assess their feasibility and environmental impact. Overcoming these challenges will unlock advancements in sustainable and innovative solutions across various sectors.

Acknowledgments

This work was supported by the Xiamen University Malaysia Research Fund (XMUMRF) (XMUMRF/2019-C3/IENG/0010).

References

Abidov, M., Ramazanov, Z., Seifulla, R., Grachev, S., 2010. The effects of Xanthigen ™ in the weight management of obese premenopausal women with non-alcoholic fatty liver disease and normal liver fat. Diabetes, Obesity and Metabolism 12 (1), 72–81. Available at https://doi.org/10.1111/j.1463-1326.2009.01132.x.

Asla, A., Martínez-Rico, Ó., Otero, P., Domínguez, Á., González, B., 2023. Water decolorization using tuned ternary deep eutectic solvents. Journal of Molecular Liquids 381, 121832. Available at https://doi.org/10.1016/j.molliq.2023.121832.

Bajkacz, S., Adamek, J., 2018. Development of a method based on natural deep eutectic solvents for extraction of flavonoids from food samples. Food Analytical Methods 11 (5), 1330–1344. Available at https://doi.org/10.1007/s12161-017-1118-5.

Barzegar, R. Extraction of Astaxanthin from H. pluvialis Using Deep Eutectic Solvents. https://ir.lib.uwo.ca/etd/5952/

Che Zain, M.S., Yeoh, J.X., L., S.Y., Shaari, K., 2021. Physicochemical Properties of Choline Chloride-Based Natural Deep Eutectic Solvents (NaDES) and Their Applicability for Extracting Oil Palm Flavonoids. Sustainability 13 (23), 12981. Available at https://doi.org/10.3390/su132312981.

Chen, Z., Bai, X., Lusi, A., Wan, C., 2018. High-Solid Lignocellulose Processing Enabled by Natural Deep Eutectic Solvent for Lignin Extraction and Industrially Relevant Production of Renewable Chemicals. ACS Sustainable Chemistry & Engineering 6 (9), 12205–12216. Available at https://doi.org/10.1021/acssuschemeng.8b02541.

Choi, Y.H., van Spronsen, J., Dai, Y., Verberne, M., Hollmann, F., Arends, I.W.C.E., et al., 2011. Are natural deep eutectic solvents the missing link in understanding cellular metabolism and physiology? Plant Physiology 156 (4), 1701–1705. Available at https://doi.org/10.1104/pp.111.178426, http://www.plantphysiol.org/content/156/4/1701.full.pdf+html.

Couteau, C., Coiffard, L., 2018. Microalgal application in cosmetics. Microalgae in Health and Disease Prevention. Elsevier, France, pp. 317–323. Available at https://doi.org/10.1016/B978-0-12-811405-6.00015-3, http://www.sciencedirect.com/science/book/9780128114056.

Dai, Y., van Spronsen, J., Witkamp, G.J., Verpoorte, R., Choi, Y.H., 2013. Natural deep eutectic solvents as new potential media for green technology. Analytica Chimica Acta 766, 61–68. Available at https://doi.org/10.1016/j.aca.2012.12.019.

Dai, Y., Witkamp, G.J., Verpoorte, R., Choi, Y.H., 2015. Tailoring properties of natural deep eutectic solvents with water to facilitate their applications. Food Chemistry 187, 14–19. Available at https://doi.org/10.1016/j.foodchem.2015.03.123, www.elsevier.com/locate/foodchem.

Das, A.K., Sharma, M., Mondal, D., Prasad, K., 2016. Deep eutectic solvents as efficient solvent system for the extraction of κ-carrageenan from Kappaphycus alvarezii. Carbohydrate Polymers 136, 930–935. Available at https://doi.org/10.1016/j.carbpol.2015.09.114.

Delgado-Mellado, N., Larriba, M., Navarro, P., Rigual, V., Ayuso, M., García, J., Rodríguez, F., 2018. Thermal stability of choline chloride deep eutectic solvents by TGA/FTIR-ATR analysis. Journal of Molecular Liquids 260, 37–43. Available at https://doi.org/10.1016/j.molliq.2018.03.076.

Doldolova, K., Bener, M., Lalikoğlu, M., Aşçı, Y.S., Arat, R., Apak, R., 2021. Optimization and modeling of microwave-assisted extraction of curcumin and antioxidant compounds from turmeric by using natural deep eutectic solvents. Food Chemistry 353. Available at https://doi.org/10.1016/j.foodchem.2021.129337, www.elsevier.com/locate/foodchem.

El Achkar, T., Greige-Gerges, H., Fourmentin, S., 2021. Basics and properties of deep eutectic solvents: a review. Environmental Chemistry Letters 19 (4), 3397–3408. Available at https://doi.org/10.1007/s10311-021-01225-8.

Fan, C., Liu, Y., Shan, Y., Cao, X., 2022. A priori design of new natural deep eutectic solvent for lutein recovery from microalgae. Food Chemistry 376, 131930. Available at https://doi.org/10.1016/j.foodchem.2021.131930.

Gille, A., Stojnic, B., Derwenskus, F., Trautmann, A., Schmid-Staiger, U., Posten, C., et al., 2019. A lipophilic fucoxanthin-rich *Phaeodactylum tricornutum* extract ameliorates effects of diet-induced obesity in C57BL/6J mice. Nutrients 11 (4). Available at https://doi.org/10.3390/nu11040796, https://www.mdpi.com/2072-6643/11/4/796/pdf.

Gutiérrez, M.C., Ferrer, M.L., Mateo, C.R., Monte, F.D., 2009. Freeze-drying of aqueous solutions of deep eutectic solvents: a suitable approach to deep eutectic suspensions of self-assembled structures. Langmuir: the ACS Journal of Surfaces and Colloids 25 (10), 5509–5515. Available at https://doi.org/10.1021/la900552bSpain, http://pubs.acs.org/doi/pdfplus/10.1021/la900552b.

Huang, Y., Feng, F., Chen, Z.G., Wu, T., Wang, Z.H., 2018. Green and efficient removal of cadmium from rice flour using natural deep eutectic solvents. Food Chemistry 244,

260–265. Available at https://doi.org/10.1016/j.foodchem.2017.10.060, www.elsevier.com/locate/foodchem.

Jaumaux, P., Liu, Q., Zhou, D., Xu, X., Wang, T., Wang, Y., et al., 2020. Deep-eutectic-solvent-based self-healing polymer electrolyte for safe and long-life lithium-metal batteries. Angewandte Chemie International Edition 59 (23), 9134–9142. Available at https://doi.org/10.1002/anie.202001793.

Jhong, H.-R., Wong, D.S.-H., Wan, C.-C., Wang, Y.-Y., Wei, T.-C., 2009. A novel deep eutectic solvent-based ionic liquid used as electrolyte for dye-sensitized solar cells. Electrochemistry Communications 11 (1), 209–211. Available at https://doi.org/10.1016/j.elecom.2008.11.001.

Kalhor, P., Ghandi, K., 2019. Deep eutectic solvents for pretreatment, extraction, and catalysis of biomass and food waste. Molecules (Basel, Switzerland) 24 (22), 4012. Available at https://doi.org/10.3390/molecules24224012.

Kao-ian, W., Pornprasertsuk, R., Thamyongkit, P., Maiyalagan, T., Kheawhom, S., 2019. Rechargeable zinc-ion battery based on choline chloride-urea deep eutectic solvent. Journal of the Electrochemical Society 166 (6), A1063. Available at https://doi.org/10.1149/2.0641906jes.

Karimi, M., Dadfarnia, S., Shabani, A.M.H., Tamaddon, F., Azadi, D., 2015. Deep eutectic liquid organic salt as a new solvent for liquid-phase microextraction and its application in ligandless extraction and preconcentraion of lead and cadmium in edible oils. Talanta 144, 648–654. Available at https://doi.org/10.1016/j.talanta.2015.07.021, https://www.journals.elsevier.com/talanta.

Kovács, A., Neyts, E.C., Cornet, I., Wijnants, M., Billen, P., 2020. Modeling the physicochemical properties of natural deep eutectic solvents. ChemSusChem 13 (15), 3789–3804. Available at https://doi.org/10.1002/cssc.202000286, http://onlinelibrary.wiley.com/journal/10.1002/(ISSN)1864-564X.

Kranz, M., Hofmann, T., 2018. Food-Grade Synthesis of Maillard-Type Taste Enhancers Using Natural Deep Eutectic Solvents (NADES). Molecules 23 (2), 261. Available at https://doi.org/10.3390/molecules23020261.

Lee, Y.R., Row, K.H., 2016. Comparison of ionic liquids and deep eutectic solvents as additives for the ultrasonic extraction of astaxanthin from marine plants. Journal of Industrial and Engineering Chemistry 39, 87–92. Available at https://doi.org/10.1016/j.jiec.2016.05.014, http://www.sciencedirect.com/science/journal/1226086X.

Lellis, B., Fávaro-Polonio, C.Z., Pamphile, J.A., Polonio, J.C., 2019. Effects of textile dyes on health and the environment and bioremediation potential of living organisms. Biotechnology Research and Innovation 3 (2), 275–290. Available at https://doi.org/10.1016/j.biori.2019.09.001.

Li, G., Dai, Y., Wang, X., Row, K.H., 2019. Molecularly imprinted polymers modified by deep eutectic solvents and ionic liquids with two templates for the simultaneous solid-phase extraction of fucoidan and laminarin from marine kelp. Analytical Letters 52 (3), 511–525. Available at https://doi.org/10.1080/00032719.2018.1471697, www.tandf.co.uk/journals/titles/00032719.asp.

Liu, Y., An, D., Xiao, Q., Chen, F., Zhang, Y., Weng, H., et al., 2022. A novel κ-carrageenan extracting process with calcium hydroxide and carbon dioxide. Food Hydrocolloids 127, 107507. Available at https://doi.org/10.1016/j.foodhyd.2022.107507.

Lochhead, R.Y., 2017. The use of polymers in cosmetic products. Cosmetic Science and Technology: Theoretical Principles and Applications. Elsevier Inc., United States, pp. 171–221. Available at http://www.sciencedirect.com/science/book/9780128020050 10.1016/B978-0-12-802005-0.00013-6.

Lores, H., Romero, V., Costas, I., Bendicho, C., Lavilla, I., 2017. Natural deep eutectic solvents in combination with ultrasonic energy as a green approach for solubilisation of proteins: application to gluten determination by immunoassay. Talanta 162, 453–459. Available at https://doi.org/10.1016/j.talanta.2016.10.078.

Mamilla, Jhansi L.K., Novak, Uroš, Grilc, Miha, Likozar, Blaž, 2019. Natural deep eutectic solvents (DES) for fractionation of waste lignocellulosic biomass and its cascade conversion to value-added bio-based chemicals. Biomass and Bioenergy 120, 417–425. Available at https://doi.org/10.1016/j.biombioe.2018.12.002.

Martins, M.A.R., Pinho, S.P., Coutinho, J.A.P., 2019. Insights into the nature of eutectic and deep eutectic mixtures. Journal of Solution Chemistry 48 (7), 962–982. Available at https://doi.org/10.1007/s10953-018-0793-1, http://www.kluweronline.com/issn/0095-9782.

Martins, R., Mouro, C., Pontes, R., Nunes, J., Gouveia, I., 2023. Natural Deep Eutectic Solvent Extraction of Bioactive Pigments from Spirulina platensis and Electrospinning Ability Assessment. Polymers 15 (6), 1574. Available at https://doi.org/10.3390/polym15061574.

Maugeria, Z., Domínguez de María, P., 2012. Novel choline-chloride-based deep-eutectic-solvents with renewable hydrogen bond donors: levulinic acid and sugar-based polyols. RSC Adv. 2 (2), 421–425. Available at https://doi.org/10.1039/c1ra00630d.

Mehdi, M., Hussain, N., Khatri, M., Hakro, R.A., Ghaffar, A., Khatri, Z., et al., 2022. Salts and water-free dyeing of cellulose nanofibers using novel green deep eutectic solvents: isotherm, kinetics, and thermodynamic studies. Journal of Applied Polymer Science 139 (23). Available at https://doi.org/10.1002/app.52279, http://onlinelibrary.wiley.com/journal/10.1002/(ISSN)1097-4628.

Meneses, L., Santos, F., Gameiro, A.R., Paiva, A., Duarte, A.R.C., 2019. Preparation of binary and ternary deep eutectic systems. Journal of Visualized Experiments 2019 (152). Available at https://doi.org/10.3791/60326, https://www.jove.com/pdf/60326.

Ming, J., Guo, J., Xia, C., Wang, W., Alshareef, H.N., 2019. Zinc-ion batteries: materials, mechanisms, and applications. Materials Science and Engineering R: Reports 135, 58–84. Available at https://doi.org/10.1016/j.mser.2018.10.002.

Mišan, A., Nađpal, J., Stupar, A., Pojić, M., Mandić, A., Verpoorte, R., et al., 2020. The perspectives of natural deep eutectic solvents in agri-food sector. Critical Reviews in Food Science and Nutrition 60 (15), 2564–2592. Available at https://doi.org/10.1080/10408398.2019.1650717, https://www.tandfonline.com/loi/bfsn20.

Morais, T., Cotas, J., Pacheco, D., Pereira, L., 2021. Seaweeds compounds: an ecosustainable source of cosmetic ingredients? Cosmetics 8 (1), 8. Available at https://doi.org/10.3390/cosmetics8010008.

Muhammad, G., Wang, J., Xiong, W., Lv, Y., Zhang, S., Zhao, A., Jahanbakhsh-Bonab, P., Solovchenko, A., Xu, J., Alam, Md.A., 2022. Polyol based deep eutectic solvent-assisted pretreatment for enhanced lutein extraction from Chlorella pyrenoidosa. Journal of Molecular Liquids 368, 120775. Available at https://doi.org/10.1016/j.molliq.2022.120775.

New, E.K., Wu, T.Y., Tien Loong Lee, C.B., Poon, Z.Y., Loow, Y.-L., Wei Foo, L.Y., Procentese, A., Siow, L.F., Teoh, W.H., Nik Daud, N.N., Jahim, J.Md., Mohammad, A.W., 2019. Potential use of pure and diluted choline chloride-based deep eutectic solvent in delignification of oil palm fronds. Process Safety and Environmental Protection 123, 190–198. Available at https://doi.org/10.1016/j.psep.2018.11.015.

Nie, J., Chen, D., Lu, Y., 2020. Deep Eutectic Solvents Based Ultrasonic Extraction of Polysaccharides from Edible Brown Seaweed Sargassum horneri. Journal of Marine Science and Engineering 8 (6), 440. Available at https://doi.org/10.3390/jmse8060440.

Obluchinskaya, E.D., Daurtseva, A.V., Pozharitskaya, O.N., Flisyuk, E.V., Shikov, A.N., 2019. Natural deep eutectic solvents as alternatives for extracting phlorotannins from brown algae. Pharmaceutical Chemistry Journal 53 (3), 243–247. Available at https://doi.org/10.1007/s11094-019-01987-0.

Obluchinskaya, E.D., Pozharitskaya, O.N., Zakharova, L.V., Daurtseva, A.V., Flisyuk, E.V., Shikov, A.N., 2021. Efficacy of natural deep eutectic solvents for extraction of hydrophilic and lipophilic compounds from *Fucus vesiculosus*. Molecules (Basel, Switzerland) 26 (14). Available at https://doi.org/10.3390/molecules26144198, https://www.mdpi.com/1420-3049/26/14/4198/pdf.

Paiva, A., Craveiro, R., Aroso, I., Martins, M., Reis, R.L., Duarte, A.R.C., 2014. Natural deep eutectic solvents - solvents for the 21st century. ACS Sustainable Chemistry and Engineering 2 (5), 1063–1071. Available at https://doi.org/10.1021/sc500096j, http://pubs.acs.org/journal/ascecg.

Pawar, S.S., Athalye, A., Adivarekar, R.V., 2021. Solvent assisted dyeing of silk fabric using deep eutectic solvent as a swelling agent. Fibers and Polymers 22 (2), 405–411. Available at https://doi.org/10.1007/s12221-021-0142-7, http://link.springer.com/journal/volumesAndIssues/12221.

Piemontese, L., Perna, F.M., Logrieco, A., Capriati, V., Solfrizzo, M., 2017. Deep eutectic solvents as novel and effective extraction media for quantitative determination of Ochratoxin A in wheat and derived products. Molecules (Basel, Switzerland) 22 (1). Available at https://doi.org/10.3390/molecules22010121, http://www.mdpi.com/1420-3049/22/1/121/pdf.

Popovic, B.M., Micic, N., Potkonjak, A., Blagojevic, B., Pavlovic, K., Milanov, D., et al., 2022. Novel extraction of polyphenols from sour cherry pomace using natural deep eutectic solvents – ultrafast microwave-assisted NADES preparation and extraction. Food Chemistry 366. Available at https://doi.org/10.1016/j.foodchem.2021.130562, www.elsevier.com/locate/foodchem.

Procentese, A., Raganati, F., Olivieri, G., Russo, M.E., Rehmann, L., Marzocchella, A., 2018. Deep Eutectic Solvents pretreatment of agro-industrial food waste. Biotechnology for Biofuels 11 (1). Available at https://doi.org/10.1186/s13068-018-1034-y.

Procentese, A., Raganati, F., Olivieri, G., Russo, M.E., Rehmann, L., Marzocchella, A., 2017. Low-energy biomass pretreatment with deep eutectic solvents for bio-butanol production. Bioresource Technology 243, 464–473. Available at https://doi.org/10.1016/j.biortech.2017.06.143.

Rebocho, S., Mano, F., Cassel, E., Anacleto, B., Bronze, M.doR., Paiva, A., Duarte, A.R.C., 2022. Fractionated extraction of polyphenols from mate tea leaves using a combination of hydrophobic/ hydrophilic NADES. Current Research in Food Science 5, 571–580. Available at https://doi.org/10.1016/j.crfs.2022.03.004.

Saravana, P.S., Cho, Y.N., Woo, H.C., Chun, B.S., 2018. Green and efficient extraction of polysaccharides from brown seaweed by adding deep eutectic solvent in subcritical water hydrolysis. Journal of Cleaner Production 198, 1474–1484. Available at https://doi.org/10.1016/j.jclepro.2018.07.151, https://www.journals.elsevier.com/journal-of-cleaner-production.

Satlewal, A., Agrawal, R., Bhagia, S., Sangoro, J., Ragauskas, A.J., 2018. Natural deep eutectic solvents for lignocellulosic biomass pretreatment: recent developments, challenges and novel opportunities. Biotechnology Advances 36 (8), 2032–2050. Available at https://doi.org/10.1016/j.biotechadv.2018.08.009, www.elsevier.com/inca/publications/store/5/2/5/4/5/5/index.htt.

Shafie, M.H., Kamal, M.L., Zulkiflee, F.F., Hasan, S., Uyup, N.H., Abdullah, S., et al., 2022. Application of carrageenan extract from red seaweed (Rhodophyta) in cosmetic products: a review. Journal of the Indian Chemical Society 99 (9). Available at https://doi.org/10.1016/j.jics.2022.100613, https://www.sciencedirect.com/journal/journal-of-the-indian-chemical-society.

Shang, X.-c., Chu, D., Zhang, J.-x., Zheng, Y.-f., Li, Y., 2021. Microwave-assisted extraction, partial purification and biological activity in vitro of polysaccharides from bladder-wrack (Fucus vesiculosus) by using deep eutectic solvents. Separation and Purification Technology 259, 118169. Available at https://doi.org/10.1016/j.seppur.2020.118169.

Smith, E.L., Abbott, A.P., Ryder, K.S., 2014. Deep Eutectic Solvents (DESs) and Their Applications. Chemical Reviews 114 (21), 11060–11082. Available at https://doi.org/10.1021/cr300162p.

Vigier, K.D.O., Chatel, G., Jérôme, F., 2015. Contribution of deep eutectic solvents for biomass processing: opportunities, challenges, and limitations. ChemCatChem 7 (8), 1250–1260. Available at https://doi.org/10.1002/cctc.201500134.

Wan Mahmood, W.M.A., Lorwirachsutee, A., Theodoropoulos, C., Gonzalez-Miquel, M., 2019. Polyol-based deep eutectic solvents for extraction of natural polyphenolic antioxidants from *Chlorella vulgaris*. ACS Sustainable Chemistry and Engineering 7 (5), 5018–5026. Available at https://doi.org/10.1021/acssuschemeng.8b05642, http://pubs.acs.org/journal/ascecg.

Wei, Z.-F., Wang, X.-Q., Peng, X., Wang, W., Zhao, C.-J., Zu, Y.-G., Fu, Y.-J., 2015. Fast and green extraction and separation of main bioactive flavonoids from Radix Scutellariae. Industrial Crops and Products 63, 175–181. Available at https://doi.org/10.1016/j.indcrop.2014.10.013.

Wils, L., Leman-Loubière, C., Bellin, N., Clément-Larosière, B., Pinault, M., Chevalier, S., et al., 2021. Natural deep eutectic solvent formulations for spirulina: preparation, intensification, and skin impact. Algal Research 56, 102317. Available at https://doi.org/10.1016/j.algal.2021.102317.

Wu, J., Liang, Q., Yu, X., Qiu-Feng, L., Ma, L., Qin, X., et al., 2021. Deep eutectic solvents for boosting electrochemical energy storage and conversion: a review and perspective. Advanced Functional Materials 31 (22). Available at https://doi.org/10.1002/adfm.202011102, http://onlinelibrary.wiley.com/journal/10.1002/(ISSN)1616-3028.

Xu, Dingmeng, Chow, J., Weber, C.C., Packer, M.A., Baroutian, S., Shahbaz, K., 2022. Evaluation of deep eutectic solvents for the extraction of fucoxanthin from the alga Tisochrysis lutea – COSMO-RS screening and experimental validation. Journal of Environmental Chemical Engineering 10 (5), 108370. Available at https://doi.org/10.1016/j.jece.2022.108370.

Xu, G.-C., Ding, J.-C., Han, R.-Z., Dong, J.-J., Ni, Y., 2016. Enhancing cellulose accessibility of corn stover by deep eutectic solvent pretreatment for butanol fermentation. Bioresource Technology 203, 364–369. Available at https://doi.org/10.1016/j.biortech.2015.11.002.

Yang, Z., 2018. Natural Deep Eutectic Solvents and Their Applications in Biotechnology. Advances in Biochemical Engineering/Biotechnology,Application of Ionic Liquids in Biotechnology. Springer International Publishing, pp. 31–59. Available at https://doi.org/10.1007/10_2018_67.

Ye, X.Y., Zhu, E.Q., Wang, D.W., Yang, J., Yang, H.Y., Shi, Z.J., 2022. Cationic functionalized bamboo fibers with spinnable and antibacterial properties prepared in chlorocholine chloride/urea deep eutectic solvent. Industrial Crops and Products 188. Available at https://doi.org/10.1016/j.indcrop.2022.115607, www.elsevier.com/inca/publications/store/5/2/2/8/2/5.

Zhao, J., Zhang, J., Yang, W., Chen, B., Zhao, Z., Qiu, H., Dong, S., Zhou, X., Cui, G., Chen, L., 2019. "Water-in-deep eutectic solvent" electrolytes enable zinc metal anodes for rechargeable aqueous batteries. Nano Energy 57, 625–634. Available at https://doi.org/10.1016/j.nanoen.2018.12.086.

CHAPTER 4

Algae as enzyme biofactories

Pei Fang Sim[1], Gou Rean Wong[1], Der Jiun Ooi[1], and Fong Fong Liew[1,2]
[1]Preclinical Sciences Department, Faculty of Dentistry, MAHSA University, Bandar Saujana Putra, Jenjarom, Selangor, Malaysia
[2]Department of Pharmaceutical Life Sciences, Faculty of Pharmacy, Universiti Malaya, Kuala Lumpur, Malaysia

4.1 Introduction

The global industrial enzyme market has experienced unprecedented growth across multiple sectors, including pharmaceuticals, biofuels, food and beverage processing, detergents, and textiles. Current market projections indicate the industry will reach approximately USD 15.5 billion by 2030, with growth primarily fueled by expanding applications in baking, brewing, and advanced bioprocessing technologies (GVR Catalysts & Enzymes Research Team, 2023). Enzymes have become indispensable in modern food processing. Integrating enzymes in fruit juice production significantly enhances extraction efficiency while improving both flavor profiles and nutritional retention. The technology enables processors to maximize yield from raw materials while maintaining product quality standards. On the other hand, the wine sector has also embraced enzymatic solutions to address critical production challenges. These biocatalysts improve wine clarity, stabilize color compounds, and preserve the complex organoleptic properties that define wine quality. This application demonstrates enzymes' ability to enhance traditional processes while maintaining product authenticity (Cosme et al., 2023). This trend reflects growing awareness of food security and safety considerations, where enzymes play a crucial role in extending shelf life, improving nutritional value, and ensuring consistent product quality (Mohammad et al., 2024).

Enzymes are traditionally extracted from diverse biological sources, including animal tissues like calf stomach, microorganisms, and plant materials. The use of advanced biotechnologies has since expanded production options to include transgenic microorganisms and engineered plant systems. Commercially, enzyme production predominantly relies on microbial platforms that utilize specialized enzyme-producing yeasts and bacterial systems. While these established methods have driven industry

Applied Phycology
DOI: https://doi.org/10.1016/B978-0-443-24758-3.00008-3

growth, they face significant technological and economic constraints that limit their scalability and sustainability (Ndochinwa et al., 2024). One of the most significant obstacles is economic barriers that are primarily driven by high fermentation costs as a direct result of complex nutrient requirements and energy-intensive processing. This severely impacts the profitability of large-scale enzyme manufacturing. To tackle this, industries look toward utilizing alternative low-cost microbial solutions. Nevertheless, while bacterial systems are cost-effective, they frequently lead to the production of insoluble or improperly folded eukaryotic proteins. This, in turn, reduces enzyme activity and overall yield. Additionally, environmental constraints and resource limitations also pose ongoing challenges for expanding conventional production systems to meet growing global demand (Ndochinwa et al., 2024; Fasim et al., 2021).

Algae-based enzyme production has since demonstrated a paradigm shift in biotechnology by offering sustainable solutions to conventional production limitations. The algal kingdom spans two primary categories: microalgae (microscopic, unicellular, or simple multicellular organisms) and macroalgae (large, complex multicellular seaweeds). This taxonomic diversity translates into remarkable metabolic versatility, enabling the production of biofuels, nutraceuticals, pharmaceuticals, and industrial enzymes within a single biological platform. In addition, algae encompass an extraordinarily diverse group of photosynthetic organisms inhabiting both freshwater and marine ecosystems. These photosynthetic platforms enable light-driven biosynthesis, which dramatically reduces dependence on external carbon sources and associated production costs. Specifically, recent advances in algal biotechnology have demonstrated their potential in producing high-value enzymes, including cellulases for biomass conversion, lipases for biotransformations, and therapeutic proteins for pharmaceutical applications (Goshtasbi et al., 2023). These achievements demonstrate the practical viability of algal enzyme production systems.

Unlike heterotrophic microbes that demand complex nutrient inputs, algae require minimal resources and can be cultivated on nonarable land. This feature effectively eliminates competition with food production systems. Their rapid growth rates, low resource demands, and adaptability to various cultivation systems further enhance the environmental sustainability of algae-based enzyme production. This inherent ecological adaptability and metabolic plasticity make algae ideal candidates for scalable biomanufacturing operations (Tahir et al., 2024). Beyond their sustainable cultivation profile, algae possess sophisticated eukaryotic posttranslational modification machinery. This ability enables the production of complex enzymes that bacterial systems cannot properly synthesize. This is particularly vital for enzymes requiring specific folding patterns and posttranslational modifications (PTMs) to achieve optimal activity (Xie et al., 2023). The transition to algae-based enzyme production systems represents more than just a technological advancement. It signifies a fundamental shift toward sustainable, scalable biotechnology that addresses both economic and environmental challenges facing the enzyme industry. This book chapter will further explore how algal platforms offer a viable pathway to meet growing industrial demands while simultaneously minimizing environmental impact and resource consumption.

4.2 Biology of algae

4.2.1 Types of algae

Algae are fundamentally categorized into two distinct groups: microalgae (unicellular or colonial organisms) and macroalgae (multicellular, seaweed-like structures). Both microalgae and macroalgae possess unique biological attributes that determine their effectiveness as enzyme biofactories. This taxonomic distinction has profound implications for biotechnological applications and production scalability. Macroalgae, including species like *Chlorophyta, Rhodophyta, and Phaeophyta*, offer substantial biomass yields and environmental resilience but present significant limitations for enzyme production. Their complex multicellular tissue structures, slower growth rates (weeks to months for substantial biomass accumulation), and limited genetic engineering tools make them less suitable for controlled enzyme production systems (Pérez-Lloréns et al., 2012). However, macroalgae may serve specialized niche applications where their unique biochemical properties provide specific advantages. Specifically, multiple enzymes have been isolated from the epiphytic bacteria and fungi associated with macroalgae hosts (Li et al., 2022; Sarasan et al., 2021; Vega-Portalatino et al., 2023).

On the other hand, microalgae represent the preferred platform for enzyme production due to their exceptional biological characteristics. Species such as *Chlamydomonas* sp. and *Nannochloropsis* sp. demonstrate rapid growth rates, with doubling times measured in hours to days, exceptional scalability in both bioreactors and open pond systems, as well as remarkable genetic tractability that facilitates metabolic engineering (Stavridou et al., 2024). These characteristics position microalgae as ideal candidates for producing industrially relevant enzymes, including amylases for starch processing and proteases for biotechnological applications. The superior genetic tractability of microalgae, enhanced by advanced molecular tools, including clustered regularly interspaced short palindromic repeats (CRISPR)-based editing systems, establishes them as the primary choice for enzyme production platforms. Recent advances in genome editing have also dramatically improved the precision and efficiency of genetic modifications in these organisms (Ambily et al., 2025).

4.2.2 Cellular and genetic features

Algal cells share typical eukaryotic features. The cellular architecture of algae provides an optimal foundation for efficient enzyme synthesis and secretion. This includes the presence of chloroplasts where photosynthesis occurs, along with an extensive endoplasmic reticulum (ER) system to facilitate proper protein folding, PTMs, and glycosylation. All of these processes are critical for producing functional, industrially active enzymes. This eukaryotic cellular machinery enables algae to overcome the significant limitations of bacterial expression systems, particularly the production of complex proteins requiring specific folding patterns and modifications (Xie et al., 2023; Mathieu-Rivet, 2020).

From a genetic perspective, model microalgae such as *Chlamydomonas reinhardtii* possess compact, well-characterized genomes (approximately 120 Mb) with thoroughly mapped regulatory elements that enable precise control of gene expression patterns. Molecular advancements have established *Chlamydomonas* as a viable host for biotechnological exploitation, with robust nuclear transgene expression now possible due to several impressive improvements made in recent years. Recent developments in CRISPR-Cas (Clustered Regularly Interspaced Short Palindromic Repeats and CRISPR-Associated) technology have further revolutionized genetic engineering capabilities in these organisms, with efficient homology-directed methods for knock-in mutagenesis now available through the delivery of CRISPR-Cas ribonucleoproteins and linear double-stranded DNA donors via electroporation. These methods allow smooth integration of fusion tags and sequence modifications of proteins to boost enzyme yields without the need for preceding mutant lines (Zhang et al., 2021).

4.2.3 Photosynthesis and carbon fixation capabilities

Photosynthesis represents the fundamental advantage of microalgae-based enzyme production systems. This process enables these organisms to harness solar energy for driving both biomass accumulation and protein synthesis processes. The photosynthetic machinery converts light energy into chemical energy, providing the metabolic foundation for sustained enzyme production without requiring external organic carbon sources. The Calvin-Benson cycle, serving as the primary carbon fixation pathway in these organisms, efficiently converts atmospheric CO_2 into organic compounds that fuel all metabolic processes essential for enzyme synthesis. This autotrophic capability distinguishes algae from heterotrophic production systems that depend on expensive organic substrates, providing significant economic advantages for large-scale enzyme manufacturing (Prasad et al., 2021).

Microalgae demonstrate superior photosynthetic efficiency compared to terrestrial plants. This capability supports rapid biomass accumulation that directly correlates with enhanced enzyme production capacity. As such, microalgae emerge as sustainability champions with applications spanning food, feed production, biofuels, cosmetics, pharmaceuticals, and environmental remediation, positioning them as key players in achieving over 50% of UN Sustainable Development Goals simultaneously.

Specific metabolic engineering strategies, including overexpression of key enzymes such as ribulose-1,5-bisphosphate carboxylase/oxygenase (RuBisCO) and optimization of carbon flux pathways, have demonstrated significant improvements in carbon dioxide fixation rates and corresponding enzyme yields. These approaches enable fine-tuning of metabolic pathways to maximize enzyme production while maintaining cellular health and growth rates (Prasad et al., 2021).

By utilizing atmospheric or industrial carbon dioxide waste streams as primary carbon sources, microalgae-based enzyme production systems could contribute to greenhouse gas reduction while generating valuable biotechnological products. This dual benefit of carbon sequestration and enzyme production creates a sustainable, environmentally beneficial production paradigm that addresses both industrial needs and climate change mitigation. Essentially, the integration of photosynthetic capability with minimal nutrient requirements forms the basis for microalgae's exceptional potential as eco-friendly enzyme biofactories. This combination of sustainability, efficiency, and scalability positions algae-based systems as transformative platforms for next-generation biotechnological production (Selvam et al., 2024).

4.3 Genetic engineering of algae

Microalgae are increasingly recognized as valuable platforms in bioengineering due to their rapid growth rates, high photosynthetic efficiency, and capacity to synthesize a wide range of high-value biomolecules. Genetic manipulation of microalgae can be achieved through transformation of the nuclear, chloroplast, or mitochondrial genomes. Each of these approaches offers distinct advantages and technical limitations.

4.3.1 Transformation techniques

4.3.1.1 Nuclear transformation

Nuclear transformation is the most commonly used approach in microalgal genetic engineering. This method typically involves random integration of transgenes via nonhomologous end joining or homologous recombination. Among the species frequently utilized for nuclear transformation, *C. reinhardtii* is particularly prominent due to its well-characterized nuclear genome. These features establish *C. reinhardtii* as an ideal model organism and highlight Cre-based cell engineering as a promising method for generating microalgal strains capable of stable foreign gene expression (Shirakawa et al., 2021). Alternatively, the diatom Phaeodactylum *tricornutum* can be genetically transformed through techniques such as biolistics and electroporation. These techniques enable the introduction of foreign DNA into the nuclear genome (Ifuku and Yan, 2019). More recently, the CRISPR-Cas9 system coupled with proteolistic delivery offers another DNA-free method for nuclear transformation in diatoms by directly introducing ribonucleoprotein (RNP) complexes. This strategy offers several advantages, like reduced off-target mutations, transient Cas9 activity, and prevention of random transgene insertions (Russo et al., 2022). Another notable microalgal genus is *Nannochloropsis*. *Nannochloropsis* has also demonstrated high efficiency in nuclear transformation and is particularly valuable for lipid metabolic engineering, thereby supporting advancements in biofuel production (Li et al., 2014; Ye et al., 2024).

4.3.1.2 Chloroplast transformation

Chloroplast transformation exploits the high copy number of the chloroplast genome and its reduced susceptibility to gene silencing, thereby facilitating stable and high-level transgene integration. *C. reinhardtii* is particularly well suited for chloroplast engineering due to its efficient homologous recombination machinery. This feature is crucial for precise and stable integration of foreign genes into the chloroplast genome (Larrea-Alvarez and Purton, 2020). Furthermore, *Dunaliella salina* has also been successfully utilized in metabolic engineering studies. This is significant in highlighting concurrent adaptability to high-salinity environments and the ability to produce commercially relevant compounds like carotenoids (Chen et al., 2023; Borowitzka, 2013). Compared with nuclear transformation, chloroplast transformation offers several distinct advantages in plant and algal systems. These include predictable transgene expression, reduced position effects, and enhanced genetic stability. Successful chloroplast engineering requires careful vector design, typically incorporating homologous flanking sequences and selectable markers to facilitate targeted genome integration. A range of DNA delivery methods has since been developed for chloroplast transformation, including biolistic particle delivery, polyethylene glycol (PEG)-mediated transformation, carbon nanotube-based carriers, UV-laser microbeam techniques, and Agrobacterium-mediated approaches (Kumar and Ling, 2021).

4.3.1.3 Mitochondrial transformation

Although mitochondrial transformation remains relatively uncommon due to substantial technical constraints, successful attempts have been reported in *C. reinhardtii*. Specifically, this highlights a potential avenue for further research and technological development. Targeted genetic modification of the mitochondrial genome could enable precise control over organelle-specific functions, which then enhances selected phenotypic traits and metabolic pathways (Bonnefoy and Remacle, 2023). While practical biotechnological applications of mitochondrial transformation in microalgae remain limited, ongoing exploration in this area may unlock novel opportunities for strain optimization and functional studies.

The expanding range of genetic transformation strategies in microalgae reflects the increasing sophistication and versatility of algal genetic engineering. The demonstrated efficiency of nuclear transformation has established *C. reinhardtii* and *P. tricornutum* as key organisms for genetic modification, while progress in chloroplast and mitochondrial transformation methods signals a promising future for microalgal engineering (Larrea-Alvarez and Purton, 2020; Hu and Pan, 2019).

In microalgal genetic engineering, successful transgene expression critically depends on the effective design and deployment of promoters and expression vectors. The development of strong and tunable promoters, coupled with optimized vectors, is crucial for maximizing transgene output. Such advancements are central to biotechnological applications, including biofuel production, bioremediation, and the biosynthesis of high-value metabolites and enzymes (Gan and Maggs, 2017).

4.3.2 Promoters for gene expression

4.3.2.1 Constitutive promoters

Constitutive promoters enable continuous expression of target genes across different growth conditions. For instance, the HSP70A/RBCS2 promoter in *C. reinhardtii* has been reported to provide strong expression, making it suitable for applications requiring consistent gene output. This combination has been shown to counteract gene silencing and enhance expression (Schroda et al., 2000, 2002). Similarly, the fucoxanthin chlorophyll protein (FCP) promoter in *P. tricornutum* is associated with light-harvesting complexes that can enhance gene expression linked to photosynthetic efficiency. The expression of FCP genes themselves is often highly regulated by light intensity and quality, making their promoters valuable for applications where strong, light-responsive gene output is desired (Eilers et al., 2016). Additionally, the development of advanced genetic engineering toolkit for *Nannochloropsis oceanica*, an oleaginous microalga rich in eicosapentaenoic acid (EPA), has enabled efficient constitutive expression of multiple transgenes. Using a gene-stacking strategy driven by strong promoters, researchers achieved the coordinated overexpression of several fatty acid desaturases, resulting in a substantial increase in long-chain polyunsaturated fatty acids, particularly EPA, in genetically modified strains (Rasala et al., 2014).

4.3.2.2 Inducible promoters

Inducible promoters enable precise temporal control of gene expression in response to environmental stimuli or chemical cues. For example, the NIT1 promoter in *C. reinhardtii* is nitrate-inducible. This makes it valuable for experiments where gene expression is controlled based on nutrient availability. Researchers may activate or repress target genes through controlled manipulation of nitrate levels in the growth medium (Koblenz and Lechtreck, 2005; Loppes et al., 1999). Similarly, the CYC6 promoter in *C. reinhardtii* is responsive to copper availability. It could be used to modulate metabolic pathways in a copper-dependent manner, thus enabling dynamic control of gene expression through changes in copper concentration (Ferrante et al., 2008). In *Volvox carteri*, the aryl sulfatase (ARS) gene is strongly induced under conditions of sulfur deprivation (Hallmann and Sumper, 1994). This makes its promoter an excellent tool for inducible gene expression and has the potential to be developed as a robust sulfur-responsive inducible system. Specifically, this mechanism for environmentally controlled gene expression in Volvox is comparable to the nitrate-responsive NIT1 system in *Chlamydomonas* (von der Heyde et al., 2023).

4.3.2.3 Vectors and selection markers

The genetic transformation of microalgae often employs specific vectors designed to facilitate the incorporation and expression of transgenes. The *pChlamy* series of shuttle vectors is commonly used in *C. reinhardtii*, while the *pPha-T1* vector is tailored for use in *P. tricornutum*. These vector systems enable the selection of successfully transformed cells through the incorporation of antibiotic resistance markers (Zaslavskaia et al., 2000; von der Heyde et al., 2015). These vectors typically include essential components necessary for replication and integration within the host genome, along with features promoting efficient transcription.

The adoption of CRISPR-based genome-editing strategies has further enriched the toolkit for genetic engineering. For instance, the plasmid Bluescript KS - nourseothricin resistance *(pKS-NatR)* plasmid in *Nannochloropsis* has been utilized to achieve targeted gene disruption and precise genome editing using the CRISPR-Cas9 system. This enables precise control over genetic modifications (Naduthodi et al., 2021; Guihéneuf et al., 2016). In parallel, transgene constructs often include selectable markers such as antibiotic resistance genes like bleomycin and aphVIII, which facilitate the identification and isolation of successfully transformed cells (Suttangkakul et al., 2019). As an alternative to antibiotic selection, auxotrophic markers could also be used. The use of NIT1 for nitrate uptake complementation offers a more sustainable selection strategy for genetically modified microalgal strains (Ross et al., 2024).

Overall, continued advances in microalgal genetic engineering are closely linked to the strategic selection of promoters, vectors, and selection systems. The integration of robust constitutive and inducible promoters with efficient vector platforms highlights the versatility and scalability of microalgae as biotechnological hosts. With an expanding range of molecular tools, significant improvements in the production of biofuels, enzymes, and other high-value metabolites are anticipated. This eventually reinforces the role of microalgae as sustainable and economically viable biological resources.

4.3.3 Synthetic biology tools in algal systems

The integration of advanced genetic tools has fundamentally transformed microalgal synthetic biology. Precise genome editing and increasingly sophisticated metabolic pathway engineering are now available. Technologies like CRISPR-Cas9, modular cloning systems, and biosensors have substantially expanded the biotechnological potential of microalgae, supporting applications ranging from biofuel production, high-value metabolite synthesis, to environmental sensing and monitoring (Jagadevan et al., 2018; Sreenikethanam et al., 2022).

Among these tools, CRISPR-Cas9, including emerging base-editing variants, has emerged as a highly effective platform for targeted genetic modification in microalgae. In *C. reinhardtii*, CRISPR-mediated disruption of lipid biosynthesis genes, such as *DGTT1,* has been shown to enhance lipid accumulation and subsequent biofuel applications (Shin et al., 2016; Lee et al., 2022). The Cas9 ribonucleoprotein (RNP) delivery system has also significantly increased lipid production in *Nannochloropsis gaditana,* while gene knock-ins via homology-directed repair (HDR) in *P. tricornutum* have enabled precise insertion of transgenes for complex metabolic pathway engineering (Naduthodi et al., 2019; Verruto et al., 2018; Moosburner et al., 2020; Qi, 2016).

4.3.3.1 Modular cloning techniques

Modular cloning methods, like Golden Gate assembly and the modular cloning (MoClo) system, enable the efficient and standardized assembly of multiple genetic components into complex expression constructs. These methodologies have been utilized in *Synechocystis* sp. *PCC 6803* to generate modular genetic parts for the systematic construction of engineered metabolic pathways; this has led to enhanced metabolic flux and overall pathway efficiency (Vasudevan et al., 2019; Englund et al., 2015). Similarly, *C. reinhardtii* has benefited from the development of standardized BioBrick-compatible vectors that facilitate the rapid assembly and stable integration of multigene constructs. This modular design framework is particularly advantageous for synthetic biology applications that require coordinated expression of multiple genes and fine-tuning of complex metabolic networks (Crozet et al., 2018).

4.3.3.2 Biosensors and optogenetics

The incorporation of biosensors in microalgal systems, particularly in *C. reinhardtii*, has enabled real-time monitoring of environmental pollutants and cellular responses. Among these, fluorescent reporter-based biosensors have proven especially effective for the rapid and sensitive detection of heavy metals, such as cadmium. It could potentially serve as a powerful tool for environmental monitoring and biosafety assessment (Iwai et al., 2014; Sahabudin et al., 2024).

In parallel, *P. tricornutum* has been engineered with light-switchable gene circuits, allowing precise control of transgene expression in response to defined light stimuli. Such systems are particularly valuable for dynamic gene regulation in metabolic engineering, as temporal control over pathway activity is essential (Kassaw et al., 2022). More broadly, the application of optogenetics, especially those based on photoreceptors such as cryptochromes, enables spatially and temporally resolved modulation of gene expression. While fully engineered optogenetic circuits require further research, integrating such systems could enable fine-tuned control of metabolic flux and cellular behavior in response to experimental or environmental cues (Kassaw et al., 2022; Ye and Fussenegger, 2019).

Collectively, advances in synthetic biology are rapidly reshaping the landscape of microalgal engineering. These tools enable the creation of customized microalgal strains for diverse applications, from sustainable bioenergy and industrial production to bio-sensing and environmental solutions. All of these functions reinforce the role of microalgae as versatile and future-ready biotechnological platforms.

4.4 Enzyme expression and secretion

This section will delve into key target enzymes produced by various algal species (see Table 4.1), their potential applications, and the challenges associated with protein processing and secretion mechanisms.

4.4.1 Target enzymes and their applications

4.4.1.1 Hydrolytic enzymes

Hydrolytic enzymes, such as cellulases and lipases, play significant roles in industrial processes like biofuel production and waste degradation (Manisha, 2017; Chandra et al., 2020). The green alga *C. reinhardtii* is particularly noted for its ability to express cellulase, which is crucial for the breakdown of cellulose for biofuel applications (Bellido-Pedraza et al., 2024). Additionally, the diatom *P. tricornutum* produces lipases that can be valuable in the bioremediation of lipid-based pollutants (Prestegard et al., 2016). These enzymes facilitate the degradation of complex organic materials, thus enhancing both energy recovery and environmental cleanup efforts.

Table 4.1 Types of algae derived enzymes and their functions.

No	Enzyme type	Enzyme	Algae species	Function	References
1	Lyase	Fatty acid photodecarboxylase	*Chlorella variabilis* NC64A	Drives the blue-light-mediated decarboxylation of fatty acids into alkanes (biofuel precursors).	Moulin et al. (2021)
2	Lyase	Ulvan lyase	*Ulva* sp., *Enteromorpha*, *Chaetomorpha*, *Codium*, and *Caulerpa*	Depolymerizes ulvan, facilitating cell-wall remodeling and the release of oligosaccharides. Also shows potential as an antidiabetic agent.	Gajanayaka et al. (2024); Romero (2023)
3	Lyase	Alginate lyase	Various algae	Degrades alginate into oligosaccharides, with applications in biofuel, food, and pharmaceutical industries.	Zhang et al. (2021)
4	Lyase	Carbonic anhydrase	*Chlamydomonas reinhardtii*, *Phaeodactylum tricornutum*, and various microalgae	Catalyzes the rapid interconversion of carbon dioxide and water into bicarbonate and protons. Relevant for CO_2 capture and utilization, biomineralization, and enhanced oil recovery.	Harada et al. (2005); DiMario et al. (2018)
5	Hydrolase	L-Asparaginase	*Chlamydomonas* sp., *Chlorella vulgaris*	A potential therapeutic enzyme with anticancer properties.	Ebrahiminezhad et al. (2014)

6	Hydrolase	β-Porphyranase	*Chondrus crispus*	Cleaves porphyran, a polysaccharide found in red algal cell walls.	Manat et al. (2022)
7	Hydrolase	Cyanophycinase	*Synechocystis* sp. PCC 6803	Degrades cyanophycin (nitrogen storage polymer).	Richter et al. (1999); Watzer and Forchhammer (2018)
8	Hydrolase	α-Amylase	*Chlamydomonas reinhardtii*	Hydrolyzes starch and plays a crucial role in carbohydrate metabolism.	Scholtysek et al. (2024); Torresi et al. (2025)
9	Hydrolase	Lipases	*Arthrospira*, *Chlorella*, various algae	Hydrolyzes lipids and phospholipids, proving useful in cell disruption and biodiesel processing. It catalyzes the hydrolysis of triacylglycerols into fatty acids and glycerol, and also performs esterification and transesterification in low water activity environments.	Ben Hlima et al. (2021); Demir and Tükel (2010)
10	Hydrolase	Nitrilase	Various algae	Catalyzes the hydrolysis of nitriles into their corresponding carboxylic acids and ammonia. Holds potential for bioremediation and the synthesis of fine chemicals.	Bhatt et al. (2021); Howden and Preston (2009)

(*Continued*)

Table 4.1 (Continued)

No	Enzyme type	Enzyme	Algae species	Function	References
11	Hydrolase	Exo-polygalacturonase	*Chlorella vulgaris*	Degrades pectin, which is used for the clarification of fruit juice.	Uzuner et al. (2022); Swain et al. (2009)
12	Hydrolase	Endoglucanases, exoglucanases, and β-glucosidases	*Chlamydomonas reinhardtii, Gonium pectorale*, and various algae	Breaks down cellulose into simpler sugars (glucose, cellobiose). Applications include biofuel production (saccharification of lignocellulosic biomass), the textile industry (biostoning, biopolishing), paper and pulp, animal feed, and pharmaceuticals.	Poidevin et al. (2013); Ejaz et al. (2021); Paul(2021)
13	Hydrolase	Serine proteases, metalloproteases, cysteine proteases, aspartic proteases	*Spirulina platensis, Arthrospira platensis, Halothece* sp. PCC7418, various algae, and cyanobacteria	Catalyzes the hydrolysis of peptide bonds in proteins. Employed in food processing (e.g., meat tenderization, dairy, baking), detergent formulations, pharmaceuticals (as therapeutic enzymes), and bioremediation.	López-Otín and Bond (2008); Patipong et al. (2021); Elleuch et al. (2021)
14	Hydrolase	Phytase	*Chlamydomonas reinhardtii* and various algae	Hydrolyzes phytate (an antinutritional factor in plant-based feeds) to release inorganic phosphate. Utilized in animal feed to improve nutrient absorption and reduce phosphorus pollution.	Erpel et al. (2016); Yoon et al. (2011)

Table 4.1 (Continued)

No	Enzyme type	Enzyme	Algae species	Function	References
15	Oxidoreductase	[FeFe]-hydrogenase, [NiFe]-hydrogenase	*Chlamydomonas reinhardtii*, various green algae, and cyanobacteria	Catalyzes the reversible oxidation of molecular hydrogen. There is significant interest in its role in biotechnological biohydrogen production.	Kosourov et al. (2021); Hippler and Khosravitabar (2024)
16	Oxidoreductase	Aldose reductase	*Ecklonia stolonifera*, *Saccharina japonica*	Catalyzes the reduction of aldoses (sugars) to their corresponding sugar alcohols. Inhibition of aldose reductase is a therapeutic target for diabetic complications.	Swain et al. (2009); Sharifuddin et al. (2015)
17	Oxidoreductase	Phenol-oxidases/ Laccase-like	*Chlorella* sp., *Tetracystis aeria*, *Scenedesmus* spp.	Oxidize phenolic pollutants; show potential in bioremediation and cosmetics.	Otto and Schlosser (2014); Otto et al. (2015); Carbone et al. (2018)
18	Oxidoreductase	Superoxide dismutase, catalase, and peroxidase	*Porphyridium cruentum*, *Ulva*, *Chaetomorpha spp.*, *Chlorella*, *Scenedesmus*	Act as antioxidant enzymes, scavenging reactive oxygen species. Used in cosmetics and for stress protection.	Flórez-Fernández et al. (2023); Zakeri et al. (2020); Ameri et al. (2020); Dummermuth et al. (2003)

4.4.1.2 Therapeutic enzymes

Therapeutic enzymes are another area of interest, particularly for applications in biopharmaceuticals and low-cost therapeutics. *D. salina* has been investigated for its potential in oral vaccine delivery, which is a novel approach to immunization (Gamberi et al., 2024). Such capabilities are particularly aligned with the growing demand for cost-effective and sustainable therapeutic production systems.

The enzyme for medical application derived from algae is the widely studied L-asparaginase, that was first purified from *Chlamydomonas* sp. (Paul, 1982). The L-asparaginase enzyme is used for the treatment of acute lymphoblastic leukemia, acute myeloid leukemia, and non-Hodgkin's lymphoma. This enzyme break downs L-asparagine (an essential amino acid for leukemic cells), which in turn depletes the body's supply of asparagine and subsequently causes starvation to the cancer cells (Batool et al., 2016).

4.4.1.3 Oxidoreductases

Oxidoreductases, including laccases and hydrogenases, represent an important class of enzymes with broad industrial and environmental relevance. Chlorella *vulgaris* is recognized for its production of laccases. The enzymes have been widely explored for applications in dye degradation and bioremediation due to their capacity to oxidize a wide range of recalcitrant organic pollutants (Abd Ellatif et al., 2021; Hultberg et al., 2019). Similarly, the cyanobacterium *Synechococcus elongatus* is capable of producing bidirectional hydrogenases. Serving a central role in hydrogen metabolism, hydrogenases have since received considerable interest for photobiological biohydrogen production (Kosourov et al., 2021; Carrieri et al., 2008). Overall, these microalgal and cyanobacterial oxidoreductases highlight promising approaches for integrating biological systems into bioremediation and green energy production. Specifically, ongoing research focuses on enzyme optimization, heterologous expression, and process engineering to improve efficiency and scalability.

4.4.1.4 Challenges in protein processing and secretion

Despite the numerous advantages of algae as platforms for enzyme production, several challenges remain. In particular, algal proteins often exhibit complex and species-specific glycosylation patterns. This can interfere with proper protein folding, stability, and activity of the heterologously expressed proteins (Rasala and Mayfield, 2015; Banerjee and Ward, 2022). Furthermore, the efficiency of protein secretion mechanisms varies widely among algal species, potentially limiting the yield of extracellular enzymes required for large-scale industrial applications. Structural barriers such as rigid cell walls further complicate enzyme extraction. Cell disruption and lysis strategies that balance both efficient biomass breakdown and the preservation of enzyme activity are necessitated (Ramos-Martinez et al., 2017; Rosales-Mendoza et al., 2020).

Nevertheless, algae remain a highly promising platform for the production of a diverse array of enzymes with significant industrial, medical, and environmental applications. Hydrolytic enzymes from *C. reinhardtii* and *P. tricornutum* enhance biofuel processing and waste degradation. Therapeutic enzymes from *D. salina* and *N. gaditana* provide innovative solutions in biopharmaceuticals, while oxidoreductases from *C. vulgaris* and *S. elongatus* contribute to bioremediation and clean energy initiatives. Despite challenges in protein processing and secretion persisting, continued advances in genetic engineering, strain optimization, and downstream processing are expected to overcome these limitations. These developments will facilitate the full realization of algal biofactories as sustainable and economically viable enzyme production systems.

4.4.2 Protein folding and posttranslational modifications

Eukaryotic algae serve as valuable platforms for enzyme production due to their ability to perform essential PTMs that are critical for the correct folding, stability, and functionality of proteins. Major PTMs include glycosylation, disulfide bond formation, and other covalent modifications that are often absent or inefficient in prokaryotic expression systems (Banerjee and Ward, 2022). This section explores the specific requirements and challenges associated with these modifications in various algal species.

4.4.2.1 Glycosylation

Glycosylation involves the enzymatic attachment of carbohydrate moieties to proteins. This PTM plays a pivotal role in determining protein conformation, stability, and functional performance. *C. reinhardtii* is known for its capacity to perform high-mannose glycosylation, a common form of N-glycosylation found in eukaryotes that supports correct protein folding and stability (Mathieu-Rivet et al., 2014). However, this glycosylation pattern differs from human-compatible N-glycan structures and therefore limits the therapeutic applicability of recombinant proteins produced via this system. To address this limitation, glycoengineering strategies have been developed to remodel the algal glycosylation pathway. One effective strategy involves the targeted knockout of xylosyltransferase genes in *C. reinhardtii*. This reduces the incorporation of plant-specific glycan residues and shifts the glycan profile toward a more human-like configuration. This engineering approach has been shown to improve glycosylation patterns and enhance the therapeutic potential of algal-derived recombinant proteins (Oltmanns et al., 2020; Lucas et al., 2020; Leprovost et al., 2024).

4.4.2.2 Disulfide bond formation

Disulfide bonds are critical for the structural stability and functional integrity of many enzymes, particularly those that are secreted extracellularly (Hatahet et al., 2014). In eukaryotic algae, the ER provides an effective environment for the formation of disulfide bonds due to its specialized folding machinery. The coexpression of molecular chaperones, such as Binding Protein (BiP) and Protein Disulfide Isomerase (PDI), can further enhance the folding efficiency of complex enzymes by facilitating proper formation of disulfide bonds and promoting the correct assembly of multimodal proteins. Strategic utilization of these chaperones can therefore significantly improve both the yield and enzymatic activity of heterologously expressed proteins in algal expression systems (Hatahet et al., 2014; Elshobary et al., 2025; Lin, 2024; Mayer et al., 2000).

4.4.2.3 Proteolytic degradation

Challenges may also arise from proteolytic degradation that can substantially reduce protein yield and stability (Rosales-Mendoza et al., 2012; Ng et al., 2020). Being a well-studied microalga, *Haematococcus pluvialis* exhibits elevated endogenous protease activity that compromises protein integrity and limits accumulation of heterologous products (Ng et al., 2020; Saei et al., 2012). To mitigate this problem, several strategies have been proposed. These include targeting recombinant proteins to subcellular compartments such as vacuoles, where proteolytic activity may be reduced. Development and utilization of protease-deficient strains have also been proposed (Marin Viegas et al., 2017). These approaches aim to create a more favorable intracellular environment for enzyme accumulation and stability, thus enhancing protein stability and overall production efficiency.

Supported by their intrinsic eukaryotic cellular machinery, eukaryotic algae offer unique capabilities for producing enzymes with appropriate folding and complex PTMs. These features make algae especially suitable for expressing complex therapeutic proteins and industrial enzymes that require eukaryotic-like processing, unlike prokaryotic systems (e.g., bacteria) that often lack such modifications (Banerjee and Ward, 2022; Mathieu-Rivet et al., 2014; Specht et al., 2010). However, key challenges remain. Achieving human-compatible glycosylation profiles and managing proteolytic degradation necessitate continued research and innovative engineering strategies. Recent and ongoing advances in genetic and synthetic biology are progressively addressing these hurdles (Specht et al., 2010; Vingiani et al., 2019). These innovations enhance protein yield, quality, and compatibility, thereby reinforcing microalgae's potential as sustainable, low-cost, and scalable platforms for biopharmaceuticals and industrial enzymes.

4.4.3 Secretion pathways in algae

Efficient recovery of algal-produced enzymes can be significantly improved through the exploitation of native and engineered secretion pathways. Secretion of recombinant proteins into the extracellular medium eliminates the need for cell disruption. This strategy preserves protein integrity, simplifies downstream processing, and often enhances overall yields. This section outlines the major secretion pathways utilized by various algal species and discusses strategies to optimize these pathways for enzyme production.

4.4.3.1 Default secretory pathway

The default secretory pathway (also known as constitutive secretory pathway) in eukaryotic cells involves the translocation of proteins into the ER, subsequent processing through the Golgi apparatus, and final secretion into the extracellular environment. In *C. reinhardtii*, this pathway has been successfully harnessed for the secretion of recombinant proteins (Wang et al., 2022). Notably, the secretion of the yellow fluorescent protein Venus was achieved using the native gametolysin signal sequence. Further enhancement was obtained by appending synthetic glycomodules $(SP)_n$ (n=10 or 20) to the Venus protein. This modification resulted in up to a 12-fold increase in secretion yield (reaching approximately 15 mg/L), while simultaneously improving resistance to proteolytic degradation. The $(SP)_n$-tagged proteins underwent glycosylation, likely initiated within the ER, highlighting the role of glycomodules in facilitating protein folding, stability, and secretion (Ramos-Martinez et al., 2017). On the other hand, the diatom *P. tricornutum* also exhibits inherent capacity for efficient protein secretion, highlighting its suitability as a host for the production of recombinant and therapeutic proteins (Molino et al., 2018). In both *C. reinhardtii* and *P. tricornutum*, secretion efficiency can be further enhanced by fusing target proteins with native or optimized signal peptides. These signal sequences facilitate accurate targeting of recombinant proteins to the secretory machinery, thereby improving transport through the ER–Golgi pathway and increasing extracellular protein accumulation (Lauersen, 2019; Rasala et al., 2010). All in all, these strategies have proven effective in optimizing secretion yields and advancing the use of microalgae as platforms for recombinant enzyme production.

4.4.3.2 Unconventional protein secretion

Unconventional protein secretion (UPS) refers to a group of mechanisms by which proteins are transported out of the cell independently of the classical ER–Golgi secretory pathway. In contrast to the conventional pathway, which requires an N-terminal signal peptide to direct proteins into the ER, UPS pathways enable the extracellular release of proteins lacking signal peptides or those that are incompatible with canonical secretion. Although UPS has been extensively characterized in mammalian and yeast systems, its study in algae remains relatively nascent. Nevertheless, emerging evidence suggests that UPS plays important roles in algal physiology, particularly under conditions of cellular stress. As photosynthetic organisms, algae are frequently exposed to rapidly fluctuating environmental conditions, including changes in light intensity, temperature, salinity, and nutrient availability. Such stressors can activate UPS pathways, facilitating the rapid secretion of proteins involved in stress adaptation, cell–cell communication, or extracellular remodeling without reliance on the slower ER–Golgi trafficking route (Rabouille, 2017).

UPS pathways are commonly categorized into four major types based on their transport mechanisms. Type I UPS involves direct translocation of proteins across plasma membrane through transient pores, whereas Type II UPS relies on ATP-binding cassette (ABC) transporters. Type III UPS employs vesicular intermediates to deliver cytosolic proteins to the extracellular space, while Type IV UPS enables Golgi bypass for ER-derived transmembrane proteins under stress conditions or within specialized cellular compartments such as cilia. Unlike Types I–III that primarily secrete leaderless cytosolic proteins lacking signal peptides, Type IV supports the export of ER-derived proteins that circumvent the canonical secretory route. In addition, Types I and II enable direct plasma membrane translocation, whereas Types III and IV rely on vesicle-mediated trafficking, either through encapsulation of soluble cargo or integration of membrane proteins into vesicular membranes (Rabouille et al., 2012; Noh et al., 2022).

Although the molecular components and regulatory networks governing UPS in algae have yet to be fully elucidated, it is increasingly recognized that algal cells likely employ UPS or analogous mechanisms to secrete enzymes, signaling molecules, and stress-response proteins that are unsuitable for classical ER–Golgi secretion. Further characterization of UPS pathways in algal systems may, therefore, unlock new opportunities to enhance extracellular enzyme recovery and broaden the utility of microalgae as biotechnological production platforms.

4.4.3.3 Engineered secretion

Several algal species, including *C. reinhardtii* and *P. tricornutum*, have also demonstrated significant potential as engineered protein secretion platforms due to their well-developed intrinsic eukaryotic machinery. Recent advancements in algal genetic engineering have enabled targeted modifications to enhance specific steps within the secretory pathway to improve extracellular protein yield. One such strategy focuses on optimizing vesicle trafficking between the ER and the Golgi apparatus. This is a process mediated by coat protein complex II (COPII) vesicles. Eukaryotic cells rely on COPII-mediated vesicular transport for protein secretion from the ER to the Golgi, with the small GTPase Sar1 playing a central regulatory role. While Sar1 has been extensively characterized in yeast, animals, and plants, its functional role in algae remains unexplored. A previous study identified a Sar1 homolog (CrSar1) in *C. reinhardtii* and showed that expression of a dominant-negative mutant (CrSar1DN) markedly inhibits protein secretion. Contrary to observations in other eukaryotic systems, CrSar1DN expression did not disrupt ER–Golgi structure but still supports a conserved and essential role for Sar1 in algal COPII-mediated secretory transport (Barlowe, 1995; Chung et al., 2024).

4.5 Cultivation and bioprocessing

As photosynthetic and fast-growing microorganisms, algae are capable of thriving in diverse environments that are often unsuitable for conventional agricultural crops. They can be cultivated in freshwater, seawater, and wastewater systems, where they efficiently utilize available nutrients as carbon and energy sources. This versatility makes algae particularly attractive for sustainable bioprocessing applications. To achieve optimal algal growth and productivity, several physicochemical parameters must be carefully regulated (Fig. 4.1).

These include nutrient composition, light intensity, photoperiod, pH, salinity, aeration, and temperature. Precise control of these variables is essential to maximize biomass accumulation and enhance the yield of target products, such as enzymes or metabolites. The two common types of algal growth systems known are the open culture system, open raceway ponds (ORPs), and closed culture system (photobioreactors). Each system offers distinct advantages and limitations in terms of scalability, cost, contamination risk, and process control (Ugwu et al., 2008; Zohir et al., 2023).

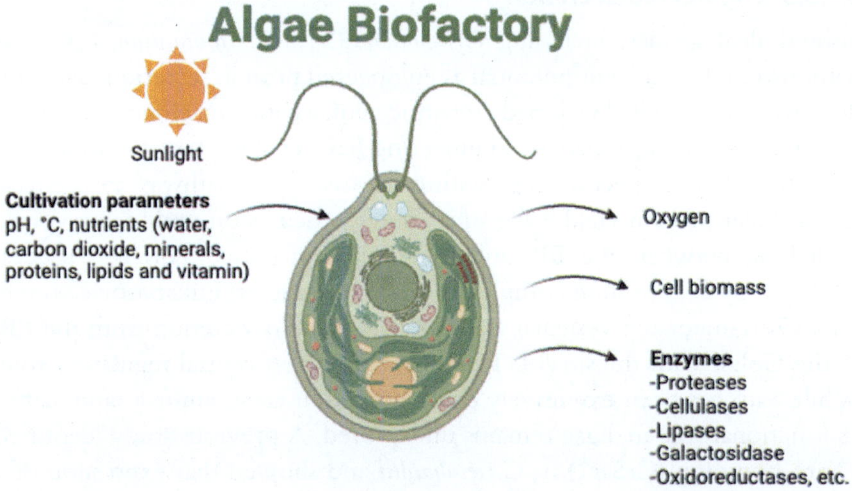

Figure 4.1 ***Schematic illustration of an algae biofactory using* Chlamydomonas *sp. as a model organism.*** Algal cells are cultivated in an aqueous medium under sunlight, with growth regulated through optimized cultivation parameters, including pH, temperature, carbon dioxide supply, and nutrient availability. Through photosynthetic metabolism, algae convert these inputs into oxygen and biomass while simultaneously producing a range of value-added products, including industrial enzymes (e.g., proteases, cellulases, lipases, galactosidases, and oxidoreductases), thereby functioning as sustainable biofactories for biotechnological applications.

4.5.1 Open raceway pond

ORPs are a type of artificial cultivation system typically designed in lopped, oval, or circular configurations resembling a racetrack, from which their name is derived. They are widely used for large-scale algal cultivation due to their shallow configuration (generally 15–20 cm in depth) that maximizes light penetration and supports efficient photosynthetic activity within the culture (Sirajunnisa and Surendhiran, 2016). ORPs are equipped with paddlewheels that provide continuous mixing, generating a unidirectional flow that prevents sedimentation, promotes uniform nutrient distribution, and enhances gas exchange (Fig. 4.2). One of the primary advantages of ORPs is their relatively low construction and operational costs compared with closed photobioreactor systems. However, these systems also present notable limitations, like limited process control, high water consumption, and increased susceptibility to contamination by unwanted microorganisms due to their open exposure to the atmosphere. Additionally, environmental factors such as temperature fluctuations and evaporation can further complicate process consistency (Egbo et al., 2018).

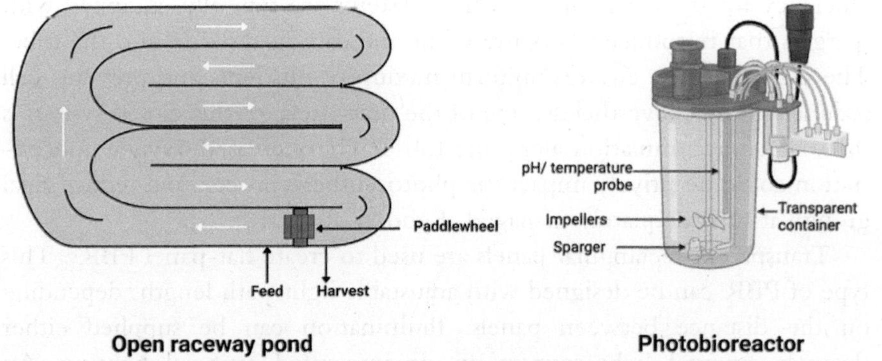

Figure 4.2 ***Schematic illustration of algal cultivation systems of an open raceway pond (ORP) (open system) and a stirred-tank photobioreactor (closed system).*** The ORP is exposed to the surrounding environment and is equipped with a motorized paddlewheel to facilitate continuous mixing and circulation of algal cells. In contrast, the stirred-tank photobioreactor is a closed, tightly sealed system that incorporates spargers and impellers for controlled aeration and mixing, along with integrated pH and temperature probes for precise monitoring and regulation of algal growth conditions.

4.5.2 Photobioreactors

Photobioreactors (PBRs) are closed cultivation systems constructed using transparent materials to allow light penetration while isolating the culture from the external environment. This closed design significantly reduces the risk of contamination compared with open systems. PBRs are commonly used in laboratory- and pilot-scale cultivation, where precise control of growth conditions is essential for experimental studies and optimization of cultivation parameters to achieve maximum biomass or product yield. The three common PBR configurations include stirred tank, tubular, and flat-panel designs.

Stirred-tank PBRs are particularly suited for laboratory-scale cultivation. Agitation is achieved mechanically using impeller. This provides efficient mixing, heat transfer, and mass transfer. These characteristics make stirred-tank PBRs suitable for controlled cultivation and the production of high-value algal products. However, their application at large scales is limited by high operational costs. The potential for shear stress caused by mechanical agitation may also damage sensitive algal cells (Xu et al., 2009).

Tubular PBRs consist of cylindrical tubes arranged to provide a high surface area-to-volume ratio to optimize light exposure and photosynthetic efficiency for the algal culture. These systems are typically equipped with spargers that introduce a mixture of air and carbon dioxide into the tube. The use of sparger ensures uniform mixing of nutrients and prevents cell sedimentation. Nevertheless, one of the drawbacks of this closed system is the oxygen accumulation along the tubes. The increased oxygen concentration could negatively impact the photosynthesis process and reduce algal growth if not adequately managed (Egbo et al., 2018).

Transparent rectangular panels are used to create flat-panel PBRs. This type of PBR can be designed with adjustable light-path lengths depending on the distance between panels. Illumination can be supplied either through external light sources or via integrated artificial lighting. An innovative variation of this concept is the "Liquid Tree." Briefly, it is a microalgae-based photobioreactor designed to mimic trees and installed in urban or indoor environments. This system aims to convert carbon dioxide and water into oxygen, thereby contributing to improved air quality and urban sustainability (Dhar et al., 2023).

In addition to these three main configurations, bag PBRs are also used due to their low setup cost and closed-system design, which helps prevent contamination. These systems utilize transparent polyethylene bags mounted on supporting frames and are aerated using pumps. The main limitation, however, is that the bags are prone to leakage and need to be replaced periodically. The long-term scalability of their use is therefore restricted (Chanquia et al., 2022).

Finally, hybrid cultivation systems that integrate both closed and open approaches have gained increasing attention. In such systems, algae are initially cultured in closed PBRs to achieve high cell density while minimizing contamination. Once sufficient biomass is obtained, the culture is transferred to open systems to promote the accumulation of target metabolites, such as lipids or bioactive compounds. This hybrid strategy combines the advantages of both systems and offers a practical route for large-scale algal bioprocessing (Zohir et al., 2023; Tan et al., 2016).

4.5.3 Optimization of growth conditions

Although algae can be cultivated under relatively low nutrient conditions, growth rates, and biomass productivity vary widely among species and can be significantly influenced by culture conditions. Key parameters affecting algal growth include mineral nutrient composition, carbon dioxide availability, temperature, pH, light intensity, and aeration (Munir et al., 2015). In PBR systems, mechanical factors such as continuous mixing, gas bubble size and distribution, as well as mass transfer efficiency further influence growth performance and biomass accumulation (Schenk et al., 2008).

Temperature is a critical determinant of algal growth. The optimum temperature range is typically between 16°C and 25°C, but it also depends significantly on the species. Growth rates generally decline when temperatures exceed 30°C due to thermal stress and enzyme inactivation (Munir et al., 2015). Similarly, the pH of the growth medium greatly affects algal metabolism. Most species exhibit optimal growth within a pH range of 7.0–8.0. In contrast, more extreme pH conditions, such as pH 5.0 or pH 9.0, have been shown to severely inhibit or completely obstruct algal growth (Cerón Garcí et al., 2000; Zhu et al., 2002).

Light availability and quality are also critical factors governing algal productivity. Munir and coworkers (2015) reported that algae grown under controlled fluorescent lighting often exhibit higher growth rates than those exposed to direct sunlight. This effect might be attributable to photoinhibition caused by excessive light intensity under natural sunlight. The high-light intensity damages photosynthetic pigments and reduces growth efficiency (Munir et al., 2015; Hu et al., 2008; Janssen et al., 1999; Sharma et al., 2008). In addition, continuous aeration and effective mixing are equally essential for maintaining optimal growth conditions. Constant aeration ensures an adequate carbon dioxide supply, prevents cell sedimentation, and promotes uniform distribution of nutrients and light. The absence of sufficient mixing and aeration has been associated with significant cell mortality and reduced biomass yields (Chen, 1996; Ip and Chen, 2005).

4.5.4 Downstream processing and enzyme recovery

Algal biomass represents a valuable and sustainable resource for the production of a wide range of bioproducts, including animal feed, protein tablets, biofuels, and industrial enzymes. Efficient recovery of intracellular proteins requires a multistep downstream processing workflow, beginning with solid–liquid separation of the harvested biomass (Rahman et al., 2022). In particular, the choice of separation and extraction strategy depends on whether the target proteins are located intracellularly or secreted extracellularly. This section focuses specifically on intracellular protein recovery.

Initial biomass harvesting is typically achieved through sedimentation, filtration, centrifugation, flotation, or flocculation, depending on algal size, density, and cultivation conditions (Milledge and Heaven, 2013). Following harvesting, intracellular protein extraction can proceed via either dry or wet processing routes (Rajauria et al., 2021). Dry routes involve dehydration of biomass through sun drying, drum drying, or freeze drying prior to extraction (Show et al., 2015). In contrast, wet routes bypass drying and instead rely on immediate cell disruption to release intracellular products. Wet processing is generally preferred for high-value proteins and enzymes, as these biomolecules are often sensitive to heat and oxidative damage (Kim et al., 2013; Gangl et al., 2015).

Cell disruption is a critical step for solubilizing intracellular proteins and may be achieved through mechanical, chemical, or biological methods. Mechanical disruption applies physical forces to break the algal cell wall and membrane, commonly using bead milling or high-speed homogenization. These techniques are widely adopted in food, chemical, and cosmetic industries due to their scalability, operational simplicity, and effectiveness across multiple algal species (Quiroz-Arita et al., 2022).

Ultrasonication is another mechanical approach based on cavitation-induced shear forces, offering rapid processing and high extraction efficiency. However, its high energy demand and species-specific effectiveness limit its industrial scalability. Nonmechanical approaches include chemical and biological methods. Organic solvents such as methanol, acetone, and phenols have been employed for pigment extraction (Alam et al., 2018; Garcia-Perez et al., 2022). Similarly, mixtures of nonpolar and polar solvents are commonly used to extract lipids and carbohydrates from algal biomass (Li et al., 2014; Ryckebosch et al., 2012).

Chemical pretreatment may also involve the addition of acids (e.g., hydrochloric, sulfuric, or phosphoric acid), bases (e.g., sodium hydroxide, ammonia, or lime), or salts to facilitate cell-wall disruption. Acid hydrolysis has been shown to enhance cellulose breakdown during biomass processing while requiring lower energy input compared to enzymatic hydrolysis (Dong et al., 2016). However, chemical methods are also frequently associated with drawbacks such as corrosiveness, toxicity, and increased operational costs due to high-temperature requirements.

Biological disruption using enzymes offers a more selective and energy-efficient alternative. Enzymatic treatments are highly effective but require careful selection of enzymes based on biomass composition and target product, often necessitating high enzyme-to-substrate ratios that increase production costs. For example, enzymatic extraction of proteins from *Palmaria palmata* using xylanase and cellulase yielded protein levels tenfold higher than those obtained via mechanical extraction (Joubert and Fleurence, 2008). Likewise, cellulase- and lysozyme-assisted extraction also significantly enhanced protein recovery from *C. vulgaris* compared with bead milling or sonication (Barbarino and Lourenço, 2005).

Following cell disruption and solubilization, removal of insoluble cell debris is achieved using filtration or centrifugation. Subsequent downstream processing aims to concentrate proteins, remove excess water, and eliminate contaminants such as nucleic acids, polysaccharides, and non-target proteins. Membrane-based filtration techniques employing semi-permeable membranes enable size- and molecular-weight-based separation and are particularly advantageous for protein recovery due to their nonthermal and environmentally friendly nature.

Chemical precipitation is another scalable protein concentration method. Chemical agents such as ammonium sulfate, polyethylene glycol, ethanol, or acid–base treatments are commonly used. For instance, combined alkaline extraction (pH 12) and acid precipitation (pH 2) of

Saccharina latissima biomass has been reported to achieve a maximum protein recovery of 34.5% (Vilg and Undeland, 2017).

The extent of purification required ultimately depends on the intended application. High-resolution purification techniques such as chromatography are essential for therapeutic enzymes and pharmaceutical applications, whereas moderate purification is sufficient for animal feed, food ingredients, or bulk industrial enzymes. An efficient initial purification strategy is three-phase partitioning (TPP), which combines ammonium sulfate and *t*-butanol to simultaneously concentrate and partially purify proteins. In TPP, proteins form a distinct intermediate phase between the aqueous and organic layers, while nonpolar and polar contaminants partition into the upper solvent and lower aqueous phases, respectively (Harde and Singhal, 2012). Key parameters influencing TPP efficiency include ammonium sulfate concentration, solvent type, incubation time, pH, solid loading, and slurry-to-solvent ratio (Vetal and Rathod, 2015).

4.6 Current research and developments

The eukaryotic nature of algae, coupled with their diverse metabolic pathways such as photosynthesis and lipid synthesis, and their ability to perform PTMs, offers significant opportunities for the discovery of novel catalytic activities. These features not only deepen our understanding of fundamental enzyme functions but also open new avenues for innovative industrial and biomedical applications (El-Fakharany et al., 2024). Examples include Acyl-CoA diacylglycerol acyltransferase (DGAT) from *P. tricornutum,* which exhibits distinct catalytic capabilities in fatty acid synthesis. It plays a pivotal role in the biotechnological production of unsaturated fatty acids and other high-value fatty acids, thereby reducing dependence on marine-derived sources, such as fish oil (Klińska-Bąchor, 2024).

In parallel, substantial progress has been made in algal genetic engineering, driven by the development of diverse transformation techniques, expression vectors, modular cloning platforms (e.g., MoClo toolkits), promoters, transcription factors, and genome-editing technologies such as zinc-finger nucleases (ZFNs), transcription activator-like nucleases (TALENs), and CRISPR-Cas9. These advances have enabled the generation of novel algal strains with enhanced or rewired metabolic capabilities. Ongoing efforts continue to focus on expanding the molecular toolkit for metabolic engineering with the aim of improving product yields and enabling the biosynthesis of new bioproducts suitable for commercialization (Sproles et al., 2021).

One notable example is the development of the Chlamys Sequence Optimizer (CSO). It is a computational tool that facilitates the rational design of heterologous genes for chloroplast expression in the model microalga *C. reinhardtii*. This tool has significantly enhanced the predictability and efficiency of chloroplast transgene expression (Weiner, 2020). Complementing this approach, Shahar et al. (2021) demonstrated the use of orthogonal regulatory systems by introducing nucleus-encoded chloroplast gene activators from maize (PPR10) and *Arabidopsis* to drive plastid transgene expression in *C. reinhardtii* (Shahar et al., 2021). Collectively, these innovations lay the groundwork for tunable, host-independent expression systems and further unlock the potential of microalgal chloroplasts as versatile biofactories.

In the area of genome editing, significant progress has been achieved since the first successful transient application of CRISPR-Cas9 in *C. reinhardtii* in 2014 (Jiang et al., 2014). Although early implementations faced challenges such as Cas9-associated toxicity, subsequent methodological refinements have enabled the successful generation of stable mutant strains across multiple microalgal species (Sproles et al., 2021). Specifically, CRISPR-Cas9-mediated mutagenesis in *Dunaliella tertiolecta* produced strains with higher specific growth rates under medium- and high-light conditions compared with the wild type (Kim et al., 2017). In *C. reinhardtii*, targeted knockout of the zeaxanthin epoxidase gene resulted in increased zeaxanthin content and productivity (Baek et al., 2018). Similarly, CRISPR-Cas9 disruption of the ADP-glucose pyrophosphorylase (AGP) gene in *Tetraselmis* sp. led to significantly enhanced lipid accumulation, highlighting the promise of metabolic engineering in marine microalgae for biofuel production (Chang et al., 2020).

4.7 Challenges and future prospects

4.7.1 Limitations and challenges

The transition from laboratory-scale studies to industrial enzyme biofactories is a complex and multifaceted process, requiring the resolution of several critical challenges. Among these, maintaining enzyme stability during prolonged operation and optimizing cultivation conditions for large-scale biomass production represent two of the most significant barriers to commercial deployment.

Preserving high catalytic efficiency under industrial conditions that are comparable to those achieved in controlled laboratory environments, along with long-term enzyme stability, is essential for commercial success. Current strategies to enhance enzyme robustness include immobilization on solid supports and covalent binding techniques, both of which have demonstrated substantial improvements in sustaining enzymatic activity over extended periods (El-Fakharany et al., 2024). A notable example is the enzyme precipitate coating (EPC) protocol. Through the covalent attachment of enzymes to electrospun polymer nanofibers, followed by controlled precipitation and cross-linking, this approach generates a highly stable enzyme matrix that preserves structural integrity while effectively minimizing metal ion leaching and enzyme deactivation (Jun et al., 2020).

Although microalgae are capable of thriving in diverse and even extreme environmental conditions, efficient biomass production and downstream processing at industrial scale depend critically on the careful selection of suitable microalgal species (Hernández et al., 2017). This selection process must balance biological performance, process robustness, and economic feasibility. Among the many variables involved, optimization of culture conditions remains one of the most challenging aspects of scale-up and is a key determinant of process success. Achieving reproducible, high-density biomass production requires precise control of multiple physicochemical parameters, including light, temperature, nutrient supply, gas exchange, and hydrodynamics, to enable scalable and cost-effective bioprocessing (Ru et al., 2020). An overview of preferred culture conditions is summarized in Table 4.2.

Recent advances in cultivation technologies have begun to address these challenges, particularly through the development of advanced photobioreactor designs that provide precise control over environmental parameters while minimizing the risk of contamination (Barboza-Rodríguez et al., 2024). As described in earlier sections, the deployment of genetically engineered algal strains with enhanced growth rates and productivity also offers considerable potential for further improving biomass and enzyme yields. Nevertheless, the application of genetically modified microalgae at industrial scale necessitates careful consideration of regulatory frameworks and biosafety measures to prevent the unintended release of modified organisms into natural ecosystems.

4.7.2 The future prospects: sustainability and economic considerations

Algal cultivation systems are generally more environmentally sustainable than conventional agricultural or microbial fermentation platforms, as they

Table 4.2 The preferred culture conditions of algae in controlled environment.

Cultural conditions	Preferences	References
Carbon source	Grow autotrophically using CO_2 or mixotrophically with organic carbon sources like glucose	Daliry et al. (2017); Liang et al. (2009)
Light	Optimal intensity is around 100–200 μmol photons/m^2/s, with red and yellow light promoting faster growth	Hultberg et al. (2014); Zhao and Su (2014)
pH	Varies but generally falls between 7.5 and 10.5 (depending on species)	Rachlin and Grosso (1991); Gong et al. (2014)
Temperature	Optimum growth occurs between 15°C and 30°C, inhibit growth at temperature above 35°C	Kumar et al. (2010); Chinnasamy et al. (2009)
Salinity	Some species tolerate salinity up to 0.5 M NaCl	Rai et al. (2015)
Aeration and mixing	Gentle agitation prevents cell damage while improving gas exchange and nutrient distribution	Daliry et al. (2017)

require fewer resources, particularly regarding land use and freshwater consumption (El-Fakharany et al., 2024). Closed photobioreactor systems, which operate under tightly controlled environmental conditions, enable higher photosynthetic efficiency and achieve substantially greater biomass densities than traditional agricultural systems (Adeniyi et al., 2018). In addition, the integration of wastewater as a nutrient source further enhances the sustainability and cost-effectiveness of algal cultivation by reducing freshwater demand and lowering nutrient input costs (Adeniyi et al., 2018; Roy et al., 2022). Beyond biomass production, algae have attracted considerable attention for their broader environmental applications, including roles as bioindicators and biosensors, sources of renewable biofuels, and agents for the removal of organic and inorganic pollutants (Roy et al., 2022).

Microbial enzymes play a central role in a wide range of biotechnological applications spanning nutrition, industrial processing, and medicine. In the context of algal bioprocessing, cultivated algal biomass offers additional value beyond enzyme production, as it contains a diverse array of high-value components, such as essential amino acids, polyunsaturated fatty acids, carbohydrates, pigments, and other bioactive compounds (El-Fakharany et al., 2024). When combined with reduced media costs, lower requirements for external carbon sources, decreased contamination risks, and simplified downstream processing and formulation, algal-based enzyme manufacturing has emerged as a rapidly growing global industry with substantial environmental, health, and economic implications (Rasala and Mayfield, 2015). The market for algal-derived enzymes is expected to expand further in the coming years, in line with projections estimating that the global industrial enzyme market will reach USD 15.5 billion by 2034 (El-Fakharany et al., 2024). Given the numerous advantages of microalgal enzymes over conventional microbial counterparts, there is strong justification for continued investment in microalgal research and development.

As illustrated throughout this chapter, algae represent a versatile and sustainable biological platform for enzyme production in the modern biotechnological landscape. Ongoing advances in genetic engineering, synthetic biology, and bioprocess optimization continue to expand their functionality and economic viability as enzyme biofactories. As research matures and market demand for algal-derived enzymes increases, the integration of algae-based production systems into commercial pipelines has the potential to substantially reduce manufacturing costs and environmental impact. Continued interdisciplinary collaboration across molecular biology, process engineering, regulatory science, and industrial biotechnology will be critical to overcoming existing challenges and fully realizing the biotechnological potential of algae in enzyme manufacturing.

References

Abd Ellatif, S., El-Sheekh, M.M., Senousy, H.H., 2021. Role of microalgal ligninolytic enzymes in industrial dye decolorization. International Journal of Phytoremediation 41–52. Available at https://doi.org/10.1080/15226514.2020.1789842. Available from: https://www.tandfonline.com/loi/bijp20.

Adeniyi, O.M., Azimov, U., Burluka, A., 2018. Algae biofuel: current status and future applications. Renewable and Sustainable Energy Reviews 90, 316–335. Available at https://doi.org/10.1016/j.rser.2018.03.067. Available from: https://www.journals.elsevier.com/renewable-and-sustainable-energy-reviews.

Alam, T., Najam, L., Harrasi, A.A., 2018. Extraction of natural pigments from marine algae. Journal of Agricultural & Marine Sciences (JAMS).

Ambily, B., Limna Mol, V.P., Sini, H., Nevin, K.G., 2025. CRISPR-based microalgal genome editing and the potential for sustainable aquaculture: a comprehensive review. Journal of Applied Phycology 37 (1), 265–285. Available at https://doi.org/10.1007/s10811-024-03376-x.

Ameri, M., Baron-Sola, A., Khavari-Nejad, R.A., Soltani, N., Najafi, F., Bagheri, A., et al., 2020. Aluminium triggers oxidative stress and antioxidant response in the microalgae Scenedesmus sp. Journal of Plant Physiology 246-247, 153114. Available at https://doi.org/10.1016/j.jplph.2020.153114.

Baek, K., Yu, J., Jeong, J., Sim, S.J., Bae, S., Jin, E.S., 2018. Photoautotrophic production of macular pigment in a Chlamydomonas reinhardtii strain generated by using DNA-free CRISPR-Cas9 RNP-mediated mutagenesis. Biotechnology and Bioengineering 115 (3), 719–728. Available at https://doi.org/10.1002/bit.26499.

Banerjee, A., Ward, V., 2022. Production of recombinant and therapeutic proteins in microalgae. Current Opinion in Biotechnology 78, 102784. Available at https://doi.org/10.1016/j.copbio.2022.102784.

Barbarino, E., Lourenço, S.O., 2005. An evaluation of methods for extraction and quantification of protein from marine macro- and microalgae. Journal of Applied Phycology 17 (5), 447–460. Available at https://doi.org/10.1007/s10811-005-1641-4.

Barboza-Rodríguez, R., Rodríguez-Jasso, R.M., Rosero-Chasoy, G., Rosales Aguado, M.L., Ruiz, H.A., 2024. Photobioreactor configurations in cultivating microalgae biomass for biorefinery. Bioresource Technology 394, 130208. Available at https://doi.org/10.1016/j.biortech.2023.130208.

Barlowe, C., 1995. COPII: a membrane coat that forms endoplasmic reticulum-derived vesicles. FEBS Letters 369 (1), 93–96. Available at https://doi.org/10.1016/0014-5793(95)00618-j.

Batool, T., Makky, E.A., Jalal, M., Yusoff, M.M., 2016. A comprehensive review on l-asparaginase and its applications. Applied Biochemistry and Biotechnology 178 (5), 900–923. Available at https://doi.org/10.1007/s12010-015-1917-3.

Bellido-Pedraza, C.M., Torres, M.J., Llamas, A., 2024. The microalgae chlamydomonas for bioremediation and bioproduct production. Cells 13 (13). Available at https://doi.org/10.3390/cells13131137. Available from: https://www.mdpi.com/journal/cells.

Ben Hlima, H., Dammak, M., Karray, A., Drira, M., Michaud, P., Fendri, I., et al., 2021. Molecular and structural characterizations of lipases from chlorella by functional genomics. Marine Drugs 19 (2), 70. Available at https://doi.org/10.3390/md19020070.

Bhatt, A., Prajapati, D., Gupte, A., 2021. Current status and future of nitrile catalysis using key nitrilases enzymes and their biotechnological impact. Open Biotechnology Journal 15 (1), 71–81. Available at https://doi.org/10.2174/1874070702115010071. Available from: https://openbiotechnologyjournal.com/VOLUME/15/PAGE/71/.

Bonnefoy, N., Remacle, C., 2023. Biolistic transformation of Chlamydomonas reinhardtii and Saccharomyces cerevisiae mitochondria. Methods in Molecular Biology 2615, 345–364. Available at https://doi.org/10.1007/978-1-0716-2922-2_24. Available from: https://www.springer.com/series/7651.

Borowitzka, M.A., 2013. High-value products from microalgae-their development and commercialisation. Journal of Applied Phycology 25 (3), 743–756. Available at https://doi.org/10.1007/s10811-013-9983-9.

Carbone, D.A., Gargano, I., Chiaiese, P., Pollio, A., Marotta, R., Olivieri, G., et al., 2018. Scenedesmus vacuolatus cultures for possible combined laccase-like phenoloxidase activity and biodiesel production. Annals of Microbiology 68 (1), 9–15. Available at https://doi.org/10.1007/s13213-017-1309-8. Available from: http://springerlink.com/content/121707/.

Carrieri, D., Ananyev, G., Garcia Costas, A.M., Bryant, D.A., Dismukes, G.C., 2008. Renewable hydrogen production by cyanobacteria: nickel requirements for optimal hydrogenase activity. International Journal of Hydrogen Energy 33 (8), 2014–2022. Available at https://doi.org/10.1016/j.ijhydene.2008.02.022.

Cerón Garcí, M.C., Fernández Sevilla, J.M., Acién Fernández, F.G., Molina Grima, E., García Camacho, F., 2000. Mixotrophic growth of Phaeodactylum tricornutum on glycerol: growth rate and fatty acid profile. Journal of Applied Phycology 12 (3-5), 239–248. Available at https://doi.org/10.1023/a:1008123000002.

Chandra, P., Enespa, Singh, R., Arora, P.K., 2020. Microbial lipases and their industrial applications: a comprehensive review. Microbial Cell Factories 19 (1). Available at https://doi.org/10.1186/s12934-020-01428-8.

Chang, K.S., Kim, J., Park, H., Hong, S.-J., Lee, C.-G., Jin, E.S., 2020. Enhanced lipid productivity in AGP knockout marine microalga Tetraselmis sp. using a DNA-free CRISPR-Cas9 RNP method. Bioresource Technology 303, 122932. Available at https://doi.org/10.1016/j.biortech.2020.122932.

Chanquia, S.N., Vernet, G., Kara, S., 2022. Photobioreactors for cultivation and synthesis: specifications, challenges, and perspectives. Engineering in Life Sciences 22 (12), 712–724. Available at https://doi.org/10.1002/elsc.202100070. Available from: http://onlinelibrary.wiley.com/journal/10.1002/(ISSN)1618-2863.

Chen, F., 1996. High cell density culture of microalgae in heterotrophic growth. Trends in Biotechnology 14 (11), 421–426. Available at https://doi.org/10.1016/0167-7799(96)10060-3.

Chen, H.-H., Liang, M.-H., Ye, Z.-W., Zhu, Y.-H., Jiang, J.-G., Fuente-Nunez, C.de la, 2023. Engineering the β-carotene metabolic pathway of microalgae dunaliella to confirm its carotenoid synthesis pattern in comparison to bacteria and plants. Microbiology Spectrum 11 (2). Available at https://doi.org/10.1128/spectrum.04361-22.

Chinnasamy, S., Ramakrishnan, B., Bhatnagar, A., Das, K., 2009. Biomass production potential of a wastewater alga Chlorella vulgaris ARC 1 under elevated levels of CO2 and temperature. International Journal of Molecular Sciences 10 (2), 518–532. Available at https://doi.org/10.3390/ijms10020518.

Chung, K.P., Frieboese, D., Waltz, F., Engel, B.D., Bock, R., 2024. Identification and characterization of the COPII vesicle-forming GTPase Sar1 in Chlamydomonas. Plant Direct 8 (6) Available from. Available at https://doi.org/10.1002/pld3.614, onlinelibrary.wiley.com/journal/24754455.

Cosme, F., Inês, A., Vilela, A., 2023. Microbial and commercial enzymes applied in the beverage production process. Fermentation 9 (4), 385. Available at https://doi.org/10.3390/fermentation9040385.

Crozet, P., Navarro, F.J., Willmund, F., Mehrshahi, P., Bakowski, K., Lauersen, K.J., et al., 2018. Birth of a photosynthetic chassis: a MoClo toolkit enabling synthetic biology in the microalga Chlamydomonas reinhardtii. ACS Synthetic Biology 7 (9), 2074–2086. Available at https://doi.org/10.1021/acssynbio.8b00251.

Daliry, S., Hallajisani, A., Mohammadi Roshandeh, J., Nouri, H., Golzary, A., 2017. Investigation of optimal condition for Chlorella vulgaris microalgae growth. Global Journal of Environmental Science and Management 3 (2), 217–230. Available at https://doi.org/10.22034/gjesm.2017.03.02.010. Available from: http://www.gjesm.net/article_22943_4886dbd2ff922a98ef31e68a80c9f2e6.pdf.

Demir, B.S., Tükel, S.S., 2010. Purification and characterization of lipase from Spirulina platensis. Journal of Molecular Catalysis B: Enzymatic 64 (3-4), 123–128. Available at https://doi.org/10.1016/j.molcatb.2009.09.011.

Dhar, A., Dey, S., Sarkar, S., 2023. Liquid trees: a novel approach for air pollution mitigation. Vigyan Varta 4.

DiMario, R.J., Machingura, M.C., Waldrop, G.L., Moroney, J.V., 2018. The many types of carbonic anhydrases in photosynthetic organisms. Plant Science 268, 11–17. Available at https://doi.org/10.1016/j.plantsci.2017.12.002. Available from: http://www.journals.elsevier.com/plant-science/.

Dong, T., Van Wychen, S., Nagle, N., Pienkos, P.T., Laurens, L.M.L., 2016. Impact of biochemical composition on susceptibility of algal biomass to acid-catalyzed pretreatment for sugar and lipid recovery. Algal Research 18, 69–77. Available at https://doi.org/10.1016/j.algal.2016.06.004.

Dummermuth, A.L., Karsten, U., Fisch, K.M., König, G.M., Wiencke, C., 2003. Responses of marine macroalgae to hydrogen-peroxide stress. Journal of Experimental Marine Biology and Ecology 289 (1), 103–121. Available at https://doi.org/10.1016/s0022-0981(03)00042-x.

Ebrahiminezhad, A., Rasoul-Amini, S., Ghoshoon, M.B., Ghasemi, Y., 2014. Chlorella vulgaris, a novel microalgal source for l-asparaginase production. Biocatalysis and Agricultural Biotechnology 3 (2), 214–217. Available at https://doi.org/10.1016/j.bcab.2013.10.005.

Egbo, M.K., Okoani, A.O., Okoh, I.E., 2018. Photobioreactors for microalgae cultivation–an overview. International Journal of Scientific Engineering and Research 9 (11), 65–74.

Eilers, U., Bikoulis, A., Breitenbach, J., Büchel, C., Sandmann, G., 2016. Limitations in the biosynthesis of fucoxanthin as targets for genetic engineering in Phaeodactylum tricornutum. Journal of Applied Phycology 28 (1), 123–129. Available at https://doi.org/10.1007/s10811-015-0583-8.

Ejaz, U., Sohail, M., Ghanemi, A., 2021. Cellulases: from bioactivity to a variety of industrial applications. Biomimetics 6 (3) Available from. Available at https://doi.org/10.3390/biomimetics6030044, ww.mdpi.com/journal/biomimetics.

El-Fakharany, E.M., Saleh, A.K., El-Maradny, Y.A., El-Sayed, M.H., Alali, I., Alsirhani, A.M., et al., 2024. Comprehensive insight into recent algal enzymes production and purification advances: toward effective commercial applications: a review. International Journal of Biological Macromolecules 283 (Pt 2), 137783. Available at https://doi.org/10.1016/j.ijbiomac.2024.137783.

Elleuch, J., Hadj Kacem, F., Ben Amor, F., Hadrich, B., Michaud, P., Fendri, I., et al., 2021. Extracellular neutral protease from Arthrospira platensis: production, optimization and partial characterization. International Journal of Biological Macromolecules 167, 1491–1498. Available at https://doi.org/10.1016/j.ijbiomac.2020.11.102.

Elshobary, M.E., Abo-Shanab, W.A., Ende, S.S.W., Alquraishi, M., El-Shenody, R.A., 2025. Optimizing Phaeodactylum tricornutum cultivation: integrated strategies for enhancing biomass, lipid, and fucoxanthin production. Biotechnology for Biofuels and Bioproducts 18 (1). Available at https://doi.org/10.1186/s13068-024-02602-5. Available from: https://biotechnologyforbiofuels.biomedcentral.com/.

Englund, E., Andersen-Ranberg, J., Miao, R., Hamberger, B., Lindberg, P., 2015. Metabolic engineering of Synechocystis sp. PCC 6803 for production of the plant diterpenoid manoyl oxide. ACS Synthetic Biology 4 (12), 1270–1278. Available at https://doi.org/10.1021/acssynbio.5b00070.

Erpel, F., Restovic, F., Arce-Johnson, P., 2016. Development of phytase-expressing chlamydomonas reinhardtii for monogastric animal nutrition. BMC Biotechnology 16 (1) Available from. Available at https://doi.org/10.1186/s12896-016-0258-9, http://www.biomedcentral.com/bmcbiotechnol/.

Fasim, A., More, V.S., More, S.S., 2021. Large-scale production of enzymes for biotechnology uses. Current Opinion in Biotechnology 69, 68–76. Available from. Available at https://doi.org/10.1016/j.copbio.2020.12.002, http://www.elsevier.com/locate/copbio.

Ferrante, P., Catalanotti, C., Bonente, G., Giuliano, G., El-Shemy, H.A., 2008. An optimized, chemically regulated gene expression system for Chlamydomonas. PLoS One 3 (9), e3200. Available at https://doi.org/10.1371/journal.pone.0003200.

Flórez-Fernández, N., Rodríguez-Coello, A., Latire, T., Bourgougnon, N., Torres, M.D., Buján, M., et al., 2023. Anti-inflammatory potential of ulvan. International Journal of Biological Macromolecules 253, 126936. Available at https://doi.org/10.1016/j.ijbiomac.2023.126936.

Gajanayaka, N.D., Jo, E., Bandara, M.S., Marasinghe, S.D., Bawkar, C., Lee, Y.J., et al., 2024. Characterisation of high alkaline-tolerant novel ulvan lyase from Pseudoalteromonas agarivorans: potential applications of enzyme derived oligo-ulvan as anti-diabetic agent. Marine Drugs 22 (12) Available from. Available at https://doi.org/10.3390/md22120577, http://www.mdpi.com/journal/marinedrugs.

Gamberi, C., Leverette, C.L., Davis, A.C., Ismail, M., Piccialli, I., Borbone, N., et al., 2024. Oceanic breakthroughs: marine-derived innovations in vaccination, therapy, and immune health. Vaccines 12 (11), 1263. Available at https://doi.org/10.3390/vaccines12111263.

Gangl, D., Zedler, J.A.Z., Rajakumar, P.D., Martinez, E.M.R., Riseley, A., Włodarczyk, A., et al., 2015. Biotechnological exploitation of microalgae. Journal of Experimental Botany 66 (22), 6975–6990. Available at https://doi.org/10.1093/jxb/erv426.

Gan, S.Y., Maggs, C.A., 2017. Random mutagenesis and precise gene editing technologies: applications in algal crop improvement and functional genomics. European Journal of Phycology 52 (4), 466–481. Available at https://doi.org/10.1080/09670262.2017.1358827. Available from: https://www.tandfonline.com/loi/tejp20.

Garcia-Perez, P., Lourenço-Lopes, C., Silva, A., Pereira, A., Fraga-Corral, M., Zhao, C., et al., 2022. Pigment composition of nine brown algae from the Iberian Northwestern coastline: influence of the extraction solvent. Marine Drugs 20 (2), 113. Available at https://doi.org/10.3390/md20020113.

Gong, Q., Feng, Y., Kang, L., Luo, M., Yang, J., 2014. Effects of light and ph on cell density of Chlorella vulgaris. Energy Procedia 61, 2012–2015. Available at https://doi.org/10.1016/j.egypro.2014.12.064.

Goshtasbi, H., Okolodkov, Y.B., Movafeghi, A., Awale, S., Safary, A., Barar, J., et al., 2023. Harnessing microalgae as sustainable cellular factories for biopharmaceutical production. Algal Research 74, 103237. Available at https://doi.org/10.1016/j.algal.2023.103237.

GVR Catalysts & Enzymes Research Team. 2023. Grand View Research Enzymes market size, share & trends analysis report by type (industrial, specialty), by product (carbohydrase, proteases), by source (plants, animals), by region, and segment forecasts, 2025 - 2033 (Report ID: 978-1-68038-022-4). https://www.grandviewresearch.com/industry-analysis/enzymes-industry

Guihéneuf, F., Khan, A., Tran, L.S.P., 2016. Genetic engineering: a promising tool to engender physiological, biochemical, and molecular stress resilience in green microalgae. Frontiers in Plant Science 7, 2016. Available at https://doi.org/10.3389/fpls.2016.00400. Available from: http://journal.frontiersin.org/article/10.3389/fpls.2016.00400/pdf.

Hallmann, A., Sumper, M., 1994. An inducible arylsulfatase of Volvox carteri with properties suitable for a reporter-gene system: purification, characterization and molecular cloning. European Journal of Biochemistry 221 (1), 143–150. Available at https://doi.org/10.1111/j.1432-1033.1994.tb18723.x.

Harada, H., Nakatsuma, D., Ishida, M., Matsuda, Y., 2005. Regulation of the expression of intracellular β -carbonic anhydrase in response to CO2 and light in the marine diatom Phaeodactylum tricornutum. Plant Physiology 139 (2), 1041–1050. Available at https://doi.org/10.1104/pp.105.065185.

Harde, S.M., Singhal, R.S., 2012. Extraction of forskolin from Coleus forskohlii roots using three phase partitioning. Separation and Purification Technology 96, 20–25. Available at https://doi.org/10.1016/j.seppur.2012.05.017.

Hatahet, F., Boyd, D., Beckwith, J., 2014. Disulfide bond formation in prokaryotes: history, diversity and design. Biochimica et Biophysica Acta - Proteins and Proteomics 1844 (8), 1402–1414. Available at https://doi.org/10.1016/j.bbapap.2014.02.014. Available from: http://www.elsevier.com/locate/bbapap.

Hernández, E., Lobato-Benítez, C., Hernández, C., 2017. Algal Enzymes, Biotechnological Potential Uses: A Review 1–15.

von der Heyde, E.L., Klein, B., Abram, L., 2015. The inducible nitA promoter provides a powerful molecular switch for transgene expression in Volvox carteri. BMC Biotechnology 15, 1–13.

von der Heyde, B., von der Heyde, E.L., Hallmann, A., 2023. Cell type-specific promoters of Volvox carteri for molecular cell biology studies. Molecular Cell Biology Studies. Genes 14 (7).

Hippler, M., Khosravitabar, F., 2024. Light-driven H2 production in Chlamydomonas reinhardtii: lessons from engineering of photosynthesis. Plants 13 (15), 2114. Available at https://doi.org/10.3390/plants13152114.

Howden, A.J.M., Preston, G.M., 2009. Nitrilase enzymes and their role in plant-microbe interactions. Microbial Biotechnology 2 (4), 441–451. Available at https://doi.org/10.1111/j.1751-7915.2009.00111.xUnitedKingdom. Available from: http://www3.interscience.wiley.com/cgi-bin/fulltext/122328411/PDFSTART.

Hultberg, M., Bodin, H., Birgersson, G., 2019. Impact on wastewater quality of biopellets composed of chlorella vulgaris and aspergillus niger and lipid content in the harvested biomass. Journal of Water Resource and Protection 11 (07), 831–843. Available at https://doi.org/10.4236/jwarp.2019.117050.

Hultberg, M., Jönsson, H.L., Bergstrand, K.-J., Carlsson, A.S., 2014. Impact of light quality on biomass production and fatty acid content in the microalga Chlorella vulgaris. Bioresource Technology 159, 465–467. Available at https://doi.org/10.1016/j.biortech.2014.03.092.

Hu, H., Pan, Y., 2019. Electroporation Transformation Protocol for Phaeodactylum tricornutum. Springer Science and Business Media LLC, pp. 163–167. Available at 10.1007/978-1-4939-9740-4_17.

Hu, Q., Sommerfeld, M., Jarvis, E., Ghirardi, M., Posewitz, M., Seibert, M., et al., 2008. Microalgal triacylglycerols as feedstocks for biofuel production: perspectives and advances. The Plant Journal 54 (4), 621–639. Available at https://doi.org/10.1111/j.1365-313x.2008.03492.x.

Ifuku, K., Yan, D., 2019. Efficient Transformation of the Diatoms Phaeodactylum tricornutum by Multipulse Electroporation. Science and Business Media LLC, pp. 169–174. Available at 10.1007/978-1-4939-9740-4_18.

Ip, P.F., Chen, F., 2005. Production of astaxanthin by the green microalga Chlorella zofingiensis in the dark. Process Biochemistry 40 (2), 733–738. Available at https://doi.org/10.1016/j.procbio.2004.01.039.

Iwai, M., Ikeda, K., Shimojima, M., Ohta, H., 2014. Enhancement of extraplastidic oil synthesis in C hlamydomonas reinhardtii using a type-2 diacylglycerol acyltransferase with a phosphorus starvation–inducible promoter. Plant Biotechnology Journal 12 (6), 808–819. Available at https://doi.org/10.1111/pbi.12210.

Jagadevan, S., Banerjee, A., Banerjee, C., Guria, C., Tiwari, R., Baweja, M., et al., 2018. Recent developments in synthetic biology and metabolic engineering in microalgae towards biofuel production. Biotechnology for Biofuels 11 (1). Available at https://doi.org/10.1186/s13068-018-1181-1.

Janssen, M., Kuijpers, T.C., Veldhoen, B., Ternbach, M.B., Tramper, J., Mur, L.R., et al., 1999. Specific growth rate of Chlamydomonas reinhardtii and Chlorella sorokiniana under medium duration light/dark cycles: 13–87 s. Journal of Biotechnology 70, 323–333. Available at 10.1016/s0079-6352(99)80124-6.

Jiang, W., Brueggeman, A.J., Horken, K.M., Plucinak, T.M., Weeks, D.P., 2014. Successful transient expression of Cas9 and single guide RNA genes in Chlamydomonas reinhardtii. Eukaryotic Cell 13 (11), 1465–1469. Available at https://doi.org/10.1128/ec.00213-14.

Joubert, Y., Fleurence, J., 2008. Simultaneous extraction of proteins and DNA by an enzymatic treatment of the cell wall of Palmaria palmata (Rhodophyta). Journal of Applied Phycology 20 (1), 55–61. Available at https://doi.org/10.1007/s10811-007-9180-9.

Jun, S.H., Yang, J., Jeon, H., Kim, H.S., Pack, S.P., Jin, E., et al., 2020. Stabilized and immobilized carbonic anhydrase on electrospun nanofibers for enzymatic CO2 conversion and utilization in expedited microalgal growth. Environmental Science and Technology 54 (2), 1223–1231. Available at https://doi.org/10.1021/acs.est.9b05284. Available from: http://pubs.acs.org/journal/esthag.

Kassaw, T.K., Paton, A.J., Peers, G., 2022. Episome-based gene expression modulation platform in the model diatom Phaeodactylum tricornutum. ACS Synthetic Biology 11 (1), 191–204. Available at https://doi.org/10.1021/acssynbio.1c00367. Available from: http://pubs.acs.org/journal/asbcd6.

Kim, M., Ahn, J., Jeon, H., Jin, E.S., 2017. Development of a Dunaliella tertiolecta strain with increased zeaxanthin content using random mutagenesis. Marine Drugs 15 (6), 189. Available at https://doi.org/10.3390/md15060189.

Kim, J., Yoo, G., Lee, H., Lim, J., Kim, K., Kim, C.W., et al., 2013. Methods of downstream processing for the production of biodiesel from microalgae. Biotechnology Advances 31 (6), 862–876. Available at https://doi.org/10.1016/j.biotechadv.2013.04.006.

Klińska-Bąchor, S., 2024. Biochemical characterization of acyl-CoA: diacylglycerol acyltransferase2 from the diatom Phaeodactylum tricornutum and its potential effect on LC-PUFAs biosynthesis in planta. BMC Plant Biology 24 (1).

Koblenz, B., Lechtreck, K.F., 2005. The NIT1 promoter allows inducible and reversible silencing of centrin in Chlamydomonas reinhardtii. Eukaryotic Cell 4 (11), 1959–1962. Available at https://doi.org/10.1128/EC.4.11.1959-1962.2005.

Kosourov, S., Böhm, M., Senger, M., Berggren, G., Stensjö, K., Mamedov, F., et al., 2021. Photosynthetic hydrogen production: Novel protocols, promising engineering approaches and application of semi-synthetic hydrogenases. Physiologia Plantarum 173 (2), 555–567. Available at https://doi.org/10.1111/ppl.13428.

Kumar, A., Ergas, S., Yuan, X., Sahu, A., Zhang, Q., Dewulf, J., et al., 2010. Enhanced CO2 fixation and biofuel production via microalgae: recent developments and future directions. Trends in Biotechnology 28 (7), 371–380. Available at https://doi.org/10.1016/j.tibtech.2010.04.004.

Kumar, A.U., Ling, A.P.K., 2021. Gene introduction approaches in chloroplast transformation and its applications. Journal of Genetic Engineering and Biotechnology 19 (1). Available at https://doi.org/10.1186/s43141-021-00255-7. Available from: https://jgeb.springeropen.com/.

Larrea-Alvarez, M., Purton, S., 2020. Multigenic engineering of the chloroplast genome in the green alga Chlamydomonas reinhardtii. Microbiology 166 (6), 510–515. Available at https://doi.org/10.1099/mic.0.000910.

Lauersen, K.J., 2019. Eukaryotic microalgae as hosts for light-driven heterologous isoprenoid production. Planta 249 (1), 155–180. Available from: Available at https://doi.org/10.1007/s00425-018-3048-x, link.springer.de/link/service/journals/00425/index.htm.

Lee, Y.Y., Park, R., Miller, S.M., Li, Y., 2022. Genetic compensation of triacylglycerol biosynthesis in the green microalga Chlamydomonas reinhardtii. Plant Journal 111 (4),

1069–1080. Available at https://doi.org/10.1111/tpj.15874. Available from: http://onlinelibrary.wiley.com/journal/10.1111/(ISSN)1365-313X.

Leprovost, S., Plasson, C., Balieu, J., Walet-Balieu, M.-L., Lerouge, P., Bardor, M., et al., 2024. Fine-tuning the N -glycosylation of recombinant human erythropoietin using Chlamydomonas reinhardtii mutants. Plant Biotechnology Journal 22 (11), 3018–3027. Available at https://doi.org/10.1111/pbi.14424.

Liang, Y., Sarkany, N., Cui, Y., 2009. Biomass and lipid productivities of Chlorella vulgaris under autotrophic, heterotrophic and mixotrophic growth conditions. Biotechnology Letters 31 (7), 1043–1049. Available at https://doi.org/10.1007/s10529-009-9975-7.

Y., Lin, et al., 2024. Protein Disulfide Isomerase and Its Potential Function on Endoplasmic Reticulum Quality Control in Diatom Phaeodactylum tricornutum. Phyton 93 (1), 137.

Li, F., Gao, D., Hu, H., 2014. High-efficiency nuclear transformation of the oleaginous marine Nannochloropsis species using PCR product. Bioscience, Biotechnology and Biochemistry 78 (5), 812–817. Available at https://doi.org/10.1080/09168451.2014.905184. Available from: http://www.tandfonline.com/doi/pdf/10.1080/09168451.2014.905184.

Li, J., Gu, X., Zhang, Q., Fu, L., Tan, J., Zhao, L., 2022. Biochemical characterization of a carrageenase, Car1383, derived from associated bacteria of antarctic macroalgae. Frontiers in Microbiology 13. Available at https://doi.org/10.3389/fmicb.2022.851182.

Loppes, R., Radoux, M., Ohresser, M.C.P., Matagne, R.F., 1999. Transcriptional regulation of the Nia1 gene encoding nitrate reductase in Chlamydomonas reinhardtii: Effects of various environmental factors on the expression of a reporter gene under the control of the Nia1 promoter. Plant Molecular Biology 41 (5), 701–711. Available at https://doi.org/10.1023/A:1006381527119.

López-Otín, C., Bond, J.S., 2008. Proteases: multifunctional enzymes in life and disease. Journal of Biological Chemistry 283 (45), 30433–30437. Available at https://doi.org/10.1074/jbc.R800035200. Available from: http://www.jbc.org/cgi/reprint/283/45/30433.

Lucas, P.L., Mathieu-Rivet, E., Song, P.C.T., Oltmanns, A., Loutelier-Bourhis, C., Plasson, C., et al., 2020. Multiple xylosyltransferases heterogeneously xylosylate protein N-linked glycans in Chlamydomonas reinhardtii. Plant Journal 102 (2), 230–245. Available at https://doi.org/10.1111/tpj.14620. Available from: http://onlinelibrary.wiley.com/journal/10.1111/(ISSN)1365-313X.

Manat, G., Fanuel, M., Jouanneau, D., Jam, M., Mac-Bear, J., Rogniaux, H., et al., 2022. Specificity of a β-porphyranase produced by the carrageenophyte red alga Chondrus crispus and implications of this unexpected activity on red algal biology. Journal of Biological Chemistry 298 (12), 102707. Available at https://doi.org/10.1016/j.jbc.2022.102707.

Manisha, S.K.Y., 2017. Technological advances and applications of hydrolytic enzymes for valorization of lignocellulosic biomass. Bioresource Technology 245, 1727–1739. Available at https://doi.org/10.1016/j.biortech.2017.05.066. Available from: http://www.elsevier.com/locate/biortech.

Marin Viegas, V.S., Ocampo, C.G., Petruccelli, S., 2017. Vacuolar deposition of recombinant proteins in plant vegetative organs as a strategy to increase yields. Bioengineered 8 (3), 203–211. Available at https://doi.org/10.1080/21655979.2016.1222994. Available from: http://www.tandfonline.com/loi/kbie20.

Mathieu-Rivet, E., 2020. N-and O-glycosylation pathways in the microalgae polyphyletic group. Frontiers in Plant Science 11.

Mathieu-Rivet, E., Kiefer-Meyer, M.-C., Vanier, G., Ovide, C.Ã.©ment, Burel, C., Lerouge, P., et al., 2014. Protein N-glycosylation in eukaryotic microalgae and its impact on the production of nuclear expressed biopharmaceuticals. Frontiers in Plant Science 5. Available at https://doi.org/10.3389/fpls.2014.00359.

Mayer, M., Kies, U., Kammermeier, R., Buchner, J., 2000. BiP and PDI cooperate in the oxidative folding of antibodies in vitro. Journal of Biological Chemistry 275 (38), 29421–29425. Available at https://doi.org/10.1074/jbc.m002655200.

Milledge, J.J., Heaven, S., 2013. A review of the harvesting of micro-algae for biofuel production. Reviews in Environmental Science and Biotechnology 12 (2), 165–178. Available at https://doi.org/10.1007/s11157-012-9301-z.

Mohammad, Z.H., Ahmad, F., Ibrahim, S.A., 2024. Biotechnology approaches to food security: risks and solutions. Microbial Biotechnology in the Food Industry: Advances, Challenges, and Potential Solutions. Springer International Publishing, United States, pp. 1–13.. Available from: https://link.springer.com/book/10.1007/978-3-031-51417-3. Available at 10.1007/978-3-031-51417-3_1.

Molino, J.V.D., de Carvalho, J.C.M., Mayfield, S.P., 2018. Comparison of secretory signal peptides for heterologous protein expression in microalgae: Expanding the secretion portfolio for Chlamydomonas reinhardtii. PLoS One 13 (2). Available at https://doi.org/10.1371/journal.pone.0192433. Available from: http://journals.plos.org/plosone/article/file?id=10.1371/journal.pone.0192433&type=printable.

Moosburner, M.A., Gholami, P., McCarthy, J.K., Tan, M., Bielinski, V.A., Allen, A.E., 2020. Multiplexed knockouts in the model diatom phaeodactylum by episomal delivery of a selectable Cas9. Frontiers in Microbiology 11. Available at https://doi.org/10.3389/fmicb.2020.00005. Available from: https://www.frontiersin.org/journals/microbiology#.

Moulin, S.L.Y., Beyly-Adriano, A., Cuiné, S., Blangy, S., Légeret, B., Floriani, M., et al., 2021. Fatty acid photodecarboxylase is an ancient photoenzyme that forms hydrocarbons in the thylakoids of algae. Plant Physiology 186 (3), 1455–1472. Available at https://doi.org/10.1093/plphys/kiab168.

Munir, N., Imtiaz, A., Sharif, N., Naz, S., 2015. Optimization of growth conditions of different algal strains and determination of their lipid contents. Journal of Animal and Plant Sciences 25 (2), 546–553. Available from: http://thejaps.org.pk/docs/v-25-02/30.pdf.

Naduthodi, M.I.S., Mohanraju, P., Südfeld, C., D'Adamo, S., Barbosa, M.J., Van Der Oost, J., 2019. CRISPR-Cas ribonucleoprotein mediated homology-directed repair for efficient targeted genome editing in microalgae Nannochloropsis oceanica IMET1. Biotechnology for Biofuels 12 (1). Available at https://doi.org/10.1186/s13068-019-1401-3. Available from: http://www.biotechnologyforbiofuels.com/.

Naduthodi, M.I.S., Südfeld, C., Avitzigiannis, E.K., Trevisan, N., van Lith, E., Alcaide Sancho, J., et al., 2021. Comprehensive genome engineering toolbox for microalgae Nannochloropsis oceanica based on CRISPR-Cas systems. ACS Synthetic Biology 10 (12), 3369–3378. Available at https://doi.org/10.1021/acssynbio.1c00329. Available from: http://pubs.acs.org/journal/asbcd6.

Ndochinwa, O.G., Wang, Q.-Y., Amadi, O.C., Nwagu, T.N., Nnamchi, C.I., Okeke, E.S., et al., 2024. Current status and emerging frontiers in enzyme engineering: an industrial perspective. Heliyon 10 (11), e32673. Available at https://doi.org/10.1016/j.heliyon.2024.e32673.

Ng, I.S., Keskin, B.B., Tan, S.I., 2020. A critical review of genome editing and synthetic biology applications in metabolic engineering of microalgae and cyanobacteria. Biotechnology journal 15 (8). Available at https://doi.org/10.1002/biot.201900228. Available from: http://onlinelibrary.wiley.com/journal/10.1002/(ISSN)1860-7314.

Noh, S.H., Kim, Y.J., Lee, M.G., 2022. Autophagy-related pathways in vesicular unconventional protein secretion. Frontiers in Cell and Developmental Biology 10.

Available at https://doi.org/10.3389/fcell.2022.892450. Available from: https://www.frontiersin.org/journals/cell-and-developmental-biology#.

Oltmanns, A., Hoepfner, L., Scholz, M., Zinzius, K., Schulze, S., Hippler, M., 2020. Novel insights into N-glycan fucosylation and core xylosylation in C. reinhardtii. Frontiers in Plant Science 10. Available at https://doi.org/10.3389/fpls.2019.01686.

Otto, B., Beuchel, C., Liers, C., Reisser, W., Harms, H., Schlosser, D., 2015. Laccase-like enzyme activities from chlorophycean green algae with potential for bioconversion of phenolic pollutants. FEMS Microbiology Letters 362 (11). Available at https://doi.org/10.1093/femsle/fnv072.

Otto, B., Schlosser, D., 2014. First laccase in green algae: purification and characterization of an extracellular phenol oxidase from Tetracystis aeria. Planta 240 (6), 1225–1236. Available from. Available at https://doi.org/10.1007/s00425-014-2144-9, link.springer.de/link/service/journals/00425/index.htm.

Patipong, T., Kageyama, H., Waditee-Sirisattha, R., 2021. Insights into the phylogeny and transcriptional response of serine proteases in a halotolerant cyanobacterium Halothece sp. PCC7418. Plant Signaling & Behavior 16 (9), 1913556. Available at https://doi.org/10.1080/2021.1913556.

Paul, J.H., 1982. Isolation and characterization of a Chlamydomonas l -asparaginase. Biochemical Journal 203 (1), 109–115. Available at https://doi.org/10.1042/bj2030109.

Paul, M., 2021. Microbial cellulases–an update towards its surface chemistry, genetic engineering and recovery for its biotechnological potential. Bioresource Technology 340.

Pérez-Lloréns, J.L., Benítez, E., Vergara, J.J., Berges, J.A., 2012. Characterization of proteolytic enzyme activities in macroalgae. European Journal of Phycology 38 (1), 55–64. Available at https://doi.org/10.1080/0967026031000096254.

Poidevin, L., Feliu, J., Doan, A., Berrin, J.-G., Bey, M., Coutinho, P.M., et al., 2013. Insights into exo- and endoglucanase activities of family 6 glycoside hydrolases from Podospora anserina. Applied and Environmental Microbiology 79 (14), 4220–4229. Available at https://doi.org/10.1128/aem.00327-13.

Prasad, R., Gupta, S.K., Shabnam, N., Oliveira, C.Y.B., Nema, A.K., Ansari, F.A., et al., 2021. Role of microalgae in global CO2 sequestration: physiological mechanism, recent development, challenges, and future prospective. Sustainability 13 (23), 13061. Available at https://doi.org/10.3390/su132313061.

Prestegard, S., Erga, S., Steinrücken, P., Mjøs, S., Knutsen, G., Rohloff, J., 2016. Specific Metabolites in a Phaeodactylum tricornutum Strain Isolated from Western Norwegian Fjord Water. Marine Drugs 14 (1), 9. Available at https://doi.org/10.3390/md14010009.

Qi, W., 2016. High-efficiency CRISPR/Cas9 multiplex gene editing using the glycine tRNA-processing system-based strategy in maize. BMC Biotechnology 16, 1–8.

Quiroz-Arita, C., Shinde, S., Kim, S., Monroe, E., George, A., Quinn, J., et al., 2022. Bioproducts from high-protein algal biomass: an economic and environmental sustainability review and risk analysis. Sustainable Energy & Fuels 6 (10), 2398–2422. Available at https://doi.org/10.1039/d1se01230d.

Rabouille, C., Malhotra, V., Nickel, W., 2012. Diversity in unconventional protein secretion. Journal of Cell Science 125 (22), 5251–5255. Available at https://doi.org/10.1242/jcs.103630Netherlands. Available from: http://jcs.biologists.org/content/125/22/5251.full.pdf.

Rabouille, C., 2017. Pathways of unconventional protein secretion. Trends in Cell Biology 27 (3), 230–240. Available at https://doi.org/10.1016/j.tcb.2016.11.007.

Rachlin, J.W., Grosso, A., 1991. The effects of pH on the growth of Chlorella vulgaris and its interactions with cadmium toxicity. Archives of Environmental Contamination and Toxicology 20 (4), 505–508. Available at https://doi.org/10.1007/BF01065839.

Rahman, M.M., Hosano, N., Hosano, H., 2022. Recovering microalgal bioresources: a review of cell disruption methods and extraction technologies. Molecules 27 (9). Available at https://doi.org/10.3390/molecules27092786. Available from: https://www.mdpi.com/1420-3049/27/9/2786/pdf.

Rai, M.P., Gautom, T., Sharma, N., 2015. Effect of salinity, pH, light intensity on growth and lipid production of microalgae for bioenergy application. OnLine Journal of Biological Sciences 15 (4), 260–267. Available at https://doi.org/10.3844/ojbsci.2015.260.267. Available from: http://thescipub.com/pdf/10.3844/ojbsci.2015.260.267.

Rajauria, G., Yuan, Y.V., Lee, D.-J., 2021. Advances in drying and milling technologies for algae. Recent Advances in Micro and Macroalgal Processing: Food and Health Perspectives. Wiley, pp. 72–95. Available at 10.1002/9781119542650.ch3.

Ramos-Martinez, E.M., Fimognari, L., Sakuragi, Y., 2017. High-yield secretion of recombinant proteins from the microalga Chlamydomonas reinhardtii. Plant Biotechnology Journal 15 (9), 1214–1224. Available at https://doi.org/10.1111/pbi.12710. Available from: http://www.wiley.com/bw/journal.asp?ref=1467-7644.

Rasala, B.A., Chao, S.-S., Pier, M., Barrera, D.J., Mayfield, S.P., Chen, W.N., 2014. Enhanced genetic tools for engineering multigene traits into green algae. PLoS One 9 (4), e94028. Available at https://doi.org/10.1371/journal.pone.0094028.

Rasala, B.A., Mayfield, S.P., 2015. Photosynthetic biomanufacturing in green algae; Production of recombinant proteins for industrial, nutritional, and medical uses. Photosynthesis Research 123 (3), 227–239. Available at https://doi.org/10.1007/s11120-014-9994-7.

Rasala, B.A., Muto, M., Lee, P.A., Jager, M., Cardoso, R.M.F., Behnke, C.A., et al., 2010. Production of therapeutic proteins in algae, analysis of expression of seven human proteins in the chloroplast of Chlamydomonas reinhardtii. Plant Biotechnology Journal 8 (6), 719–733. Available at https://doi.org/10.1111/j.1467-7652.2010.00503.x. Available from: http://onlinelibrary.wiley.com/journal/10.1111/(ISSN)1467-7652.

Richter, R., Hejazi, M., Kraft, R., Ziegler, K., Lockau, W., 1999. Cyanophycinase, a peptidase degrading the cyanobacterial reserve material multi-L-arginyl-poly-L-aspartic acid (cyanophycin). Molecular cloning of the gene of Synechocystis sp. PCC 6803, expression in Escherichia coli, and biochemical characterization of the purified enzyme. European Journal of Biochemistry 263 (1), 163–169. Available at https://doi.org/10.1046/j.1432-1327.1999.00479.x.

Romero, M., 2023. Enzyme-assisted extraction of ulvan from the green macroalgae ulva fenestrata. Molecules 28 (19).

Rosales-Mendoza, S., Paz-Maldonado, L.M.T., Soria-Guerra, R.E., 2012. Chlamydomonas reinhardtii as a viable platform for the production of recombinant proteins: current status and perspectives. Plant Cell Reports 31 (3), 479–494. Available at https://doi.org/10.1007/s00299-011-1186-8.

Rosales-Mendoza, S., Solís-Andrade, K.I., Márquez-Escobar, V.A., González-Ortega, O., Bañuelos-Hernandez, B., 2020. Current advances in the algae-made biopharmaceuticals field. Expert Opinion on Biological Therapy 20 (7), 751–766. Available at https://doi.org/10.1080/0.1739643.

Ross, I.L., Le, H.P., Budiman, S., Xiong, D., Hemker, F., Millen, E.A., et al., 2024. A cyclical marker system enables indefinite series of oligonucleotide-directed gene editing in Chlamydomonas reinhardtii. Plant Physiology 196 (4), 2330–2345. Available at https://doi.org/10.1093/plphys/kiae427. Available from: https://academic.oup.com/plphys/article/196/4/2330/7740471.

Roy, A., Gogoi, N., Yasmin, F., Farooq, M., 2022. The use of algae for environmental sustainability: trends and future prospects. Environmental Science and Pollution Research 29 (27), 40373–40383. Available at https://doi.org/10.1007/s11356-022-19636-7.

Russo, M.T., Santin, A., Rogato, A., Ferrante, M.I., 2022. Optimized proteolistic protocol for the delivery of the Cas9 Protein in Phaeodactylum tricornutum. Methods in Molecular Biology 2498, 327–336. Available at https://doi.org/10.1007/978-1-0716-2313-8_18. Available from: http://www.springer.com/series/7651.

Ru, I.T.K., Sung, Y.Y., Jusoh, M., Wahid, M.E.A., Nagappan, T., 2020. Chlorella vulgaris: a perspective on its potential for combining high biomass with high value bioproducts. Applied Phycology 1 (1), 2–11. Available at https://doi.org/10.1080/26388081.2020.1715256. Available from: https://www.tandfonline.com/journals/tapy20.

Ryckebosch, E., Muylaert, K., Foubert, I., 2012. Optimization of an analytical procedure for extraction of lipids from microalgae. Journal of the American Oil Chemists' Society 89 (2), 189–198. Available at https://doi.org/10.1007/s11746-011-1903-z.

Saei, A.A., Ghanbari, P., Barzegari, A., 2012. Haematococcus as a promising cell factory to produce recombinant pharmaceutical proteins. Molecular Biology Reports 39 (11), 9931–9939. Available at https://doi.org/10.1007/s11033-012-1861-z.

Sahabudin, E., Kubo, S., Yuzir, M.A.M., Othman, N.'azizi, Nadia Md Akhir, F., Suzuki, K., et al., 2024. The cadmium tolerance and bioaccumulation mechanism of Tetratostichococcus sp. P1: insight from transcriptomics analysis. Bioengineered 15 (1). Available at https://doi.org/10.1080/21655979.2024.2314888.

Sarasan, M., Job, N., Puthumana, J., Ravinesh, R., Kachiprath, B., Philip, R., 2021. Enzyme profiling of macroalgal endophytes: An attempt to uncover the arsenal of novel biocatalysts. Journal of the Marine Biological Association of India 62 (2), 112–116. Available at https://doi.org/10.6024/jmbai.2020.62.2.2183-14.

Schenk, P.M., Thomas-Hall, S.R., Stephens, E., Marx, U.C., Mussgnug, J.H., Posten, C., et al., 2008. Second generation biofuels: high-efficiency microalgae for biodiesel production. BioEnergy Research 1 (1), 20–43. Available at https://doi.org/10.1007/s12155-008-9008-8.

Scholtysek, L., Poetsch, A., Hofmann, E., Hemschemeier, A., 2024. The activation of Chlamydomonas reinhardtii alpha amylase 2 by glutamine requires its N-terminal aspartate kinase–chorismate mutase–tyrA (ACT) domain. Plant Direct 8 (6). Available at https://doi.org/10.1002/pld3.609.

Schroda, M., Beck, C.F., Vallon, O., 2002. Sequence elements within an HSP70 promoter counteract transcriptional transgene silencing in Chlamydomonas. Plant Journal 31 (4), 445–455. Available at https://doi.org/10.1046/j.1365-313X.2002.01371.x.

Schroda, M., Blöcker, D., Beck, C.F., 2000. The HSP70A promoter as a tool for the improved expression of transgenes in Chlamydomonas. Plant Journal 21 (2), 121–131. Available at https://doi.org/10.1046/j.1365-313X.2000.00652.x.

Selvam, J., Mal, J., Singh, S., Yadav, A., Giri, B.S., Pandey, A., et al., 2024. Bioprospecting marine microalgae as sustainable bio-factories for value-added compounds. Algal Research 79, 103444. Available at https://doi.org/10.1016/j.algal.2024.103444.

Shahar, N., Elman, T., Williams-Carrier, R., Ben-Zvi, O., Yacoby, I., Barkan, A., 2021. Use of plant chloroplast RNA-binding proteins as orthogonal activators of chloroplast transgenes in the green alga Chlamydomonas reinhardtii. Algal Research 60, 102535. Available at https://doi.org/10.1016/j.algal.2021.102535.

Sharifuddin, Y., Chin, Y.-X., Lim, P.-E., Phang, S.-M., 2015. Potential bioactive compounds from seaweed for diabetes management. Marine Drugs 13 (8), 5447–5491. Available at https://doi.org/10.3390/md13085447.

Sharma, Y.C., Singh, B., Upadhyay, S.N., 2008. Advancements in development and characterization of biodiesel: a review. Fuel 87 (12), 2355–2373. Available at https://doi.org/10.1016/j.fuel.2008.01.014.

Shin, S.E., Lim, J.M., Koh, H.G., Kim, E.K., Kang, N.K., Jeon, S., et al., 2016. CRISPR/Cas9-induced knockout and knock-in mutations in Chlamydomonas reinhardtii. Scientific Reports 6 Available from. Available at https://doi.org/10.1038/srep27810, www.nature.com/srep/index.html.

Shirakawa, K., Kawabe, Y., Huang, G., Ito, A., Kamihira, M., Shimada, M., 2021. Targeted gene integration into nuclear genome of microalgae using Cre/ loxP recombination system. MATEC Web of Conferences 333, 07003. Available at https://doi.org/10.1051/matecconf/202133307003.

Show, K.Y., Lee, D.J., Tay, J.H., Lee, T.M., Chang, J.S., 2015. Microalgal drying and cell disruption - recent advances. Bioresource Technology 184, 258–266. Available from. Available at https://doi.org/10.1016/j.biortech.2014.10.139, www.elsevier.com/locate/biortech.

Sirajunnisa, A.R., Surendhiran, D., 2016. Algae – a quintessential and positive resource of bioethanol production: a comprehensive review. Renewable and Sustainable Energy Reviews 66, 248–267. Available at https://doi.org/10.1016/j.rser.2016.07.024. Available from: https://www.journals.elsevier.com/renewable-and-sustainable-energy-reviews.

Specht, E., Miyake-Stoner, S., Mayfield, S., 2010. Micro-algae come of age as a platform for recombinant protein production. Biotechnology Letters 32 (10), 1373–1383. Available at https://doi.org/10.1007/s10529-010-0326-5.

Sproles, A.E., Fields, F.J., Smalley, T.N., Le, C.H., Badary, A., Mayfield, S.P., 2021. Recent advancements in the genetic engineering of microalgae. Algal Research 53, 102158. Available at https://doi.org/10.1016/j.algal.2020.102158.

Sreenikethanam, A., Raj, S., Poornachandar Gugulothu, R.B.J., Bajhaiya, A.K., 2022. Genetic engineering of microalgae for secondary metabolite production: recent developments, challenges, and future prospects. Frontiers in Bioengineering and Biotechnology 10. Available at https://doi.org/10.3389/fbioe.2022.836056.

Stavridou, E., Karapetsi, L., Nteve, G.M., Tsintzou, G., Chatzikonstantinou, M., Tsaousi, M., et al., 2024. Landscape of microalgae omics and metabolic engineering research for strain improvement: an overview. Aquaculture 587, 740803. Available at https://doi.org/10.1016/j.aquaculture.2024.740803.

Suttangkakul, A., Sirikhachornkit, A., Juntawong, P., Puangtame, W., Chomtong, T., Srifa, S., et al., 2019. Evaluation of strategies for improving the transgene expression in an oleaginous microalga Scenedesmus acutus. BMC Biotechnology 19 (1). Available at https://doi.org/10.1186/s12896-018-0497-z.

Swain, M.R., Kar, S., Ray, R.C., 2009. Exo-polygalacturonase production by Bacillus subtilis CM5 in solid state fermentation using cassava bagasse. Brazilian Journal of Microbiology 40 (3), 636–648. Available at https://doi.org/10.1590/S1517-83822009000300028.

Tahir, F., Ashfaq, H., Khan, A.Z., Amin, M., Akbar, I., Malik, H.A., et al., 2024. Emerging trends in algae farming on non-arable lands for resource reclamation, recycling, and mitigation of climate change-driven food security challenges. Reviews in Environmental Science and Bio/Technology 23 (3), 869–896. Available at https://doi.org/10.1007/s11157-024-09697-0.

Tan, C.H., Chen, C.Y., Show, P.L., Ling, T.C., Lam, H.L., Lee, D.J., et al., 2016. Strategies for enhancing lipid production from indigenous microalgae isolates. Journal of the Taiwan Institute of Chemical Engineers 63, 189–194. Available at https://doi.org/10.1016/j.jtice.2016.02.034. Available from: http://www.elsevier.com/wps/find/journaldescription.cws_home/715607/description#description.

Torresi, F., Carrillo, J.B., Gomez-Casati, D.F., Busi, M.V., Martín, M., 2025. Enhancing starch levels, granule size and phosphate content in Chlamydomonas reinhardtii through overexpression of ChlreSEX4. Plant Science 352, 112360. Available at https://doi.org/10.1016/j.plantsci.2024.112360.

Ugwu, C.U., Aoyagi, H., Uchiyama, H., 2008. Photobioreactors for mass cultivation of algae. Bioresource Technology 99 (10), 4021–4028. Available at https://doi.org/10.1016/j.biortech.2007.01.046.

Uzuner, S., Evrendilek, G.A., Kurhan, S., 2022. Production of microalgae-derived exo-polygalacturonase for fruit juice clarification. Frontiers in Food Science and Technology 2 Available from. Available at https://doi.org/10.3389/frfst.2022.951609, www.frontiersin.org/journals/food-science-and-technology.

Vasudevan, R., Gale, G.A.R., Schiavon, A.A., Puzorjov, A., Malin, J., Gillespie, M.D., et al., 2019. CyanoGate: a modular cloning suite for engineering cyanobacteria based on the plant MoClo syntax. Plant Physiology 180 (1), 39–55. Available at https://doi.org/10.1104/pp.18.01401.

Vega-Portalatino, E.J., Rosales-Cuentas, M.M., Valdiviezo-Marcelo, J., Arana-Torres, N.M., Espinoza-Espinoza, L.A., Moreno-Quispe, L.A., et al., 2023. Antimicrobial and production of hydrolytic enzymes potentials of bacteria and fungi associated with macroalgae and their applications: a review. Frontiers in Marine Science 10. Available at https://doi.org/10.3389/fmars.2023.1174569. Available from: https://www.frontiersin.org/journals/marine-science#.

Verruto, J., Francis, K., Wang, Y., Low, M.C., Greiner, J., Tacke, S., et al., 2018. Unrestrained markerless trait stacking in Nannochloropsis gaditana through combined genome editing and marker recycling technologies. Proceedings of the National Academy of Sciences 115 (30). Available at https://doi.org/10.1073/pnas.1718193115.

Vetal, M.D., Rathod, V.K., 2015. Three phase partitioning a novel technique for purification of peroxidase from orange peels (Citrus sinenses). Food and Bioproducts Processing 94, 284–289. Available at https://doi.org/10.1016/j.fbp.2014.03.007. Available from: http://www.elsevier.com/wps/find/journaldescription.cws_home/713883/description#description.

Vilg, J.V., Undeland, I., 2017. pH-driven solubilization and isoelectric precipitation of proteins from the brown seaweed Saccharina latissima—effects of osmotic shock, water volume and temperature. Journal of Applied Phycology 29 (1), 585–593. Available at https://doi.org/10.1007/s10811-016-0957-6.

Vingiani, G.M., De Luca, P., Ianora, A., Dobson, A.D.W., Lauritano, C., 2019. Microalgal enzymes with biotechnological applications. Marine Drugs 17 (8), 459. Available at https://doi.org/10.3390/md17080459.

Wang, Q., Zhuang, J., Ni, S., Luo, H., Zheng, K., Li, X., et al., 2022. Overexpressing CrePAPS polyadenylate activity enhances protein translation and accumulation in Chlamydomonas reinhardtii. Marine Drugs 20 (5), 276. Available at https://doi.org/10.3390/md20050276.

Watzer, B., Forchhammer, K., 2018. Cyanophycin synthesis optimizes nitrogen utilization in the unicellular cyanobacterium Synechocystis sp. strain PCC 6803. Applied and Environmental Microbiology 84 (20). Available at https://doi.org/10.1128/AEM.01298-18. Available from: https://aem.asm.org/content/aem/84/20/e01298-18.full.pdf.

Weiner, I., 2020. CSO–A sequence optimization software for engineering chloroplast expression in Chlamydomonas reinhardtii. Algal Research.

Xie, Z., He, J., Peng, S., Zhang, X., Kong, W., 2023. Biosynthesis of protein-based drugs using eukaryotic microalgae. Algal Research 74, 103219. Available at https://doi.org/10.1016/j.algal.2023.103219.

Xu, L., Weathers, P.J., Xiong, X.-R., Liu, C.-Z., 2009. Microalgal bioreactors: challenges and opportunities. Engineering in Life Sciences 9 (3), 178–189. Available at https://doi.org/10.1002/elsc.200800111.

Ye, H., Fussenegger, M., 2019. Optogenetic medicine: synthetic therapeutic solutions precision-guided by light. Cold Spring Harbor Perspectives in Biology. **11** 9. Available at https://doi.org/10.1101/cshperspect.a034371. Available from: http://perspectivesinmedicine.cshlp.org/content/9/9/a034371.full.pdf.

Ye, Y., Liu, M., Yu, L., Sun, H., Liu, J., 2024. Nannochloropsis as an emerging algal chassis for light-driven synthesis of lipids and high-value products. Marine Drugs 22 (2), 54. Available at https://doi.org/10.3390/md22020054.

Yoon, S.M., Kim, S.Y., Li, K.F., Yoon, B.H., Choe, S., Kuo, M.M.C., 2011. Transgenic microalgae expressing Escherichia coli AppA phytase as feed additive to reduce phytate excretion in the manure of young broiler chicks. Applied Microbiology and Biotechnology 91 (3), 553–563. Available at https://doi.org/10.1007/s00253-011-3279-2.

Zakeri, H.A., Razak, S.Z., Hasmudin, H.H., 2020. Antioxidative enzymes activity of chlorella vulgaris (umt-m1) in response to ic50 of pb2+ and hg2+. Malaysian Applied Biology 49 (4), 63–70. Available from. Available at https://doi.org/10.55230/mabjournal.v49i4.1568, http://www.mabjournal.com/images/49_4_December_2020/49_04_08.pdf.

Zaslavskaia, L.A., Lippmeier, J.C., Kroth, P.G., Grossman, A.R., Apt, K.E., 2000. Transformation of the diatom Phaeodactylum tricornutum (Bacillariophyceae) with a variety of selectable marker and reporter genes. Journal of Phycology 36 (2), 379–386. Available at https://doi.org/10.1046/j.1529-8817.2000.99164.x.

Zhang, M.P., Wang, M., Wang, C., 2021. Nuclear transformation of Chlamydomonas reinhardtii: A review. Biochimie 181, 1–11. Available from. Available at https://doi.org/10.1016/j.biochi.2020.11.016, www.elsevier.com/locate/biochimie.

Zhao, B., Su, Y., 2014. Process effect of microalgal-carbon dioxide fixation and biomass production: a review. Renewable and Sustainable Energy Reviews 31, 121–132. Available at https://doi.org/10.1016/j.rser.2013.11.054.

Zhu, M., Zhou, P.P., Yu, L.J., 2002. Extraction of lipids from Mortierella alpina and enrichment of arachidonic acid from the fungal lipids. Bioresource Technology 84 (1), 93–95. Available at https://doi.org/10.1016/S0960-8524(02)00028-7.

Zohir, W.F., Makhlof, M.E.M., Abdallah, A.M., El-Sheekh, M.M., 2023. Algae Cultivation Systems. Springer Science and Business Media LLC, pp. 11–41. Available at 10.1007/978-3-031-42026-9_2.

CHAPTER 5

Microalgal based bioactive compounds for medical and pharmaceutical aspects

Noor Haziqah Kamaludin[1,2], Nur Azalina Suzianti Feisal[3,4], and Tengku Nilam Baizura Tengku Ibrahim[2,5]
[1]Center for Environmental Health and Safety Study, Faculty of Health Sciences, Universiti Teknologi MARA, Puncak Alam, Selangor, Malaysia
[2]Occupational Health and Safety Risk Management (OHSeRM) Research Initiative Group, Faculty of Health Sciences, Universiti Teknologi MARA, Puncak Alam, Selangor, Malaysia
[3]MSU Centre for Climate Resilience and Strategy (m-CREST), Management and Science University, Shah Alam, Selangor, Malaysia
[4]Department of Diagnostic and Allied Health Science, Faculty of Health and Life Sciences, Management and Science University, Shah Alam, Selangor, Malaysia
[5]Department of Environmental Health, Faculty of Health Sciences, Universiti Teknologi MARA Cawangan Pulau Pinang, Bertam, Pulau Pinang, Malaysia

5.1 Introduction

Algae have gained growing attention as a versatile biological resource in pharmaceutical and health sciences. Their broad metabolic capacity enables the production of diverse compounds, including sulfated polysaccharides, pigments, phenolic derivatives, and fatty acids that support a range of therapeutic activities (Bratchkova and Kroumov, 2020). Many of these metabolites exhibit antiinflammatory, antioxidant, antiviral, antibacterial, and anticancer properties, positioning algae as a valuable foundation for natural product–based drug development (Saide et al., 2021). Recent work has further highlighted the potential of microalgae to generate high-value metabolites with direct applications in healthcare and pharmaceutical industries (Ibrahim et al., 2023). In a period marked by rising antimicrobial resistance and chronic disease burdens, algae offer promising opportunities for discovering new and effective bioactive agents.

Beyond their role as sources of bioactive molecules, algae also contribute significantly to advances in pharmaceutical formulation and drug delivery. Biopolymers such as alginate, carrageenan, and agar are widely used in controlled-release and encapsulation systems due to their biocompatibility and favorable physicochemical characteristics (Hsu et al., 2024). Recent progress in polymer science and nanotechnology has

Applied Phycology
DOI: https://doi.org/10.1016/B978-0-443-24758-3.00012-5

strengthened the potential of algal-derived materials for targeted delivery and stabilization of therapeutic compounds (Venkatesan et al., 2016), highlighting their relevance in both drug development and pharmaceutical engineering.

Algae also play an important role at the interface between nutrition and medicine through their use in nutraceutical and functional formulations. Species rich in antioxidants, essential fatty acids, vitamins, polysaccharides, and other bioactive components are increasingly incorporated into products designed to support metabolic regulation, immune modulation, and overall physiological resilience (Patel et al., 2021). In particular, macro- and microalgal polysaccharides and related fractions have been recognized for their prebiotic potential and their capacity to modulate gut microbiota, a mechanism increasingly linked to the prevention and adjunct management of cardiometabolic, inflammatory, and gastrointestinal disorders (Patel et al., 2021). The convergence of nutritional and therapeutic functions positions algal ingredients as promising components in broader medical and pharmaceutical strategies, particularly those targeting disease prevention and long-term health maintenance.

Emerging research on neuroprotective properties further expands the biomedical relevance of algae. Compounds such as phlorotannins, carotenoids, and sulfated polysaccharides have demonstrated potential to modulate oxidative stress, reduce neuroinflammation, and support neuronal function in preclinical studies (Hannan et al., 2020). Although still developing, these findings suggest new possibilities for algae-derived interventions targeting neurodegenerative diseases (ND). As algae move toward broader clinical and commercial applications, evaluating their safety, toxicology, and regulatory status remains essential. Species variability, cultivation conditions, and extraction methods can influence chemical composition and safety margins, underscoring the need for standardized assessments and robust quality control (Gao et al., 2024). This chapter aims to provide an integrated understanding of the medical and pharmaceutical significance of algae by examining their bioactive potential, roles in drug formulation and delivery, development as nutraceuticals, emerging neuroprotective functions, and the essential safety and regulatory considerations that guide their application in contemporary healthcare and biomedicine.

5.2 Algae-derived bioactive compounds for therapeutic use

Algae represent a unique reservoir of bioactive molecules that have attracted considerable attention in pharmaceutical research due to their structural diversity and multifunctional properties (Bouafir et al., 2025). These organisms synthesize compounds such as sulfated polysaccharides, pigments, polyphenols, peptides, and lipids, which exhibit a wide range of biological activities, including antiviral, antibacterial, antifungal, anticancer, antiinflammatory, and antioxidant effects (Osathanunkul et al., 2025). Their ecological advantages, such as rapid growth in seawater and the ability to sequester carbon, make algae a sustainable source for drug discovery and the development of nutraceuticals (Goshtasbi et al., 2023).

Among the most studied algal metabolites are fucoidan, phycocyanin, astaxanthin, and various polyphenols (Pakan et al., 2025). Fucoidan, a sulfated polysaccharide derived from brown algae, is recognized for its anticoagulant, antiviral, and anticancer properties, which act through mechanisms such as the inhibition of angiogenesis and modulation of immune responses (Jin et al., 2021). Phycocyanin, a pigment–protein complex from cyanobacteria, demonstrates strong antioxidant and antiinflammatory activity, with emerging evidence of its role in apoptosis induction in cancer cells (Sabat et al., 2025). Astaxanthin, a carotenoid from microalgae such as *Haematococcus pluvialis*, is recognized for its potent antioxidant capacity, offering neuroprotective and cardioprotective benefits (Donoso et al., 2021). Algal polyphenols further contribute to reducing oxidative stress and inflammation, supporting chronic disease management (Pereira, Cotas and Valado, 2024). These compounds influence critical biological pathways, including oxidative-stress regulation, immune modulation, microbial adhesion, and tumor signaling. Their multifunctionality positions algae-derived bioactives as promising candidates for therapeutic applications ranging from infectious disease control to chronic disease prevention.

5.2.1 Antiviral, antibacterial, and antifungal properties

Marine algae produce sulfated polysaccharides such as fucoidan, porphyran, and carrageenan, which exhibit strong antiviral properties (Table 5.1). These compounds act primarily by blocking viral attachment and entry into host cells through electrostatic interactions with viral envelope proteins and host cell receptors (Dong et al., 2025). For instance, carrageenan-based nasal sprays have been tested against respiratory viruses, including influenza and SARS-CoV-2, showing reduced viral load and severity (Kumar et al., 2022). Fucoidan has demonstrated inhibitory effects on herpes simplex virus and dengue virus by interfering with replication enzymes and

Table 5.1 Algae-derived compounds and their antimicrobial mechanisms.

Compound	Source	Activity type	Mechanism of action	Target pathogens	References
Fucoidan	Brown algae (*Fucus*, *Laminaria*)	Antiviral	Blocks viral attachment and replication, interfere with viral enzymes, immune modulation	Herpes simplex virus, dengue virus, influenza, SARS-CoV-2	Musarra-Pizzo et al. (2021); Homaeigohar, Liu and Elbahri (2023)
Porphyran	Red algae (*Porphyra*)	Antiviral	Electrostatic interaction with viral envelope proteins prevents entry	HIV, respiratory viruses	Pereira and Valado (2025); Sharma et al. (2025)
Carrageenan	Red algae (*Kappaphycus*, *Gigartina*)	Antiviral	Forms protective barrier on mucosal surfaces, inhibits viral binding	HPV, influenza, SARS-CoV-2	Kumar et al. (2022)
Phycocyanin	Cyanobacteria (*Spirulina*)	Antibacterial/antifungal	Antioxidant and antiinflammatory, disrupts biofilm formation, inhibits oxidative stress	*Staphylococcus aureus*, *Candida albicans*	López 2019; Abdel-Moneim et al. (2022)
Hassallidin	Cyanobacteria	Antifungal	Disrupt fungal cell membranes, antibiofilm activity	*Candida* spp., *Aspergillus* spp.	Shariati et al. (2022)
Algal poly-phenols	Green/brown algae	Antibacterial/antifungal	Inhibit microbial adhesion and quorum sensing, antioxidant activity, and enzyme inhibition	Multidrug-resistant bacteria, fungal pathogens	Wong Chin et al. (2021)
Astaxanthin	Microalgae (*Haematococcus pluvialis*)	Antiviral (adjunct)	Immunomodulation, reduces oxidative stress, synergistic with antiviral drugs	Viral infections (adjunct therapy), opportunistic bacteria	Peinsipp et al. (2025)

enhancing immune responses through cytokine modulation (Ziyaei et al., 2022). Recent reviews emphasize that these polysaccharides can also stimulate innate immunity, making them promising candidates for broad-spectrum antiviral formulations. However, challenges such as low oral bioavailability and scale-up for pharmaceutical-grade purity remain key hurdles for clinical translation (Salehi & Rashidinejad, 2025).

Beyond polysaccharides, certain algal-derived proteins and pigments have shown antiviral potential. For example, phycocyanin from cyanobacteria has been reported to inhibit viral replication indirectly by reducing oxidative stress and modulating immune signaling pathways. Similarly, algal polyphenols exhibit virucidal activity by destabilizing viral envelopes and interfering with nucleic acid synthesis. These mechanisms highlight the multifaceted nature of algal compounds, which can act at different stages of the viral life cycle. Recent research also explores the incorporation of algal bioactives into nanocarrier systems to overcome bioavailability limitations. Fucoidan-based nanoparticles have been developed for targeted delivery of antiviral drugs, enhancing therapeutic efficacy while reducing systemic toxicity. Such innovations underscore the potential of algae-derived compounds not only as standalone antivirals but also as functional excipients in advanced drug formulations.

Microalgae and cyanobacteria are known to produce a wide spectrum of bioactive metabolites, including peptides, alkaloids, terpenoids, polyphenols, and various fatty acids that can act against bacterial and fungal pathogens. These substances work through several mechanisms, damaging microbial cell membranes, blocking essential protein synthesis pathways, and interrupting quorum-sensing signals that regulate pathogen virulence. Interestingly, microalgae exposed to environmental stressors such as elevated salinity or intense UV light often show stronger antibacterial properties, suggesting that extreme habitats may serve as valuable sources for discovering new antimicrobial agents.

One of the persistent challenges in clinical microbiology is the ability of pathogens to form biofilms, which shield them from many conventional antibiotics. Certain algal-derived molecules, such as phycocyanin, hassallidin, and polyphenolic compounds with tannin-like behavior, have demonstrated the capacity to disrupt biofilm formation by preventing early surface attachment and interfering with matrix development. Several studies also note that combining algal extract with existing antibiotics can enhance antimicrobial effectiveness, particularly against multidrug-resistant organisms like *Staphylococcus aureus* and *Candida albicans*. To support pharmaceutical development, advanced analytical tools such as high-performance liquid chromatography (HPLC), liquid chromatography-mass spectrometry (LC-MS), Fourier transform infrared (FTIR), and nuclear magnetic resonance (NMR) spectroscopy are used to isolate, identify, and characterize these active constituents and to refine extraction methods.

Although algae produce many useful health-related compounds, turning these substances into real medical products is not always straightforward. Some key molecules, like fucoidan and phycocyanin, do not absorb well when taken by mouth because they are large and easily broken down in the body. Many algae pigments, such as phycocyanin and astaxanthin, also break down quickly if they are not stored properly, so they need special protective packaging or coating to keep them stable and effective (Table 5.2). Another major challenge is consistency: different algae species, or even the same species grown under different conditions, can produce varying amounts and qualities of these compounds, making it difficult to meet strict regulatory standards.

To overcome these issues, researchers are turning to modern delivery technologies such as nanoparticles, liposomes, and nanoemulsions. These systems can improve how well the compounds dissolve, help them reach specific parts of the body, and release them slowly over time. There is also growing interest in adding algae-derived molecules to functional foods, supplements, and skin-care products to support general health. Scientists are further exploring how these natural compounds can work together with existing medicines to fight drug-resistant infections and enhance cancer treatments. With better extraction methods and more sustainable production approaches, algae-based products are expected to play a larger role in the future of medicine and health care.

5.2.2 Notable algal bioactive compounds and their therapeutic roles

Phycocyanin is a water-soluble pigment–protein complex commonly produced by cyanobacteria, particularly *Spirulina platensis*. It is well known for its strong antioxidant properties, largely due to its ability to scavenge reactive oxygen species (ROS) and limit lipid peroxidation. By reducing oxidative stress, phycocyanin helps protect cells from damage associated with chronic conditions such as diabetes, neurodegenerative disorders, and various cancers stated in various study as in Table 5.3. In addition to its

Table 5.2 Clinical applications and formulation strategies of algal bioactive compounds.

Compound	Therapeutic application	Formulation strategy	Advantages	References
Fucoidan	Antiviral (Herpes Simplex Virus (HSV), dengue), anticancer adjunct	Nanoparticles, hydrogels, injectable gels	Improves bioavailability, enables targeted drug delivery	Jia et al. (2025); Lee et al. (2025)
Carrageenan	Antiviral (influenza, Human Papillomavirus (HPV), SARS-CoV-2)	Nasal sprays, topical gels	Forms protective barrier on mucosa, easy administration	Unger-Manhart et al. (2024); Sonvico et al. (2023); Hebar et al. (2015)
Porphyran	Antiviral (HIV, respiratory viruses)	Oral capsules, functional foods	Enhances immune response, convenient nutraceutical use	Skenderidou, Leontopoulos and Skenderidis (2025); Jung et al. (2020)
Phycocyanin	Antiinflammatory, antioxidant, anticancer	Liposomes, microencapsulation, tablets	Protect pigment stability, controlled release	Soliman et al. (2024); Fernandes et al. (2023)
Astaxanthin	Neuroprotective, cardioprotective, antiinflammatory	Soft gels, nanoemulsions, oral supplements	Improves solubility and absorption, sustained release	Oninku et al. (2025)
Algal polyphenols	Antibacterial, antifungal, gut health	Functional beverages, polymeric coatings	Supports microbiome modulation, food-grade applications	Liu et al. (2025); Törős et al. (2023)
Hassallidin	Antifungal (*Candida, Aspergillus*)	Topical creams, biofilm-targeted gels	Direct application to infection site, antibiofilm effect	Lin et al. (2023)

Table 5.3 Potential algae on application in neurodegenerative disease (ND), modulation of inflammation, and oxidative-stress pathways.

Potential algae/ compound	Application in ND	Modulation of inflammation	Oxidative-stress pathways	Preclinical evidence	Citation
Eisenia bicyclis dieckol (phlorotannin)	Neuroprotective properties reported in context of radical scavenging and neuroprotection (suggested relevance to ND).	Upregulates HO-1 (heme oxygenase-1) contributing to antiinflammatory effects; reduces lipopolysaccharide (LPS)-induced cytokines.	Radical-scavenging activity; attenuates reactive oxygen species (ROS) in macrophage/cell models.	In vitro: LPS-stimulated murine macrophages; cell assays.	Yayeh et al. (2014)
Fucoxanthin (brown algae carotenoid)	Reviewed as having neuroprotective potential via antioxidant/ antiinflammatory actions—proposed for ND applications.	Inhibits inflammatory mediators (NO, PGE2) and NF-κB signaling in macrophages.	Strong antioxidant activity; reduces ROS and increases antioxidant enzyme activity (reviewed).	Review of in vitro and in vivo (preclinical) studies across multiple models.	Kim and Pangestuti (2011)
Agarum clathratum extracts (red/ brown seaweed)	Not specifically ND-focused; cell death/ cytotoxicity work (cancer). ND application not primary.	Reported modulation of apoptotic and inflammatory markers in cell lines (general antiinflammatory compounds identified).	Some antioxidant assays reported.	In vitro (ovarian cancer cells)—not ND model.	Kim et al. (2022)

Red-alga oligosaccharide—*agaropentaose* (from agarose)	Direct ND application protects SH-SY5Y neuronal cells against 6-OHDA—model for Parkinson's disease (PD).	Modulates NF-κB and p38 mitogen-activated protein kinase (MAPK) pathways (reduces inflammatory signaling).	Protects against oxidative injury and mitochondrial dysfunction.	In vitro: SH-SY5Y cells exposed to 6-OHDA (neurotoxin)—cell viability, ROS, signaling assays.	Ye et al. (2019)
Bifurcaria bifurcata (brown seaweed)—extracts/isolated terpenoids (eleganolone, fucosterol)	ND relevance: review/experimental work shows neuroprotective and antiinflammatory potential—tested in cellular PD models.	Reduces LPS-induced inflammation in macrophage models; downregulates inflammatory mediators.	Antioxidant capacity assays (2,2-diphenyl-1-picrylhydrazyl (DPPH), Oxygen Radical Absorbance Capacity (ORAC), etc.) and mitochondrial protective effects described.	In vitro neuroprotection (SH-SY5Y with 6-OHDA) and macrophage LPS models; some fractionation/compound assays.	Silva et al. (2022)
Compounds from Bifurcaria/related brown algae (eleganolone, fucosterol)—(pharmacological research paper)	Evaluated for PD-related models; neuroprotective candidates proposed.	Antiinflammatory actions shown (reduced cytokine release in macrophage models).	Antioxidant/neuroprotective markers improved in cell assays.	In vitro ND models (6-OHDA on neuronal cells) and LPS-stimulated macrophages.	Silva et al. (2021)

(*Continued*)

Table 5.3 (Continued)

Potential algae/ compound	Application in ND	Modulation of inflammation	Oxidative-stress pathways	Preclinical evidence	Citation
Ulva conglobata (green seaweed)	Neuroprotective and antiinflammatory effects in hippocampal and microglial cells—potential relevance to Alzheimer's disease (AD)/ neuroinflammation.	Decreased microglial activation and inflammatory mediators in cell studies.	Shows antioxidant activity and protection against glutamate toxicity.	In vitro: murine hippocampal neurons and microglial cells; cell assays.	Jin et al. (2006)
Microalgae (carotenoid-enriched extracts) (diverse species)	Studied for antiinflammatory and neuroprotective potency—carotenoids highlighted for ND relevance.	Carotenoid-rich extracts inhibit inflammatory mediators and NF-κB signaling.	Strong antioxidant properties (carotenoids reduce ROS); neuroprotective in cellular assays.	In vitro screening of microalgae extracts; mechanistic assays.	Gallego et al. (2022)
Palmaria palmata (red seaweed/dulse)	2025 Frontiers paper: cognitive and antiinflammatory effects in a rodent	Reduced Iba-1 expression (microglial activation), modulated CD4+	Histology and biomarkers show reduced neuroinflammation/ oxidative damage.	In vivo: mouse model (behavioral tests, histology,	Yousof et al. (2025)

	model (implications for neurodegeneration/cognitive dysfunction).	cells, and cyclic Adenosine 3', 5'-Monophosphate *response element binding protein* (CREB) signaling—antiinflammatory immunomodulation.		immunohistochemistry).	
Spirulina platensis (cyanobacterium/microalga)	Neuroprotective in 6-OHDA Parkinson's model; suggested for ND due to antiinflammatory actions.	Antiinflammatory effects associated with decreased proinflammatory markers in brain tissue.	Antioxidant activity (reduced oxidative markers in brain) implicated in neuroprotection.	In vivo: 6-OHDA rodent model of PD—behavioral and biochemical end points.	Lima et al. (2017)
Sargassum serratifolium (brown seaweed)—ethanolic extract	Not directly ND in title but antiinflammatory compounds identified that could be neuroprotective.	Inhibits LPS-induced inflammatory responses in macrophages (NF-κB/iNOS/cytokines).	Reported antioxidant components (meroterpenoids) that scavenge ROS.	In vitro: LPS-stimulated peritoneal macrophages; component ID.	Joung et al., (2016)
Some brown marine algae (general extract)	Reported antiinflammatory activities—possible ND	Inhibits iNOS/NO and inflammatory signaling (IκBα phosphorylation	Antioxidant effects measured (assays reported).	In vitro: LPS-stimulated Radial	Hwang, Oh and Lim (2014)

(*Continued*)

Table 5.3 (Continued)

Potential algae/ compound	Application in ND	Modulation of inflammation	Oxidative-stress pathways	Preclinical evidence	Citation
	relevance (inflammation is ND risk factor).	inhibition reported).		Adenocarcinoma/ macrophage Williams (RAW) 264.7 macrophages.	
Fucoxanthin (brown algae)	Fucoxanthin has been evaluated for antiinflammatory and antioxidant effects—implication for ND protection.	Inhibits NO and PGE2 production in LPS-stimulated macrophages; modulates NF-κB.	Antioxidant/radical-scavenging activity; protective in oxidative-stress assays.	In vitro: RAW 264.7 macrophage cells; various biochemical assays.	Heo et al. (2010)
Red-algae secondary metabolites (various)	Reported to reduce neurodegenerative biomarkers in preclinical models—proposed neuroprotective uses.	Modulate MAPK/ NF-κB and NLRP3 inflammasome pathways (mechanistic study).	Reduce oxidative markers and upregulate antioxidant responses in tissues.	In vitro mechanistic assays and some in vivo histopathology reported (paper: American Chemical Society Omega).	Nabil-Adam, Ashour and Shreadah (2023)

Neorhodomela aculeata (red alga) methanolic extract	Neuroprotective potential noted in hippocampal and microglial cell models—ND relevance.	Reduced microglial activation and inflammatory markers in hippocampal/ microglial cells.	Antioxidant activity evidenced in cell assays.	In vitro: hippocampal neurons and microglial cells.	Lim et al. (2006)
Undaria pinnatifida (wakame) glycoprotein	Anti-Alzheimer's and antiinflammatory activities reported (food-research study suggests ND-relevant activity).	Suppresses inflammatory mediators relevant to neuroinflammation.	Antioxidant properties and protection in oxidative-stress assays reported.	In vitro and ex vivo assays; assays relevant to AD (enzyme/ inflammatory endpoints).	Rafiquzzaman et al. (2015)
Various seaweeds (*Sargassum muticum, Codium tomentosum, Saccorhiza polyschides*)	Seaweeds screened for neuroprotective potential in human cellular stress models—several species showed protection.	Many extracts reduced inflammatory markers in stressed cells (NF-κB, iNOS pathways).	Antioxidant, mitochondrial protection, and reduced caspase-3 activation noted.	In vitro: human cell stress models (dopamine toxicity, oxidative stress).	Silva et al. (2020)
North Bornean seaweeds (*Kappaphycus alvarezii, Eucheuma denticulatum,*	Review lists neuroprotective and antiinflammatory activities of	Multiple mechanisms across species: COX/iNOS/	Antioxidant and radical-scavenging properties summarized.	Review of in vitro and in vivo preclinical	Shah et al. (2022)

(*Continued*)

Table 5.3 (Continued)

Potential algae/ compound	Application in ND	Modulation of inflammation	Oxidative-stress pathways	Preclinical evidence	Citation
Halymenia durvillaei (Rhodophyta), *Caulerpa lentillifera, Caulerpa racemosa* (Chlorophyta), *Dictyota dichotoma, Sargassum polycystum* (Ochrophyta))	several local seaweeds (ND relevance).	NF-κB/MAPK discussed.		evidence across species.	
Phlorotannins (eckol, dieckol) from *Ecklonia cava*	Neuroprotective → reduced Aβ25–35-induced cytotoxicity in PC12 cells (relevance to AD).	Inhibited proinflammatory cytokines (TNF-α, IL-1β, PGE_2); antineuroinflammatory effects.	Suppressed intracellular ROS, preserved mitochondrial function, reduced caspase activation.	In vitro: PC12 neuronal cell model of Aβ(25–35) toxicity.	Lee et al. (2019)
Multiple microalgae species—various bioactives (carotenoids, lipids, pigments)	Review: microalgae as sources of neuroprotective compounds; potential therapeutic candidates for ND.	Carotenoids/ polysaccharides modulate NF-κB, reduce cytokine production (reviewed).	Strong antioxidant potential (carotenoids, phycobiliproteins)—ROS scavenging, antioxidant enzyme upregulation.	Review summarizing many in vitro and in vivo preclinical studies.	Silva et al. (2022)

Microalgae-derived compounds (review)	Directly addresses microalgae as therapeutic candidates for ND (review of mechanisms and candidates).	Reviews inhibition of inflammatory cascades (TLR4, NF-κB, MAPK) by microalgal bioactives.	Antioxidant pathways (Nrf2/HO-1) and mitochondrial protection summarized.	Review: compiles in vitro and in vivo preclinical evidence across species.	Parameswari & Lakshmi (2022)
Seaweed bioactives (review): antimicrobial/ pharmacological	Review: lists pharmacological promise of seaweed extracts (includes neuroprotection potential).	Highlights antiinflammatory metabolites (phenolics, polysaccharides) and their mechanisms.	Reviews antioxidant compounds and scavenging activity across species.	Review (prospects and pharmacology)—no single preclinical experiment.	Sneha et al. (2024)
Marine algal metabolites (review)	Explicit review of marine algal metabolites in ND—synthesizes recent preclinical evidence and mechanisms.	Summarizes microglial modulation, decreased proinflammatory cytokines, NF-κB, and NLRP3 modulation.	Emphasizes Nrf2/HO-1 activation, ROS reduction, and mitochondrial protection.	Review summarizing in vitro and in vivo rodent work.	Naik et al. (2024)
Bioactives from marine macroalgae (review)	Review focused on marine macroalgae compounds for prevention/ treatment of ND.	Describes antineuroinflammatory candidates (fucoidans, phlorotannins) and signaling targets.	Discusses antioxidant defenses (Nrf2, Superoxide Dismutase (SOD), catalase) and antioxidative roles.	Review paper covering numerous preclinical reports.	Leal et al. (2025)

(*Continued*)

Table 5.3 (Continued)

Potential algae/ compound	Application in ND	Modulation of inflammation	Oxidative-stress pathways	Preclinical evidence	Citation
Zonarol (sesquiterpene hydroquinone) from *Dictyopteris undulata*	Pharmacokinetics study—supports translational use; zonarol has reported antioxidant/ antiinflammatory properties relevant to ND.	Shown previously to reduce inflammatory markers in preclinical studies (literature summarized).	Antioxidant activity; study provides protein kinase data to support dosing in vivo.	In vivo: mouse pharmacokinetics (support future ND preclinical models).	Fei et al. (2023)
Indole derivatives from *Sargassum thunbergii* spp.	Natural indole derivatives with bioactivity; potential neuroprotective leads (antioxidant/ antiinflammatory).	Reported antiinflammatory and antibacterial activities (mechanistic assays).	Antioxidant assays reported, supporting neuroprotective potential.	Compound isolation + in vitro pharmacology; not specific ND animal models.	Kang et al. (2017)
Fucoxanthin and fucoxanthinol (from *Undaria pinnatifida*, diatoms)	Demonstrated neuroprotective effects in neuronal cultures (attenuate oxidative stress; ND relevant).	Antiinflammatory via reduced microglial activation and inhibition of NF-κB signaling (fucoxanthinol studies).	Activates PI3K/Akt; upregulates Nrf2/HO-1; reduces intracellular ROS and apoptosis.	In vitro: primary/ immortalized neuronal cells and BV-2 microglia; some in vivo rodent support in literature.	Mohibbullah et al. (2018)

Spirulina/Arthrospira bioactives (phycocyanin, C-PC)	Reviews/evidence that Spirulina compounds can prevent progression of neurological disease; C-phycocyanin inhibits amyloid and α-synuclein aggregation in some studies.	Inhibits microglial activation, lowers proinflammatory gene expression (TLR4-mediated modulation reported).	Antioxidant, ROS scavenging, reduces oxidative damage in brain tissue; supports mitochondrial health.	In vitro microglial assays; in vivo rodent PD/KA/α-synuclein models showing neuroprotection.	Elkader et al. (2024)
Spirulina acetone extract (*Arthrospira platensis*)	Demonstrated ability to impede LPS-triggered neuroinflammation in microglia—implication for ND.	Decreased microglial activation and proinflammatory cytokine release (in vitro).	Reported antioxidant markers and attenuation of LPS-induced ROS in microglia.	In vitro: BV-2/microglia LPS activation model; pre and posttreatment experiments.	Piovan et al. (2021)
Arthrospira (Spirulina) *platensis* (dietary Spirulina)	In vivo neuroprotection: Spirulina diet reduced lesion size and microglial activation in α-synuclein PD model.	Reduced microglial activation (major histocompatibility complex class II (MHCII), Iba1), decreased inflammatory responses in brain.	Associated with reduced oxidative markers and better neuronal survival.	In vivo: rat α-synuclein model of PD.	Pérez-Juárez et al. (2016)

(*Continued*)

Table 5.3 (Continued)

Potential algae/ compound	Application in ND	Modulation of inflammation	Oxidative-stress pathways	Preclinical evidence	Citation
Fucoxanthin (brown algae carotenoid)	Fucoxanthin protects hippocampal neurons from H_2O_2-induced apoptosis—ND relevance for oxidative-stress related neuronal death.	Indirect antiinflammatory effects via signaling cascades (PI3K/Akt modulation).	Activates PI3K/Akt and inhibits extracellular signal kinase (ERK)pathway; reduces ROS and apoptosis in neuronal cells.	In vitro: SH-SY5Y neuronal cell model of H_2O_2-induced oxidative injury.	Yu et al. (2017)
Plastoquinones from *Sargassum serratifolium*	Enzyme inhibitory bioactives (PTP1B, α-glucosidase)—-metabolic and neuroprotective implications discussed.	Reported antiinflammatory/antioxidant potential; mechanism not ND-specific but supportive.	Antioxidant assays and enzymatic modulation reported.	In vitro enzymatic assays and compound isolation.	Ali et al. (2017)
Ecklonia cava extract (phlorotannin-rich)	Demonstrated protection against oxidative stress–induced neuronal damage in PC12 cells—ND relevance.	Reduced apoptotic signaling and inflammatory markers in neuronal cells.	Antioxidant, antiapoptotic (mitochondrial protection) in PC12 oxidative-stress assays.	In vitro: PC12 neuronal cells exposed to oxidative stress.	Shin et al. (2021)

Fucoxanthinol (from diatom/seaweed)	Fucoxanthinol shown to ameliorate neuroinflammatory responses in BV-2 microglia—ND-relevant antineuroinflammatory agent.	Suppresses proinflammatory mediators in LPS-stimulated microglia; modulates NF-κB and related signaling.	Enhances Nrf2 nuclear translocation; upregulates HO-1 and NQO1; reduces ROS in microglia.	In vitro: BV-2 microglial LPS model; mechanistic cellular assays.	Li et al. (2020)
Diketopiperazines/ marine-derived cyclic dipeptides	Antiinflammatory molecules with potential for neuroinflammatory modulation (indirect ND relevance).	Inhibit LPS/ polyphosphate-mediated inflammatory responses; reduce adhesion molecules and leukocyte migration.	Antioxidant status not primary endpoint but antiinflammatory action reduces oxidative sequelae.	In vitro human endothelial cells and mice (sepsis/ polyphosphate models).	Jeong et al. (2016)
C-Phycocyanin (C-PC) from Spirulina	Demonstrated modulation of microglial activation via TLR4—neuroprotective potential for ND by dampening neuroinflammation.	Acts through TLR4-mediated signaling to reduce proinflammatory gene expression in microglia.	Reduces ROS generation in activated microglia; supports antioxidant defenses indirectly.	In vitro: BV-2/ microglial cells; mechanistic TLR4 assays.	Piovan et al. (2022)

(*Continued*)

Table 5.3 (Continued)

Potential algae/ compound	Application in ND	Modulation of inflammation	Oxidative-stress pathways	Preclinical evidence	Citation
Marine diketopiperazines/ marine-derived small molecules	Several marine-derived small molecules show antiinflammatory/ neuroprotective leads (potential ND relevance).	Inhibit endothelial and inflammatory pathways (LPS/ polyP models)—may reduce neurovascular inflammation.	Some antioxidant activity reported depending on compound class.	In vitro human endothelial and mouse models (sepsis/ inflammation assays).	Lee et al. (2016)
Dietary Spirulina (whole biomass)	In vivo evidence of neuroprotection in α-synuclein PD model: reduced dopaminergic cell loss and microglial activation.	Decreased microglial activation and MHC II expression; lowered brain inflammatory response.	Improved neuronal survival associated with reduced oxidative damage markers.	In vivo: rat α-synuclein Parkinson's model (dietary supplementation).	Pabon et al. (2012)
Spirulina/C-phycocyanin (cell-based studies)	In vitro evidence that Spirulina and C-PC reduce microglial cytotoxicity and inflammatory gene expression—ND relevance.	Inhibited expression of inflammation-related genes in BV-2 microglial cells; lowered NO and cytokines.	Reduced ROS/cytotoxicity in microglial culture models.	In vitro: BV-2 microglia studies; gene expression and cytotoxicity assays.	Chen et al. (2012)

antioxidant activity, phycocyanin also demonstrates notable antiinflammatory effects by suppressing pathways involving NF-κB and COX-2, which in turn lowers the production of proinflammatory cytokines and minimizes tissue inflammation.

Emerging evidence shows that phycocyanin may exert anticancer effects by triggering apoptosis in cancer cells through caspase activation and disruption of mitochondrial membranes. It has also been associated with protective benefits for the nervous system and liver, highlighting its potential for use in functional foods and therapeutic products. Despite these advantages, the compound is sensitive to heat and light, which can reduce its effectiveness. As a result, modern encapsulation approaches, such as liposomal delivery or microencapsulation, are increasingly used to maintain its stability during processing, storage, and administration .

Fucoidan is a sulfated polysaccharide commonly obtained from brown seaweeds such as *Fucus vesiculosus* and *Laminaria japonica*. It has attracted considerable attention due to its wide range of biological activities, including antiviral, anticancer, anticoagulant, and immune-enhancing effects. One of its key anticancer actions involves suppressing angiogenesis by disrupting VEGF-related signaling, which limits the formation of new blood vessels needed for tumor growth and spread. Fucoidan also shows strong antiviral potential by blocking viral enzymes and preventing viruses from attaching to and entering host cells, with reported effects against herpes simplex, dengue, and various influenza strains.

In terms of immune support, fucoidan can activate natural killer (NK) cells and macrophages, strengthening the body's overall defense mechanisms. Advances in formulation science have led to the integration of fucoidan into nanoparticles, hydrogels, and other delivery systems to improve its absorption and to target specific tissues more effectively. However, its clinical translation is still challenged by differences in structure and purity that arise from using various seaweed species and extraction methods, making standardization an important ongoing focus.

5.3 Algae-based nutraceuticals and functional foods

The expanding interest in algae as sources of therapeutic metabolites has naturally extended into their utilization within nutraceuticals and functional foods. Many of the pharmacologically active compounds previously discussed, such as carotenoids, phycobiliproteins, long-chain omega-3 polyunsaturated fatty acids, vitamins, and sulfated polysaccharides, also constitute the biochemical foundation of algae-based preventive health interventions. Their high nutrient density and bioactive complexity contribute to a profile capable of modulating key metabolic, inflammatory, and oxidative pathways relevant to disease prevention (Ahmed et al., 2024; Lafarga, Acién-Fernández and Garcia-Vaquero, 2020). Such characteristics place algae in a category that sits between conventional dietary ingredients and biofunctional agents, creating growing interest in their integration into evidence-based preventive nutrition models. As illustrated in Fig. 5.1, algae-derived bioactives contribute to multiple human health functions and underpin a diverse and expanding range of nutraceutical and functional-food applications.

Evidence for metabolic health benefits associated with algal bioactives continues to accumulate, yet it demonstrates considerable heterogeneity across studies. Fucoxanthin extracted from brown algae influences adipocyte metabolism through UCP1 upregulation and enhanced mitochondrial β-oxidation, although outcomes from human trials vary due to differences in fucoxanthin purity, esterification state, and formulation stability (Lee et al., 2020; Mikami and Hosokawa, 2013). Phlorotannins and sulfated polysaccharides demonstrate inhibitory activity against

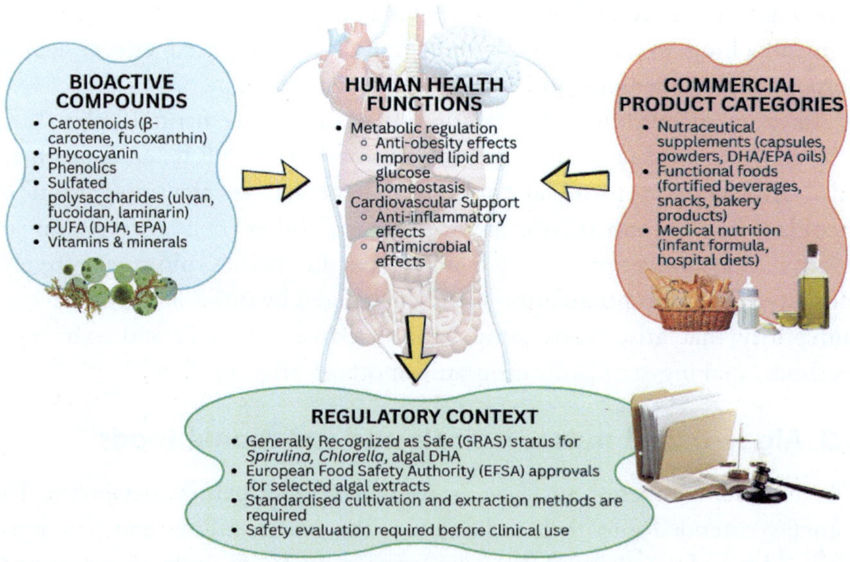

Figure 5.1 Overview of algal nutraceutical functions, including roles in lifestyle disease prevention, immune and antioxidant modulation, and their incorporation into regulated commercial products.

α-glucosidase, α-amylase, and DPP-IV, lending support to their potential role in postprandial glycaemic regulation (Zaharudin, Salmeán and Dragsted, 2018). However, molecular-weight variability and susceptibility to enzymatic degradation in the gastrointestinal tract result in inconsistent bioaccessibility. Microalgal docosahexaenoic acid (DHA) shows promise through inflammation-resolving pathways and lipid-profile modulation, yet comparative studies with fish-derived DHA indicate that differences in triglyceride structure and oxidative resistance may influence physiological effectiveness (Lv and Xu, 2022).

Immunomodulatory and antioxidant effects attributed to algae involve mechanisms that are highly dependent on compound integrity and matrix interactions (Heo, Kim and Cho, 2026). Phycocyanin, ulvan, astaxanthin, and other pigments exert well-documented radical-scavenging, antiinflammatory, and macrophage-regulating activities, though their stability is challenged by thermal processing, pH variation, oxygen exposure, and light sensitivity (Coulombier, Jauffrais and Lebouvier, 2021; Pereira, Cotas and Valado, 2024). Incorporation into functional foods therefore raises formulation-specific barriers, particularly when used in beverages, fermented matrices, or high-shear processing conditions. Cell-wall architecture further influences the bioavailability of encapsulated bioactives, the recalcitrant trilaminar sheath of Chlorella species often requires mechanical or enzymatic disruption to enhance nutrient release, while certain red algae exhibit mucilage layers that complicate extraction yields (Guil-Guerrero and Prates, 2025; Kumari, Garima and Bharadvaja, 2023).

Regulatory acceptance of selected microalgae such as Spirulina and Chlorella with generally recognized as safe (GRAS) designation reflects progress in establishing their safety, but does not fully resolve concerns associated with production variability. Biomass composition changes substantially in response to culture parameters including nutrient limitation, salinity, light regime, and harvest timing, all of which alter pigment concentration, lipid profiles, and polysaccharide structures (Čabarkapa et al., 2022). The potential for heavy metal accumulation, cyanotoxin contamination, or microbial proliferation in poorly controlled systems necessitates rigorous quality-control frameworks involving contaminant screening, chemical fingerprinting, and standardized extraction protocols (AlFadhly et al., 2022; Čabarkapa et al., 2022). Industrial-scale production of algal DHA illustrates the influence of strain selection, fermentation optimization, and antioxidant stabilization strategies on final product quality, especially for applications in infant formula and medical nutrition (Eilam et al., 2023).

As interest in algae-based nutraceuticals intensifies, unresolved scientific challenges continue to shape the field (Tiwari and Garcia-Vaquero, 2022). Issues such as low oral bioavailability, compound instability during storage and processing, species-specific variability, extraction-dependent differences in chemical profiles, and inconsistent clinical evidence limit the predictability of therapeutic effectiveness. Harmonized regulatory standards, improved delivery systems including microencapsulation, emulsification, and nanoformulation, and advanced analytical benchmarking remain essential for ensuring reproducibility and efficacy across commercial products (Tan et al., 2023). The evolving understanding of algal metabolite chemistry, combined with improvements in cultivation and downstream processing, provides a foundation for expanding their integration into medical nutrition and functional-food systems (Tiwari and Garcia-Vaquero, 2022).

5.4 Neuroprotective and antiinflammatory potential of algae

Marine algae, which are derived bioactives, have emerged as a captivating materials, progressively explored, and potentially useful for ND intervention. Algae are among the most convincing therapeutic actions in several studies for disease-oriented evidence related to Parkinson's disease (PD) and Alzheimer's disease (AD), where oxidative stress, chronic neuroinflammation, and protein aggregation converge as central pathogenic mechanisms. Agaropentaose derived from red algae protected dopaminergic neuronal cells against 6-hydroxydopamine-induced cytotoxicity, a canonical PD model, by restoring cellular redox balance and suppressing apoptotic signaling (Ye et al., 2019). *S. platensis* and its bioactive pigments, particularly C-phycocyanin, showed consistent protection of nigrostriatal dopaminergic neurons and improved behavioral outcomes in toxin-induced PD in animal model studies (Lima et al., 2017; Pabón et al., 2012). These findings are remarkable because they go beyond in vitro screening and demonstrate functional neuroprotection in vivo.

For AD-related damage, phlorotannins like dieckol and eckol have shown protective effects against toxicity from amyloid-β, which ties into the synaptic issues and neuron loss seen in AD (Yayeh et al., 2014; Jeong et al., 2016). Meanwhile, fucoxanthin and its breakdown product fucoxanthinol consistently support neuron survival amid oxidative and inflammatory stress, pointing to potential benefits for both AD and Parkinson's (Kim and Pangestuti, 2011; Heo et al., 2010). That said, a key drawback is that many of the algae highlighted in the review were not tested directly in models of neurodegeneration. Their links to AD or PD mostly come from broader antioxidant or antiinflammatory effects (Gallego et al., 2022; Silva et al., 2022). This kind of extrapolation makes biological sense, but it highlights a real gap between initial lab screenings and proven benefits in specific diseases, indicating that targeted studies are needed to back up any real-world applications.

The most solid and reliable findings from these studies on how algae compounds apply to both inflammation and oxidative stress in two linked processes that fuel neurodegeneration. Almost every compound reviewed reveals the proinflammatory signals while boosting the body's own antioxidant systems. At the inflammatory level, multiple compounds inhibited nuclear factor such as kappa B signaling, mitogen-activated protein kinase cascades, and downstream cytokine production, including tumor necrosis factor-α, interleukin-1β, and interleukin-6 (Yayeh et al., 2014; Silva et al., 2021). Dieckol algae significantly induced heme oxygenase-1 expression while attenuating lipopolysaccharide-induced macrophage activation, which emphasizes a mechanism highly relevant to microglial-driven neuroinflammation (Yayeh et al., 2014). Fucoxanthin and fucoxanthinol consistently suppressed inducible nitric oxide synthase and cyclooxygenase-2 expression, reinforcing their role as broad-spectrum antiinflammatory agents (Heo et al., 2010; Lim et al., 2006).

Oxidative-stress modulation was equally prominent. The carotenoids, phlorotannins, and sulfated polysaccharides demonstrated ROS scavenging capacity, activation of nuclear factor erythroid 2-related factor 2 signaling, and preservation of mitochondrial integrity (Gallego et al., 2022). Importantly, several studies emphasized pathway-specific antioxidant regulation rather than nonspecific radical scavenging, strengthening their mechanistic relevance to chronic neurodegenerative conditions. The convergence of antiinflammatory and antioxidant mechanisms is particularly significant, as neurodegeneration is increasingly understood as a self-amplifying cycle of redox imbalance and immune dysregulation. Algae-derived compounds appear well positioned to disrupt this cycle; however, most evidence remains mechanistic rather than outcome driven.

Despite strong mechanistic coherence, the translational landscape remains challenging for preclinical research, with a predominance of in vitro models and a limited number of animal studies, especially for algae used in neurological effects. Neuronal cell lines, macrophages, and microglia dominate the experimental designs, offering valuable mechanistic insight but limited predictive validity for human disease progression (Jin et al., 2006; Ye et al., 2019). Encouragingly, a subset of studies progressed to in vivo validation, particularly for Spirulina-based interventions and selected red-algae extracts, demonstrating reductions in neuroinflammation, oxidative damage, and neuronal loss (Lima et al., 2017; Yousof et al., 2025). Moreover, pharmacokinetic and bioavailability investigations, such as those conducted for zonarol, represent an important shift toward addressing practical barriers to therapeutic development (Nabil-Adam, Ashour and Shreadah, 2023). Nevertheless, clinical evidence remains absent, and critical issues, including dose optimization, long-term safety, metabolic stability, and blood–brain barrier permeability, are largely unresolved and ongoing studies. Review-based articles within the dataset consistently acknowledge that most algae-derived compounds are still at the discovery or early preclinical stage, and caution against premature clinical extrapolation (Kim and Pangestuti, 2011; Naik et al., 2024). Bridging this gap will require standardized extracts, rigorous animal models, and early-phase human studies to determine whether the strong mechanistic promise of algae-derived bioactives can translate into effective therapies for ND.

5.5 Conclusion

Taken together, the evidence discussed in this chapter positions algae as a strategic platform for next-generation medical and pharmaceutical innovation rather than merely an auxiliary source of bioactive ingredients. By uniting high-value metabolites, tunable biomaterials for drug delivery, and nutraceutical applications that sit at the food–medicine interface, algae can support integrated strategies for disease prevention, adjunct therapy, and long-term health maintenance. At the same time, persistent challenges related to bioavailability, product standardization, safety assessment, and regulatory harmonization show that translation from laboratory promise to clinical and commercial impact will depend on coordinated advances in cultivation technologies, process engineering, analytical characterization, and well-designed clinical trials, including longitudinal safety studies and comparative effectiveness research against conventional therapies. In future, global drivers such as aging populations, the rising burden of chronic and infectious diseases, antimicrobial resistance, as well as the need for climate-resilient production systems give algae a broader public-health relevance that extends beyond niche biomedical applications. Addressing all these scientific and regulatory priorities will not only determine the credibility and uptake of algae-based interventions in mainstream healthcare, but will also shape how policymakers, industry stakeholders, and clinicians ultimately position algae within future pharmacotherapy, medical nutrition, and preventive health frameworks.

References

Abdel-Moneim, A.M.E., El-Saadony, M.T., Shehata, A.M., Saad, A.M., Aldhumri, S.A., Ouda, S.M., et al., 2022. Antioxidant and antimicrobial activities of Spirulina platensis extracts and biogenic selenium nanoparticles against selected pathogenic bacteria and fungi. Journal of Biological Sciences 29 (2), 1197–1209. Available at https://doi.org/10.1016/j.sjbs.2021.09.046, http://www.ksu.edu.sa/sites/Colleges/CollegeofScinces/BotanyDepartment/SJBS/Default.aspx.

Ahmed, N., Sheikh, M.A., Ubaid, M., Chauhan, P., Kumar, K., Choudhary, S., 2024. Comprehensive exploration of marine algae diversity, bioactive compounds, health benefits, regulatory issues, and food and drug applications. Measurement: Food 14, 100163. Available at https://doi.org/10.1016/j.meafoo.2024.100163.

AlFadhly, N.K.Z., Alhelfi, N., Altemimi, A.B., Verma, D.K., Cacciola, F., Narayanankutty, A., 2022. Trends and technological advancements in the possible food applications of spirulina and their health benefits: a review. Molecules (Basel, Switzerland) 27 (17). Available at https://doi.org/10.3390/molecules27175584, http://www.mdpi.com/journal/molecules.

Ali, Y., Kim, D.H., Seong, S.H., Kim, H.R., Jung, H.A., Choi, J.S., 2017. α-Glucosidase and protein tyrosine phosphatase 1b inhibitory activity of plastoquinones from marine brown alga sargassum serratifolium. Marine Drugs 15 (12). Available at https://doi.org/10.3390/md15120368, http://www.mdpi.com/1660-3397/15/12/368/pdf.

Bouafir, Y., Bouhenna, M.M., Nebbak, A., Belfarhi, L., Aouzal, B., Boufahja, F., Bendif, H., Bruno, M., 2025. Algal bioactive compounds: A review on their characteristics and medicinal properties. Fitoterapia 183, 106591. Available at https://doi.org/10.1016/j.fitote.2025.106591.

Bratchkova, A., Kroumov, A.D., 2020. Microalgae as producers of biologically active compounds with antibacterial, antiviral, antifungal, antialgal, antiprotozoal, antiparasitic and anticancer activity. Acta Microbiologica Bulgarica 2020 (3), 79–89. Available at actamicrobio.bg/.

Čabarkapa, I., Rakita, S., Popović, S., Tomičić, Z., Spasevski, N., Vulić, J., et al., 2022. Characterization of organic Spirulina spp. and Chlorella vulgaris as one of the most nutrient-dense food. Journal of Food Safety and Food Quality-Archiv für Lebensmittelhygiene 73 (3). Available at https://doi.org/10.31083/0003-925x-73-78.

Chen, J.C., Liu, K.S., Yang, T.J., Hwang, J.H., Chan, Y.C., Lee, I.T., 2012. Spirulina and C-phycocyanin reduce cytotoxicity and inflammation-related genes expression of microglial cells. Nutritional Neuroscience 15 (6), 252–256. Available at https://doi.org/10.1179/1476830512Y.0000000020, http://www.ingentaconnect.com/content/maney/nns/2012/00000015/00000006/art00003.

Coulombier, N., Jauffrais, T., Lebouvier, N., 2021. Antioxidant compounds from microalgae: a review. Marine Drugs 19 (10), 549. Available at https://doi.org/10.3390/md19100549.

Dong, J., Ma, H., Liu, K., Zhou, W., Liu, S., Sun, Y., Ju, W., Wang, S., 2025. Effects of *Undaria pinnatifida*-derived brown algae polysaccharide (UPS) on the nutritional composition, digestive capacity, immune performance and intestinal microbiota of juvenile sea cucumber (*Apostichopus japonicus*). PeerJ 13, e19944. Available at https://doi.org/10.7717/peerj.19944.

Donoso, A., González-Durán, J., Muñoz, A.A., González, P.A., Agurto-Muñoz, C., 2021. Therapeutic uses of natural astaxanthin: An evidence-based review focused on human clinical trials. Pharmacological Research 166, 105479. Available at https://doi.org/10.1016/j.phrs.2021.105479.

Eilam, Y., Khattib, H., Pintel, N., Avni, D., 2023. Microalgae—sustainable source for alternative proteins and functional ingredients promoting gut and liver health. Global Challenges 7 (5). Available at https://doi.org/10.1002/gch2.202200177.

Elkader, H.-T.A.E.A., Essawy, A.E., Al-Shami, A.S., 2024. Bioactive compounds of the genus Spirulina can prevent the progression of neurological diseases. Neurochemical Journal 18 (1), 47–59. Available at https://doi.org/10.1134/s1819712424010057.

Fei, J., Yamada, S., Satoh, T., Koyama, T., 2023. Pharmacokinetic analysis of zonarol, a marine algal hydroquinone, in mice using HPLC with fluorescence detection. Antibiotics 12 (6), 1013. Available at https://doi.org/10.3390/antibiotics12061013.

Fernandes, R., Campos, J., Serra, M., Fidalgo, J., Almeida, H., Casas, A., et al., 2023. Exploring the benefits of phycocyanin: from Spirulina cultivation to its widespread applications. Pharmaceuticals 16 (4). Available at https://doi.org/10.3390/ph16040592, https://www.mdpi.com/journal/pharmaceuticals.

Gallego, R., Valdés, A., Suárez-Montenegro, Z.J., Sánchez-Martínez, J.D., Cifuentes, A., Ibáñez, E., et al., 2022. Anti-inflammatory and neuroprotective evaluation of diverse microalgae extracts enriched in carotenoids. Algal Research 67. Available at https://doi.org/10.1016/j.algal.2022.102830, http://www.sciencedirect.com/science/journal/aip/22119264.

Gao, X., Wang, H., Chen, K., Guo, Y., Zhou, J., Xie, W., 2024. Toxicological and pharmacological activities, and potential medical applications, of marine algal toxins. International Journal of Molecular Sciences 25 (17), 9194. Available at https://doi.org/10.3390/ijms25179194.

Goshtasbi, H., Okolodkov, Y.B., Movafeghi, A., Awale, S., Safary, A., Barar, J., Omidi, Y., 2023. Harnessing microalgae as sustainable cellular factories for biopharmaceutical production. Algal Research 74, 103237. Available at https://doi.org/10.1016/j.algal.2023.103237.

Guil-Guerrero, J.L., Prates, J.A.M., 2025. Microalgae bioactives for functional food innovation and health promotion. Foods 14 (12). Available at https://doi.org/10.3390/foods14122122, www.mdpi.com/journal/foods.

Hannan, M.A., Dash, R., Haque, M.N., Mohibbullah, M., Sohag, A.A.M., Rahman, M.A., et al., 2020. Neuroprotective potentials of marine algae and their bioactive metabolites: pharmacological insights and therapeutic advances. Marine Drugs 18 (7). Available at https://doi.org/10.3390/md18070347, https://www.mdpi.com/1660-3397/18/7/347/pdf.

Hebar, A., Koller, C., Seifert, J.-M., Chabicovsky, M., Bodenteich, A., Bernkop-Schnürch, A., et al., 2015. Non-clinical safety evaluation of intranasal iota-carrageenan. PLoS One 10 (4), e0122911. Available at https://doi.org/10.1371/journal.pone.0122911.

Heo, Y., Kim, M.Y., Cho, J.Y., 2026. Chlorella vulgaris, a representative edible algae as integrative and alternative medicine. Integrative Medicine Research 15 (1). Available at https://doi.org/10.1016/j.imr.2025.101228, https://www.sciencedirect.com/science/journal/22134220.

Heo, S.J., Yoon, W.J., Kim, K.N., Ahn, G.N., Kang, S.M., Kang, D.H., et al., 2010. Evaluation of anti-inflammatory effect of fucoxanthin isolated from brown algae in lipopolysaccharide-stimulated RAW 264.7 macrophages. Food and Chemical Toxicology 48 (8-9), 2045–2051. Available at https://doi.org/10.1016/j.fct.2010.05.003.

Homaeigohar, S., Liu, X., Elbahri, M., 2023. Antiviral polysaccharide and antiviral peptide delivering nanomaterials for prevention and treatment of SARS-CoV-2 caused COVID-19 and other viral diseases. Journal of Controlled Release 358, 476–497. Available at https://doi.org/10.1016/j.jconrel.2023.05.010.

Hsu, C.Y., Allela, O.Q.B., Hussein, A.M., Mustafa, M.A., Kaur, M., Alaraj, M., et al., 2024. Recent advances in polysaccharide-based drug delivery systems for cancer therapy: a comprehensive review. Artificial Cells, Nanomedicine, and Biotechnology 52 (1), 564–586. Available at https://doi.org/10.1080/21691401.2024.2436350, http://www.tandfonline.com/loi/ianb20#.VmugQbfovcs.

Hwang, J.H., Oh, Y.S., Lim, S.B., 2014. Anti-inflammatory activities of some brown marine algae in LPS-stimulated RAW 264.7 cells. Food Science and Biotechnology 23 (3), 865–871. Available at https://doi.org/10.1007/s10068-014-0116-2, http://www.springerlink.com/content/n3031612t262/.

Ibrahim, T.N.B.T., Feisal, N.A.S., Kamaludin, N.H., Cheah, W.Y., How, V., Bhatnagar, A., et al., 2023. Biological active metabolites from microalgae for healthcare and pharmaceutical industries: a comprehensive review. Bioresource Technology 372. Available at https://doi.org/10.1016/j.biortech.2023.128661, www.elsevier.com/locate/biortech.

Jeong, S., Ku, S.K., Min, G., Choi, H., Park, D.H., Bae, J.S., 2016. Suppressive effects of three diketopiperazines from marine-derived bacteria on polyphosphate-mediated septic responses. Chemico-Biological Interactions 257, 61–70. Available at https://doi.org/10.1016/j.cbi.2016.07.032, www.elsevier.com/locate/chembioint.

Jia, H., Li, Y., Zheng, Y., Wang, H., Zhao, F., Yang, X., et al., 2025. Recent advances in fucoidan-based improved delivery systems: structure, carrier types and biomedical applications. Carbohydrate Polymers 352, 123183. Available at https://doi.org/10.1016/j.carbpol.2024.123183.

Jin, D.Q., Lim, C.S., Sung, J.Y., Choi, H.G., Ha, I., Han, J.S., 2006. Ulva conglobata, a marine algae, has neuroprotective and anti-inflammatory effects in murine hippocampal and microglial cells. Neuroscience Letters 402 (1-2), 154–158. Available at https://doi.org/10.1016/j.neulet.2006.03.068.

Jin, X., Gong, S., Yang, B., Wu, J., Li, T., Wu, H., Wu, H., Xiang, W., 2021. Transcriptomic analysis for phosphorus limitation-induced β-glucans accumulation in Chlorella sorokiniana SCSIO 46784 during the early phase of growth. Algal Research 54, 102208. Available at https://doi.org/10.1016/j.algal.2021.102208.

Jung, S.J., Jang, H.Y., Jung, E.S., Noh, S.O., Shin, S.W., Ha, K.C., et al., 2020. Effects of porphyra tenera supplementation on the immune system: a randomized, double-blind, and placebo-controlled clinical trial. Nutrients 12 (6). Available at https://doi.org/10.3390/nu12061642, https://www.mdpi.com/2072-6643/12/6/1642/pdf.

Kang, M.C., Ding, Y., Kim, E.A., Choi, Y.K., De Araujo, T., Heo, S.J., et al., 2017. Indole derivatives isolated from brown alga Sargassum thunbergii inhibit adipogenesis through AMPK activation in 3T3-L1 preadipocytes. Marine Drugs 15 (4). Available at https://doi.org/10.3390/md15040119, http://www.mdpi.com/1660-3397/15/4/119/pdf.

Kim, S.K., Pangestuti, R., 2011. Biological activities and potential health benefits of fucoxanthin derived from marine brown algae. Advances in Food and Nutrition Research 64, 111–128. Available at https://doi.org/10.1016/B978-0-12-387669-0.

00009-0, http://www.elsevier.com/wps/find/bookdescription.cws_home/704624/description#description.
Kim, L., Hong, T., Ham, J., Lim, W., 2022. Effects of Agarum clathratum extract on cell death and calcium ion levels of ovarian cancer cell. Molecular & Cellular Toxicology 19 (2), 303–310. Available at https://doi.org/10.1007/s13273-022-00259-5.
Kumari, A., Garima, Bharadvaja, N., 2023. A comprehensive review on algal nutraceuticals as prospective therapeutic agent for different diseases. 3 Biotech 13 (2). Available at https://doi.org/10.1007/s13205-022-03454-2.
Kumar, A., Singh, R.P., Kumar, I., Yadav, P., Singh, S.K., Kaushalendra, Singh, P.K., et al., 2022. Algal metabolites can be an immune booster against COVID-19 pandemic. Antioxidants 11 (3). Available at https://doi.org/10.3390/antiox11030452, https://www.mdpi.com/2076-3921/11/3/452/pdf.
Lafarga, T., Acién-Fernández, F.G., Garcia-Vaquero, M., 2020. Bioactive peptides and carbohydrates from seaweed for food applications: natural occurrence, isolation, purification, and identification. Algal Research 48, 101909. Available at https://doi.org/10.1016/j.algal.2020.101909.
Leal, M.F.C., Duarte, R.J.G., Cardoso, I.L., Catarino, R.I.L., Pimenta, A.M., Souto, M.R.S., 2025. Bioactive compounds from marine macroalgae in the treatment and prevention of neurodegenerative diseases. Neurophysiology. Available at https://doi.org/10.1007/s11062-025-09953-1, https://www.springer.com/journal/11062.
Lee, W., Ku, S.K., Choi, H., Bae, J.S., 2016. Inhibitory effects of three diketopiperazines from marine-derived bacteria on endothelial protein C receptor shedding in human endothelial cells and mice. Fitoterapia 110, 181–188. Available at https://doi.org/10.1016/j.fitote.2016.03.016, www.elsevier.com/locate/fitote.
Lee, S., Youn, K., Kim, D.H., Ahn, M.R., Yoon, E., Kim, O.Y., et al., 2019. Anti-neuroinflammatory property of phlorotannins from ecklonia cava on Aβ25-35-induced damage in PC12 cells. Marine Drugs 17 (1). Available at https://doi.org/10.3390/md17010007, http://www.mdpi.com/journal/marinedrugs.
Lee, H.G., Lu, Y.A., Li, X., Hyun, J.M., Kim, H.S., Lee, J.J., et al., 2020. Anti-obesity effects of grateloupia elliptica, a red seaweed, in mice with high-fat diet-induced obesity via suppression of adipogenic factors in white adipose tissue and increased thermogenic factors in brown adipose tissue. Nutrients 12 (2). Available at https://doi.org/10.3390/nu12020308, https://www.mdpi.com/2072-6643/12/2/308/pdf.
Lee, Y., Kim, M., Kim, N., Byun, S., Seo, S., Han, J.Y., 2025. Injectable hydrogel systems for targeted drug delivery: from site-specific application to design strategy. Applied Sciences (Switzerland) 15 (21). Available at https://doi.org/10.3390/app152111599, https://www.mdpi.com/journal/applsci/.
Lima, F.A.V., Joventino, I.P., Joventino, F.P., de Almeida, A.C., Neves, K.R.T., do Carmo, M.R., et al., 2017. Neuroprotective activities of Spirulina platensis in the 6-OHDA model of Parkinson's disease are related to its anti-inflammatory effects. Neurochemical Research 42 (12), 3390–3400. Available at https://doi.org/10.1007/s11064-017-2379-5, www.wkap.nl/journalhome.htm/0364-3190.
Lim, C.S., Jin, D.Q., Sung, J.Y., Lee, J.H., Choi, H.G., Ha, I., et al., 2006. Antioxidant and anti-inflammatory activities of the methanolic extract of Neorhodomela aculeate in hippocampal and microglial cells. Biological and Pharmaceutical Bulletin 29 (6), 1212–1216. Available at https://doi.org/10.1248/bpb.29.1212, http://www.jstage.jst.go.jp/article/bpb/29/6/1212/_pdf.
Lin, S., Chen, X., Chen, H., Cai, X., Chen, X., Wang, S., 2023. The bioprospecting of microbial-derived antimicrobial peptides for sustainable agriculture. Engineering 27, 222–233. Available at https://doi.org/10.1016/j.eng.2022.08.011.
Liu, Y., Wang, M., Castagnini, J.M., Barba, F.J., Asensio-Grau, A., 2025. The impact of microalgae-derived bioactive compounds on gut microbiota modulation. Current

Opinion in Food Science 66. Available at https://doi.org/10.1016/j.cofs.2025.101365, https://www.sciencedirect.com/science/journal/22147993.

López, Y., Soto, S.M., 2019. The Usefulness of Microalgae Compounds for Preventing Biofilm Infections. Antibiotics 9 (1), 9. Available at https://doi.org/10.3390/antibiotics9010009.

Li, Y., Liu, L., Sun, P., Zhang, Y., Wu, T., Sun, H., et al., 2020. Fucoxanthinol from the diatom Nitzschia laevis ameliorates neuroinflammatory responses in lipopolysaccharide-stimulated BV-2 microglia. Marine Drugs 18 (2). Available at https://doi.org/10.3390/md18020116, https://www.mdpi.com/1660-3397/18/2/116.

Lv, W., Xu, D., 2022. Docosahexaenoic acid delivery systems, bioavailability, functionality, and applications: a review. Foods 11 (17), 2685. Available at https://doi.org/10.3390/foods11172685.

Mikami, K., Hosokawa, M., 2013. Biosynthetic pathway and health benefits of fucoxanthin, an algae-specific xanthophyll in brown seaweeds. International Journal of Molecular Sciences 14 (7), 13763–13781. Available at https://doi.org/10.3390/ijms140713763.

Mohibbullah, M., Haque, M.N., Khan, M.N.A., Park, I.S., Moon, I.S., Hong, Y.K., 2018. Neuroprotective effects of fucoxanthin and its derivative fucoxanthinol from the phaeophyte Undaria pinnatifida attenuate oxidative stress in hippocampal neurons. Journal of Applied Phycology 30 (6), 3243–3252. Available at https://doi.org/10.1007/s10811-018-1458-6, www.wkap.nl/journalhome.htm/0921-8971.

Musarra-Pizzo, M., Pennisi, R., Ben-Amor, I., Mandalari, G., Sciortino, M.T., 2021. Antiviral activity exerted by natural products against human viruses. Viruses 13 (5), 828. Available at https://doi.org/10.3390/v13050828.

Nabil-Adam, A., Ashour, M.L., Shreadah, M.A., 2023. Modulation of MAPK/NF-κB pathway and NLRP3 inflammasome by secondary metabolites from red algae: a mechanistic study. ACS Omega 8 (41), 37971–37990. Available at https://doi.org/10.1021/acsomega.3c03480, pubs.acs.org/journal/acsodf.

Naik, B., Richa, S., Bharadwaj, S., Mishra, S., Kumar, V., Kumar, V., et al., 2024. Utilizing marine algal metabolites to fight neurodegenerative diseases. Frontiers in Marine Science 11. Available at https://doi.org/10.3389/fmars.2024.1370839, https://www.frontiersin.org/journals/marine-science#.

Osathanunkul, Maslin, Thanaporn, Suebsuya, Karapetsi, Lefkothea, Nteve, Georgia Maria, Pratsinakis, Emmanouil, Stefanidou, Eleni, Lagiotis, Giorgos, Avramidou, Eleni, Zorxzobokou, Lydia, Tsintzou, Georgia, Athanasiou, Artemis, Mpelai, Sofia, Constandinidis, Constantinos, Pantiora, Panagiota, Merino, Marián, Mullor, José Luis, Dobrovic, Luka, Cerasino, Leonardo, Ogawa, Tomohisa, Tsaousi, Meropi, Rodrigues, Alexandre M.C., Cardoso, Helena, Pires, Rita, Figueiredo, Daniel, Costa, Inês F., Anjos, Catarina, Labrou, Nikolaos E., Madesis, Panagiotis, 2025. Diversity of Bioactive Compounds in Microalgae: Key Classes and Functional Applications. Marine Drugs 23 (6), 222. Available at 10.3390/md23060222.

Oninku, B., Lomas, M.W., Burr, G., Aryee, A.N.A., 2025. Astaxanthin: an overview of its sources, extraction methods, encapsulation techniques, characterization, and bioavailability. Journal of Agriculture and Food Research 21, 101869. Available at https://doi.org/10.1016/j.jafr.2025.101869.

Pabon, M.M., Jernberg, J.N., Morganti, J., Contreras, J., Hudson, C.E., Klein, R.L., et al., 2012. A spirulina-enhanced diet provides neuroprotection in an α-synuclein model of Parkinson's disease. PLoS One 7 (9). Available at https://doi.org/10.1371/journal.pone.0045256, http://www.plosone.org/article/fetchObjectAttachment.action?uri=info%3Adoi%2F10.1371%2Fjournal.pone.0045256&representation=PDF.

Pakan, P.D., Siu, A.C.W., Lee, H., Singh, M., De Rubis, G., Yeung, S., Kulkarni, M.P., Goh, B.H., Hsu, A.C., Chellappan, D., Gupta, G., Yow, Y-Y., Oliver, B.G.G., Paudel, K.R., Dua, K., 2025. Algal bioactives: Unlocking future frontiers in respiratory therapeutics. Food Bioscience 69, 106778. Available at https://doi.org/10.1016/j.fbio.2025.106778.

Patel, A.K., Singhania, R.R., Awasthi, M.K., Varjani, S., Bhatia, S.K., Tsai, M.L., et al., 2021. Emerging prospects of macro- and microalgae as prebiotic. Microbial Cell Factories 20 (1). Available at https://doi.org/10.1186/s12934-021-01601-7, http://www.microbialcellfactories.com/home/.

Peinsipp, P., Gerlza, T., Kircher, J., Zatloukal, K., Jäger, C., Pucher, P., et al., 2025. Antiviral activity of haematococcus pluvialis algae extract is not exclusively due to astaxanthin. Pathogens 14 (8). Available at https://doi.org/10.3390/pathogens14080791, http://www.mdpi.com/journal/pathogens.

Pereira, L., Cotas, J., Valado, A., 2024. Antioxidants from microalgae and their potential impact on human well-being. Exploration of Drug Science 292–321. Available at https://doi.org/10.37349/eds.2024.00048.

Pereira, L., Valado, A., 2025. Beyond nutrition: the therapeutic promise of seaweed-derived polysaccharides against bacterial and viral threats. Marine Drugs 23 (10), 407. Available at https://doi.org/10.3390/md23100407.

Piovan, A., Battaglia, J., Filippini, R., Dalla Costa, V., Facci, L., Argentini, C., et al., 2021. Pre- and early post-treatment with Arthrospira platensis (Spirulina) extract impedes lipopolysaccharide-triggered neuroinflammation in microglia. Frontiers in Pharmacology 12. Available at https://doi.org/10.3389/fphar.2021.724993.

Piovan, A., Filippini, R., Argentini, C., Moro, S., Giusti, P., Zusso, M., 2022. The effect of C-phycocyanin on microglia activation is mediated by toll-like receptor 4. International Journal of Molecular Sciences 23 (3), 1440. Available at https://doi.org/10.3390/ijms23031440.

Pérez-Juárez, A., Chamorro, G., Alva-Sánchez, C., Paniagua-Castro, N., Pacheco-Rosado, J., 2016. Neuroprotective effect of Arthrospira (Spirulina) platensis against kainic acid-neuronal death. Pharmaceutical Biology 54 (8), 1408–1412. Available at https://doi.org/10.3109/13880209.2015.1103756.

Rafiquzzaman, S.M., Kim, E.Y., Lee, J.M., Mohibbullah, M., Alam, M.B., Soo Moon, I., et al., 2015. Anti-Alzheimers and anti-inflammatory activities of a glycoprotein purified from the edible brown alga Undaria pinnatifida. Food Research International 77, 118–124. Available at https://doi.org/10.1016/j.foodres.2015.08.021, www.elsevier.com/inca/publications/store/4/2/2/9/7/0.

Sabat, S., Patra, S., Swain, S., Bej, S., Bishoyi, A., Sahoo, C.R., Padhy, R.N., 2025. Phycocompounds from Cyanobacteria: Exploring Synergistic Effects with Conventional Anticancer and Antimicrobial Properties. ACS Omega 10 (23), 23957–23980. Available at https://doi.org/10.1021/acsomega.5c03526.

Saide, A., Martínez, K.A., Ianora, A., Lauritano, C., 2021. Unlocking the health potential of microalgae as sustainable sources of bioactive compounds. International Journal of Molecular Sciences 22 (9), 4383. Available at https://doi.org/10.3390/ijms22094383.

Salehi, M., Rashidinejad, A., 2025. Multifaceted roles of plant-derived bioactive polysaccharides: A review of their biological functions, delivery, bioavailability, and applications within the food and pharmaceutical sectors. International Journal of Biological Macromolecules 290, 138855. Available at https://doi.org/10.1016/j.ijbiomac.2024.138855.

Shah, M.D., Maran, B.A.V., Shaleh, S.R.M., Zuldin, W.H., Gnanaraj, C., Yong, Y.S., 2022. Therapeutic potential and nutraceutical profiling of north bornean seaweeds: a review. Marine Drugs 20 (2). Available at https://doi.org/10.3390/md20020101, https://www.mdpi.com/1660-3397/20/2/101/pdf.

Shariati, A., Didehdar, M., Razavi, S., Heidary, M., Soroush, F., Chegini, Z., 2022. Natural compounds: a hopeful promise as an antibiofilm agent against candida species. Frontiers in Pharmacology 13. Available at https://doi.org/10.3389/fphar.2022.917787.

Sharma, A., Virmani, T., Kumar, G., Singh, H., Kumar Mishra, A., Bhunia, A., et al., 2025. Exploring sustainable aquatic proteins from algae, seaweed, water lentils for biological characteristics, and commercialization. Applied Food Research 5 (2), 101413. Available at https://doi.org/10.1016/j.afres.2025.101413.

Shin, Y.S., Kim, K.J., Park, H., Lee, M.G., Cho, S., Choi, S.I., et al., 2021. Effects of ecklonia cava extract on neuronal damage and apoptosis in pc-12 cells against oxidative stress. Journal of Microbiology and Biotechnology 31 (4), 581–591. Available at https://doi.org/10.4014/JMB.2012.12013, http://www.jmb.or.kr/journal/view.html?doi=10.4014/jmb.2012.12013.

Silva, J., Alves, C., Pinteus, S., Mendes, S., Pedrosa, R., 2020. Seaweeds' neuroprotective potential set in vitro on a human cellular stress model. Molecular and Cellular Biochemistry 473 (1-2), 229–238. Available at https://doi.org/10.1007/s11010-020-03824-5.

Silva, J., Alves, C., Pinteus, S., Susano, P., Simões, M., Guedes, M., Martins, A., Rehfeldt, S., Gaspar, H., Goettert, M., Alfonso, A., Pedrosa, R., 2021. Disclosing the potential of eleganolone for Parkinson's disease therapeutics: Neuroprotective and anti-inflammatory activities. Pharmacological Research 168, 105589. Available at https://doi.org/10.1016/j.phrs.2021.105589.

Silva, J., Alves, C., Susano, P., Martins, A., Pinteus, S., Gaspar, H., Alfonso, A., Pedrosa, R., 2022. Exploring marine resources against neurological disorders– the neuroprotective and anti-inflammatory potential of the brown seaweed Bifurcaria bifurcata. Journal of Applied Phycology 34 (5), 2671–2688. Available at https://doi.org/10.1007/s10811-022-02794-z.

Skenderidou, I., Leontopoulos, S., Skenderidis, P., 2025. Functional food ingredients enhancing immune health: a systematic review. International Journal of Molecular Sciences 26 (17), 8408. Available at https://doi.org/10.3390/ijms26178408.

Sneha, P., Vijayakumar, S., Devadharshini, D., Wadaan, M.A., Srinivasan, P., Vidhya, E., 2024. Potential Benefits of Bioactive Compounds from Seaweed Extracts: Their Prospects and Pharmacological Promises. Waste and Biomass Valorization 16 (7), 3413–3425. Available at https://doi.org/10.1007/s12649-024-02833-5.

Soliman, T.N., Negm El-Dein, A., Abd Al-Daim, S., Allayeh, A., Awad, H., Flefil, N.S., 2024. Characterization of C-phycocyanin antioxidant, anti-inflammatory, anti-tumour, and anti-HCoV-229E activities and encapsulation for implementation in an innovative functional yogurt. Heliyon 10 (11), e31642. Available at https://doi.org/10.1016/j.heliyon.2024.e31642.

Sonvico, F., Colombo, G., Quarta, E., Guareschi, F., Banella, S., Buttini, F., et al., 2023. Nasal delivery as a strategy for the prevention and treatment of COVID-19. Expert Opinion on Drug Delivery 20 (8), 1115–1130. Available at https://doi.org/10.1080/17425247.2023.2263363.

Tan, K.Y., Low, S.S., Manickam, S., Ma, Z., Banat, F., Munawaroh, H.S.H., et al., 2023. Prospects of microalgae in nutraceuticals production with nanotechnology applications. Food Research International 169. Available at https://doi.org/10.1016/j.foodres.2023.112870, www.elsevier.com/inca/publications/store/4/2/2/9/7/0.

Tiwari, B.K., Garcia-Vaquero, M., 2022. Nutraceuticals from algae: current view and prospects from a esearch perspective. MDPI, Ireland. Marine Drugs 20 (11). Available at https://doi.org/10.3390/md20110671, http://www.mdpi.com/journal/marinedrugs.

Törős, G., El-Ramady, H., Prokisch, J., Velasco, F., Llanaj, X., Nguyen, D.H.H., Peles, F., 2023. Modulation of the Gut Microbiota with Prebiotics and Antimicrobial Agents from Pleurotus ostreatus Mushroom. Foods 12 (10), 2010. Available at https://doi.org/10.3390/foods12102010.

Unger-Manhart, N., Morokutti-Kurz, M., Zieglmayer, P., Lemell, P., Savli, M., Zieglmayer, R., et al., 2024. Carrageenan-containing nasal spray alleviates allergic symptoms in participants with grass pollen allergy: a randomized, controlled, crossover clinical trial. International Journal of General Medicine Volume 17, 419–428. Available at https://doi.org/10.2147/ijgm.s447359.

Venkatesan, J., Anil, S., Kim, S.K., Shim, M.S., 2016. Seaweed polysaccharide-based nanoparticles: preparation and applications for drug delivery. Polymers 8 (2). Available at https://doi.org/10.3390/polym8020030, http://www.mdpi.com/2073-4360/8/2/30/pdf.

Wong Chin, J.M., Puchooa, D., Bahorun, T., Jeewon, R., 2021. Antimicrobial properties of marine fungi from sponges and brown algae of Mauritius. Mycology 12 (4), 231–244. Available at https://doi.org/10.1080/21501203.2021.1895347.

Yayeh, T., Im, E.J., Kwon, T.H., Roh, S.S., Kim, S., Kim, J.H., et al., 2014. Hemeoxygenase 1 partly mediates the anti-inflammatory effect of dieckol in lipopolysaccharide stimulated murine macrophages. International Immunopharmacology 22 (1), 51–58. Available at https://doi.org/10.1016/j.intimp.2014.06.009, www.elsevier.com/locate/intimp.

Ye, Q., Wang, W., Hao, C., Mao, X., 2019. Agaropentaose protects SH-SY5Y cells against 6-hydroxydopamine-induced neurotoxicity through modulating NF-κB and p38MAPK signaling pathways. Journal of Functional Foods 57, 222–232. Available at https://doi.org/10.1016/j.jff.2019.04.017.

Yousof, S.M., Alghamdi, B.S., Alqurashi, T., Alam, M.Z., Tash, R., El-Fadeal, N.M.A., et al., 2025. Cognitive and anti-inflammatory effects of Palmaria palmata in a schizophrenia mouse model: insights into CREB signaling, Iba-1 expression, and CD4+ cell modulation. Frontiers in Neuroscience 19. Available at https://doi.org/10.3389/fnins.2025.1551764, https://www.frontiersin.org/journals/neuroscience#.

Yu, J., Lin, J.-J., Yu, R., He, S., Wang, Q.-W., Cui, W., et al., 2017. Fucoxanthin prevents H 2 O 2 -induced neuronal apoptosis via concurrently activating the PI3-K/Akt cascade and inhibiting the ERK pathway. Food & Nutrition Research 61 (1), 1304678. Available at https://doi.org/10.1080/16546628.2017.1304678.

Zaharudin, N., Salmeán, A.A., Dragsted, L.O., 2018. Inhibitory effects of edible seaweeds, polyphenolics and alginates on the activities of porcine pancreatic α-amylase. Food Chemistry 245, 1196–1203. Available at https://doi.org/10.1016/j.foodchem.2017.11.027, www.elsevier.com/locate/foodchem.

Ziyaei, K., Ataie, Z., Mokhtari, M., Adrah, K., Daneshmehr, M.A., 2022. An insight to the therapeutic potential of algae-derived sulfated polysaccharides and polyunsaturated fatty acids: Focusing on the COVID-19. International Journal of Biological Macromolecules 209, 244–257. Available at https://doi.org/10.1016/j.ijbiomac.2022.03.063.

CHAPTER 6

Algae in biomedical and nutraceutical applications

Vasumathi Krishnan[1,2], Sankaran Krishnamoorthy[3], Rahul Goswami[1,4], Natanamurugaraj Govindan[4], Nadana Raja Vadivu Ganapathy[5], and Chin-Tsan Wang[1,6]

[1]Department of Mechanical and Electromechanical Engineering, National Ilan University, Yilan City, Taiwan
[2]Department of Biotechnology, Kalasalingam Academy of Research and Education, Virudhunagar, Tamil Nadu, India
[3]Department of Biotechnology, Vignan's Foundation for Science, Technology and Research, Guntur, Andhra Pradesh, India
[4]Department of Botany and Microbiology, Hemvati Nandan Bahuguna Garhwal University, Srinagar, Uttarakhand, India
[5]Department of Biotechnology and Chemical Engineering, Manipal University, Jaipur, Rajasthan, India
[6]Department of Chemical Engineering, Indian Institute of Technology Guwahati, Assam, India

6.1 Introduction

Nutrition is increasingly recognized as a central determinant of long-term health, disease prevention, and overall quality of life. Beyond meeting basic dietary requirements, contemporary nutrition science emphasizes the role of functional foods and bioactive compounds in supporting physiological processes associated with chronic, metabolic, and age-related disorders (Ali et al., 2025). At the same time, modern food systems face growing pressures arising from environmental degradation, resource scarcity, and the rising prevalence of lifestyle-related diseases. These interconnected challenges have driven sustained interest in alternative biological resources that support both human health and sustainable production paradigms (Yang et al., 2024; Yusof et al., 2025).

Within this broader context, algae have attracted increasing attention as versatile biological platforms for food, nutraceutical, and biomedical applications. Unlike terrestrial crops, algae can be cultivated in diverse environments, including nonarable land, saline water, and wastewater systems, thereby reducing competition for fertile soil and freshwater resources (Tahir et al., 2024; Yang et al., 2024). High photosynthetic efficiency, rapid growth rates, and year-round productivity further enhance their potential as scalable and resource-efficient bioresources (Ali et al., 2025). In addition, algal cultivation contributes to carbon dioxide assimilation through photosynthesis, linking biomass production to environmental management and climate-mitigation strategies (Daneshvar and Wicker, 2022).

Applied Phycology
DOI: https://doi.org/10.1016/B978-0-443-24758-3.00013-7

A defining characteristic of algae is their exceptional metabolic diversity. Algal biomass encompasses a wide array of primary and secondary metabolites, including proteins, polyunsaturated fatty acids (PUFAs), complex polysaccharides, pigments, vitamins, minerals, and low-molecular-weight bioactive compounds (Yusof et al., 2025). Many of these constituents are directly relevant to human health, supporting physiological functions such as antioxidant defense, immune modulation, lipid metabolism, and inflammatory regulation (Tan et al., 2020; Skjånes et al., 2021). As a result, algae occupy a distinctive position at the interface of nutrition and biomedicine, serving both as nutrient-dense food ingredients and as sources of biologically active compounds with preventive and therapeutic relevance.

Interest in algal proteins as alternatives to conventional plant- and animal-derived proteins reflects both nutritional and sustainability considerations, particularly their favorable amino-acid profiles and high protein content (Ali et al., 2025). Similarly, algal lipids—especially omega-3 PUFAs—have gained prominence for their established roles in cardiovascular and neurological health (Skifa et al., 2025). Photosynthetic pigments such as astaxanthin and β-carotene are widely valued for their antioxidant properties and are incorporated into dietary supplements, functional foods, and cosmetic formulations. Beyond these well-established compound classes, a broader range of algal-derived bioactives has been reported to influence immune responses, inflammatory pathways, and metabolic regulation, with relevance to conditions such as obesity, diabetes, and cardiovascular disease (Skjånes et al., 2021).

Despite these advantages, incorporating algae into functional foods and preventive healthcare frameworks poses several scientific, technological, and regulatory challenges. The biochemical composition of algal biomass is species-dependent and strongly influenced by cultivation conditions, resulting in variability in metabolite profiles and functional performance (Lauersen, 2019; Fabris et al., 2020; Khan et al., 2018; Vavitsas et al., 2018). Structural features such as rigid or recalcitrant cell walls can limit the bioaccessibility of intracellular compounds, while processing and formulation strategies influence stability and bioavailability (Duan et al., 2024). In addition, safety considerations—including the potential accumulation of heavy metals, toxins, or environmental contaminants—necessitate controlled cultivation, rigorous quality assurance, and regulatory oversight to ensure consumer protection (Singh et al., 2021).

From a systems perspective, algae are increasingly integrated into circular-economy and biorefinery concepts in which biomass production is coupled with wastewater treatment, nutrient recovery., and the sequential extraction of high-value compounds (Ali et al., 2025). Such integrated approaches enhance resource efficiency and economic feasibility while reducing environmental burdens. Residual biomass can be further valorized across multiple downstream pathways, reinforcing the role of algae as multifunctional platforms rather than single-product resources (Laurens et al., 2017). Nevertheless, translating these concepts into standardized and widely accepted products requires coordinated advances in cultivation engineering, downstream processing, regulatory harmonization, and evidence-based health validation (Yusof et al., 2025).

Against this backdrop, this chapter examines algae as bioresources for biomedical and nutraceutical applications, with emphasis on the nature and functional relevance of algal metabolites, their roles in human health, and the factors governing their production and utilization. By situating algal biotechnology within broader nutritional, biomedical, and sustainability frameworks, the chapter provides an integrated overview of key opportunities and constraints relevant to the use of algae in preventive healthcare and functional-food systems. The following section introduces the major metabolite classes produced by algae and outlines how cultivation and processing conditions shape their functional availability.

6.2 Algal metabolites

Algae constitute a biologically diverse and sustainable reservoir of primary and secondary metabolites with relevance across nutritional, biomedical, and industrial domains (Kaushik et al., 2023; Yang et al., 2023). Their broad distribution across freshwater, marine, and extreme environments has driven the evolution of flexible metabolic networks capable of synthesizing compounds essential for energy storage, cellular structure, stress tolerance, and ecological defense. This section presents algal metabolites from a functional perspective, emphasizing their roles as bioresources shaped by species characteristics, environmental conditions, and cultivation strategies (Yang et al., 2024; Kaur et al., 2022).

6.2.1 Major categories of algal metabolites

The biochemical composition of algal biomass is inherently variable and reflects genetic background, environmental conditions, and cultivation regime (Bhatnagar et al., 2024). From a functional perspective, algal metabolites can be grouped into proteins and amino acids; lipids and fatty acids; carbohydrates and polysaccharides; pigments; and a diverse range of secondary metabolites. The hierarchical relationships among these metabolite classes are illustrated schematically in Fig. 6.1, which provides a structural overview of primary and secondary algal metabolites.

6.2.1.1 Proteins and amino acids

Algae are recognized as high-quality protein sources, with protein content typically accounting for approximately 30% of dry biomass and reaching 55%–60% in cyanobacteria such as *Spirulina* (Bhatnagar et al., 2024). Several microalgae, including *Chlorella* species, provide balanced profiles of essential amino acids, supporting their use in dietary supplements and functional food formulations (Bhatnagar et al., 2024). In addition to nutritional proteins, algae synthesize functionally specialized proteins with biomedical relevance. Phycobiliproteins such as phycocyanin and phycoerythrin function as natural pigments and fluorescent probes and also exhibit antioxidant and antiinflammatory properties (Bhatnagar et al., 2024). The effective utilization of algal proteins depends on digestibility, cell wall structure, and downstream processing, which together determine protein accessibility and bioavailability (Green et al., 2022).

6.2.1.2 Lipids and fatty acids

Algal lipids include triacylglycerols, phospholipids, and glycolipids that contribute to membrane integrity, energy storage, and stress adaptation (Bhatnagar et al., 2024). Of particular interest are PUFAs such as

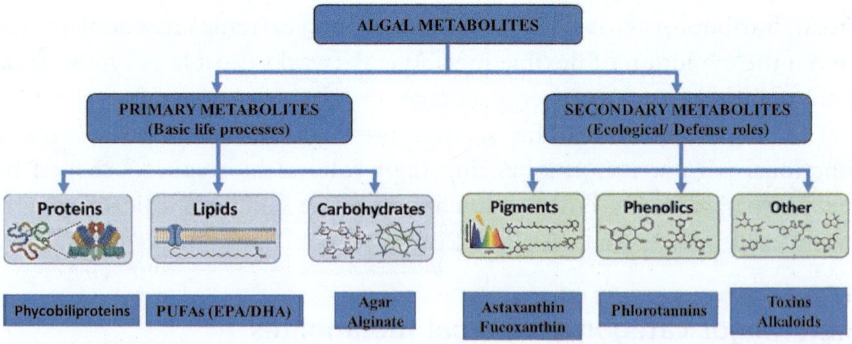

Figure 6.1 Hierarchical classification of primary and secondary algal metabolites.

eicosapentaenoic acid (EPA) and docosahexaenoic acid (DHA), which support cardiovascular health, neurological development, and inflammatory regulation (Dyall, 2015; Yang et al., 2023). Total lipid content typically ranges from 20% to 40% of dry biomass and may increase under stress conditions such as nutrient limitation (Kumar et al., 2022). In biorefinery systems, algal lipids support both bulk applications and high-value nutraceutical products, while additional lipid-derived compounds, including sterols and terpenes, further expand functional potential (Bhatnagar et al., 2024).

6.2.1.3 Carbohydrates and polysaccharides

Algae synthesize storage carbohydrates and structurally diverse polysaccharides that are often taxonomically specific (Moreira et al., 2023; Muhamad et al., 2019; Waoo et al., 2025). Examples include agar, carrageenan, and porphyran in red algae; alginate, fucoidan, and laminarin in brown algae; and ulvan in green algae (Moreira et al., 2023; Waoo et al., 2025).

These polysaccharides exhibit anticoagulant, antiviral, immunomodulatory, and anticancer activities and are applied in nutraceuticals, drug-delivery systems, and tissue engineering (Moreira et al., 2023; Muhamad et al., 2019; Waoo et al., 2025; Rahmati et al., 2019). Algal β-glucans also serve as prebiotics that support gut health and metabolic regulation (Rahmati et al., 2019). Biological performance depends on molecular weight, sulfation pattern, and extraction methodology (Jagtap et al., 2025).

6.2.1.4 Pigments

Photosynthetic pigments produced by algae include chlorophylls, carotenoids, and phycobiliproteins, which function in light harvesting and photoprotection (Bhatnagar et al., 2024). Many algal pigments also exhibit antioxidant activity and contribute to cellular protection against oxidative stress (Bhatnagar et al., 2024; Sharma et al., 2023). Carotenoids such as β-carotene, astaxanthin, and fucoxanthin are of particular biomedical interest. Astaxanthin from *Haematococcus pluvialis* exhibits high antioxidant activity, while fucoxanthin from brown algae has demonstrated anticancer potential (Bhatnagar et al., 2024; Sharma et al., 2023).

6.2.1.5 Other secondary metabolites

Algae also produce secondary metabolites, including phenolics, alkaloids, volatile organic compounds, and bioactive peptides that function in ecological defense (Yang et al., 2023; Wu et al., 2023). Many of these compounds exhibit antioxidant, antimicrobial, antiproliferative, and cyto-protective activities, underscoring their biomedical relevance (Yang et al., 2023; Kaushik et al., 2023; Sharma et al., 2023).

6.2.2 Functional mapping of algal metabolites to biomedical applications

The biomedical relevance of algal metabolites becomes clearer when their biochemical composition is considered alongside their biological functions and application contexts. Although metabolite classes are often described as separate categories, their significance in health-related applications lies in the functional roles they play within physiological systems. To provide an integrated overview, Table 6.1 presents the major algal metabolite classes, representative compounds, their principal biological functions, and associated biomedical applications.

It illustrates how structurally diverse metabolites, including carotenoids, phycobiliproteins, PUFAs, sulfated polysaccharides, peptides, and phenolic compounds, contribute to shared biological processes such as redox regulation, immune modulation, inflammation control, and tissue repair. In addition, emerging applications in wound healing, tissue regeneration, and metabolic regulation demonstrate the expanding scope of algal bioactives beyond conventional nutraceutical uses. This functional mapping highlights the breadth of biomedical potential associated with algal-derived compounds.

6.2.3 Factors influencing metabolite production

The biosynthesis and accumulation of algal metabolites are governed by dynamic interactions between environmental conditions and cultivation strategies (Yang et al., 2023; Bhatnagar et al., 2024). Environmental parameters such as light intensity and spectral quality, temperature, nutrient availability, salinity, pH, and carbon dioxide concentration exert strong control over metabolic fluxes and resulting metabolite profiles (Bhatnagar et al., 2024). Variations in these factors can alter the relative allocation of cellular resources toward proteins, lipids, carbohydrates, pigments, or secondary metabolites.

Nutrient availability is a particularly influential driver of metabolite production. Nutrient limitation, especially nitrogen deprivation, is widely used to stimulate lipid accumulation and secondary metabolite synthesis,

Table 6.1 Major algal metabolite classes and their principal biological functions and biomedical applications.

Algal source	Metabolites classification	Algal metabolites	Role of agents	References
Haematococcus pluvialis, *Undaria pinnatifida*	Pigments	Carotenoids (β-carotene, astaxanthin, and fucoxanthin)	**Antioxidant activity:** Potent radical scavenging; activation of the Nuclear factor erythroid 2-related factor 2 (Nrf2) pathway and upregulation of antioxidant and detoxifying enzymes.	Matin et al. (2024); Deepa and Subha (2025); Guil-Guerrero and Prates (2025); Prates (2025)
			Antiinflammatory signaling: Suppression of Nuclear factor kappa B (NF-κB) and mitogen-activated protein kinase (MAPK) pathways; reduced pro-inflammatory cytokine expression.	Matin et al. (2024); Deepa and Subha (2025)
			Anticancer activity (preclinical): Induction of caspase-dependent apoptosis, modulation of cell-cycle regulators, and inhibition of tumor-cell proliferation.	Mirza et al. (2025); Deepa and Subha (2025); Prates (2025)
Spirulina sp.	Proteins	Phycobiliproteins (phycocyanin, phycoerythrin, and allophycocyanin)	**Immunomodulation:** Suppression of pro-inflammatory cytokines (TNF-α, IL-1β, IL-6) and enhancement of	Reddy et al. (2000); Romay and Gonzalez (2003)

(*Continued*)

Table 6.1 (Continued)

Algal source	Metabolites classification	Algal metabolites	Role of agents	References
			antiinflammatory IL-10. Also inhibits Nuclear factor kappa B (NF-κB) activation and downregulates Cyclooxygenase 2 (COX-2) and inducible nitric oxide synthase (iNOS) enzymes in immune cells.	
			NF-κB/COX-2 inhibition: Selective COX-2 inhibition and downregulation of inducible nitric oxide synthase (iNOS) without affecting Cyclooxygenase 1 (COX-1).	Reddy et al. (2000); Bhat and Madyastha (2001)
			Antioxidant defense: Scavenging of reactive oxygen species (ROS)/reactive nitrogen species (RNS); phycocyanobilin acts analogously to bilirubin.	Reddy et al. (2000); Bhat and Madyastha (2001); Romay and Gonzalez (2003)
			Immune regulation: Promotion of M2 macrophage polarization, enhancement of humoral and cellular immunity, stimulation of natural killer (NK) cell	Reddy et al. (2000); Wu and Liu, 2016; Jia et al. (2025)

			activity, and activation of Nrf2-antioxidant response element (Nrf2-ARE) signaling.	
Schizochytrium sp., Nannochloropsis sp.	Lipids	(Polyunsaturated fatty acids; eicosapentaenoic acid; docosahexaenoic acid)	**Cardiovascular health:** Reduction of plasma triglycerides via inhibition of very-low-density lipoprotein (VLDL) synthesis and enhanced β-oxidation; improved endothelial function through endothelial nitric oxide synthase (eNOS) activation; and antithrombotic and antiarrhythmic effects.	Bays (2006); Bercea et al. (2021); Harris et al. (2009); Mozaffarian and Wu (2011)
			Neural health: Structural support of neuronal membranes; enhanced synaptic plasticity and neurogenesis.	Innis (2007); Bazinet and Layé (2014)
			Neuroprotection: Production of resolvins and neuroprotectin D1, reducing neuroinflammation and oxidative stress; benefits for cognition and mood.	Serhan and Chiang, 2008; Bazan (2009); Freeman et al. (2006); Su et al. (2015)

(*Continued*)

Table 6.1 (Continued)

Algal source	Metabolites classification	Algal metabolites	Role of agents	References
Fucus vesiculosus, U. pinnatifida, Chondrus crispus, Kappaphycus alvarezii	Polysaccharides	Sulfated polysaccharides: (fucoidan and carrageenan)	**Antiviral activity:** Inhibition of viral attachment and entry via interaction with viral envelope glycoproteins; efficacy reported against herpes simplex virus (HSV), human immunodeficiency virus (HIV), influenza, human papillomavirus (HPV), and Severe acute respiratory syndrome coronavirus 2 (SARS-CoV-2) (in vitro/clinical formulations).	Damonte et al. (2004); Grassauer et al. (2008); Morokutti-Kurz et al. (2021)
			Immunostimulation: Activation of macrophages, NK cells, and interferon production.	Fitton et al. (2015); Hayashi et al. (2013)
			Anticoagulant effects: Heparin-like interactions with antithrombin III; inhibition of thrombin and factor Xa.	Chevolot et al. (2001); Berteau (2003); Lichota et al. (2020); Ale et al. (2011); Chandika et al. (2022)
			Anticancer activity (preclinical): Induction of apoptosis, cell-cycle arrest,	Banack et al. (2007); Park et al. (2011); Campo et al. (2009); Yuan et al. (2006);

			immune-mediated tumor suppression, and synergistic enhancement of chemotherapeutic efficacy.	Jin et al. (2022); Taghipour et al. (2025)
Halymenia durvillei,	Phenolics and peptides	Peptide compounds	**Antimicrobial activity:** Membrane disruption, inhibition of DNA/RNA/protein synthesis, endotoxin neutralization, and quorum-sensing interference.	Patra et al. (2022); Chitthan et al. (2025); Rojas et al. (2020); Rajesh et al. (2023); Boominathan et al. (2022)
			Cytoprotection: Activation of Nrf2-antioxidant response element (Nrf2-ARE signaling, enhancement of antioxidant enzymes, mitochondrial stabilization, and preservation of epithelial barrier integrity.	Zhang et al. (2025); Sun et al. (2022); Gaviria et al. (2024)
		Phenolic compounds	**Antimicrobial effects:** Increased membrane permeability; inhibition of DNA gyrase, Adenosine triphosphate (ATP) synthase, and fungal ergosterol synthesis.	Rabea et al. (2025); Mandal and Domb (2024); Besednova et al. (2020)
			Cytoprotective activity: Nrf2 activation, Nuclear factor kappa B (NF-κB) suppression, modulation of Bcl-2/Bax	Besednova et al. (2020); Mikhailova (2023); Yarnpakdee et al. (2019)

(Continued)

Table 6.1 (Continued)

Algal source	Metabolites classification	Algal metabolites	Role of agents	References
			balance, inhibition of lipid peroxidation via metal chelation.	
		Emerging roles (wound healing, tissue regeneration, and metabolic regulation)	**Wound healing and tissue regeneration**: Hydrogel-forming polysaccharides (alginate, fucoidan, carrageenan, ulvan) support moisture retention, oxygenation, angiogenesis, and infection control; bioactive peptides and pigments promote regenerative immune responses.	Zhong et al. (2021); Kang et al. (2024); Obaíd et al. (2021); Behzadnia et al. (2024); Pandian et al. (2024)
			Metabolic regulation: Modulation of glucose and lipid metabolism via antioxidant and antiinflammatory pathways; emerging systems biology approaches enable targeted metabolite production for metabolic health.	Ibrahim et al. (2023); Enayati et al. (2022)

whereas nutrient-rich wastewater streams support high biomass productivity and enable simultaneous bioremediation (Kumar et al., 2022; Javed et al., 2024; Pierson et al., 2023). At the system and engineering level, cultivation design, strain selection, downstream processing, and emerging genetic and metabolic engineering approaches collectively determine metabolite yield, purity, and reproducibility (Kaur et al., 2022; Goold et al., 2024).

Advances in omics technologies have improved understanding of the biosynthetic pathways and regulatory networks involved in algal metabolite production (Yang et al., 2023; Goold et al., 2024). Transcriptomic studies have identified stress-responsive pathways associated with enhanced carotenoid and astaxanthin biosynthesis in species such as *H. pluvialis* and *Chlamydomonas zofingiensis* (Yang et al., 2023). Building on these insights, synthetic biology approaches aim to redesign algal genomes to enable application-specific metabolite production, thereby improving yield control and consistency (Goold et al., 2024).

Collectively, algal metabolites encompass a wide range of chemical structures and biological activities, underscoring their relevance to nutraceutical and biomedical applications. At the same time, intrinsic bioactivity alone is not sufficient to ensure practical implementation. Metabolite accessibility, production consistency, and scalability, shaped by biological characteristics and engineering constraints, ultimately determine translational utility. By integrating metabolite classification with functional mapping and production considerations, this section establishes a conceptual foundation for subsequent discussion of biomedical applications. With this framework in place, the following section examines how specific algal metabolite classes translate into defined biomedical functions and application domains.

6.3 Biomedical applications of algae

Algae represent a rich and renewable source of bioactive compounds with broad biomedical relevance (Yang et al., 2024). Increasing global health challenges, including chronic noncommunicable diseases, emerging infectious threats, and rising antimicrobial resistance, have intensified interest in therapeutic strategies that combine biological efficacy with favorable safety and sustainability profiles (Sharma et al., 2023). In this context, algal metabolites have attracted attention for their chemical diversity, multifunctional biological activities, and generally lower toxicity than many synthetic agents (Srinivasan et al., 2022).

Marine algae, in particular, synthesize structurally distinctive secondary metabolites shaped by variable and often extreme environmental conditions, resulting in compounds with potent pharmacological properties. Across biomedical contexts, algal bioactivities frequently converge on shared molecular processes, including antioxidant defense, inflammatory signaling, immune modulation, mitochondrial regulation, and cell-survival pathways. As a result, antioxidant, antiinflammatory, antimicrobial, and anticancer effects are often mechanistically interconnected. The following subsections, therefore, emphasize application-specific outcomes while maintaining a consistent mechanistic framework.

6.3.1 Anticancer activities of algal metabolites

Cancer remains a major global health burden, and current therapeutic approaches are often limited by systemic toxicity, off-target effects, and the emergence of drug resistance (Sharma et al., 2023). Algal-derived compounds have demonstrated anticancer activity through multiple complementary mechanisms, including induction of apoptosis, inhibition of tumor-cell proliferation, suppression of angiogenesis, and modulation of metastatic pathways (Sharma et al., 2023). Carotenoids, phycobiliproteins, and sulfated polysaccharides are among the most extensively investigated algal-derived compounds with anticancer activity. Fucoxanthin from brown algae induces apoptosis and cell-cycle arrest in prostate, colon, and leukemia models by modulating Akt/mechanistic target of rapamycin (mTOR) and mitogen-activated protein kinase (MAPK) signaling pathways (Sharma et al., 2023). Astaxanthin exhibits antiproliferative effects in breast and gastric cancer cells, while phycocyanin derived from cyanobacteria promotes apoptosis in chronic myeloid leukemia and other malignancies (Srinivasan et al., 2022). Sulfated polysaccharides, such as fucoidans and carrageenans, further inhibit tumor progression by regulating immune responses, angiogenesis, and apoptotic signaling, supporting their potential use as adjuncts or as combinatorial agents in cancer therapy (Waoo et al., 2025; Sharma et al., 2023).

6.3.2 Antioxidant and antiinflammatory properties

Oxidative stress and chronic inflammation are central contributors to numerous pathological conditions, including cardiovascular, neurodegenerative, metabolic, and oncological disorders (Srinivasan et al., 2022). Algae are abundant sources of antioxidant compounds, including carotenoids, phenolics, vitamins, and phycobiliproteins, which collectively support antioxidant defense and cellular protection. Astaxanthin, particularly from *H. pluvialis*, exhibits exceptionally high antioxidant capacity, exceeding that of vitamin E and β-carotene, and protects cellular membranes and nucleic acids from oxidative damage (Kaur et al., 2022). In parallel, algal metabolites exert antiinflammatory effects by suppressing key mediators, such as TNF-α and IL-6, and inhibiting NF-κB signaling pathways. Fucoidans and phycocyanin have therefore been explored for the management of chronic inflammatory and autoimmune conditions (Srinivasan et al., 2022).

6.3.3 Antimicrobial and antiviral effects

The rapid emergence of antimicrobial resistance has intensified interest in alternative bioactive agents derived from natural sources. Algae synthesize a wide range of antimicrobial compounds with activity against bacteria, fungi, and viruses. Sulfated polysaccharides, including fucoidans, carrageenans, and ulvans, exhibit antiviral activity by interfering with viral attachment, entry, and replication processes (Waoo et al., 2025). These compounds have demonstrated inhibitory effects against enveloped viruses such as Herpes simplex virus (HSV), Human immunodeficiency virus (HIV), and Severe acute respiratory syndrome coronavirus 2 (SARS-CoV-2). In addition, algal extracts and metabolites exhibit antibacterial activity against multidrug-resistant strains, including methicillin-resistant *Staphylococcus aureus* (MRSA), through mechanisms such as membrane disruption, enzyme inhibition, and oxidative imbalance. Taken together, these findings highlight algae as a valuable source of antimicrobial and antiviral compounds at a time when conventional therapeutic pipelines are under increasing pressure from resistance (Zuorro et al., 2024).

6.3.4 Immunomodulatory effects

Algal polysaccharides play a significant role in immune regulation by stimulating both innate and adaptive immune responses (Muhamad et al., 2019; Waoo et al., 2025). Compounds such as fucoidan and β-glucans activate macrophages, natural killer (NK) cells, and lymphocytes, thereby enhancing host defense against infections and tumor development. Their prebiotic properties further contribute to immune homeostasis by modulating the composition of the gut microbiota (Moreira et al., 2023). From an applied perspective, these immunomodulatory effects provide a rationale for incorporating selected algal polysaccharides into immunonutrition strategies and adjunct formulations to support immune resilience (Hwang et al., 2022).

6.3.5 Neuroprotective potential

Neurodegenerative disorders are characterized by oxidative stress, chronic inflammation, mitochondrial dysfunction, and synaptic impairment. Algal metabolites such as astaxanthin, β-carotene, and omega-3 PUFAs target these pathological processes and exhibit neuroprotective effects (Yang et al., 2023). DHA, a major structural component of neuronal membranes, supports synaptic plasticity and cognitive function, while antioxidant carotenoids mitigate neuronal oxidative damage. The combined contributions of membrane support via DHA and oxidative protection via carotenoids underpin growing interest in algal-derived ingredients for neuroprotection-oriented nutrition and healthy aging applications (Weiser et al., 2016).

6.3.6 Tissue engineering and regenerative medicine

Algal polysaccharides such as alginate, agar, and carrageenan exhibit high biocompatibility and tunable mechanical properties, supporting their use in tissue engineering and regenerative medicine (Moreira et al., 2023; Rahmati et al., 2019). Alginate hydrogels are widely applied in cell encapsulation, wound healing, and cartilage regeneration, while agar and carrageenan are employed in scaffold fabrication and controlled drug-delivery systems. Chemical and structural modification of these polymers enables tailored interactions with cells, bioactive molecules, and growth factors, thereby enhancing tissue regeneration and repair outcomes (Muhamad et al., 2019).

6.3.7 Green synthesis of nanoparticles and drug delivery

Algae also provide environmentally benign routes for synthesizing metallic nanoparticles by using naturally occurring reducing and capping agents present in algal extracts (Chaudhary et al., 2020). These biogenic nanoparticles exhibit improved biocompatibility and reduced toxicity compared with conventionally synthesized counterparts, supporting applications in targeted drug delivery, diagnostics, antimicrobial therapy, and cancer treatment.

Integration of biogenic nanoparticles with algal polysaccharides enables controlled drug release and enhanced therapeutic performance, further expanding the biomedical utility of algal-derived materials (Moreira et al., 2023). To visually summarize the breadth of biomedical applications discussed in this section, Fig. 6.2 illustrates the roles of key algal bioactive compounds, including astaxanthin, fucoidan, alginate, phycocyanin, and PUFAs, in the prevention and treatment of neurological disorders, cancer, tissue damage, inflammation, and cardiovascular diseases (Yang et al., 2023).

Collectively, the diverse bioactivities of algal metabolites underscore their promise as sustainable resources for biomedical innovation. Advances in omics technologies and metabolic engineering are expected to refine the targeted production of high-value compounds (Yang et al., 2023; Ahmad et al., 2022). Translation into clinical and pharmaceutical practice will depend on rigorous preclinical validation, well-designed clinical studies,

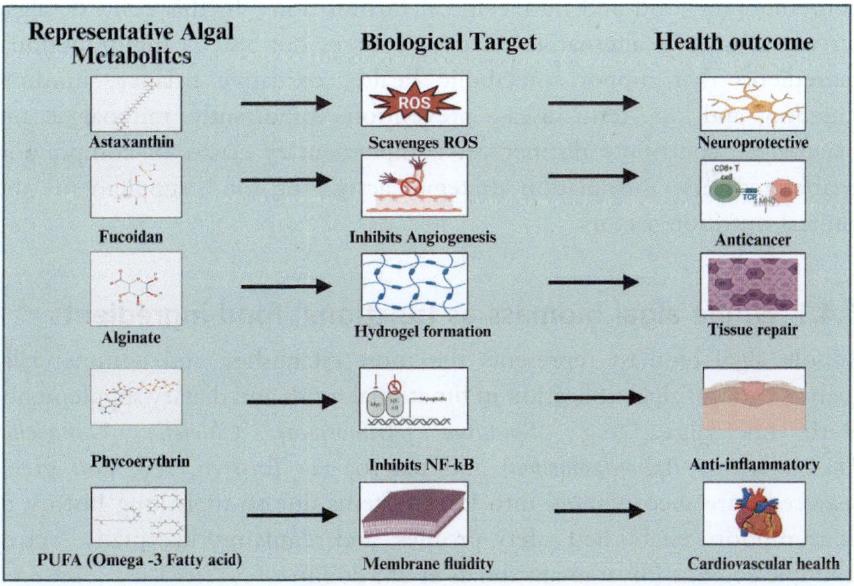

Figure 6.2 Representative algal metabolites, their primary biological targets, and associated health outcomes. PubChem CID No: Astaxanthin (5281224), Fucoidan (92023653), Alginate (91666324), Phycoerythrin (238) and Omega-3 Fatty Acid (56842239).

and scalable production strategies to ensure safety, efficacy, and reproducibility.

6.4 Functional foods and dietary supplements

Algae, encompassing both microalgae and macroalgae, are increasingly recognized as strategic bioresources for developing functional foods and dietary supplements due to their high nutritional density, biochemical diversity, and broad spectrum of health-promoting properties. Unlike conventional plant- or animal-derived ingredients, algal biomass integrates proteins, essential fatty acids, polysaccharides, pigments, vitamins, minerals, and specialized bioactive compounds within compact, highly productive biological systems. Their rapid growth rates, ability to utilize nonarable land, and compatibility with controlled cultivation and marine harvesting position algae as sustainable contributors to global nutritional security and preventive healthcare strategies.

A growing body of experimental, clinical, and industrial evidence supports the incorporation of algal biomass and derived fractions into contemporary food and nutraceutical formulations. In this context, algae serve not only as alternative nutrient sources but also as multifunctional ingredients that support metabolic health, oxidative balance, immune function, and long-term disease prevention. Importantly, microalgae and macroalgae contribute distinct yet complementary classes of compounds, enabling diverse formulation strategies across the food, supplement, and clinical nutrition sectors.

6.4.1 Whole algal biomass as functional food ingredients

Whole algal biomass represents the most established and commercially mature form of algal utilization in functional foods and dietary supplements. Both microalgae (e.g., *Spirulina [Arthrospira], Chlorella, Dunaliella, Haematococcus, Aphanizomenon*) and macroalgae (brown, red, and green seaweeds) are incorporated into food systems due to their long history of consumption, established safety profiles, and regulatory acceptance across multiple regions (Sivaramakrishnan et al., 2025).

Unlike isolated compounds, whole-biomass formulations deliver complex molecular matrices in which proteins, lipids, polysaccharides, pigments, and micronutrients coexist and may exert additive or synergistic physiological effects (Guil-Guerrero and Prates, 2025). This integrative nutritional profile underpins their classification as functional foods rather than as single-nutrient supplements.

Among microalgae, Spirulina is one of the most widely consumed species and is valued for its high protein content, favorable amino-acid composition, and abundance of micronutrients (Ahmad et al., 2023). It is incorporated into powders, tablets, bakery products, beverages, dairy formulations, and snack foods, facilitating routine dietary intake. Clinical and observational studies associate Spirulina supplementation with improvements in lipid metabolism and metabolic markers relevant to cardiovascular disease and type 2 diabetes (Gouveia et al., 2007; Carvalho et al., 2017), effects commonly attributed to its protein quality and bioactive pigments such as C-phycocyanin (Bishop and Zubeck, 2012).

Chlorella is similarly marketed as whole biomass and is characterized by a dense nutritional composition that includes proteins, chlorophyll, dietary fiber, and a wide spectrum of vitamins (Naguib, 2000). Reported benefits include gastrointestinal support, blood-pressure regulation, and cardioprotective effects, although outcomes vary depending on strain selection, dosage, and processing conditions (Nicoletti, 2016; Kumari et al., 2023).

Macroalgae are commonly consumed as whole foods or minimally processed ingredients, particularly within East Asian dietary traditions. Brown seaweeds (e.g., Laminaria, Undaria, and Fucus) provide dietary fiber and bioactive polysaccharides such as alginate and fucoidan, while red seaweeds (e.g., Porphyra and Gracilaria) contribute proteins, minerals, and sulfated galactans, including agar and carrageenan. These seaweeds are increasingly incorporated into functional foods due to their contributions to satiety, glycemic regulation, and modulation of gut health (Zang et al., 2023).

Some algae are cultivated primarily for targeted metabolite recovery rather than direct whole-biomass consumption. *H. pluvialis* exemplifies this approach as the dominant natural source of astaxanthin, a carotenoid with exceptionally strong antioxidant capacity (Naguib, 2000; Yuan et al., 2011). Human intervention studies indicate good tolerability and suggest protective roles in visual, neurological, and age-associated contexts of oxidative stress (Satoh et al., 2009). Overall, while whole-biomass algal products offer nutritional completeness and accessibility, variability in biochemical composition, limitations in bioavailability associated with rigid cell walls, and processing-dependent stability increasingly motivate selective fractionation, enrichment, or formulation–optimization strategies (Sanjeewa et al., 2023; Magnabosco et al., 2025).

6.4.2 Algal-derived omega-3 fatty acids in functional nutrition

Algae represent the primary biological source of omega-3 PUFAs in aquatic ecosystems and provide a sustainable alternative to fish-derived oils. DHA and EPA play essential roles in cardiovascular health, neurodevelopment, and the regulation of inflammation (Norzagaray-Valenzuela et al., 2017). Concerns related to overfishing, marine ecosystem degradation, and contaminant accumulation in fish oils have further increased interest in algal-derived omega-3 production (Topuz, 2016).

Several microalgal species, including Schizochytrium, Nannochloropsis, *Crypthecodinium cohnii*, and *Phaeodactylum tricornutum*, are efficient producers of DHA and EPA and can be cultivated at an industrial scale using photobioreactors, heterotrophic fermentation, or hybrid systems (Norzagaray-Valenzuela et al., 2017). Compared with fish oils, algal omega-3 products offer higher purity, reduced risk of heavy-metal contamination, and suitability for vegetarian and vegan consumers (Zhu et al., 2020).

Algal-derived DHA and EPA are now widely incorporated into functional foods, infant formulas, and dietary supplements, supporting cardiovascular, cognitive, and inflammatory health outcomes across different life stages (Innes and Calder, 2020; Bhat et al., 2023; De Bhowmick et al., 2023; Beuzenberg et al., 2014). Their increasing adoption reflects a broader transition toward sustainable lipid sources with established physiological relevance.

6.4.3 Antioxidant and metabolic health nutraceuticals

Algal pigments and phenolic compounds constitute an important class of nutraceutical ingredients relevant to oxidative balance, metabolic regulation, and lifestyle-related health maintenance. Both microalgae and macroalgae produce carotenoids, including astaxanthin, fucoxanthin, β-carotene, and lutein, which protect cellular components from oxidative damage and are associated with reduced risk of chronic disorders such as cardiovascular disease, diabetes, neurodegenerative conditions, and cancer (Swapnil et al., 2021).

Beyond antioxidant activity, algal carotenoids influence lipid metabolism, insulin sensitivity, mitochondrial efficiency, and energy homeostasis. Astaxanthin has been associated with enhanced mitochondrial performance and physical endurance; fucoxanthin promotes fat oxidation and thermogenesis; while β-carotene and lutein support immune competence and ocular health. In macroalgae, additional antioxidant and metabolic effects arise from phlorotannins and sulfated polysaccharides, which contribute to glycemic control and modulation of inflammation (Gunathilaka et al., 2020).

These multifunctional properties underpin the growing incorporation of algal bioactives into functional foods, fortified beverages, dietary supplements, and lifestyle nutrition products. Collectively, they illustrate how algae, across both microalgal and macroalgal systems, have progressed from niche supplements to integrated platforms for functional nutrition.

6.5 Safety, toxicology, and regulatory considerations

The successful integration of algae into functional foods, dietary supplements, and biomedical applications depends not only on demonstrated bioactivity but also on well-defined safety profiles, toxicological evaluations, and regulatory compliance. Although many algal species have long histories of dietary use, safety outcomes are strongly influenced by species selection, cultivation environments, processing practices, and consumption patterns. This section outlines key safety considerations for algal consumption and summarizes the regulatory frameworks governing algal-derived products in major global markets.

6.5.1 Safety and toxicological aspects

6.5.1.1 Historical use and general safety profile of edible algae

Numerous edible algae, including microalgae such as Spirulina (Arthrospira) and Chlorella, as well as macroalgae such as kelps (Laminaria, Undaria), red seaweeds (Porphyra), and green seaweeds (Ulva), have been consumed for centuries, particularly within East Asian and coastal dietary traditions. This extensive history of consumption, supported by experimental and clinical investigations, underpins their general recognition as safe when produced or harvested under controlled conditions.

Regulatory authorities such as the US Food and Drug Administration (FDA) and the European Food Safety Authority (EFSA) have approved selected algal species, including Spirulina, Chlorella, and *H. pluvialis*, for use in foods and dietary supplements, subject to compliance with defined quality and production standards. Several macroalgal species are also authorized as traditional foods or food ingredients in multiple jurisdictions, reflecting their established consumption history.

From a nutritional perspective, edible algae provide dense sources of essential amino acids, vitamins, minerals, dietary fiber, and antioxidant compounds. Under optimal conditions, Spirulina and Chlorella typically contain 50%–70% protein by dry weight, with amino acid profiles comparable to those of conventional protein sources (Mosibo et al., 2024; Wani et al., 2023). Macroalgae contribute lower protein levels but are rich in polysaccharides, trace minerals, including iodine, iron, and calcium, and bioactive sulfated compounds. Collectively, these components support the reported benefits in cardiovascular health, cognitive function, immune modulation, and inflammatory regulation (Gao et al., 2024; Martínez-Ruiz et al., 2025).

6.5.1.2 Toxicological risks associated with algal consumption

Despite these advantages, algal safety is strongly influenced by environmental exposure and production control. A primary toxicological concern for both microalgae and macroalgae is the bioaccumulation of heavy metals. Algal cells readily absorb elements such as aluminum, cadmium, lead, and mercury from the surrounding water and growth media. Chronic exposure to these metals may result in toxicity affecting the digestive, hematopoietic, cardiovascular, respiratory, and nervous systems, with some metals exhibiting carcinogenic potential under prolonged exposure (Gnonsoro et al., 2022).

Empirical studies indicate that Spirulina and Chlorella can accumulate metals such as lead and cadmium at concentrations that exceed those in their cultivation environments when grown in contaminated water sources (Faruque et al., 2024; Malletzidou et al., 2025). Similarly, macroalgae harvested from polluted coastal zones may accumulate heavy metals and other marine contaminants, underscoring the importance of monitoring harvest sites and ensuring traceability (Rakib et al., 2021).

Another safety consideration is iodine overload, particularly associated with brown macroalgae that naturally contain high iodine concentrations. Excessive iodine intake can disrupt thyroid hormone synthesis and regulation, potentially leading to hypo or hyperthyroidism when consumption exceeds recommended limits (Singh et al., 2021). This risk is especially relevant for individuals with preexisting thyroid disorders and highlights the importance of portion control, product labeling, and consumer guidance.

Cyanotoxins represent an additional hazard primarily associated with certain cyanobacterial strains. Some cyanobacteria produce hepatotoxic compounds such as microcystins, which may cause liver damage and systemic toxicity upon ingestion. Although commercially cultivated edible strains are generally nontoxic, contamination by toxin-producing species can occur in poorly managed open systems, reinforcing the need for strain authentication and routine toxin surveillance (Gnonsoro et al., 2022).

6.5.1.3 Cultivation control, strain selection, and dosage considerations

Accordingly, the safety profile of edible algae is closely linked to strain selection, cultivation strategy, and harvesting practices. Controlled cultivation systems, particularly closed or semi-closed photobioreactors used for microalgae, substantially reduce the risk of contamination by heavy metals, pharmaceuticals, pesticides, and other environmental pollutants. In contrast, open-pond systems and wild-harvested macroalgae are more vulnerable to contamination from wastewater inputs, agricultural runoff, and industrial effluents.

Selecting strains with low toxin production and low metal accumulation potential, combined with rigorous quality-control protocols, is essential for regulatory compliance and consumer protection. For macroalgae, responsible site selection, seasonal harvesting strategies, and post-harvest processing controls are equally critical (Nicoletti, 2016).

Consumption-related risks are also dose-dependent. Moderate intake of algae-based products is generally considered safe and beneficial, whereas excessive or prolonged consumption may increase the risk of heavy-metal bioaccumulation, iodine overload, or inadvertent exposure to toxins. Vulnerable populations, including children, pregnant women, older adults, and individuals with chronic health conditions, require particular caution. Long-term safety assessments should therefore integrate estimated daily intake (EDI), tolerable weekly intake (TWI), and bioaccumulation potential to inform evidence-based consumption guidelines (Almomani and Bhosale, 2021). Beyond product-level toxicology, regulatory and quality-assurance frameworks ultimately determine whether algae-based foods and nutraceuticals can achieve large-scale adoption and sustained consumer trust.

6.5.2 Regulatory and quality considerations

Regulatory treatment of edible algae and algal-derived products varies across global markets, reflecting differences in dietary traditions, regulatory philosophies, and risk-assessment approaches. In the United States, widely consumed microalgae such as Spirulina, Chlorella, and *H. pluvialis* are recognized as Generally Recognized as Safe (GRAS) by the FDA, permitting their use in foods and dietary supplements under defined conditions. Depending on processing methods, concentration, and intended use, algal ingredients may also fall under the Food, Drug, and Cosmetic Act or require individual GRAS evaluations.

In the European Union, a more precautionary approach is applied through the Novel Foods Regulation (EU) 2015/2283, which mandates premarket authorization for foods not widely consumed before May 15, 1997. While Spirulina and Chlorella are exempt due to their established consumption history, other microalgae, macroalgae, and concentrated extracts require a comprehensive safety and toxicological assessment prior to authorization. EFSA has approved selected species, including *H. pluvialis* and *Tetraselmis chuii*, as novel foods, while additional species and high-potency extracts remain subject to further evaluation.

In Asian markets, particularly Japan and South Korea, both microalgae and macroalgae are integrated within long-standing functional-food regulatory systems, often through positive lists or structured approval pathways for health claims. Nevertheless, regulatory stringency and claim authorization requirements vary substantially across regions. Within the EU, health claims for algal products are governed by Regulation (EC) No. 1924/2006, which requires robust scientific substantiation, frequently including human intervention studies (Bigliardi and Galati, 2013).

Clear and accurate labeling is central to consumer safety and informed decision-making across algal-based products. Labels should specify the nutritional composition, iodine content (where relevant), potential heavy-metal levels, recommended daily intake, and any applicable warnings or contraindications. Routine testing for contaminants, including heavy metals, cyanotoxins, pesticide residues, and microbial pathogens, is essential to ensure compliance. Continuous monitoring of metal concentrations in widely consumed products such as Spirulina, Chlorella, and edible seaweeds is particularly important.

As algae-based functional foods and nutraceuticals continue to expand globally, harmonized production standards and robust quality-assurance systems become increasingly critical. Controlled cultivation, traceable harvesting, standardized processing, and batch-level testing are essential for minimizing contamination and ensuring consistent bioactive profiles (Sharma et al., 2024; Kapoor et al., 2024). Certification schemes, including organic, non-GMO, and sustainability labels, further enhance consumer confidence and market acceptance, while emerging eco-labels emphasize responsible marine and biotechnological production practices (Zhong et al., 2023).

Collectively, harmonized regulation, transparent labeling, stringent quality control, and credible certification systems are essential for ensuring the safety, efficacy, and long-term success of algae-based products.

6.6 Challenges and future perspectives

The transition of algal systems from experimental platforms to reliable contributors within functional food, nutraceutical, and biomedical sectors has progressed substantially over the past decade. At the current stage of development, the remaining challenges primarily concern system integration, economic alignment, and functional performance rather than fundamental biological feasibility. For both microalgae and macroalgae, these challenges define the operational space within which algal biotechnology must evolve toward scalable, reproducible, and broadly accepted applications. In this context, interpretation of reported bioactivities requires careful consideration of the underlying evidence base. Although many algal bioactivities are supported by in vitro and animal studies, translation to human health outcomes remains dependent on compound type, dosage, and formulation, and will benefit from further validation through well-designed clinical investigations (Trigo et al., 2023).

6.6.1 Technical and economic challenges

Scaling algal production from laboratory and pilot systems to industrial operation requires coordinated control of biological productivity, environmental conditions, and resource efficiency (Vasumathi et al., 2012). For microalgae, open cultivation systems such as raceway ponds remain attractive due to their relatively low capital costs and operational simplicity. However, productivity and metabolite consistency are strongly affected by environmental variability, seasonal fluctuations, and susceptibility to biological contamination (Piyatilleke et al., 2025).

Closed photobioreactor systems offer greater control over light distribution, temperature, gas exchange, and nutrient delivery, enabling more consistent production of high-value bioactive compounds. These benefits are offset by higher capital and energy costs, making economic viability closely dependent on product value rather than bulk biomass output (Ali et al., 2025). Consequently, hybrid cultivation strategies that combine open and closed systems, supported by real-time sensing, automation, and data-driven process control, are increasingly adopted to balance scalability and control (Zhu et al., 2020).

For macroalgae, technical challenges are shaped by geographic location, seasonal growth patterns, oceanographic conditions, and exposure to environmental contaminants. Although seaweed farming typically requires lower energy inputs than photobioreactor-based microalgal cultivation, variability in biomass composition and supply stability remain key constraints for downstream processing and standardized product development (Laurens et al., 2017).

Beyond cultivation, downstream processing represents a major cost contributor across algal systems. Harvesting, dewatering, and extraction are energy-intensive, and conventional separation methods such as centrifugation and filtration face limitations in scalability and sustainability (Musa et al., 2019). As a result, alternative approaches, including bioflocculation, autoflocculation, and hybrid separation strategies, are being explored to reduce energy demand while maintaining product quality (Abu-Shamleh and Najjar, 2020). Advances in extraction technologies, such as pulsed electricfield treatment, enzyme-assisted extraction, and microwave-assisted extraction, further support the selective recovery of bioactive compounds while preserving their functional integrity.

Inherent variability in algal biomass composition presents an additional challenge. Protein content, lipid profiles, and bioactive metabolite concentrations depend on species selection, cultivation regime, and environmental parameters such as light intensity and nutrient availability (González-González and de-Bashan, 2021). For nutraceutical and biomedical applications, this variability necessitates robust process standardization. Genetic and metabolic engineering, combined with real-time monitoring and predictive modeling, is increasingly applied to improve production stability and reproducibility (Yusof et al., 2025). Although algal systems offer sustainability advantages, assessment of net environmental benefits at scale requires consideration of energy inputs, infrastructure requirements, and life-cycle impacts (Braud et al., 2023).

6.6.2 Bioavailability and product development

Bioaccessibility and bioavailability are key determinants of the functional performance of algal-derived ingredients in nutritional and biomedical applications. Bioaccessibility refers to the proportion of nutrients or bioactive compounds released from the algal matrix during gastrointestinal digestion, whereas bioavailability describes the fraction absorbed and made systemically available for physiological action (Demarco et al., 2022). In algal systems, both parameters are strongly influenced by cell-wall architecture, processing methods, and formulation strategies, thereby linking biomass structure directly to product efficacy.

Although proteins, lipids, and secondary metabolites from microalgae and macroalgae exhibit high intrinsic nutritional and therapeutic potential, structural features such as rigid microalgal cell walls or complex polysaccharide matrices in macroalgae can limit digestibility and compound release (Ali et al., 2025). Accordingly, processing is increasingly viewed as an integral design step rather than a corrective measure. Physical and biochemical treatments, including enzymatic hydrolysis, ultrasound-assisted disruption, and ionic-liquid-based processing, have been applied to improve nutrient accessibility and absorption (Machado et al., 2022), while selective extraction and matrix modification enhance compound stability during processing and storage (Qing et al., 2022).

Advanced formulation strategies further support the effective use of algal bioactives. Encapsulation and controlled-delivery systems improve stability, bioavailability, and targeted release of carotenoids, omega-3 fatty acids, peptides, and other sensitive compounds. Techniques such as spray drying, freeze-drying, and nanostructured carrier systems reduce oxidative degradation and gastrointestinal losses, thereby enhancing functional performance in foods, supplements, and biomedical materials (Cotas et al., 2025; Yusof et al., 2025).

6.6.3 Future research directions and industrial opportunities

Future developments in algal biotechnology are likely to depend on improved integration across biological, engineering, and application domains. Genetic and metabolic engineering approaches, including CRISPR–Cas9-based genome editing, offer tools for enhancing protein production, stress tolerance, and targeted metabolite synthesis while remaining compatible with safety and regulatory requirements (Skifa et al., 2025). These approaches support the development of algal biomass tailored for specific nutraceutical and biomedical uses.

Integration of algal cultivation into circular biorefinery systems represents another important direction. Coupling microalgal production with wastewater treatment enables nutrient recovery and biomass generation, while similar circular strategies are emerging for macroalgae through integrated multitrophic aquaculture and coastal biorefineries (Zayed et al., 2024). Such systems improve resource efficiency and allow for the sequential recovery of multiple value-added products.

Looking ahead, advances in nutrigenomics, digital health tools, and formulation science may enable more targeted applications of algal ingredients in functional foods and nutraceuticals. As cultivation and processing technologies mature, both microalgae and macroalgae are likely to assume broader roles in sustainable, health-oriented food systems (Yusof et al., 2025).

6.7 Conclusions and outlook

Algae, including both microalgae and macroalgae, constitute a biologically diverse resource relevant to biomedical, nutraceutical, and functional-food applications. Their metabolic capacity enables the production of proteins, PUFAs, carotenoids, polysaccharides, vitamins, minerals, and other bioactive metabolites associated with antioxidant, antiinflammatory, antimicrobial, immunomodulatory, and cytoprotective activities relevant to human health.

Recent advances in cultivation, processing, metabolic engineering, and formulation have expanded algal utilization beyond traditional whole-biomass consumption to the targeted use of defined bioactive ingredients (Papadaki et al., 2024). This transition has facilitated the incorporation of algal-derived compounds into functional foods, dietary supplements, infant nutrition, cosmeceuticals, and emerging biomedical applications.

From a sustainability perspective, algal systems offer advantages such as high productivity, the ability to utilize nonarable land and saline or wastewater resources, and compatibility with circular biorefinery concepts. However, challenges related to compositional variability, scale-up, and limited long-term toxicological and clinical evidence remain. Addressing these issues will require coordinated advances in process standardization, regulatory alignment, and evidence generation. As these areas progress, algae are expected to play an increasingly important role in health-oriented and sustainable bioeconomic strategies.

AI disclosure

During the preparation of this work, the author(s) used ChatGPT to organize and improve the language. After using this tool/service, the author(s) reviewed and edited the content as needed and take(s) full responsibility for the content of the publication.

References

Abu-Shamleh, A., Najjar, Y.S.H., 2020. Optimization of mechanical harvesting of microalgae by centrifugation for biofuels production. Biomass and Bioenergy 143, 105877. Available at https://doi.org/10.1016/j.biombioe.2020.105877.

Ahmad, A., Banat, F., Taher, H., 2022. Algal Biotechnology: Integrated Algal Engineering for Bioenergy, Bioremediation, and Biomedical Applications. Elsevier, United Arab Emirates, pp. 1–453. Available at https://www.sciencedirect.com/book/9780323904766, http://doi.org/10.1016/C2020-0-03239-1.

Ahmad, A.M.R., Intikhab, A., Zafar, S., Farooq, U., Shah, H.B.U., Akram, S., et al., 2023. Spirulina, an FDA-approved functional food: worth the hype? Cellular and Molecular Biology 69 (1), 137–144. Available at https://doi.org/10.14715/cmb/2022.69.1.24, http://www.cellmolbiol.org/index.php/CMB/issue/view/268.

Ale, M.T., Mikkelsen, J.D., Meyer, A.S., 2011. Important determinants for fucoidan bioactivity: a critical review of structure-function relations and extraction methods for fucose-containing sulfated polysaccharides from brown seaweeds. Marine Drugs 9 (10), 2106–2130. Available at https://doi.org/10.3390/md9102106, http://www.mdpi.com/1660–3397/9/10/2106/pdf.

Ali, S.S., Al-Tohamy, R., Al-Zahrani, M., Schagerl, M., Kornaros, M., Sun, J., 2025. Advancements and challenges in microalgal protein production: a sustainable alternative to conventional protein sources. Microbial Cell Factories 24 (1). Available at https://doi.org/10.1186/s12934-025-02685-1, https://microbialcellfactories.biomedcentral.com/.

Almomani, F., Bhosale, R.R., 2021. Bio-sorption of toxic metals from industrial wastewater by algae strains Spirulina platensis and Chlorella vulgaris: application of isotherm, kinetic models and process optimization. Science of The Total Environment 755 (Pt 2), 142654. Available at https://doi.org/10.1016/j.scitotenv.2020.142654.

Banack, S.A., Johnson, H.E., Cheng, R., Cox, P.A., 2007. Production of the neurotoxin BMAA by a marine cyanobacterium. Marine Drugs 5 (4), 180–196. Available at https://doi.org/10.3390/md504180, http://www.mdpi.net/marinedrugs/papers/md504180.pdf.

Bays, H., 2006. Clinical overview of omacor: a concentrated formulation of omega-3 polyunsaturated fatty acids. The American Journal of Cardiology 98 (4), 71–76. Available at https://doi.org/10.1016/j.amjcard.2005.12.029.

Bazan, N.G., 2009. Cellular and molecular events mediated by docosahexaenoic acid-derived neuroprotectin D1 signaling in photoreceptor cell survival and brain protection. Prostaglandins, Leukotrienes, and Essential Fatty Acids 81 (2–3), 205–211. Available at https://doi.org/10.1016/j.plefa.2009.05.024.

Bazinet, R.P., Layé, S., 2014. Polyunsaturated fatty acids and their metabolites in brain function and disease. Nature Reviews. Neuroscience 15 (12), 771–785. Available at https://doi.org/10.1038/nrn3820, http://www.nature.com/nrn/.

Behzadnia, A., Moosavi-Nasab, M., Oliyaei, N., 2024. Anti-biofilm activity of marine algae-derived bioactive compounds. Frontiers in Microbiology 15. Available at https://doi.org/10.3389/fmicb.2024.1270174.

Bercea, C.I., Cottrell, G.S., Tamagnini, F., McNeish, A.J., 2021. Omega-3 polyunsaturated fatty acids and hypertension: a review of vasodilatory mechanisms of docosahexaenoic acid and eicosapentaenoic acid. British Journal of Pharmacology 178 (4), 860–877. Available at https://doi.org/10.1111/bph.15336, http://onlinelibrary.wiley.com/journal/10.1111/(ISSN)1476–5381.

Berteau, O., 2003. Sulfated fucans, fresh perspectives: structures, functions, and biological properties of sulfated fucans and an overview of enzymes active toward this class of polysaccharide. Glycobiology 13 (6), 29 R. Available at https://doi.org/10.1093/glycob/cwg058.

Besednova, N.N., Andryukov, B.G., Zaporozhets, T.S., Kryzhanovsky, S.P., Kuznetsova, T.A., Fedyanina, L.N., et al., 2020. Algae polyphenolic compounds and modern antibacterial strategies: current achievements and immediate prospects. Biomedicines 8 (9). Available at https://doi.org/10.3390/BIOMEDICINES8090342, https://res.mdpi.com/d_attachment/biomedicines/biomedicines-08-00342/article_deploy/biomedicines-08-00342.pdf.

Beuzenberg, V., Smith, K.F., Packer, M.A., 2014. Isolation and characterisation of halo-tolerant Dunaliella strains from Lake Grassmere/Kapara Te Hau, New Zealand. New Zealand Journal of Botany 52 (1), 136–152. Available at https://doi.org/10.1080/0028825x.2013.859627.

Bhat, V.B., Madyastha, K.M., 2001. Scavenging of peroxynitrite by phycocyanin and phycocyanobilin from Spirulina platensis: protection against oxidative damage to DNA. Biochemical and Biophysical Research Communications 285 (2), 262–266. Available at https://doi.org/10.1006/bbrc.2001.5195.

Bhat, S., Sarkar, S., Zaffar, D., Dandona, P., Kalyani, R.R., 2023. Omega-3 fatty acids in cardiovascular disease and diabetes: a review of recent evidence. Current Cardiology Reports 25 (2), 51–65. Available at https://doi.org/10.1007/s11886-022-01831-0, https://www.springer.com/journal/11886.

Bhatnagar, P., Gururani, P., Parveen, A., Gautam, P., Chandra Joshi, N., Tomar, M.S., et al., 2024. Algae: a promising and sustainable protein-rich food ingredient for bakery and dairy products. Food Chemistry 441. Available at https://doi.org/10.1016/j.foodchem.2023.138322, https://www.sciencedirect.com/science/journal/03088146.

Bigliardi, B., Galati, F., 2013. Innovation trends in the food industry: the case of functional foods. Trends in Food Science & Technology 31 (2), 118–129. Available at https://doi.org/10.1016/j.tifs.2013.03.006.

Bishop, W.M., Zubeck, H.M., 2012. Evaluation of microalgae for use as nutraceuticals and nutritional supplements. Journal of Nutrition & Food Sciences 2 (5). Available at https://doi.org/10.4172/2155-9600.1000147.

Boominathan, R., Devanesan, S., AlSalhi, M.S., Balasubramanian, A., Alkhalid, I.Z., Paul, P., et al., 2022. Quorum quenching action of marine red alga Halemenia durvillei on biofilm forming Gram negative bacterial isolates from contact lens. Algal Research 64. Available at https://doi.org/10.1016/j.algal.2022.102693, http://www.sciencedirect.com/science/journal/aip/22119264.

Braud, L., McDonnell, K., Murphy, F., 2023. Environmental life cycle assessment of algae systems: critical review of modelling approaches. Renewable and Sustainable Energy Reviews 179, 113218. Available at https://doi.org/10.1016/j.rser.2023.113218.

Campo, V.L., Kawano, D.F., Silva, D.B.D., Carvalho, I., 2009. Carrageenans: biological properties, chemical modifications and structural analysis - a review. Carbohydrate Polymers 77 (2), 167–180. Available at https://doi.org/10.1016/j.carbpol.2009.01.020.

Carvalho, L.F., Moreira, J.B., Oliveira, M.S., Costa, J.A.V., 2017. Novel food supplements formulated with Spirulina to meet athletes' needs. Brazilian Archives of Biology and Technology 60. Available at https://doi.org/10.1590/1678-4324-2017160656, http://www.scielo.br/scielo.php?script=sci_issues&pid=1516-8913&lng=en&nrm=iso.

Chandika, P., Tennakoon, P., Kim, T.H., Kim, S.C., Je, J.Y., Kim, J.I., et al., 2022. Marine biological macromolecules and chemically modified macromolecules; potential anticoagulants. Marine Drugs 20 (10). Available at https://doi.org/10.3390/md20100654, http://www.mdpi.com/journal/marinedrugs.

Chaudhary, R., Nawaz, K., Khan, A.K., Hano, C., Abbasi, B.H., Anjum, S., 2020. An overview of the algae-mediated biosynthesis of nanoparticles and their biomedical applications. Biomolecules 10 (11), 1–35. Available at https://doi.org/10.3390/biom10111498, https://www.mdpi.com/2218-273X/10/11/1498/pdf.

Chevolot, L., Mulloy, B., Ratiskol, J., Foucault, A., Colliec-Jouault, S., 2001. A disaccharide repeat unit is the major structure in fucoidans from two species of brown algae. Carbohydrate Research 330 (4), 529–535. Available at https://doi.org/10.1016/s0008–6215(00)00314-1.

Chitthan, S.V., Rasheed, A.S., Muniyandi, M., Dhanasekaran, D., Thajuddin, N., 2025. Bioactive peptides from microalgae and cyanobacteria and their possible mechanisms of action. International Journal of Peptide Research and Therapeutics 31 (5). Available at https://doi.org/10.1007/s10989-025-10746-7, https://www.springer.com/journal/10989.

Cotas, J., Lourenço, M., Figueirinha, A., Valado, A., Pereira, L., 2025. Seaweed polysaccharides: a rational approach for food safety studies. Marine Drugs 23 (11), 412. Available at https://doi.org/10.3390/md23110412.

Damonte, E., Matulewicz, M., Cerezo, A., 2004. Sulfated seaweed polysaccharides as antiviral agents. Current Medicinal Chemistry 11 (18), 2399–2419. Available at https://doi.org/10.2174/0929867043364504.

Daneshvar, E., Wicker, R.J., Show, P.L., Bhatnagar, A., 2022. Biologically-mediated carbon capture and utilization by microalgae towards sustainable CO2 biofixation and biomass valorization – a review. Chemical Engineering Journal 427. Available at https://doi.org/10.1016/j.cej.2021.130884, www.elsevier.com/inca/publications/store/6/0/1/2/7/3/index.htt.

De Bhowmick, G., Guieysse, B., Everett, D.W., Reis, M.G., Thum, C., 2023. Novel source of microalgal lipids for infant formula. Trends in Food Science and Technology 135, 1–13. Available at https://doi.org/10.1016/j.tifs.2023.03.012, http://www.elsevier.com/wps/find/journaldescription.cws_home/601278/description#description.

Deepa, L., Subha, T.S., 2025. Bioactivities and potential application of indian red and brown algae in pharmaceutics. International Journal of Advanced Research 13 (8), 1220–1246. Available at https://doi.org/10.21474/ijar01/21613.

Demarco, M., Oliveira de Moraes, J., Matos, Â.P., Derner, R.B., de Farias Neves, F., Tribuzi, G., 2022. Digestibility, bioaccessibility and bioactivity of compounds from algae. Trends in Food Science and Technology 121, 114–128. Available at https://doi.org/10.1016/j.tifs.2022.02.004, http://www.elsevier.com/wps/find/journaldescription.cws_home/601278/description#description.

Duan, X., Xie, C., Hill, D.R.A., Barrow, C.J., Dunshea, F.R., Martin, G.J.O., et al., 2024. Bioaccessibility, bioavailability and bioactivities of carotenoids in microalgae: a review. Food Reviews International 40 (1), 230–259. Available at https://doi.org/10.1080/87559129.2023.2165095, www.tandf.co.uk/journals/titles/87559129.asp.

Dyall, S.C., 2015. Long-chain omega-3 fatty acids and the brain: a review of the independent and shared effects of EPA, DPA and DHA. Frontiers in Aging Neuroscience 7, 125532.

Enayati, A., Ghojoghnejad, M., Roufogalis, B.D., Maollem, S.A., Sahebkar, A., 2022. Impact of phytochemicals on PPAR receptors: implications for disease treatments. PPAR Research 2022. Available at https://doi.org/10.1155/2022/4714914, http://www.hindawi.com/journals/ppar/.

Fabris, M., Abbriano, R.M., Pernice, M., Sutherland, D.L., Commault, A.S., Hall, C.C., et al., 2020. Emerging technologies in algal biotechnology: toward the establishment of a sustainable, algae-based bioeconomy. Frontiers in Plant Science 11. Available at https://doi.org/10.3389/fpls.2020.00279, https://www.frontiersin.org/journals/plant-science.

Faruque, M.O., Uddin, S., Hossain, M.M., Hossain, S.M.Z., Shafiquzzaman, M., Razzak, S.A., 2024. A comprehensive review on microalgae-driven heavy metals removal from industrial wastewater using living and nonliving microalgae. Journal of Hazardous

Materials Advances 16. Available at https://doi.org/10.1016/j.hazadv.2024.100492, https://www.sciencedirect.com/science/journal/27724166.
Fitton, J., Stringer, D., Karpiniec, S., 2015. Therapies from fucoidan: an update. Marine Drugs 13 (9), 5920–5946. Available at https://doi.org/10.3390/md13095920.
Freeman, M.P., Hibbeln, J.R., Wisner, K.L., Davis, J.M., Mischoulon, D., Peet, M., et al., 2006. Omega-3 fatty acids: evidence basis for treatment and future research in psychiatry. Journal of Clinical Psychiatry 67 (12), 1954–1967. Available at https://doi.org/10.4088/JCP.v67n1217, www.psychiatrist.com.
Gao, L., Qin, Y., Zhou, X., Jin, W., He, Z., Li, X., et al., 2024. Microalgae as future food: rich nutrients, safety, production costs and environmental effects. Science of The Total Environment 927, 172167. Available at https://doi.org/10.1016/j.scitotenv.2024.172167.
Gaviria, Y.S., Zapata, J.E., Miedes, D., Alegría, A., Cilla, A., 2024. Cytoprotective effects of antioxidant peptides from red californian worm (Eisenia foetida) hydrolysate on differentiated caco-2 cells. Nutrients 16 (21). Available at https://doi.org/10.3390/nu16213654, http://www.mdpi.com/journal/nutrients/.
Gnonsoro, U.P., Ake Assi, Y.E.D., Sangare, N.S., Kouakou, Y.U., Trokourey, A., 2022. Health risk assessment of heavy metals (Pb, Cd, Hg) in hydroalcoholic gels of Abidjan, Côte d'Ivoire. Biological Trace Element Research 200 (5), 2510–2518. Available at https://doi.org/10.1007/s12011-021-02822-y, http://www.springer.com/humana+press/journal/12011.
González-González, L.M., de-Bashan, L.E., 2021. Toward the enhancement of microalgal metabolite production through microalgae–bacteria consortia. Biology 10 (4), 282. Available at https://doi.org/10.3390/biology10040282.
Goold, H.D., Moseley, J.L., Lauersen, K.J., 2024. The synthetic future of algal genomes. Cell Genomics 4 (3). Available at https://doi.org/10.1016/j.xgen.2024.100505, https://www.sciencedirect.com/science/journal/2666979X.
Gouveia, L., Batista, A.P., Miranda, A., Empis, J., Raymundo, A., 2007. Chlorella vulgaris biomass used as colouring source in traditional butter cookies. Innovative Food Science and Emerging Technologies 8 (3), 433–436. Available at https://doi.org/10.1016/j.ifset.2007.03.026.
Grassauer, A., Weinmuellner, R., Meier, C., Pretsch, A., Prieschl-Grassauer, E., Unger, H., 2008. Iota-Carrageenan is a potent inhibitor of rhinovirus infection. Virology Journal 5 (1), 107. Available at https://doi.org/10.1186/1743-422X-5-107.
Green, A., Blattmann, C., Chen, C., Mathys, A., 2022. The role of alternative proteins and future foods in sustainable and contextually-adapted flexitarian diets. Trends in Food Science & Technology 124, 250–258. Available at https://doi.org/10.1016/j.tifs.2022.03.026.
Guil-Guerrero, J.L., Prates, J.A.M., 2025. Microalgae bioactives for functional food innovation and health promotion. Foods 14 (12). Available at https://doi.org/10.3390/foods14122122, www.mdpi.com/journal/foods.
Gunathilaka, T.L., Samarakoon, K., Ranasinghe, P., Peiris, L.D.C., 2020. Antidiabetic potential of marine brown algae - a mini review. Journal of Diabetes Research 2020. Available at https://doi.org/10.1155/2020/1230218, http://www.hindawi.com/journals/jdr/.
Harris, W.S., Mozaffarian, D., Rimm, E., Kris-Etherton, P., Rudel, L.L., Appel, L.J., et al., 2009. O-6 F acids and risk for cardiovascular disease. Circulation 119 (6), 902–907. Available at https://doi.org/10.1161/circulationaha.108.191627.
Hayashi, K., Lee, J.B., Nakano, T., Hayashi, T., 2013. Anti-influenza A virus characteristics of a fucoidan from sporophyll of Undaria pinnatifida in mice with normal and compromised immunity. Microbes and Infection 15 (4), 302–309. Available at https://doi.org/10.1016/j.micinf.2012.12.004.
Hwang, J., Yadav, D., Lee, P.C.W., Jin, J.-O., 2022. Immunomodulatory effects of polysaccharides from marine algae for treating cancer, infectious disease, and

inflammation. Phytotherapy Research 36 (2), 761–777. Available at https://doi.org/10.1002/ptr.7348.

Ibrahim, T.N.B.T., Feisal, N.A.S., Kamaludin, N.H., Cheah, W.Y., How, V., Bhatnagar, A., et al., 2023. Biological active metabolites from microalgae for healthcare and pharmaceutical industries: a comprehensive review. Bioresource Technology 372. Available at https://doi.org/10.1016/j.biortech.2023.128661, www.elsevier.com/locate/biortech.

Innes, J.K., Calder, P.C., 2020. Marine omega-3 (N-3) fatty acids for cardiovascular health: an update for 2020. International Journal of Molecular Sciences 21 (4). Available at https://doi.org/10.3390/ijms21041362, https://www.mdpi.com/1422-0067/21/4/1362/pdf.

Innis, S.M., 2007. Dietary (n-3) fatty acids and brain development. Journal of Nutrition 137 (4), 855–859. Available at https://doi.org/10.1093/jn/137.4.855, http://jn.nutrition.org/cgi/reprint/137/4/855.

Jagtap, A.S., Singh, A., Jain, A., 2025. Extraction, structural characterization, and bioactivity of laminarin and fucoidan from brown macroalgae Chorda filum. Carbohydrate Research 554, 109551. Available at https://doi.org/10.1016/j.carres.2025.109551.

Javed, M.U., Mukhtar, H., Zieniuk, B., Rashid, U., 2024. Algal-based hollow fiber membrane bioreactors for efficient wastewater treatment: a comprehensive review. Fermentation 10 (3), 131. Available at https://doi.org/10.3390/fermentation10030131.

Jia, N., Zhang, S., Chen, R., He, X., Dai, C., El-Seedi, H.R., et al., 2025. Immunomodulatory functions of algal bioactive compounds. Critical Reviews in Food Science and Nutrition 65 (30), 7133–7150. Available at https://doi.org/10.1080/10408398.2025.2460634, https://www.tandfonline.com/loi/bfsn20.

Jin, J.-O., Yadav, D., Madhwani, K., Puranik, N., Chavda, V., Song, M., 2022. Seaweeds in the oncology arena: anti-cancer potential of fucoidan as a drug—a review. Molecules (Basel, Switzerland) 27 (18), 6032. Available at https://doi.org/10.3390/molecules27186032.

Kang, Y., Xu, L., Dong, J., Yuan, X., Ye, J., Fan, Y., et al., 2024. Programmed microalgae-gel promotes chronic wound healing in diabetes. Nature Communications 15 (1). Available at https://doi.org/10.1038/s41467-024-45101-9.

Kapoor, D.U., Kukkar, M.R., Gaur, M., Prajapati, B.G., Suttiruengwong, S., Sriamornsak, P., 2024. Algae as third-generation materials: exploring the emerging role in pharmaceutical applications. Materials Today Sustainability 27. Available at https://doi.org/10.1016/j.mtsust.2024.100935, https://www.sciencedirect.com/science/journal/25892347.

Kaur, A., Kaur, G., Reetu, Rai, M.P., 2022. Algae cultivation for biomedical applications: current scenario and future direction. Algal Biotechnology: Integrated Algal Engineering for Bioenergy, Bioremediation, and Biomedical Applications. Elsevier, India, pp. 283–303. Available at https://www.sciencedirect.com/book/9780323904766, https://doi.org/10.1016/B978-0-323-90476-6.00009-1.

Kaushik, A., Sangtani, R., Parmar, H.S., Bala, K., 2023. Algal metabolites: paving the way towards new generation antidiabetic therapeutics. Algal Research 69, 102904. Available at https://doi.org/10.1016/j.algal.2022.102904.

Khan, M.I., Shin, J.H., Kim, J.D., 2018. The promising future of microalgae: current status, challenges, and optimization of a sustainable and renewable industry for biofuels, feed, and other products. Microbial Cell Factories 17 (1). Available at https://doi.org/10.1186/s12934-018-0879-x, http://www.microbialcellfactories.com/home/.

Kumar, L., Anand, R., Shah, M.P., Bharadvaja, N., 2022. Microalgae biodiesel: a sustainable source of energy, unit operations, technological challenges, and solutions. Journal of Hazardous Materials Advances 8, 100145. Available at https://doi.org/10.1016/j.hazadv.2022.100145.

Kumari, A., Garima, Bharadvaja, N., 2023. A comprehensive review on algal nutraceuticals as prospective therapeutic agent for different diseases. 3 Biotech 13 (2). Available at https://doi.org/10.1007/s13205-022-03454-2.

Lauersen, K.J., 2019. Eukaryotic microalgae as hosts for light-driven heterologous isoprenoid production. Planta. 249 (1), 155–180. Available at https://doi.org/10.1007/s00425-018-3048-x, link.springer.de/link/service/journals/00425/index.htm.

Laurens, L.M.L., Markham, J., Templeton, D.W., Christensen, E.D., Van Wychen, S., Vadelius, E.W., et al., 2017. Development of algae biorefinery concepts for biofuels and bioproducts; a perspective on process-compatible products and their impact on cost-reduction. Energy and Environmental Science 10 (8), 1716–1738. Available at https://doi.org/10.1039/c7ee01306j, http://www.rsc.org/Publishing/Journals/EE/About.asp.

Lichota, A., Szewczyk, E.M., Gwozdzinski, K., 2020. Factors affecting the formation and treatment of thrombosis by natural and synthetic compounds. International Journal of Molecular Sciences 21 (21), 7975. Available at https://doi.org/10.3390/ijms21217975.

Machado, L., Carvalho, G., Pereira, R.N., 2022. Effects of innovative processing methods on microalgae cell wall: prospects towards digestibility of protein-rich biomass. Biomass 2 (2), 80–102. Available at https://doi.org/10.3390/biomass2020006.

Magnabosco, C., Santaniello, G., Romano, G., 2025. Microalgae: a promising source of bioactive polysaccharides for biotechnological applications. Molecules (Basel, Switzerland) 30 (9), 2055. Available at https://doi.org/10.3390/molecules30092055.

Malletzidou, L., Kyratzopoulou, E., Kyzaki, N., Nerantzis, E., Kazakis, N.A., 2025. Towards the sustainable removal of heavy metals from wastewater using arthrospira platensis: a laboratory-scale approach in the context of a green circular economy. Applied Sciences 15 (2), 791. Available at https://doi.org/10.3390/app15020791.

Mandal, M.K., Domb, A.J., 2024. Antimicrobial activities of natural bioactive polyphenols. Pharmaceutics 16 (6). Available at https://doi.org/10.3390/pharmaceutics16060718, http://www.mdpi.com/journal/pharmaceutics.

Martínez-Ruiz, F.E., Andrade-Bustamante, G., Holguín-Peña, R.J., Renganathan, P., Gaysina, L.A., Sukhanova, N.V., et al., 2025. Microalgae as functional food ingredients: nutritional benefits, challenges, and regulatory considerations for safe consumption. Biomass 5 (2), 25. Available at https://doi.org/10.3390/biomass5020025 2673-8783.

Matin, M., Koszarska, M., Atanasov, A.G., Król-Szmajda, K., Jóźwik, A., Stelmasiak, A., et al., 2024. Bioactive potential of algae and algae-derived compounds: focus on antiinflammatory, antimicrobial, and antioxidant effects. Molecules (Basel, Switzerland) 29 (19), 4695. Available at https://doi.org/10.3390/molecules29194695.

Mikhailova, E.O., 2023. Selenium nanoparticles: green synthesis and biomedical application. Molecules (Basel, Switzerland) 28 (24), 8125. Available at https://doi.org/10.3390/molecules28248125.

Mirza, S.S., Akbar, S., Ijaz, M.U., Morowvat, M.H., Ishaque, A., Fatima, K., 2025. Nutritional health connection of algae and its pharmaceutical value as anticancer and antioxidant constituents of drugs. Recent Patents on Biotechnology 19 (1), 19–34. Available at https://doi.org/10.2174/0118722083287672240321081428, https://www.eurekaselect.com/article/139408.

Moreira, J.B., Santos, T.D., Cruz, C.G., Silveira, J.T.D., Carvalho, L.F.D., Morais, M.G.D., et al., 2023. Algal polysaccharides-based nanomaterials: general aspects and potential applications in food and biomedical fields. Polysaccharides 4 (4), 371–389. Available at https://doi.org/10.3390/polysaccharides4040022, https://www.mdpi.com/journal/polysaccharides.

Morokutti-Kurz, M., Fröba, M., Graf, P., Große, M., Grassauer, A., Auth, J., et al., 2021. Iota-carrageenan neutralizes SARS-CoV-2 and inhibits viral replication in vitro. PLoS One 16 (2), e0237480. Available at https://doi.org/10.1371/journal.pone.0237480.

Mosibo, O.K., Ferrentino, G., Udenigwe, C.C., 2024. Microalgae proteins as sustainable ingredients in novel foods: recent developments and challenges. Foods 13 (5). Available at https://doi.org/10.3390/foods13050733, www.mdpi.com/journal/foods.

Mozaffarian, D., Wu, J.H.Y., 2011. Omega-3 fatty acids and cardiovascular disease. Journal of the American College of Cardiology 58 (20), 2047–2067. Available at https://doi.org/10.1016/j.jacc.2011.06.063.

Muhamad, I.I., Zulkifli, N., Selvakumaran, S.A.P., Md Lazim, N.A., 2019. Bioactive algal-derived polysaccharides: multi-functionalization, therapeutic potential and biomedical applications. Current Pharmaceutical Design 25 (11), 1147–1162. Available at https://doi.org/10.2174/1381612825666190618152133, http://www.eurekaselect.com/172752/article.

Musa, M., Doshi, A., Brown, R., Rainey, T.J., 2019. Microalgae dewatering for biofuels: a comparative techno-economic assessment using single and two-stage technologies. Journal of Cleaner Production 229, 325–336. Available at https://doi.org/10.1016/j.jclepro.2019.05.039, https://www.journals.elsevier.com/journal-of-cleaner-production.

Naguib, Y.M.A., 2000. Antioxidant activities of astaxanthin and related carotenoids. Journal of Agricultural and Food Chemistry 48 (4), 1150–1154. Available at https://doi.org/10.1021/jf991106k.

Nicoletti, M., 2016. Microalgae nutraceuticals. Foods 5 (3), 54. Available at https://doi.org/10.3390/foods5030054.

Norzagaray-Valenzuela, C.D., Valdez-Ortiz, A., Shelton, L.M., Jiménez-Edeza, M., Rivera-López, J., Valdez-Flores, M.A., et al., 2017. Residual biomasses and protein hydrolysates of three green microalgae species exhibit antioxidant and anti-aging activity. Journal of Applied Phycology 29 (1), 189–198. Available at https://doi.org/10.1007/s10811-016-0938-9, www.wkap.nl/journalhome.htm/0921-8971.

Obaíd, M.L., Camacho, J.P., Brenet, M., Corrales-Orovio, R., Carvajal, F., Martorell, X., et al., 2021. A first in human trial implanting microalgae shows safety of photosynthetic therapy for the effective treatment of full thickness skin wounds. Frontiers in Medicine 8. Available at https://doi.org/10.3389/fmed.2021.772324, journal.frontiersin.org/journal/medicine.

Pandian, N., Chaurasia, R., Chatterjee, S., Biswas, B., Patra, P., Tiwari, A., et al., 2024. Roadmap of algal autotrophic tissue engineering in the avenue of regenerative wound therapy. Materials Advances 5 (19), 7516–7533. Available at https://doi.org/10.1039/d4ma00492b.

Papadaki, S., Tricha, N., Panagiotopoulou, M., Krokida, M., 2024. Innovative bioactive products with medicinal value from microalgae and their overall process optimization through the implementation of life cycle analysis—an overview. Marine Drugs 22 (4), 152.

Park, H.S., Kim, G.Y., Nam, T.J., Deuk Kim, N., Hyun Choi, Y., 2011. Antiproliferative activity of fucoidan was associated with the induction of apoptosis and autophagy in AGS human gastric cancer cells. Journal of Food Science 76 (3), T77. Available at https://doi.org/10.1111/j.1750-3841.2011.02099.x, http://onlinelibrary.wiley.com/journal/10.1111/(ISSN)1750-3841.

Patra, A., Das, J., Agrawal, N.R., Kushwaha, G.S., Ghosh, M., Son, Y.O., 2022. Marine antimicrobial peptides-based strategies for tackling bacterial biofilm and biofouling challenges. Molecules (Basel, Switzerland) 27 (21). Available at https://doi.org/10.3390/molecules27217546, http://www.mdpi.com/journal/molecules.

Pierson, J., Makkena, G.R., Kumar, S., Kumar, V., Vivekanand, V., Husain, H., et al., 2023. Modeling the production of microalgal biomass in large water resource recovery facilities and its processing into various commodity bioproducts. Fermentation 9 (10). Available at https://doi.org/10.3390/fermentation9100909, www.mdpi.com/journal/fermentation.

Piyatilleke, S., Thevarajah, B., Nimarshana, P.H.V., Ariyadasa, T.U., 2025. Microalgal biofuels: challenges and prospective in the framework of circular bioeconomy. Energy Nexus 17, 100338. Available at https://doi.org/10.1016/j.nexus.2024.100338.

Prates, J.A.M., 2025. Applications of bioactive compounds from marine microalgae in health, cosmetics, and functional foods. Applied Sciences 15 (11), 6144. Available at https://doi.org/10.3390/app15116144.

Qing, R., Hao, S., Smorodina, E., Jin, D., Zalevsky, A., Zhang, S., 2022. Protein design: from the aspect of water solubility and stability. Chemical Reviews 122 (18), 14085–14179. Available at https://doi.org/10.1021/acs.chemrev.1c00757.

Rabea, E.Y., Mahmoud, E.D., Mohamed, N.K., Ansary, E.R., Alrouby, M.R., Shehata, R.R., et al., 2025. Potential of venom-derived compounds for the development of new antimicrobial agents. Toxins 17 (5). Available at https://doi.org/10.3390/toxins17050238, http://www.mdpi.com/journal/toxins/.

Rahmati, M., Alipanahi, Z., Mozafari, M., 2019. Emerging biomedical applications of algal polysaccharides. Current Pharmaceutical Design 25 (11), 1335–1344. Available at https://doi.org/10.2174/1381612825666190423160357.

Rajesh, A.T., Beegum, S., Omgy, N., Lakshmi, S., Chellapandian, H., Jeyachandran, S., et al., 2023. Immunomodulators to Prevent Diseases and Minimize Antimicrobial Use Immunomodulators in Aquaculture and Fish Health. CRC Press, India, pp. 59–75. Available at http://www.tandfebooks.com/doi/book/10.1201/9781003361183, http://doi.org/10.1201/9781003361183-7.

Rakib, M.R.J., Jolly, Y.N., Dioses-Salinas, D.C., Pizarro-Ortega, C.I., De-la-Torre, G.E., Khandaker, M.U., et al., 2021. Macroalgae in biomonitoring of metal pollution in the Bay of Bengal coastal waters of Cox's Bazar and surrounding areas. Scientific Reports 11 (1). Available at https://doi.org/10.1038/s41598-021-99750-7, www.nature.com/srep/index.html.

Reddy, C.M., Bhat, V.B., Kiranmai, G., Reddy, M.N., Reddanna, P., Madyastha, K.M., 2000. Selective inhibition of cyclooxygenase-2 by C-phycocyanin, a biliprotein from Spirulina platensis. Biochemical and Biophysical Research Communications 277 (3), 599–603. Available at https://doi.org/10.1006/bbrc.2000.3725, http://www.sciencedirect.com/science/journal/0006291X.

Rojas, V., Rivas, L., Cárdenas, C., Guzmán, F., 2020. Cyanobacteria and eukaryotic microalgae as emerging sources of antibacterial peptides. Molecules (Basel, Switzerland) 25 (24), 5804. Available at https://doi.org/10.3390/molecules25245804.

Romay, C., Gonzalez, R., Ledon, N., Remirez, D., Rimbau, V., 2003. C-Phycocyanin: a biliprotein with antioxidant, anti-inflammatory and neuroprotective effects. Current Protein & Peptide Science 4 (3), 207–216. Available at https://doi.org/10.2174/1389203033487216.

Sanjeewa, K.K.A., Herath, K.H.I.N.M., Kim, Y.S., Jeon, Y.J., Kim, S.K., 2023. Enzyme-assisted extraction of bioactive compounds from seaweeds and microalgae. TrAC - Trends in Analytical Chemistry 167. Available at https://doi.org/10.1016/j.trac.2023.117266, www.elsevier.com/locate/trac.

Satoh, A., Tsuji, S., Okada, Y., Murakami, N., Urami, M., Nakagawa, K., et al., 2009. Preliminary clinical evaluation of toxicity and efficacy of a new astaxanthin-rich Haematococcus pluvialis extract. Journal of Clinical Biochemistry and Nutrition 44 (3), 280–284. Available at https://doi.org/10.3164/jcbn.08-238.

Serhan, C.N., Chiang, N., Van Dyke, T.E., 2008. Resolving inflammation: dual antiinflammatory and pro-resolution lipid mediators. Nature Reviews Immunology 8 (5), 349–361. Available at https://doi.org/10.1038/nri2294.

Sharma, R., Mondal, A.S., Trivedi, N., 2023. Anticancer potential of algae-derived metabolites: recent updates and breakthroughs. Future Journal of Pharmaceutical Sciences 9 (1). Available at https://doi.org/10.1186/s43094-023-00492-2.

Sharma, R., Solanki, P., Chaudhary, M., Gupta, N., Kaur, P., 2024. Unveiling the potential of microalgae for bioplastic production from wastewater – current trends, innovations, and future prospects. Biotechnology for Sustainable Materials 1 (1). Available at https://doi.org/10.1186/s44316-024-00010-1.

Singh, D.V., Bhat, R.A., Upadhyay, A.K., Singh, R., Singh, D.P., 2021. Microalgae in aquatic environs: a sustainable approach for remediation of heavy metals and emerging contaminants. Environmental Technology and Innovation 21. Available at https://doi.org/10.1016/j.eti.2020.101340, http://www.journals.elsevier.com/environmental-technology-and-innovation/.

Sivaramakrishnan, R., Kanwal, S., Incharoensakdi, A., Nirmal, N., Srimongkol, P., 2025. Exploring the nutraceutical and functional food potential of microalgae: implications for health and sustainability. Journal of Agriculture and Food Research 22, 102148. Available at https://doi.org/10.1016/j.jafr.2025.102148.

Skifa, I., Chauchat, N., Cocquet, P.-H., Guer, Y.L., 2025. Microalgae cultivation in raceway ponds: advances, challenges, and hydrodynamic considerations. EFB Bioeconomy Journal 5, 100073. Available at https://doi.org/10.1016/j.bioeco.2024.100073.

Skjånes, K., Aesoy, R., Herfindal, L., Skomedal, H., 2021. Bioactive peptides from microalgae: focus on anti-cancer and immunomodulating activity. Physiologia Plantarum 173 (2), 612–623. Available at https://doi.org/10.1111/ppl.13472.

Srinivasan, S., Arockiasamy, P., Gideon, D.A., Sekaran, S., Arumugasamy, H., Ragupathi, N.K.D., 2022. Immunomodulatory algal metabolites for alleviating inflammation and cancer. Handbook of Oxidative Stress in Cancer: Therapeutic Aspects Volume 1. Springer Singapore, India, pp. 443–463. Available at http://doi.org/10.1007/978-981-16-5422-0_34, https://link.springer.com/book/10.1007/978-981-16-5422-0.

Su, K.P., Matsuoka, Y., Pae, C.U., 2015. Omega-3 polyunsaturated fatty acids in prevention of mood and anxiety disorders. Clinical Psychopharmacology and Neuroscience 13 (2), 129–137. Available at https://doi.org/10.9758/cpn.2015.13.2.129, http://www.cpn.or.kr/journal/download_pdf.php?doi=10.9758/cpn.2015.13.2.129.

Sun, K.L., Gao, M., Wang, Y.Z., Li, X.R., Wang, P., Wang, B., 2022. Antioxidant peptides from protein hydrolysate of marine red algae Eucheuma cottonii: preparation, identification, and cytoprotective mechanisms on H2O2 oxidative damaged HUVECs. Frontiers in Microbiology 13. Available at https://doi.org/10.3389/fmicb.2022.791248, https://www.frontiersin.org/journals/microbiology#

Swapnil, P., Meena, M., Singh, S.K., Dhuldhaj, U.P., Harish, Marwal, A., 2021. Vital roles of carotenoids in plants and humans to deteriorate stress with its structure, biosynthesis, metabolic engineering and functional aspects. Current Plant Biology 26. Available at https://doi.org/10.1016/j.cpb.2021.100203, https://www.journals.elsevier.com/current-plant-biology.

Taghipour, R., Hassannia, H., Arabnozari, H., Enderami, S.E., Habibi, E., 2025. Natural Polysaccharides in Breast Cancer: Fucoidan's Role in Enhancing Cisplatin Cytotoxicity and Reducing Chemotherapy Resistance. IntechOpen. Available at http://doi.org/10.5772/intechopen.1007395.

Tahir, F., Ashfaq, H., Khan, A.Z., Amin, M., Akbar, I., Malik, H.A., et al., 2024. Emerging trends in algae farming on non-arable lands for resource reclamation, recycling, and mitigation of climate change-driven food security challenges. Reviews in Environmental Science and Biotechnology 23 (3), 869–896. Available at https://doi.org/10.1007/s11157-024-09697-0, https://www.springer.com/journal/11157.

Tan, J.S., Lee, S.Y., Chew, K.W., Lam, M.K., Lim, J.W., Ho, S.H., et al., 2020. A review on microalgae cultivation and harvesting, and their biomass extraction processing using ionic liquids. Bioengineered 11 (1), 116–129. Available at https://doi.org/10.1080/21655979.2020.1711626, http://www.tandfonline.com/loi/kbie20.

Topuz, O.K., 2016. Algal oil: a novel source of omega-3 fatty acids for human nutrition. Scientific Bulletin Series F. Biotechnologies 20.

Trigo, J.P., Palmnäs-Bédard, M., Juanola, M.V.L., Undeland, I., 2023. Effects of whole seaweed consumption on humans: current evidence from randomized-controlled intervention trials, knowledge gaps, and limitations. Frontiers in Nutrition 10. Available at https://doi.org/10.3389/fnut.2023.1226168, journal.frontiersin.org/journal/nutrition.

Vasumathi, K.K., Premalatha, M., Subramanian, P., 2012. Parameters influencing the design of photobioreactor for the growth of microalgae. Renewable and Sustainable Energy Reviews 16 (7), 5443–5450. Available at https://doi.org/10.1016/j.rser.2012.06.013, https://www.sciencedirect.com/science/article/pii/S1364032112003930.

Vavitsas, K., Fabris, M., Vickers, C., 2018. Terpenoid metabolic engineering in photosynthetic microorganisms. Genes 9 (11), 520. Available at https://doi.org/10.3390/genes9110520.

Wani, S.A., Elshikh, M.S., Al-Wahaibi, M.S., Naik, H.R., 2023. Functional Foods: Technological Challenges and Advancement in Health Promotion. CRC Press, India, pp. 1–339. Available at https://www.taylorfrancis.com/books/edit/10.1201/9781003315100/functional-foods-sajad-ahmad-wani-mohamed-elshikh-mona-al-wahaibi-haroon-rashid-naik?context=ubx&refId=54d5528e-c8f5-4478-8ef6-fa243e1cab6a, http://doi.org/10.1201/9781003315100.

Waoo, A.A., Singh, S., Chandra, N., Mishra, S., Singh, M.P., Mal, J., et al., 2025. Comprehensive review on the biomedical applications of marine algal polysaccharides. Phycology 5 (4). Available at https://doi.org/10.3390/phycology5040080, https://www.mdpi.com/journal/phycology.

Weiser, M., Butt, C., Mohajeri, M., 2016. Docosahexaenoic acid and cognition throughout the lifespan. Nutrients 8 (2), 99. Available at https://doi.org/10.3390/nu8020099.

Wu, Q., Liu, L., Miron, A., Klímová, B., Wan, D., Kuča, K., 2016. The antioxidant, immunomodulatory, and anti-inflammatory activities of Spirulina: an overview. Archives of Toxicology 90 (8), 1817–1840. Available at https://doi.org/10.1007/s00204-016-1744-5.

Wu, W., Li, S., Xie, P., Li, X., Chang, H., Ho, S.-H., 2023. Algal-fungal interactions and biomass production in wastewater treatment: current status and future perspectives. Algal Research 70, 103021. Available at https://doi.org/10.1016/j.algal.2023.103021.

Yang, Y., Hassan, S.H.A., Awasthi, M.K., Gajendran, B., Sharma, M., Ji, M.K., et al., 2023. The recent progress on the bioactive compounds from algal biomass for human health applications. Food Bioscience 51. Available at https://doi.org/10.1016/j.fbio.2022.102267, http://www.journals.elsevier.com/food-bioscience/.

Yang, S., Wang, Y., Wang, J., Cheng, K., Liu, J., He, Y., et al., 2024. Microalgal protein for sustainable and nutritious foods: a joint analysis of environmental impacts, health benefits and consumer's acceptance. Trends in Food Science & Technology 143, 104278. Available at https://doi.org/10.1016/j.tifs.2023.104278.

Yarnpakdee, S., Benjakul, S., Senphan, T., 2019. Antioxidant activity of the extracts from freshwater macroalgae (Cladophora glomerata) grown in Northern Thailand and its preventive effect against lipid oxidation of refrigerated eastern little tuna slice. Turkish Journal of Fisheries and Aquatic Sciences 19 (3), 209–219. Available at https://doi.org/10.4194/1303-2712-v19_03_04, http://www.trjfas.org/pdf.php?&id=1321.

Yuan, H., Song, J., Li, X., Li, N., Dai, J., 2006. Immunomodulation and antitumor activity of κ-carrageenan oligosaccharides. Cancer Letters 243 (2), 228–234. Available at https://doi.org/10.1016/j.canlet.2005.11.032.

Yuan, J.P., Peng, J., Yin, K., Wang, J.H., 2011. Potential health-promoting effects of astaxanthin: a high-value carotenoid mostly from microalgae. Molecular Nutrition and Food Research 55 (1), 150–165. Available at https://doi.org/10.1002/mnfr.201000414.

Yusof, Z., Tong, Y.W., Selvarajoo, K., Parakh, S.K., Foo, S.C., 2025. Overcoming challenges in microalgal bioprocessing through data-driven and computational approaches. Current Opinion in Food Science 61. Available at https://doi.org/10.1016/j.cofs.2024.101253, https://www.sciencedirect.com/science/journal/22147993.

Zang, L., Baharlooeian, M., Terasawa, M., Shimada, Y., Nishimura, N., 2023. Beneficial effects of seaweed-derived components on metabolic syndrome via gut microbiota modulation. Frontiers in Nutrition 10. Available at https://doi.org/10.3389/fnut.2023.1173225.

Zayed, A., El-Seadawy, H.M., Attia, E.Z., Rushdi, M.I., Abdelmohsen, U.R., 2024. Adopting biorefinery and a circular bioeconomy for extracting and isolating natural products from marine algae. Frontiers in Natural Products 3. Available at https://doi.org/10.3389/fntpr.2024.1425242, https://www.frontiersin.org/journals/natural-products.

Zhang, T., Ouyang, Z., Zhang, Y., Sun, H., Kong, L., Xu, Q., et al., 2025. Marine natural products in inflammation-related diseases: opportunities and challenges. Medicinal Research Reviews 45 (5), 1375–1406. Available at https://doi.org/10.1002/med.22109.

Zhong, D., Du, Z., Zhou, M., 2021. Algae: a natural active material for biomedical applications. View 2 (4). Available at https://doi.org/10.1002/viw.20200189.

Zhong, X., Wang, G., Li, F., Fang, S., Zhou, S., Ishiwata, A., et al., 2023. Immunomodulatory effect and biological significance of β-glucans. Pharmaceutics 15 (6). Available at https://doi.org/10.3390/pharmaceutics15061615, http://www.mdpi.com/journal/pharmaceutics.

Zhu, Z., Jiang, J., Fa, Y., 2020. Overcoming the biological contamination in microalgae and cyanobacteria mass cultivations for photosynthetic biofuel production. Molecules (Basel, Switzerland) 25 (22), 5220. Available at https://doi.org/10.3390/molecules25225220.

Zuorro, A., Lavecchia, R., Contreras-Ropero, J.E., Martínez, J.B.G., Barajas-Ferreira, C., Barajas-Solano, A.F., 2024. Natural antimicrobial agents from algae: current advances and future directions. International Journal of Molecular Sciences 25 (21). Available at https://doi.org/10.3390/ijms252111826, http://www.mdpi.com/journal/ijms.

CHAPTER 7

Algae biomaterials and bioengineering

Mohd. Azwan Ahmad[1,2], Huifang Xu[3], Noor Faradilla Abdullah[2], Nur Azalina Suzianti Feisal[1,2], and Nazahatul Anis Amaludin[4]

[1]MSU Centre for Climate Resilience and Strategy (m-CREST), Management and Science University, Shah Alam, Selangor, Malaysia

[2]Department of Diagnostic and Allied Health Science, Faculty of Health and Life Sciences, Management and Science University, Shah Alam, Selangor, Malaysia

[3]College of Pharmacy, Henan University of Chinese Medicine, Zhengzhou, Henan, P.R. China

[4]Faculty of Earth Science, Universiti Malaysia Kelantan, Kota Bharu, Kelantan, Malaysia

7.1 Introduction

The growing global demand for clean and safe water is severely threatened by the escalating levels of pollution in freshwater sources. Industrialization, urbanization, and agricultural expansion have significantly contributed to the discharge of harmful contaminants such as heavy metals, nutrients (e.g., nitrogen and phosphorus), synthetic dyes, and pharmaceutical residues into rivers, lakes, and groundwater (Akhtar et al., 2021). These pollutants not only degrade the quality of water but also pose serious risks to human health and aquatic ecosystems. Conventional water treatment systems, while effective to an extent, are often expensive, energy-intensive, and unable to completely remove emerging contaminants such as endocrine-disrupting compounds and microplastics (Tarigan et al., 2025). Moreover, chemical treatments used in conventional methods may produce harmful by-products, leading to secondary pollution (Altowayti et al., 2022). These limitations necessitate the search for more sustainable, cost-effective, and environmentally friendly alternatives for water purification. In many developing countries, access to advanced water treatment infrastructure remains limited. Rural and remote communities often rely on untreated or minimally treated water sources, making them vulnerable to waterborne diseases and toxic exposure (Obayomi et al., 2024). The lack of affordable and decentralized water treatment technologies contributes to persistent health disparities and inhibits socio-economic development. Additionally, the problem of eutrophication, the excessive enrichment of water bodies with nutrients, has become increasingly common due to agricultural runoff

Applied Phycology
DOI: https://doi.org/10.1016/B978-0-443-24758-3.00005-8

and domestic wastewater discharge (Akinnawo, 2023). This phenomenon leads to algal blooms, oxygen depletion, and massive die-offs of aquatic organisms, ultimately disrupting the ecological balance of freshwater ecosystems. The inefficiency in removing specific pollutants such as nitrates, phosphates, and certain heavy metals from wastewater further complicates the treatment process. Biological treatment systems that rely on microbial degradation are often limited by the sensitivity of microbial consortia to toxic pollutants and environmental stress (Kuppan et al., 2024). As a result, these systems may fail under fluctuating conditions or when confronted with complex contaminant mixtures. Rising operational costs and stringent environmental regulations also put pressure on industries and municipalities to adopt cleaner and more sustainable water treatment practices (Da Silva et al., 2025). However, innovation in green technologies remains sluggish due to insufficient awareness, limited technical capacity, and high initial investment costs. The climate crisis further exacerbates the global water challenge by increasing the frequency of droughts and reducing the availability of freshwater. This has intensified the urgency to develop alternative treatment approaches that are not only efficient but also resilient and adaptable to environmental changes. Furthermore, public concerns regarding the safety and quality of recycled or reclaimed water pose barriers to the widespread adoption of wastewater reuse (Florides et al., 2024). The social acceptability of alternative treatment solutions depends largely on the ability of technologies to consistently meet safety standards and environmental expectations.

In response to the pressing water quality challenges, algae-based treatment systems offer multiple benefits over conventional water treatment technologies, making them a viable alternative for addressing various types of pollution in an eco-friendly and cost-effective manner (Abudaqqa et al., 2024). Microalgae are microscopic photosynthetic organisms that thrive in aquatic environments and are capable of removing a wide range of contaminants from wastewater. One of the most significant advantages of microalgae is their ability to uptake nutrients such as nitrogen and phosphorus, which are primary contributors to eutrophication (Geremia et al., 2021). Through a natural assimilation process, microalgae incorporate these nutrients into their biomass, thereby reducing nutrient concentrations in treated water and mitigating the risk of algal blooms in downstream ecosystems. Moreover, microalgae demonstrate remarkable efficiency in the adsorption and bioaccumulation of heavy metals such as lead, cadmium, and mercury (Sarma et al., 2024). This capability is facilitated by functional groups on their cell walls, which interact with metal ions, leading to their removal from contaminated water. Unlike chemical precipitation or ion exchange methods, this biosorption process is environmentally benign and does not produce hazardous sludge or by-products. Algae-based systems also exhibit high scalability and flexibility. They can be integrated into existing treatment plants or deployed in decentralized units for rural and underserved areas (Omokaro et al., 2025). Low operational costs, minimal chemical usage, and the potential for solar-driven operations make these systems especially attractive for resource-limited settings where conventional technologies are impractical or too costly. Furthermore, recent advances in biotechnology and bioprocess engineering have significantly enhanced the performance and reliability of algal systems. Algae are rich in biopolymers such as polysaccharides (e.g., alginate, carrageenan, agar), proteins, lipids, and pigments, which serve as essential raw materials in the production of sustainable biomaterials (Mohan et al., 2022). These naturally derived compounds can be extracted and processed into biodegradable films, hydrogels, fibers, and composites. Such materials offer a viable alternative to petroleum-based plastics and synthetic polymers, aligning with global efforts to reduce plastic pollution and environmental degradation (Ansari et al., 2024). One of the most studied biopolymers is alginate, primarily derived from brown algae. Alginate exhibits excellent biocompatibility, gel-forming ability, and nontoxicity, making it ideal for biomedical applications such as wound dressings, tissue scaffolds, drug delivery systems, and even 3D bioprinting (Zhang et al., 2021). In tissue engineering, alginate-based scaffolds are used to support cell growth and tissue regeneration due to their ability to mimic the extracellular matrix, maintain moisture, and deliver bioactive molecules in a controlled manner (Farshidfar et al., 2023). Similarly, red algae-derived carrageenan and agar have been employed in the development of hydrogels and biomedical coatings. These materials are useful in drug encapsulation and sustained release applications, as they can protect therapeutic compounds from degradation while ensuring targeted delivery. Additionally, algae-derived materials often possess antioxidant and antimicrobial properties, further enhancing their functionality in biomedical contexts (Minhas et al., 2024). In bioengineering, algae are being explored for biosensor development and bioprinting. Algal pigments such as phycocyanin and chlorophyll exhibit fluorescence and can serve as biological indicators in sensing devices. These biosensors can detect environmental pollutants, pathogens, or physiological changes in real time, making them valuable tools in health diagnostics and environmental monitoring (Dębowski et al., 2025). Table 7.1 highlights the diverse capabilities of various algae and microalgae species in generating

Table 7.1 Types of algae in bioengineering processes.

Algae/microalgae	Types of biomaterials	Bioengineering process	Results	Reference
Chlorella vulgaris	Bioplastic (polylactic acid [PLA] composite)	Blending with PLA	Improved mechanical strength and biodegradability	Xu et al. (2024)
Spirulina platensis	Biosorbent powder	Drying and grinding for direct biosorption	92% cadmium removal at pH 5.5 within 4 h	Sayadi et al. (2019)
Nannochloropsis gaditana	Lipid-rich biomass	Genetic engineering to overexpress lipid synthesis genes	35% increase in biodiesel yield	Navarro López et al. (2015)
Synechococcus elongatus	Carbon capture strain	Genetic modification for enhanced carbonic anhydrase expression	40% higher carbon fixation under low CO_2	Kanno et al. (2017)
Chlamydomonas reinhardtii	Bioethanol feedstock	Saccharification and fermentation of starch	High ethanol conversion rate with low-carbon footprint	Banerjee et al. (2021)
Scenedesmus obliquus	Biogas substrate	Anaerobic digestion of wet biomass	Significant methane yields for renewable energy generation	Wirth et al. (2015)
Dunaliella salina	Biocomposite filter membrane	Immobilization in cellulose acetate matrix	Efficient removal of phosphate and lead from wastewater	de Souza Celente et al. (2024)
Porphyridium cruentum	Bioactive coating material	Extraction of sulfated polysaccharides for surface coatings	Antibacterial and antifouling properties	Zermeño-Pérez et al. (2024)
Botryococcus braunii	Algae-derived carbon material	Pyrolysis and carbonization of biomass	Production of carbon nanotubes and supercapacitor electrodes	Arévalo-Gallegos et al. (2018)
Anabaena variabilis	Smart biosensor component	Integration with pH-responsive polymers in sensor systems	Real-time nitrate detection with high sensitivity	Singh et al. (2021)

biomaterials through targeted bioengineering processes, emphasizing their potential in environmental and industrial applications. Species such as *Chlorella vulgaris*, *Spirulina platensis*, and *Scenedesmus obliquus* have been widely explored for their ability to produce valuable biomaterials like bioplastics, biofilms, and functional composites due to their high protein, carbohydrate, and lipid content. Through bioengineering methods such as genetic modification, cultivation optimization, and biomass immobilization, researchers have enhanced these organisms' properties to yield improved mechanical stability, pollutant adsorption, and biodegradability. For instance, *C. vulgaris* engineered with polyhydroxyalkanoate (PHA) genes can produce biodegradable plastics with desirable tensile strength. Similarly, bioengineered Spirulina has shown increased cellulose content, supporting its use in biocomposites and filters. These interventions have led to remarkable outcomes, including enhanced thermal stability, greater metal-binding capacities, and the development of eco-friendly alternatives to petrochemical-derived materials. Hence, this chapter aims to address the main advancements and applications of algae as a versatile and sustainable platform for next-generation biomaterials and bioengineering solutions. It explores the biological foundation of algal systems and delves into modern cultivation and bioprocessing strategies designed to optimize biomass yield for high-value applications. Emphasis is placed on algae-derived biopolymers, which are increasingly recognized as eco-friendly alternatives to conventional materials in fields such as material science, biomedicine, and environmental engineering.

7.2 Cultivation technologies and bioprocesses of algae

Algae have garnered increasing attention as a sustainable and versatile resource for applications in bioenergy, bioremediation, pharmaceuticals, nutraceuticals, and biomaterials (Hu et al., 2025). Their rapid growth rate, high photosynthetic efficiency, and ability to thrive in diverse environmental conditions make them ideal candidates for biomass production and biotechnological applications. The success of algae-based industries, however, largely depends on the development and optimization of cultivation technologies and bioprocessing systems that can maximize biomass yield and bioactive compound extraction in a cost-effective and scalable manner (Abdur Razzak et al., 2024). Algal cultivation can be broadly categorized into two main systems: open systems (such as raceway ponds) and closed systems (such as photobioreactors). Each system has its advantages and challenges. Open pond systems are relatively low-cost and simple to construct and operate, making them suitable for large-scale cultivation, particularly in warm climates. However, they are prone to contamination, water loss through evaporation, and fluctuations in environmental conditions (Sun et al., 2020). On the other hand, photobioreactors offer better control over cultivation parameters such as light, temperature, pH, and CO_2 concentration, resulting in higher biomass productivity and purity (Manu et al., 2024). However, they come with higher capital and operational costs. A previous study has demonstrated that closed tubular photobioreactors significantly increased biomass yield and lipid content in *C. vulgaris* compared to traditional open pond systems, highlighting the importance of controlled environments in optimizing algal growth (Usman et al., 2024). Similarly, another study has reported that hybrid systems combining open ponds with photobioreactor modules could enhance productivity while maintaining cost efficiency, offering a balanced solution for industrial applications (Yun et al., 2018). Another key aspect of algal bioprocessing is the selection of suitable strains, each with distinct metabolic profiles and environmental tolerances. Microalgae such as *Spirulina, Chlorella, Scenedesmus*, and *Nannochloropsis* have been extensively studied for their high content of proteins, lipids, and pigments (Ferreira de Oliveira and Bragotto, 2022). Strain selection must align with the intended end use of the biomass, whether for biofuel production, wastewater treatment, or pharmaceutical extraction. The downstream bioprocessing of algae involves harvesting, dewatering, cell disruption, and extraction of target compounds. Since microalgae are suspended in dilute cultures, harvesting and drying remain energy-intensive steps. Innovative techniques such as flocculation, membrane filtration, and centrifugation are being optimized to reduce energy input and cost (Zhang et al., 2019). Furthermore, integrated biorefinery models are being developed to extract multiple products such as proteins, lipids, and carbohydrates from a single biomass stream, enhancing the economic viability of algal bioprocessing.

7.2.1 Optimization of growth conditions for algae cultivation

The optimization of growth conditions is a critical step in maximizing algal biomass productivity and enhancing the yield of desired bioactive compounds (Kumari et al., 2025). Algae are photosynthetic organisms whose growth is highly influenced by external environmental factors, including light intensity, temperature, pH, carbon dioxide concentration, and nutrient availability. Each of these parameters must be carefully controlled and tailored to specific algal strains to achieve optimal cultivation outcomes (Paul et al., 2021).

Light intensity and photoperiod are among the most important factors affecting algal photosynthesis and biomass accumulation. Algae require adequate light to drive the photosynthetic process, but both insufficient and excessive light can negatively impact growth. For example, a study has demonstrated that *C. vulgaris* achieved peak productivity under a light intensity of 100 μmol photons/m^2/s, while higher intensities led to photoinhibition and oxidative stress (Esteves et al., 2025). Similarly, another report has found that light-dark cycles optimized at a 16:8 hour photoperiod significantly enhanced lipid accumulation in microalgae, which is crucial for biofuel applications (Jui et al., 2024). Temperature also plays a fundamental role in influencing metabolic rates and enzymatic activities. Most freshwater microalgae grow optimally between 20°C and 30°C. Deviations from this range can result in slower cell division or cellular damage. Studies have shown that *S. obliquus* displayed the highest growth rate and chlorophyll content at 28°C, but growth declined significantly at 35°C due to thermal stress (Kato et al., 2024). Maintaining a stable temperature environment, particularly in closed systems like photobioreactors, is therefore essential for consistent productivity, which pH levels directly affect nutrient solubility, CO_2 uptake, and cellular metabolism (Amaro et al., 2023). Most algal species prefer a pH range of 7–8 for optimal growth. An unbalanced pH can limit photosynthesis and nutrient absorption. Similar results were observed, as maintaining a controlled pH of 7.5 during the cultivation of *Nannochloropsis oculata* resulted in higher lipid yields and improved biomass stability (Cavonius et al., 2015). Automated pH monitoring and control systems are commonly employed in large-scale cultivation setups to ensure this balance is maintained. Carbon dioxide concentration is another vital factor, as CO_2 is the primary carbon source for algal photosynthesis. Supplying CO_2-enriched air to the culture medium can dramatically enhance biomass production. A study showed that supplementing algal cultures with 5% CO_2 increased the biomass of *Chlorella pyrenoidosa* by over 60% compared to ambient air conditions (Koh et al., 2023). Integration of flue gases from industrial emissions as a carbon source has also been explored as a sustainable practice, though CO_2 concentration has been moderated to avoid acidification of the medium. Nutrient availability, particularly nitrogen, phosphorus, and trace elements like iron and magnesium, is crucial for cellular growth and biochemical synthesis (Sharma et al., 2025). Nitrogen is especially important for protein formation, but its limitations can be used strategically to induce the accumulation of secondary metabolites such as lipids. For example, it has been demonstrated that nitrogen-starved cultures of *Botryococcus braunii* showed increased hydrocarbon content, useful for biofuel production (Nazloo et al., 2024).

7.2.2 Nutrient recycling in algae cultivation

Nutrient recycling is a crucial component of sustainable algae cultivation systems, aimed at reducing resource inputs, minimizing environmental impact, and improving economic feasibility (Arora et al., 2023). Since nutrients such as nitrogen (N), phosphorus (P), and trace metals are essential for algal growth, optimizing their reuse from cultivation systems can significantly enhance biomass productivity and reduce overall production costs. Efficient nutrient recycling strategies ensure continuous biomass generation while aligning with circular bioeconomy principles. In typical large-scale cultivation, a substantial amount of nutrients remains unused in the culture medium after harvesting. If not recycled, these nutrients are lost and contribute to environmental pollution when discharged. Studies have shown that reusing the spent medium or recycling nutrients from biomass residues can provide a cost-effective alternative to fresh nutrient supplementation. For example, some studies have demonstrated that reusing growth media in *C. vulgaris* cultures after partial nutrient replenishment has resulted in only a 10% reduction in biomass yield compared to fresh media, thus validating the feasibility of medium recycling (Salinas-García et al., 2025). Nitrogen and phosphorus are the primary nutrients targeted in recycling due to their high input demand and environmental significance. Nitrogen is often supplied as nitrate or ammonium, while phosphorus is usually added as phosphate salts (Liu and Zhang, 2025). However, these nutrients can precipitate or become biologically unavailable over time. To solve this, pretreatment techniques such as adjusting pH and chelation are used to maintain nutrient solubility during recycling. Furthermore, the use of bioavailable nutrient formulations has been shown to improve uptake in recycled systems. Another study has reported that supplementing recycled media with micro-chelated nutrients enhanced lipid content in *S. obliquus* without affecting growth rate (Duan et al., 2020). Another important strategy involves anaerobic digestion (AD) of harvested algal biomass, which produces nutrient-rich digestate that can be recycled as a culture medium. This closed-loop approach enables the recovery of nitrogen, phosphorus, and potassium, making it particularly attractive for integrated biorefinery systems. Previous investigations highlighted that integrating AD with algal cultivation reduced the demand for synthetic fertilizers by over 70% and improved overall process sustainability (Guenka Scarcelli et al., 2021).

However, the digestate may contain inhibitory compounds such as ammonia or heavy metals, which require dilution or treatment before reuse. Hydrothermal processing and enzymatic hydrolysis are other effective methods for recovering nutrients from algal residues. These processes break down the biomass to release intracellular nutrients, which can be returned to the cultivation cycle. This phenomenon was reported that hydrolyzed algal residues not only supported equivalent biomass growth compared to synthetic media but also improved pigment production due to the presence of organic growth promoters (Gurau et al., 2025).

Automation and monitoring tools such as sensor-based nutrient profiling are increasingly employed to optimize recycling cycles. Real-time data on nutrient concentrations allow for precise supplementation, avoiding overuse or deficiency. For example, study has demonstrated that using nutrient sensors and feedback systems improved nitrogen use efficiency by 30% in closed photobioreactors (Ugya et al., 2025). Dilution and supplementation strategies are also widely used to maintain optimal nutrient levels during recycling. Reused media are often diluted with freshwater or minimal fresh media to balance nutrient concentrations and dilute potential inhibitory compounds. Another study has shown that a 50% dilution of spent medium with fresh medium, followed by targeted supplementation of nitrogen and phosphorus, could sustain algal growth over three continuous cultivation cycles without significant productivity loss (Edmundson et al., 2017). This strategy is particularly effective in large-scale raceway ponds, where nutrient depletion and microbial contamination are common issues. In addition to primary nutrients, micronutrients and trace elements such as magnesium, iron, and zinc also play important roles in enzymatic functions and photosynthesis. Over successive reuse cycles, these elements can become limiting or imbalanced. Therefore, monitoring micronutrient concentrations is essential. Consistent trends have emphasized that incorporating trace element analysis into the recycling protocol prevents micronutrient deficiencies and stabilizes biomass quality in *Nannochloropsis* cultures (Esteves et al., 2024). Moreover, recycling efforts can be significantly enhanced by integrating algae with wastewater streams. Municipal or agricultural wastewater is rich in nitrogen and phosphorus and can be repurposed as a low-cost culture medium. However, it may also contain pathogens, heavy metals, or toxic organic compounds. Pretreatment such as sedimentation, filtration, or UV disinfection is often necessary to make wastewater suitable for algal cultivation. According to a study, the use of pretreated wastewater not only supported comparable biomass yields but also contributed to simultaneous wastewater remediation, promoting a dual-benefit system (Vavra et al., 2025). Algal-bacterial consortia are another promising approach for nutrient recycling. In such systems, bacteria degrade organic matter, releasing nutrients back into the medium, while algae absorb these nutrients for growth. This symbiotic relationship can improve nutrient turnover and reduce waste. Related studies have demonstrated that mixed cultures of *Chlorella* and nitrifying bacteria exhibited enhanced nitrogen cycling and more stable productivity over time (Abate et al., 2024). Furthermore, the use of biorefinery approaches whereby algae are processed to extract multiple high-value products such as lipids, proteins, and pigments generates nutrient-rich by-products (e.g., protein hydrolysates, lipid-extracted biomass) that can be recycled as culture media or nutrient additives. These by-products can contain amino acids, peptides, and growth-promoting factors that improve both algal growth and product quality. A similar study has demonstrated that supplementing culture medium with protein hydrolysates from previously harvested algal biomass improved growth rates and antioxidant pigment accumulation in *Spirulina* (Kaur et al., 2025). In another application, process automation and digital modeling can optimize nutrient recycling by simulating various reuse scenarios and predicting nutrient depletion rates. Advanced models incorporating parameters such as biomass growth kinetics, nutrient consumption, and metabolic shifts provide valuable insights for scheduling nutrient supplementation and media replacement. The observed behavior highlighted that predictive nutrient management systems reduced operational costs by 25% in pilot-scale photobioreactor operations (Oostlander et al., 2020).

7.3 Algae-based biopolymers for sustainable materials

The global demand for sustainable and eco-friendly materials has intensified in recent years due to growing concerns about environmental degradation, plastic pollution, and resource scarcity (Stoett et al., 2024). Traditional petroleum-based plastics, while widely used, contribute significantly to landfill accumulation and marine pollution due to their nonbiodegradable nature. As a response, biopolymers derived from renewable sources have gained increasing attention, particularly those sourced from algae. Algae-based biopolymers represent a promising class of materials that are biodegradable, renewable, and environmentally benign, offering an alternative to conventional plastics and synthetic polymers (Irfan et al., 2025). Algae, including both macroalgae (seaweeds) and microalgae, are abundant, fast-growing organisms capable of thriving in diverse environments without competing with food crops for arable land. They produce a range of biopolymers such as alginate, carrageenan, agar, ulvan, and cellulose-like materials (Sousa et al., 2025). These compounds exhibit diverse physico-chemical properties suitable for various industrial applications, including food packaging, pharmaceuticals, agriculture, and biomedical devices. Accordingly, the growing interest in algae-derived polymers is not only due to their renewable origin but also because of their functional versatility, including gel-forming ability, biocompatibility, and tunable mechanical strength (Satchanska et al., 2024). Fig. 7.1 depicts the biosynthesis of algae-based biopolymers for sustainable materials, showcasing how algal extra-cellular polymeric substances (EPS) such as alginate, carrageenan, and fucoidan are secreted and self-assemble into nanostructured matrices. These biopolymers capture and stabilize metal ions or form hydrogel networks through natural crosslinking, enabling eco-friendly fabrication of biodegradable films, encapsulants, or scaffolds.

One of the most widely utilized algae-based biopolymers is alginate, predominantly extracted from brown seaweed. Alginate forms hydrogels in the presence of divalent cations such as calcium, making it ideal for controlled drug delivery, wound dressings, and tissue engineering (Abdelbasset et al., 2022). Studies have shown that alginate-based scaffolds promote cell adhesion and proliferation while maintaining structural integrity, making them suitable for regenerative medicine (Pinto et al., 2025). Similarly, carrageenan, derived from red algae, is widely used as a thickener and stabilizer in food and pharmaceutical products. Its gelling properties and thermal stability make it a preferred material in biodegradable films and edible coatings. Microalgae, such as *Chlorella* and *Spirulina*, are also valuable sources of proteins and polysaccharides that can be processed into bioplastic precursors. Previous investigations have reported

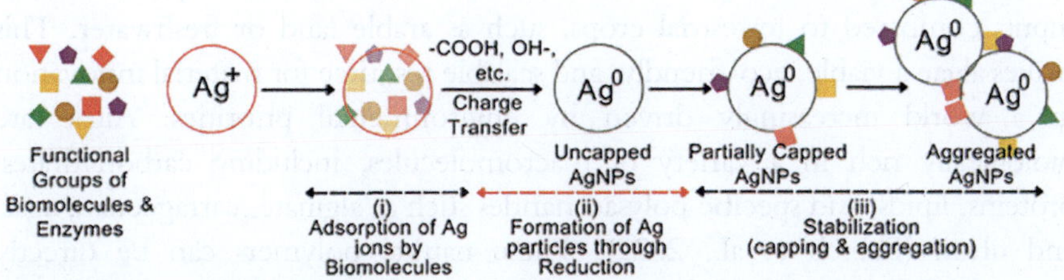

Figure 7.1 Mechanism of Ag ion reduction to Ag particles mediated by algae.

that biomass residues from microalgae, postlipid extraction, could be transformed into thermoplastic films with comparable mechanical properties to synthetic polymers (Dakhili et al., 2025). This valorization of algal waste not only reduces processing costs but also contributes to zero-waste production models in circular bioeconomy systems. Another promising development is the integration of algae-based polymers into composite materials. When blended with other biopolymers such as starch, polylactic acid (PLA), or cellulose, algae-derived materials enhance biodegradability and improve tensile strength and barrier properties (Yang et al., 2025). Related studies have demonstrated that the incorporation of algal biomass into starch-based films significantly improved water resistance and mechanical durability, expanding their potential in packaging applications (Moreira et al., 2023). Algae cultivation also aligns well with low-carbon manufacturing processes, requiring less freshwater and fertilizer compared to terrestrial crops. Moreover, algae can be grown using wastewater or CO_2-rich industrial emissions, thus contributing to environmental remediation alongside biomass production. Previous results have emphasized that this dual functionality enhances the sustainability profile of algae-based biopolymers, making them highly attractive for green manufacturing initiatives (Abuhasheesh et al., 2025).

7.3.1 Development of bioplastics and biocomposites from algal feedstock

The rising global concern over plastic pollution and fossil fuel depletion has propelled the search for sustainable and biodegradable alternatives to conventional petroleum-based plastics. In this context, algae—especially microalgae and macroalgae—have gained significant attention as a promising renewable feedstock for the development of bioplastics and biocomposites (Dang et al., 2022). These photosynthetic organisms are not only fast-growing and capable of thriving in diverse aquatic environments, but they also require minimal inputs compared to terrestrial crops, such as arable land or freshwater. This makes algae a viable, eco-friendly, and scalable resource for material innovation in a world increasingly driven by environmental priorities. Algae are biologically rich in a variety of macromolecules, including carbohydrates, proteins, lipids, and specific polysaccharides such as alginate, carrageenan, agar, and ulvan (Babich et al., 2022). These natural polymers can be directly extracted from the algal biomass and processed into films or molded into various biodegradable plastic forms. Another study has reported that alginate derived from brown algae demonstrates excellent film-forming capacity and can be chemically crosslinked to improve mechanical strength and water resistance, making it suitable for packaging applications (Metha et al., 2024). Similarly, *carrageenan*, a sulfated polysaccharide from red algae, has been widely studied for its gelling, stabilizing, and thickening properties and has been successfully used to produce edible films with biodegradable features. The inherent biodegradability of these compounds means they break down naturally in the environment, alleviating the plastic waste accumulation that plagues landfills and marine ecosystems (Elkaliny et al., 2024). One of the primary advantages of algal feedstock lies in its rich composition of polysaccharides, proteins, and lipids, which serve as precursors for bioplastic synthesis. For instance, macroalgae such as *Ulva, Gracilaria,* and *Sargassum* contain high levels of structural polysaccharides like ulvan, agar, carrageenan, and alginate. These natural polymers can be extracted and processed into bioplastics with tunable mechanical and thermal properties. Consistent trends have been identified that alginate-based films derived from brown seaweeds demonstrate excellent biocompatibility, moisture retention, and controlled degradation, making them highly suitable for wound dressing and tissue engineering applications (Janarthanan and Kumar, 2019).

Beyond the direct use of algal polymers, whole algal biomass or algal hydrolysates can also serve as a carbon-rich substrate for microbial synthesis of high-value bioplastics, such as PHAs (Chia et al., 2020). These polyesters, produced by bacteria under nutrient-limited conditions, are known for their biodegradability and biocompatibility. Another study has demonstrated the efficient conversion of hydrolyzed *Scenedesmus* biomass into PHAs, indicating that microalgae could replace more expensive substrates like glucose or sucrose in the microbial production pipeline (Ooi et al., 2023). This dual functionality of algae as both a polymer source and a feedstock for fermentation makes them uniquely valuable in the bio-based plastic industry. Research has highlighted that *Cupriavidus necator* cultivated on hydrolysates of *C. vulgaris* produced poly (3-hydroxybutyrate) (PHB) with a yield of up to 70% of cell dry weight (Rossi et al., 2025). This demonstrates the feasibility of integrating microalgae cultivation with bacterial fermentation for PHA production. Furthermore, genetic engineering efforts have enabled the direct biosynthesis of PHAs in transgenic microalgae, potentially streamlining the production process. The application of algae in biocomposites also reveals promising opportunities for sustainable material development (Barros et al., 2024). By integrating algal biomass into conventional biopolymer matrices such as PLA, polycaprolactone (PCL), or starch-based plastics, researchers have enhanced the performance, biodegradability, and cost-effectiveness of these materials. For example, it was found that incorporating Spirulina powder into PLA resulted in biocomposites with improved toughness and impact resistance, making them suitable for structural applications (Liao et al., 2023). Similarly, seaweed-derived cellulose fibers have been used as reinforcement agents in various polymer systems, increasing strength while reducing environmental impact (Tennakoon et al., 2023). In addition, algal biocomposites consist of a biopolymer matrix (e.g., PLA, starch, or PHA) reinforced with natural fibers or fillers to enhance mechanical performance. Algal residues, particularly after lipid or protein extraction, contain cellulose and other structural components that can be used as fillers or reinforcements. This evidence has shown that incorporating *S. platensis* residue into PLA matrices improved their tensile modulus and increased thermal stability while maintaining biodegradability (Bowman et al., 2025). The porous and fibrous nature of algal biomass contributes to better interfacial adhesion with the polymer matrix, a critical factor in composite performance. Moreover, the development of algae-based plastics and composites is not without challenges. The extraction and purification of biopolymers from algae can be energy-intensive and costly, requiring solvents or specific enzymes for efficient processing (Imamoglu, 2024). To address these barriers, researchers have turned to green extraction technologies, including supercritical CO_2 extraction and ultrasonic-assisted methods, which reduce environmental burden while improving yields. Another study has emphasized that enzymatic hydrolysis methods offer a cleaner and more sustainable alternative, particularly for scalable industrial applications (Dkhar et al., 2025). For example, a method was developed using a compression-molding technique to produce biocomposite sheets directly from dried *Chlorella* biomass blended with glycerol as a plasticizer (Six et al., 2024). The resulting material exhibited flexibility, water resistance, and complete biodegradability in soil within eight weeks. This approach shows the potential for low-cost, scalable production of algal biocomposites for disposable products and short-life applications.

Additionally, achieving uniform dispersion and compatibility between algal biomass and polymer matrices is critical for the mechanical integrity of the resulting composites. To this end, the use of plasticizers and compatibilizers has been explored to enhance the blending properties of algae-derived fillers. The results have shown that the inclusion of glycerol as a plasticizer significantly improves the flexibility and processability of carrageenan-based films, making them more adaptable for industrial processing (Hidayati et al., 2021). From an economic standpoint, algae hold immense potential in driving cost-effective bioplastic production, especially when integrated into existing biorefinery models (Samoraj et al., 2024). Utilizing residual biomass from algae-based fuel or pigment industries can lower raw material costs and create a diversified revenue stream. Similarly, the adoption of algae in bioplastic industries could reduce reliance on agricultural feedstocks, which are often subject to food-versus-fuel conflicts and seasonal variability (Khan et al., 2021). Furthermore, algae can be cultivated in nonarable lands and saline waters, opening avenues for countries with limited agricultural resources to participate in the green material economy (Ahmad and Ashraf, 2024). Looking forward, genetic engineering and synthetic biology may play an increasing role in optimizing algal strains for higher biopolymer content and easier extraction. Advances in strain selection and metabolic engineering could lead to algae that are custom-designed for bioplastic production, with improved yields and desirable material properties. In addition, the integration of algae-based materials with nanotechnology, such as the incorporation of algal nanocellulose or bioactive compounds, offers exciting prospects for smart and multifunctional packaging, medical devices, and environmental sensors.

7.3.2 Industrial applications and commercialization potential of algae-based biopolymers

The unique composition of algae, which is rich in polysaccharides, proteins, lipids, and bioactive compounds, makes them especially attractive as feedstock for biopolymer production with high economic and environmental value (Khanra et al., 2022). One of the most promising industrial applications of algae-based biopolymers is in the packaging industry. With the European Union's Single-Use Plastics Directive and similar regulations worldwide, demand for compostable and marine-degradable materials is rising. Alginates and agar extracted from brown and red seaweeds, respectively, have been successfully processed into flexible films and coatings that serve as moisture barriers in food packaging. Another study has reported that alginate films derived from *Laminaria hyperborea* exhibited oxygen barrier properties comparable to low-density polyethylene (LDPE), while being fully biodegradable in marine environments (Dörschmann et al., 2025). In another study, it was demonstrated that alginate-based films exhibit excellent tensile strength and moisture retention, making them suitable for biodegradable food packaging applications (Tammina et al., 2025). These biopolymers can also be infused with natural antimicrobial or antioxidant agents, enhancing shelf life and food safety. These films are particularly suited for wrapping fresh produce, where controlled permeability and biodegradability are critical. Similarly, carrageenan-based coatings have been commercialized by companies, such as CP Kelco and FMC Corporation for use in edible films that extend shelf life and reduce plastic waste (Cheng et al., 2022).

Beyond packaging, algae-based biopolymers are finding applications in agriculture and horticulture. Alginate hydrogels, for example, are used as seed coatings and soil conditioners due to their water retention and biostimulant properties (Parreidt et al., 2018). A comparable finding has been observed that seeds coated with Ulva-derived ulvan exhibited a 30% increase in germination rate and improved drought resistance in maize crops (Shefer et al., 2022). Additionally, biodegradable mulch films made from agar and PHA blends are being tested as alternatives to polyethylene mulches, which contribute significantly to agricultural plastic pollution. These films degrade naturally after the growing season, eliminating the need for removal and reducing soil contamination. In biomedical applications, algae-derived biopolymers have emerged as highly valuable due to their inherent biocompatibility, nontoxicity, and functional versatility (Beni et al., 2025). Alginates, in particular, are widely used in wound dressings, drug delivery systems, and tissue engineering scaffolds. Another study illustrated that alginate-based hydrogels loaded with therapeutic agents could control the release rate and enhance healing in chronic wounds (Silva, 2025). Similarly, carrageenan has been utilized in pharmaceutical capsules and gels for its gelling and stabilizing properties. The commercialization of such biomedical applications is already evident, with numerous FDA-approved products on the market that rely on algae-derived materials for their structural and therapeutic functions (Chinjoo and Golzary, 2025). Previous research has highlighted that porous alginate from *Macrocystis pyrifera* supported enhanced cell proliferation and vascularization in skin tissue regeneration, showing its clinical potential (Ciarlantini et al., 2025). Moreover, the antimicrobial properties of certain algal polysaccharides, such as *sulfated ulvans*, open new fields for infection-resistant medical devices. The cosmetic and personal care industries have long benefited from algal extracts, but the use of whole algae-derived biopolymers as structural ingredients in creams, lotions, and biodegradable packaging is a newer and rapidly expanding area (Pipliya et al., 2025). Seaweed-derived carrageenan and agar are now commonly used as thickening agents and film-formers in skincare formulations. More recently, the emphasis has shifted toward using these compounds not just for functionality but also for their sustainability profiles. Consistent trends have been identified that consumers increasingly prefer products with transparent sourcing and minimal environmental impact, a trend that aligns with the algae-based ingredient supply chain (Sasounian et al., 2024).

Market forecasts indicate strong growth potential for algae-based biopolymers, with projections suggesting multibillion-dollar valuations within the next decade (ElFar et al., 2021). According to recent market analysis, the global seaweed extract market alone is expected to exceed USD 9 billion by 2030, while the algae-based bioplastics segment is anticipated to grow at a compound annual growth rate (CAGR) exceeding 12% over the same period (Verma et al., 2024). This growth is driven by increased R&D investment, rising environmental awareness, and technological innovations in cultivation, extraction, and material formulation. What sets algae apart from other bio-based feedstocks is its adaptability and minimal competition with food resources. Unlike corn or sugarcane used in traditional bioplastics, algae can be cultivated in saline or wastewater, reducing land and freshwater demand (Anli Dino et al., 2025). This not only lowers production costs in the long term but also makes algae an inclusive solution for regions with limited agricultural resources. Furthermore, algae can be engineered or selectively bred to optimize the yield of specific biopolymers, increasing their value proposition in specialized applications.

7.4 Bioengineering of algal biomaterials for environmental applications

In the modern world, the application of algal biomaterials has emerged as a transformative approach to tackling critical environmental challenges, especially in areas such as water treatment, heavy metal biosorption, and carbon capture technologies. With increasing concerns about water pollution, heavy metal contamination, and atmospheric carbon dioxide levels, there is a growing interest in harnessing nature-based solutions, particularly from algae, which are fast-growing, photosynthetic organisms with high adaptability to diverse environments (Le et al., 2025). Their unique cellular composition, rich in polysaccharides, proteins, and functional groups, makes them especially suitable for environmental remediation applications. Their rapid growth rates, high surface-area-to-volume ratios, abundance of functional groups on cell walls, and ability to thrive in diverse and often extreme environments position them as ideal candidates for engineered biomaterials in eco-friendly technologies (Abdelfattah et al., 2023). With increasing industrial pollution, water scarcity, and rising atmospheric CO_2 levels, the development of scalable, low-cost, and renewable solutions is imperative, and algal-based systems are proving to be a promising frontier.

7.4.1 Algal-based biomaterials in water filtration

Algal-based biomaterials have emerged as a sustainable and effective solution in water filtration technologies, offering a green alternative to conventional, energy-intensive, and often nonrenewable filtration systems (Satpati et al., 2024). With increasing global concerns over water scarcity, pollution from industrial effluents, and contamination by nutrients and heavy metals, there is a pressing need for innovative, low-cost, and eco-friendly materials for water purification. Traditional water treatment methods, while effective to some extent, often involve high energy inputs, complex infrastructure, and the use of chemical agents that may produce secondary pollutants (Singh et al., 2023a). In this context, algae-based materials offer a unique solution through their natural biosorptive capacity, tunable biochemical composition, and compatibility with a variety of water purification systems (Kuppan et al., 2024). One of the most significant contributions of algal biomaterials lies in the removal of heavy metals from contaminated water. Industrial activities such as mining, electroplating, and battery manufacturing release toxic metals like lead (Pb^{2+}), cadmium (Cd^{2+}), chromium (Cr^{6+}), and mercury (Hg^{2+}) into aquatic environments, posing severe risks to human health and ecosystems. Accordingly, the structural complexity of algal biomass contributes significantly to its ability to adsorb a wide variety of pollutants, including heavy metals, dyes, and pathogenic microorganisms (Chugh et al., 2022). For instance, studies on brown macroalgae such as *Sargassum* and *Laminaria* have demonstrated remarkable efficiency in removing lead (Pb), cadmium (Cd), and arsenic (As) from industrial wastewater due to the presence of alginate, a naturally occurring polymer known for its ion exchange properties (Faruque et al., 2024).

One of the notable advances in this field is the development of alginate-based hydrogels and membranes for water filtration. Alginate, primarily extracted from brown algae, is widely used to fabricate biodegradable and biocompatible membranes (Siddiqui et al., 2025). These membranes can be precisely engineered to achieve desired pore sizes, porosity, and mechanical strength by adjusting the concentration of alginate, the crosslinking agent, and the gelation time. This flexibility allows researchers to tailor membrane properties for specific applications, such as the removal of bacteria, suspended solids, organic matter, and even certain dissolved heavy metals. Previous research has conducted an extensive study demonstrating that cross-linked alginate membranes showed excellent removal efficiencies for microbial contaminants and turbidity, without significantly compromising water flux (Zhao et al., 2025). Their findings show the material's capacity to balance filtration performance with operational durability, which is a crucial requirement in practical water treatment systems. One of the most appealing attributes of alginate membranes is their inherent hydrophilicity, which not only facilitates high water permeability but also helps reduce membrane fouling, one of the most persistent problems in filtration technologies (Wen et al., 2022). Hydrophilic surfaces tend to resist the adsorption of organic macromolecules and microbial adhesion, thereby maintaining filtration efficiency over longer periods and reducing the need for frequent membrane cleaning or replacement. Moreover, alginate membranes are highly compatible with biological systems, making them ideal for use in decentralized and eco-friendly water treatment setups, including those in rural or resource-limited regions. This combination of physical filtration, chemical adsorption, and biological compatibility makes alginate hydrogels especially suitable for hybrid membrane technologies, where different treatment mechanisms are integrated within a single platform (Qiao et al., 2025). Such multifunctional membranes are increasingly being explored in advanced water treatment systems designed to address complex wastewater streams, including industrial effluents and greywater. In another study, a composite alginate-activated carbon membrane was able to simultaneously remove heavy metals like lead and organic pollutants such as methylene blue dye, achieving over 90% efficiency in a single pass through the membrane (Rout et al., 2025).

Another innovative and rapidly growing area in sustainable water treatment is the incorporation of microalgal biomass directly into filter media or polymer-based composites. This approach takes advantage not only of the structural properties of biodegradable polymers but also of the natural bioactivity and adsorptive capacity of microalgae such as *Chlorella* and *Spirulina* (Wan Osman et al., 2021). These species are particularly suited for such applications due to their high surface area, functional groups on the cell wall (such as carboxyl, hydroxyl, and amine), and the presence of bioactive compounds with antimicrobial or chelating properties. Incorporating microalgae into biodegradable polymers such as PLA or polyvinyl alcohol (PVA) allows for the development of advanced biocomposite membranes that combine the strength and processability of synthetic bioplastics with the ecological and adsorptive benefits of microalgae. The resulting membranes offer several functional advantages. For instance, the inclusion of dried *Chlorella* or *Spirulina* biomass introduces reactive sites that can bind to pollutants while also enhancing the membrane's porosity and water permeability (Nath et al., 2024). These structural modifications are essential in tailoring the membranes for efficient filtration of both organic and inorganic contaminants. A notable study has highlighted the application of PLA-based composite membranes integrated with *Spirulina* biomass for water purification (Jiang et al., 2023). The study demonstrated that the membranes showed substantial removal efficiency for both ions, Cu^{+} and Zn^{2+}, from contaminated water. The enhanced adsorption was attributed to the presence of functional groups in the algal biomass that actively interact with metal ions through ion exchange and complexation mechanisms. At the same time, Zn^{2+} removal was facilitated by bioadsorption processes, which benefited from the high affinity of algal components for anionic species.

7.4.2 Biosorption of heavy metals by algal biomaterials

Heavy metal contamination in water bodies presents a significant environmental and public health concern. Metals such as lead (Pb), cadmium (Cd), mercury (Hg), chromium (Cr), and arsenic (As) are nonbiodegradable and tend to accumulate in living organisms, posing toxic effects even at low concentrations (Oladimeji et al., 2024). Traditional physicochemical methods used to remove these contaminants, including chemical precipitation, ion exchange, and membrane filtration, often involve high operational costs and generate secondary pollutants. In response, biosorption using algal biomaterials has gained considerable attention as an eco-friendly and cost-effective alternative. Algae possess a high affinity for binding heavy metal ions due to their rich composition of polysaccharides, proteins, lipids, and other functional groups that facilitate biosorption processes (Ordóñez et al., 2023). This phenomenon can be explained that *Sargassum filipendula* exhibited a maximum adsorption capacity of 132 mg/g for Pb^{2+}, outperforming many synthetic resins and activated carbons (Mohammed et al., 2024). The high density of carboxyl groups in alginates was identified as the primary factor enabling such exceptional metal uptake. Similarly, a study has highlighted that dried *Ascophyllum nodosum* biomass could remove up to 94% of lead and 87% of cadmium ions from industrial wastewater within 24 hours under optimal conditions (Sarangi et al., 2025). This high biosorption efficiency was attributed to the abundant negatively charged carboxyl groups that facilitated ion exchange and complexation.

Biosorption efficiency is highly dependent on a range of physicochemical parameters, with pH being one of the most critical factors influencing the surface charge of algal biomass and the speciation of metal ions in solution (Mohammed et al., 2025). The optimal pH for heavy metal biosorption by algal biomaterials typically ranges between 4 and 6, as this range maximizes the deprotonation of functional groups on the algal cell wall, such as carboxyl (–COOH), sulfate ($–OSO_3H$), and hydroxyl (–OH), without inducing metal precipitation (Bilal et al., 2018). At lower pH values (below 3), high concentrations of protons (H^+) dominate the solution, leading to protonation of these functional groups and creating a positively charged biomass surface. This results in electrostatic repulsion of positively charged metal ions (e.g., Cd^{2+}, Pb^{2+}, Cu^{2+}), significantly reducing biosorption capacity. As noted by a previous study, the competition between H^+ and metal cations for binding sites becomes dominant under highly acidic conditions, severely limiting metal uptake (Aslam et al., 2025). Conversely, at pH values above 6, many heavy metals begin to hydrolyze and form insoluble hydroxides (e.g., $Cd(OH)_2$, $Pb(OH)_2$, $Cr(OH)_3$), which precipitate out of solution rather than being biosorbed through specific binding mechanisms (Pohl, 2020). This phenomenon can lead to an overestimation of biosorption efficiency if not carefully controlled, as removal may be attributed to precipitation rather than true biosorption. Therefore, maintaining pH within the 4–6 window ensures that metal ions remain soluble and that algal functional groups are sufficiently deprotonated to facilitate electrostatic attraction and complexation. Temperature also plays a significant role in biosorption kinetics and thermodynamics (Ciobanu et al., 2023). While biosorption is often an exothermic process, moderate temperatures (20°C–30°C) generally enhance the diffusion rates of metal ions toward the biomass surface and increase membrane fluidity, promoting faster uptake. However, excessively high temperatures may denature biomolecules or alter cell wall integrity, reducing binding capacity. In another study, *S. platensis* exhibited maximum cadmium (Cd^{2+}) biosorption at 25°C, with efficiency declining at temperatures above 40°C due to structural degradation of surface binding sites (Al-Homaidan et al., 2015). These results attributed the high removal efficiency (92%) at optimal conditions to the synergistic effect of favorable protonation state and enhanced molecular mobility. Contact time is another crucial parameter, as biosorption typically follows a two-phase kinetic model, such as a rapid initial uptake due to abundant vacant surface sites, followed by a slower phase as sites become saturated or intracellular diffusion limits further uptake (Torres, 2020). For *Arthrospira platensis*, it was observed that equilibrium was reached within 4 hours, after which no significant increase in CO_2 uptake occurred (Kishi et al., 2020). This aligns with other studies that reported that most algal biosorbents achieve 70%–90% of maximum metal uptake within the first 60–120 minutes (Hamza et al., 2022). The initial concentration of metal ions in solution also affects biosorption performance. At low concentrations, biosorption is highly efficient due to the excess of available binding sites relative to metal ions. However, as concentration increases, the biomass reaches saturation, and removal efficiency may plateau. The maximum biosorption capacity (often reported in mg/g) is typically determined using Langmuir isotherm models, which assume monolayer adsorption on homogeneous surfaces (Acharya et al., 2024). Meanwhile, biomass pretreatment such as drying, grinding, acid/alkali washing, or heat treatment, can significantly enhance biosorption efficiency by increasing surface area, exposing hidden functional groups, or removing impurities. For instance, alkali-treated *Spirulina* showed a significant increase in Cd^{2+} uptake compared to raw biomass, as NaOH treatment removed surface proteins and lipids, exposing more carboxyl and amino groups (Bürck et al., 2024). Table 7.2 highlighted the significant role of algae and microalgae in the removal of heavy metals from contaminated water, emphasizing their effectiveness as natural biosorbents in water treatment technologies. Species such as *C. vulgaris*, *S. platensis*, *S. obliquus*, and *Cladophora glomerata* have demonstrated high affinity for binding toxic heavy metals like lead (Pb), cadmium (Cd), chromium (Cr), arsenic (As), and copper (Cu) through mechanisms involving ion exchange, complexation, and surface adsorption. These processes are facilitated by the functional groups present on algal cell walls, such as carboxyl, hydroxyl, sulfate, and phosphate groups, which actively interact with metal ions. For instance, *C. vulgaris* has been reported to remove up to 93% of Cd from aqueous solutions, while *S. platensis* showed over 90% removal efficiency for Pb and Cu. These algae are applied in various formats, including live biomass, dried powders, or immobilized in polymeric matrices, enhancing their operational feasibility in batch and continuous water treatment systems. The practical applications of these biosorbents extend to industrial effluents, mine drainage, and municipal wastewater, offering cost-effective and eco-friendly alternatives to

Table 7.2 Heavy metal removal by algae biomaterials.

Algae/microalgae	Heavy metal removal	Application	Results	Reference
Chlorella vulgaris	Lead (Pb^{2+}), cadmium (Cd^{2+})	Wastewater biosorption	Achieved 85%–90% removal efficiency at pH 5–6	Waqar et al. (2023)
Spirulina platensis	Cadmium (Cd^{2+})	Industrial effluent treatment	92% removal within 4 h at optimal pH 5.5 and 25°C	Malletzidou et al. (2025)
Scenedesmus obliquus	Chromium (Cr^{6+}), nickel (Ni^{2+})	Industrial wastewater	Reduced Cr^{6+} concentration by 88% after 6 h	Pradhan et al. (2019)
Oscillatoria sp.	Arsenic (As^{3+}), mercury (Hg^{2+})	Mining wastewater	Showed strong affinity for arsenic; >80% removal within 3 h	Thammasane and Kaosol (2018)
Leptolyngbya sp.	Copper (Cu^{2+})	Agricultural runoff remediation	76%–84% removal of Cu at pH 6	Kalita and Baruah (2024)
Nostoc commune	Lead (Pb^{2+}), chromium (Cr^{6+})	Industrial effluent treatment	Effective biosorption with over 90% Pb removal in 5 h	Lavado-Meza et al. (2023)
Ulva lactuca	Copper (Cu^{2+}), nickel (Ni^{2+})	Industrial effluent treatment	78% Cu and 66% Ni removal in batch system	Mosleh et al. (2023)
Cladophora glomerata	Chromium (Cr^{6+}), lead (Pb^{2+})	Fresh wastewater treatment	Cr^{6+} reduced to Cr^{3+} and >85% Pb removal observed	Michalak et al. (2019)
Botryococcus braunii	Cadmium (Cd^{2+}), zinc (Zn^{2+})	Acidic wastewater biosorbent	Biosorption equilibrium reached within 2 h with >80% efficiency	Areco et al. (2018)
Lyngbya majuscula	Mercury (Hg^{2+}), lead (Pb^{2+})	Contaminated wastewater treatment	Exhibited bioaccumulation and detoxification capabilities, 70%–90% removal	Ahmad (2022)

conventional chemical treatments. Moreover, the nontoxic and biodegradable nature of algal biomass ensures minimal environmental impact.

7.4.3 Algal biomaterials in carbon capture technologies

In the face of escalating climate change and rising global carbon dioxide (CO_2) levels, the search for effective carbon capture technologies has become a critical scientific and industrial priority. Among the emerging solutions, algal biomaterials have demonstrated considerable potential as bio-based platforms for capturing, utilizing, and storing atmospheric and industrial CO_2 (Ugya et al., 2024). One of the primary advantages of algal biomaterials in carbon capture lies in their high photosynthetic efficiency and adaptability to diverse environments. Microalgae such as *C. vulgaris*, *S. obliquus*, and *Nannochloropsis* spp. are capable of assimilating CO_2 directly from industrial flue gases, including those emitted by power plants, cement factories, and steel mills (Adetunji et al., 2025). These gases typically contain 10%–15% CO_2, a concentration that many algal strains can tolerate and utilize effectively. It was demonstrated that *C. vulgaris* cultivated in a flat-panel photobioreactor with simulated flue gas achieved a CO_2 fixation rate of 2.8 g/L/day, with over 85% of the supplied CO_2 being captured and converted into biomass (Li et al., 2023). This highlights the potential of algal systems to serve as biological scrubbers in industrial settings, simultaneously reducing emissions and generating valuable biomass for downstream applications. Expanding on the application of bioengineering in enhancing carbon capture efficiency, it becomes clear that genetic and biochemical modifications are pivotal in optimizing the natural abilities of algae to fix carbon. While algae inherently possess high photosynthetic efficiencies, few reports have identified several biochemical bottlenecks, particularly within the Calvin-Benson cycle, that limit their carbon fixation potential under variable environmental conditions (Zhang et al., 2024). Through the lens of synthetic biology and metabolic engineering, scientists are now tailoring algal strains to overcome these limitations, resulting in significantly improved carbon assimilation capabilities. One central target in these genetic strategies is *Rubisco* (ribulose-1, 5-bisphosphate carboxylase/oxygenase), the enzyme responsible for catalyzing the incorporation of CO_2 into organic molecules (Tommasi, 2021). Despite its importance, *Rubisco* is known for its relatively slow catalytic rate and its tendency to react with O_2 instead of CO_2, leading to photorespiration, a wasteful process that reduces carbon fixation efficiency. To address this, bioengineers have focused on increasing *Rubisco's* expression levels or replacing it with more efficient variants from other species. For instance, in a landmark study, a genetically modified strain of *Synechococcus elongatus* was engineered to express a more active form of carbonic anhydrase alongside elevated *Rubisco* activity (Zhao et al., 2024). This dual modification not only improved CO_2 conversion rates but also allowed the *cyanobacterium* to thrive under ambient or low-CO_2 conditions, a notable advantage in real-world settings where CO_2 concentrations fluctuate. Carbonic anhydrase plays a complementary role by catalyzing the reversible conversion between bicarbonate (HCO_3^-) and CO_2, thus ensuring a steady intracellular supply of CO_2 for the Calvin cycle (Supuran, 2023). This enzyme is particularly vital for photosynthetic organisms in aquatic environments, where dissolved inorganic carbon exists primarily in the form of bicarbonate. The engineered *Synechococcus* strain exhibited a 40% increase in carbon fixation compared to wild-type strains when exposed to low-CO_2 conditions, a compelling demonstration of how fine-tuning the cellular carbon concentrating mechanism (CCM) can directly translate to enhanced biomass productivity (Hu et al., 2023).

The carbon captured in algal biomass can be further utilized in multiple ways, contributing to a circular carbon economy. Unlike fossil fuel-derived carbon, which introduces ancient carbon into the modern atmosphere, algal biomass recycles contemporary CO_2, either from the air or industrial emissions, into renewable products, thereby minimizing net greenhouse gas emissions (Abraham et al., 2023). This principle of carbon reuse not only enhances climate mitigation but also creates economic value, driving the commercial viability of algal carbon capture systems. One of the most extensively studied pathways for utilizing algal-captured carbon is the production of biofuels. Microalgae such as *Chlorella, Nannochloropsis,* and *Scenedesmus* are particularly rich in lipids, carbohydrates, and proteins, which can be converted into biodiesel, bioethanol, and biogas, respectively. Biofuels derived from algal lipids are chemically similar to petroleum diesel, yet they are biodegradable and emit significantly fewer greenhouse gases during combustion. The lipid productivity of algae such as *Nannochloropsis, Chlorella, and Scenedesmus* makes them attractive candidates for biodiesel generation. It was estimated that microalgae could yield up to 136,900 L of oil per hectare annually, dwarfing traditional oilseed crops like soybean or canola (Fawcett et al., 2022). This potential, when combined with the ability to grow algae on nonarable land using wastewater or saline water, presents a compelling case for sustainable biofuel production. Similarly, a study has demonstrated that *Nannochloropsis oceanica* cultivated under high

CO_2 conditions accumulated up to 45% lipid content by dry weight, enabling a high yield of biodiesel per ton of biomass (Touliabah and Almutairi, 2021). When combusted, this biodiesel releases CO_2 that was recently captured from the atmosphere or flue gas, creating a near-carbon-neutral energy cycle. The carbon neutrality of algal-derived bioethanol stems from the fact that the CO_2 released during ethanol combustion is approximately equivalent to the CO_2 fixed by algae during photosynthesis. Thus, it represents a closed carbon cycle; carbon absorbed from the atmosphere is reused rather than contributing to net greenhouse gas emissions. This makes bioethanol from algae particularly appealing in the context of decarbonizing transportation and reducing reliance on fossil fuels. Consistent trends have been identified regarding the feasibility of extracting fermentable sugars from algae for ethanol production, with yields comparable to or exceeding those from land crops under optimized conditions (Sanz Smachetti et al., 2020). Furthermore, advancements in pretreatment technologies, such as ultrasonic-assisted extraction, steam explosion, or acid hydrolysis, have significantly improved sugar recovery from algal biomass, thereby increasing ethanol output. Through fermentation into bioethanol and AD into biogas, these biomolecules not only provide alternatives to fossil fuels but also offer a practical pathway for recycling atmospheric CO_2 into usable forms of energy (Nair et al., 2022). As technological barriers continue to fall, and with growing interest in bio-based circular economies, algal-based fuel systems are poised to play a crucial role in the future of sustainable energy.

7.5 Future prospects and emerging trends in algae bioengineering

The field of algae bioengineering is witnessing a surge of innovation and attention, driven by the urgent global need for sustainable solutions in energy, materials, and environmental remediation. Recent advances in synthetic biology, genetic engineering, and systems biology are enabling unprecedented control over algal metabolism, paving the way for next-generation bioproducts and bio-based solutions. These innovations are transforming algae from simple photosynthetic organisms into programmable "green cell factories" capable of producing fuels, pharmaceuticals, bioplastics, and high-value biomaterials with minimal environmental footprint (Einhaus et al., 2024). The convergence of algae bioengineering with smart materials and artificial intelligence (AI) is ushering in a new era of precision biotechnology, where living algal systems are dynamically monitored, optimized, and integrated into responsive, self-regulating platforms (Teng et al., 2020). Simultaneously, smart materials and AI are revolutionizing the design and control of algal systems by enabling real-time data analysis, predictive modeling, and autonomous decision-making in cultivation and bioproduction processes. This integration marks a transformative shift from static, trial-and-error approaches to intelligent, feedback-driven bioengineering.

7.5.1 Harnessing smart materials for enhanced algae-based bioengineering systems

The fusion of smart materials with algal bioengineering represents a transformative frontier in the pursuit of sustainable biotechnology and green innovation (Shelake et al., 2025). Smart materials, also known as stimuli-responsive or intelligent materials, possess the ability to alter their properties in response to external stimuli such as pH, temperature, light, magnetic fields, or specific chemical signals. When integrated into algae-based systems, these materials can significantly enhance the efficiency, functionality, and versatility of algal cultivation, processing, and application across various domains, including environmental remediation, energy production, and biomedicine. These hybrid systems leverage the natural metabolic capabilities of algae, including photosynthesis, carbon fixation, and biosorption, while enhancing performance through responsive material matrices (Guo et al., 2025). As a result, smart algal biohybrids are emerging as powerful tools for applications in environmental monitoring, targeted drug delivery, wastewater treatment, and sustainable biomanufacturing. This compatibility enables the design of smart bioreactors and cultivation systems that respond in real time to fluctuating conditions. For example, thermoresponsive films or coatings can be used inside photobioreactors to adjust thermal insulation based on temperature, helping maintain optimal growth environments for the algal culture (Armugam et al., 2024). During periods of intense sunlight, these materials may reduce heat retention to prevent overheating, while during cooler times, they retain warmth to support continued photosynthesis and biomass accumulation. For example, photochromic materials that change optical properties in response to specific wavelengths can be used to modulate light transmission in photobioreactors (Nwoba et al., 2019). A study has demonstrated that integrating spiropyran-coated films into algal reactors enabled automatic filtering of excessive blue and UV light, preventing photoinhibition in *N. oceanica* while maintaining optimal photosynthetic efficiency under high solar irradiance (Wei et al., 2020). This dynamic light regulation mimics natural shade-avoidance responses and enhances biomass productivity by up to 35% compared to static systems. In addition, conductive polymers like polyaniline or carbon nanotube-infused matrices can be embedded within algal systems to monitor physiological activity in real time. These materials change electrical conductivity in response to oxygen evolution, proton flux, or redox changes associated with photosynthesis (Riaz et al., 2022). This allows for continuous, noninvasive monitoring of algal health and productivity. The integration of such materials into bioreactor walls enables early detection of contamination or stress, triggering automated corrective actions such as gas injection or pH adjustment. Ultimately, the use of smart materials transforms static algal systems into self-optimizing environments, where both biological and material components work in tandem. This minimizes manual intervention, reduces operational costs, and increases productivity. As materials science continues to advance, especially in fields like nanotechnology, soft robotics, and functional coatings, the potential to design increasingly sophisticated, responsive, and efficient algae-based systems becomes more attainable.

7.5.2 Role of artificial intelligence in algal bioprocessing

The integration of AI into algae bioengineering is revolutionizing the way researchers design, optimize, and scale algal-based biotechnologies. As the demand for sustainable alternatives in energy, materials, and environmental remediation grows, microalgae and cyanobacteria are emerging as powerful biological platforms due to their rapid growth, high carbon fixation rates, and metabolic versatility. As algae continue to gain attention for their potential in applications ranging from biofuels and bioplastics to carbon capture and wastewater treatment, the integration of AI technologies is proving essential in optimizing each stage of algal cultivation and bioprocessing (Srimongkol et al., 2022). The convergence of biotechnology with AI algorithms is not only accelerating research and development but also enabling scalable, intelligent systems capable of adapting to real-world complexities. Traditional experimental approaches in algal biotechnology often suffer from trial-and-error methodologies, which are time-consuming and resource-intensive. However, with the aid of machine learning (ML), a subset of AI, researchers can now develop predictive models that correlate environmental variables (such as light intensity, CO_2 concentration, temperature, and nutrient availability) with algal biomass production or metabolite yields (Singh et al., 2023b). Table 7.3 outlines the application of various AI/ML algorithms in microalgae processes, highlighting their roles in optimizing growth, lipid production, nutrient management, and environmental monitoring. Algorithms like artificial neural networks (ANN), support vector machines (SVM), and random forests (RF) are used for predicting biomass, classifying strains, and identifying key nutrients. Genetic algorithms (GA) optimize metabolite production, while deep learning (DL) enables automated species recognition. k-means clustering groups growth phases, reinforcement learning (RL) controls photobioreactors dynamically, and Bayesian networks model wastewater treatment efficiency. Each application is tied to specific microalgae species and research studies, demonstrating how AI enhances precision and efficiency in algal biotechnology across diverse processes. In another study, they utilized an RF model to predict lipid productivity in *C. vulgaris* under different cultivation conditions, achieving over 92% prediction accuracy (Manochkumar et al., 2024). Such models help identify the optimal parameters for maximum yield without exhaustive experimentation. Similarly, a study was highlighted that a deep neural network trained on 18 months of cultivation data from *Nannochloropsis* cultures accurately predicted lipid productivity with 94% precision, reducing experimental trials by over 60% (Syed et al., 2024). The model identified nonintuitive synergies, such as a pulsed light regime combined with nitrogen starvation at mid-log phase, that significantly enhanced triacylglycerol (TAG) accumulation, a key precursor for biodiesel.

Another major application of AI in algal systems is the use of DL algorithms for image-based monitoring and diagnostics (Chong et al., 2024). Microscopic imaging combined with convolutional neural networks (CNNs) allows for the automated identification and classification of algal species, growth stages, and contaminations in mixed cultures. In large-scale cultivation setups, early detection of invasive species or harmful bacteria is crucial to prevent culture collapse. This observation has developed a CNN model capable of distinguishing between *Spirulina, Chlorella, and Scenedesmus* with 98% accuracy based on microscopic image inputs, facilitating real-time surveillance in photobioreactors (Otálora et al., 2021). In addition, AI-driven control systems are revolutionizing automated photobioreactor management. Smart sensors embedded within cultivation systems collect real-time data on variables such as pH, oxygen, CO_2,

Table 7.3 The application of artificial intelligence/machine learning in algae bioprocessing.

AI/ML algorithms	Microalgae species	Process	Application	Reference
Artificial neural networks	*Chlorella vulgaris*	Growth modeling	Predicting biomass productivity under varying light and nutrient conditions	Liyanaarachchi et al. (2020)
Support vector machine	*Scenedesmus obliquus*	Lipid accumulation	Classifying high-lipid strains for biodiesel production	Akenteng et al. (2025)
Random forest	*Nannochloropsis gaditana*	Nutrient optimization	Identifying key nutrient parameters affecting lipid yield	Rangel Pinto et al. (2024)
Genetic algorithms	*Dunaliella salina*	Metabolite production	Optimizing biomass production via salinity and light intensity	Lv et al. (2019)
Deep learning-convolutional neural networks	*Mixed algal consortia*	Image-based monitoring	Automated species recognition and contamination detection	Popescu et al. (2025)
k-means clustering	*Chlorella pyrenoidosa*	Process monitoring	Grouping growth phases based on spectral data	Nurlaila et al. (2021)
Reinforcement learning	*Spirulina platensis*	Photobioreactor control	Dynamic adjustment of light and CO_2 supply for maximum yield	Doan et al. (2021)
Bayesian networks	*Tetraselmis suecica*	Wastewater treatment	Modeling pollutant removal efficiency by algae	Wei et al. (2022)

nutrient levels, and turbidity. These data are processed by RL algorithms that dynamically adjust input parameters such as light intensity, aeration rate, or nutrient dosing to maintain ideal growth conditions. Similarly, an AI-integrated photobioreactor system for *Haematococcus pluvialis* cultivation achieved a 30% increase in astaxanthin productivity by continuously learning and adapting its environmental settings over multiple growth cycles (Cho et al., 2021). In other research, they have applied a constraint-based modeling approach combined with ML to identify key regulatory genes in *Chlamydomonas reinhardtii* that, when silenced, redirected carbon flux toward hydrogen production (He et al., 2018). The AI-predicted strain achieved a 2.8-fold increase in H_2 output compared to wild-type, validating the model's predictive power and drastically shortening the development cycle. In operational settings, AI is enabling real-time monitoring and autonomous control of algal bioreactors. Internet of Things (IoT) sensors continuously collect data on dissolved oxygen, chlorophyll fluorescence, turbidity, and pH, which are fed into AI systems for immediate analysis (Hippler and Khosravitabar, 2024). RL algorithms can then adjust operational parameters such as CO_2 injection, mixing speed, or light exposure on the fly to maintain optimal growth conditions. A pilot-scale study implemented an AI-controlled photobioreactor system that autonomously adapted to daily and seasonal variations in sunlight, maintaining *C. vulgaris* productivity within 5% of theoretical maximum over a six-month period (Pozzobon et al., 2024). This level of adaptive control is particularly valuable in outdoor cultivation, where environmental fluctuations can severely impact yield. Similarly, the implementation of a digital twin for a large-scale algal biomass production facility reduced downtime by 18% and increased overall yield by 12% in the first year of deployment (Sheik et al., 2024). From predictive modeling and automated control to genetic design and sustainability analysis, AI enhances nearly every facet of algal biotechnology. As the technology matures, future algal production systems are likely to become smarter, more resilient, and more sustainable, playing a vital role in addressing global challenges such as climate change, renewable energy, and circular economy development.

7.6 Conclusion

The integration of algae bioengineering with advanced technologies such as smart materials and AI represents a transformative leap toward sustainable solutions for global environmental and industrial challenges. Algae, with their rapid growth, high carbon and nutrient uptake efficiency, and metabolic diversity, have emerged as a cornerstone of next-generation biotechnologies, offering renewable alternatives across water treatment, pollution remediation, carbon capture, and bio-based material production. Their ability to thrive in diverse environments, including wastewater, seawater, and flue gas streams, enables the development of closed-loop systems that convert waste into valuable resources, aligning perfectly with circular economy principles. Algal-based biopolymers, bioplastics, and biocomposites are now being engineered to replace petroleum-derived materials, reducing plastic pollution and fossil fuel dependence while maintaining performance and biodegradability. In water filtration, algal biomaterials such as alginate-based membranes and immobilized biomass composites demonstrate exceptional efficiency in removing heavy metals, nitrates, and phosphates, offering scalable, low-energy solutions for both municipal and industrial wastewater treatment. The biosorption capacity of algal cell walls, rich in functional groups like carboxyl and sulfate, enables cost-effective and eco-friendly remediation of contaminated waters, with high reusability and minimal secondary waste. Simultaneously, algae play a pivotal role in carbon capture, naturally fixing CO_2 at rates far exceeding those of terrestrial plants; when coupled with flue gas from industrial sources, they not only mitigate emissions but also generate biomass that can be valorized into biofuels, biochemicals, and biocarbon materials, effectively closing the carbon loop. Advances in genetic and metabolic engineering have further enhanced algal strains for improved photosynthetic efficiency, stress tolerance, and targeted product synthesis, accelerating their deployment in real-world applications. The integration of smart materials, responsive to light, temperature, pH, or chemical cues, adds a dynamic dimension to algal systems, enabling self-regulating bioreactors that adapt in real time to optimize growth, prevent contamination, and maximize productivity. These intelligent platforms reduce operational inefficiencies and support autonomous control, crucial for large-scale cultivation. Complementing this, AI is revolutionizing algae bioengineering through predictive modeling, real-time monitoring, and automated decision-making. ML algorithms analyze vast datasets from omics, bioreactor sensors, and environmental conditions to forecast optimal growth parameters, detect metabolic shifts, and design high-yield strains with unprecedented speed and accuracy. AI-driven control systems enable smart algae farms that dynamically adjust lighting, nutrient supply, and harvesting cycles, minimizing human intervention and energy use.

Together, these innovations are transforming algae from simple photosynthetic organisms into programmable, responsive, and intelligent biofactories. As research progresses and interdisciplinary collaboration strengthens, the convergence of biology, materials science, and digital technologies will continue to unlock new frontiers in sustainability, positioning algae at the heart of a resilient, low-carbon, and regenerative bioeconomy.

References

Abate, R., Oon, Y.-S., Oon, Y.-L., Bi, Y., 2024. Microalgae-bacteria nexus for environmental remediation and renewable energy resources: advances, mechanisms and biotechnological applications. Heliyon 10 (10). Available at https://doi.org/10.1016/j.heliyon.2024.e31170.

Abdelbasset, W.K., Jasim, S.A., Sharma, S.K., Margiana, R., Bokov, D.O., Obaid, M.A., et al., 2022. Alginate-based hydrogels and tubes, as biological macromolecule-based platforms for peripheral nerve tissue engineering: a review. Annals of Biomedical Engineering 50 (6), 628–653. Available at https://doi.org/10.1007/s10439-022-02955-8.

Abdelfattah, A., Ali, S.S., Ramadan, H., El-Aswar, E.I., Eltawab, R., Ho, S.-H., et al., 2023. Microalgae-based wastewater treatment: mechanisms, challenges, recent advances, and future prospects. Environmental Science and Ecotechnology 13. Available at https://doi.org/10.1016/j.ese.2022.100205.

Abdur Razzak, S., Bahar, K., Islam, K.M.O., Haniffa, A.K., Faruque, M.O., Hossain, S.M.Z., et al., 2024. Microalgae cultivation in photobioreactors: sustainable solutions for a greener future. Green Chemical Engineering 5 (4), 418–439. Available at https://doi.org/10.1016/j.gce.2023.10.004.

Abraham, J., Prigiobbe, V., Abimbola, T., Christodoulatos, C., 2023. Integrating biological and chemical CO2 sequestration using green microalgae for bioproducts generation. Frontiers in Climate 4. Available at https://doi.org/10.3389/fclim.2022.949411.

Abudaqqa, W.S.K., Madhuranthakam, C.M.R., Chaalal, O., 2024. Algae based membrane bioreactors: a mini review on their progress and processes for wastewater treatment. Journal of Water Process Engineering 59. Available at https://doi.org/10.1016/j.jwpe.2024.104937.

Abuhasheesh, Y., Ghazal, A., Tang, D.Y.Y., Banat, F., Hasan, S.W., Show, P.L., 2025. Advances in Chlorella microalgae for sustainable wastewater treatment and bioproduction. Chemical Engineering Journal Advances 22. Available at https://doi.org/10.1016/j.ceja.2025.100715.

Acharya, A., Jeppu, G., Girish, C.R., Prabhu, B., Murty, V.R., Martis, A.S., et al., 2024. Adsorption of arsenic and fluoride: modeling of single and competitive adsorption systems. Heliyon 10 (11). Available at https://doi.org/10.1016/j.heliyon.2024.e31967.

Adetunji, A.I., Gumbi, S.T., Erasmus, M., 2025. Harnessing the potential of microalgae in sequestration of CO2 emissions: removal mechanisms, optimization strategies, and bioenergy production. Journal of Hazardous Materials Advances 18. Available at https://doi.org/10.1016/j.hazadv.2025.100722.

Ahmad, I.Z., 2022. The usage of Cyanobacteria in wastewater treatment: prospects and limitations. Letters in Applied Microbiology 75 (4), 718–730. Available at https://doi.org/10.1111/lam.13587.

Ahmad, A., Ashraf, S.S., 2024. Harnessing microalgae: innovations for achieving UN Sustainable Development Goals and climate resilience. Journal of Water Process Engineering 68. Available at https://doi.org/10.1016/j.jwpe.2024.106506.

Akenteng, Y.D., Chen, H., Opoku, K.N., Ullah, F., Wang, S., Kumar, S., 2025. The role of computational fluid dynamics (CFD) in phytohormone-regulated microalgae-based carbon dioxide capture technology. Sustainability (Switzerland) 17 (3). Available at https://doi.org/10.3390/su17030860.

Akhtar, N., Syakir Ishak, M.I., Bhawani, S.A., Umar, K., 2021. Various natural and anthropogenic factors responsible for water quality degradation: a review. Water (Switzerland) 13 (19). Available at https://doi.org/10.3390/w13192660.

Akinnawo, S.O., 2023. Eutrophication: causes, consequences, physical, chemical and biological techniques for mitigation strategies. Environmental Challenges 12. Available at https://doi.org/10.1016/j.envc.2023.100733.

Al-Homaidan, A.A., Alabdullatif, J.A., Al-Hazzani, A.A., Al-Ghanayem, A.A., Alabbad, A.F., 2015. Adsorptive removal of cadmium ions by Spirulina platensis dry biomass. Saudi Journal of Biological Sciences 22 (6), 795–800. Available at https://doi.org/10.1016/j.sjbs.2015.06.010.

Altowayti, W.A.H., Shahir, S., Othman, N., Eisa, T.A.E., Yafooz, W.M.S., Al-Dhaqm, A., et al., 2022. The role of conventional methods and artificial intelligence in the wastewater treatment: a comprehensive review. Processes 10 (9). Available at https://doi.org/10.3390/pr10091832.

Amaro, H.M., Salgado, E.M., Nunes, O.C., Pires, J.C.M., Esteves, A.F., 2023. Microalgae systems - environmental agents for wastewater treatment and further potential biomass valorisation. Journal of Environmental Management 337. Available at https://doi.org/10.1016/j.jenvman.2023.117678.

Ansari, M.M., Heo, Y., Do, K., Ghosh, M., Son, Y.-O., 2024. Nanocellulose derived from agricultural biowaste by-products–sustainable synthesis, biocompatibility, biomedical applications, and future perspectives: a review. Carbohydrate Polymer Technologies and Applications 8. Available at https://doi.org/10.1016/j.carpta.2024.100529.

Areco, Haug, Curutchet, 2018. Studies on bioremediation of Zn and acid waters using Botryococcus braunii. Journal of Environmental Chemical Engineering 6 (4), 3849–3859. Available at https://doi.org/10.1016/j.jece.2018.05.041.

Arévalo-Gallegos, A., Garcia-Perez, J.S., Carrillo-Nieves, D., Ramirez-Mendoza, Iqbal, H.M.N., Parra-Saldívar, R., 2018. Botryococcus braunii as a bioreactor for the production of nanoparticles with antimicrobial potentialities. International Journal of Nanomedicine 13, 5591–5604. Available at https://doi.org/10.2147/ijn.s174205.

Armugam, A., Lim, D.S.W., Riduan, S.N., Chan, S.P., Asugan, J., Zhang, Y., 2024. Structural engineering of antimicrobials for optimal broad-spectrum activity. Green Chemical Engineering 5 (4), 473–482. Available at https://doi.org/10.1016/j.gce.2023.12.001.

Arora, N., Lo, E., Legall, N., Philippidis, G.P., 2023. A critical review of growth media recycling to enhance the economics and sustainability of algae cultivation. Energies 16 (14). Available at https://doi.org/10.3390/en16145378.

Aslam, Kanwal, Javied, Nisar, Torriero, 2025. Microbial biosorption: a sustainable approach for metal removal and environmental remediation. International Journal of Environmental Science and Technology 22 (13), 13245–13276. Available at https://doi.org/10.1007/s13762-025-06611-1.

Babich, O., Sukhikh, S., Larina, V., Kalashnikova, O., Kashirskikh, E., Prosekov, A., et al., 2022. Algae: study of edible and biologically active fractions, their properties and applications. Plants 11 (6). Available at https://doi.org/10.3390/plants11060780.

Banerjee, S., Ray, A., Das, D., 2021. Optimization of Chlamydomonas reinhardtii cultivation with simultaneous CO2 sequestration and biofuels production in a biorefinery framework. Science of the Total Environment 762. Available at https://doi.org/10.1016/j.scitotenv.2020.143080.

Barros, D., Nobre, L., Bessa, J., Leite, L., Mota, C., Cunha, F., et al., 2024. Bio-inspired eco-composite materials seaweed waste integration for sustainable structural applications. Sustainability (Switzerland) 16 (24). Available at https://doi.org/10.3390/su162411051.

Beni, B.H., Falahati, M., Rostamabadi, M.M., Navid, S., Rostamabadi, H., 2025. Agar as a natural polymer: from culture media to cutting-edge biomedical applications. Carbohydrate Polymers 368. Available at https://doi.org/10.1016/j.carbpol.2025.124100.

Bilal, M., Rasheed, T., Sosa-Hernández, J.E., Raza, A., Nabeel, F., Iqbal, H.M.N., 2018. Biosorption: an interplay between marine algae and potentially toxic elements—a review. Marine Drugs 16 (2). Available at https://doi.org/10.3390/md16020065.

Bowman, S., Gu, B., Qiu, Y., Qiu, L., 2025. Making of cyanobacteria reinforced PLA composites using vat photopolymerization LCD additive manufacturing for textile accessories applications. International Journal of Biological Macromolecules 309. Available at https://doi.org/10.1016/j.ijbiomac.2025.142910.

Bürck, M., Fratelli, C., Campos Assumpção de Amarante, M., Braga, A.R.C., 2024. Unveiling the potential of Spirulina biomass—a glimpse into the future circular economy using green and blue ingredients. Biomass (Switzerland) 4 (3), 704–719. Available at https://doi.org/10.3390/biomass4030039.

Cavonius, L.R., Albers, E., Undeland, I., 2015. pH-shift processing of Nannochloropsis oculata microalgal biomass to obtain a protein-enriched food or feed ingredient. Algal Research 11, 95–102. Available at https://doi.org/10.1016/j.algal.2015.05.022.

Cheng, C., Chen, S., Su, J., Zhu, M., Zhou, M., Chen, T., et al., 2022. Recent advances in carrageenan-based films for food packaging applications. Frontiers in Nutrition 9. Available at https://doi.org/10.3389/fnut.2022.1004588.

Chia, W.Y., Ying Ying Tang, D., Khoo, K.S., Kay Lup, A.N., Chew, K.W., 2020. Nature's fight against plastic pollution: algae for plastic biodegradation and bioplastics production. Environmental Science and Ecotechnology 4. Available at https://doi.org/10.1016/j.ese.2020.100065.

Chinjoo, N., Golzary, A., 2025. Microalgae: revolutionizing skin repair and enhancement. Biotechnology Reports 47. Available at https://doi.org/10.1016/j.btre.2025.e00911.

Cho, S.J., Sung, Y.J., Lee, J.S., Yu, B.S., Sim, S.J., 2021. Robust cyst germination induction in Haematococcus pluvialis to enhance astaxanthin productivity in a semi-continuous outdoor culture system using power plant flue gas. Bioresource Technology 338. Available at https://doi.org/10.1016/j.biortech.2021.125533.

Chong, J.W.R., Khoo, K.S., Chew, K.W., Ting, H.-Y., Iwamoto, K., Ruan, R., et al., 2024. Artificial intelligence-driven microalgae autotrophic batch cultivation: a comparative study of machine and deep learning-based image classification models. Algal Research 79. Available at https://doi.org/10.1016/j.algal.2024.103400.

Chugh, M., Kumar, L., Shah, M.P., Bharadvaja, N., 2022. Algal bioremediation of heavy metals: an insight into removal mechanisms, recovery of by-products, challenges, and future opportunities. Energy Nexus 7. Available at https://doi.org/10.1016/j.nexus.2022.100129.

Ciarlantini, C., Garcia, R.B., Lacolla, E., Francolini, I., Fernández-García, M., Nuñez, C.M., et al., 2025. Nanofibrous scaffolds based on polyelectrolyte complexes of chitosan-alginate-COLLAGEN and chitosan-hyaluronic acid for tissue Engineering. Carbohydrate Polymer Technologies and Applications 11. Available at https://doi.org/10.1016/j.carpta.2025.100915.

Ciobanu, A.A., Bulgariu, D., Ionescu, I.A., Puiu, D.M., Vasile, G.G., Bulgariu, L., 2023. Evaluation of thermodynamic parameters for cu(II) ions biosorption on algae biomass and derived biochars. Symmetry 15 (8). Available at https://doi.org/10.3390/sym15081500.

Da Silva, J., Leal, E.C., Gomes, A., Gomes, P., Calheiros, D., Gonçalves, T., et al., 2025. Alginate-based hydrogels for sustained antimicrobial peptide delivery to enhance wound healing in diabetes. Biomaterials Advances 175. Available at https://doi.org/10.1016/j.bioadv.2025.214337.

Dakhili, S., Hashami, Z., Rostami, N., Moradi, S., Moslehishad, M., Shojaee-Aliabadi, S., 2025. Innovative applications of microalgae in biodegradable food packaging: a review. Journal of Agriculture and Food Research 19. Available at https://doi.org/10.1016/j.jafr.2025.101723.

Dang, B.-T., Bui, X.-T., Tran, D.P.H., Hao Ngo, H., Nghiem, L.D., Hoang, T.-K.-D., et al., 2022. Current application of algae derivatives for bioplastic production: a review. Bioresource Technology 347. Available at https://doi.org/10.1016/j.biortech.2022.126698.

de Souza Celente, G., de Cassia de Souza Schneider, R., Medianeira Rizzetti, T., Lobo, E.A., Sui, Y., 2024. Using wastewater as a cultivation alternative for microalga Dunaliella salina: potentials and challenges. Science of the Total Environment 911. Available at https://doi.org/10.1016/j.scitotenv.2023.168812.

Dębowski, M., Kazimierowicz, J., Zieliński, M., 2025. Multi-sensing monitoring of the microalgae biomass cultivation systems for biofuels and added value products synthesis—challenges and opportunities. Applied Sciences (Switzerland) 15 (13). Available at https://doi.org/10.3390/app15137324.

Dino, A., Kishore, S., Lalthazuala Rokhum, G., 2025. Bioplastic production using waste macroalgal biomass: a holistic review on challenges, prospects, economic viability and sustainability analysis. Journal of Environmental Chemical Engineering 13 (2). Available at https://doi.org/10.1016/j.jece.2025.116108.

Dkhar, D.S., Swain, R.P., Dubey, R., Patel, G.K., Chandra, P., 2025. Plant-derived enzymes as sustainable biocatalysts for biosensing and industrial applications. Industrial Crops and Products 233. Available at https://doi.org/10.1016/j.indcrop.2025.121336.

Doan, Y.T.-T., Ho, M.-T., Nguyen, H.-K., Han, H.-D., 2021. Optimization of Spirulina sp. cultivation using reinforcement learning with state prediction based on LSTM neural network. Journal of Applied Phycology 33 (5), 2733–2744. Available at https://doi.org/10.1007/s10811-021-02488-y.

Dörschmann, P., Kopplin, G., Thalenhorst, T., Seeba, C., Ullah, S.F., Srivastava, V., et al., 2025. Influence of a very high-molecular weight fucoidan from laminaria hyperborea on age-related macular degeneration-relevant pathomechanisms in ocular cell models. Marine Drugs 23 (3). Available at https://doi.org/10.3390/md23030101.

Duan, Y., Guo, X., Yang, J., Zhang, M., Li, Y., 2020. Nutrients recycle and the growth of Scenedesmus obliquus in synthetic wastewater under different sodium carbonate concentrations. Royal Society Open Science 7 (1). Available at https://doi.org/10.1098/rsos.191214.

Edmundson, Huesemann, Kruk, Lemmon, Billing, Schmidt, et al., 2017. Phosphorus and nitrogen recycle following algal bio-crude production via continuous hydrothermal liquefaction. Algal Research 26, 415–421. Available at https://doi.org/10.1016/j.algal.2017.07.016.

Einhaus, A., Baier, T., Kruse, O., 2024. Molecular design of microalgae as sustainable cell factories. Trends in Biotechnology 42 (6), 728–738. Available at https://doi.org/10.1016/j.tibtech.2023.11.010.

ElFar, O.A., Chang, C.-K., Leong, H.Y., Peter, A.P., Chew, K.W., Show, P.L., 2021. Prospects of Industry 5.0 in algae: customization of production and new advance technology for clean bioenergy generation. Energy Conversion and Management: X 10. Available at https://doi.org/10.1016/j.ecmx.2020.100048.

Elkaliny, N.E., Alzamel, N.M., Moussa, S.H., Elodamy, N.I., Madkor, E.A., Ibrahim, E.M., et al., 2024. Macroalgae bioplastics: a sustainable shift to mitigate the ecological impact of petroleum-based plastics. Polymers 16 (9). Available at https://doi.org/10.3390/polym16091246.

Esteves, A.F., Salgado, E.M., Vilar, V.J.P., Gonçalves, A.L., Pires, J.C.M., 2024. A growth phase analysis on the influence of light intensity on microalgal stress and potential biofuel production. Energy Conversion and Management 311. Available at https://doi.org/10.1016/j.enconman.2024.118511.

Esteves, M.V., Marques, D.M.C., de Almeida, J.D., Faria, N.T., Ferreira, F.C., 2025. Marine microalgae–microorganism co-cultures: an insight into nannochloropsis sp. use and biotechnological applications. Foods 14 (9). Available at https://doi.org/10.3390/foods14091522.

Farshidfar, N., Iravani, S., Varma, R.S., 2023. Alginate-based biomaterials in tissue engineering and regenerative medicine. Marine Drugs 21 (3). Available at https://doi.org/10.3390/md21030189.

Faruque, M.O., Uddin, S., Hossain, M.M., Hossain, S.M.Z., Shafiquzzaman, M., Razzak, S.A., 2024. A comprehensive review on microalgae-driven heavy metals removal from industrial wastewater using living and nonliving microalgae. Journal of Hazardous Materials Advances 16. Available at https://doi.org/10.1016/j.hazadv.2024.100492.

Fawcett, C.A., Senhorinho, G.N.A., Laamanen, C.A., Scott, J.A., 2022. Microalgae as an alternative to oil crops for edible oils and animal feed. Algal Research 64. Available at https://doi.org/10.1016/j.algal.2022.102663.

Ferreira de Oliveira, A.P., Bragotto, A.P.A., 2022. Microalgae-based products: food and public health. Future Foods 6. Available at https://doi.org/10.1016/j.fufo.2022.100157.

Florides, F., Giannakoudi, M., Ioannou, G., Lazaridou, D., Lamprinidou, E., Loukoutos, N., et al., 2024. Water reuse: a comprehensive review. Environments - MDPI 11 (4). Available at https://doi.org/10.3390/environments11040081.

Geremia, E., Ripa, M., Catone, C.M., Ulgiati, S., 2021. A review about microalgae wastewater treatment for bioremediation and biomass production—a new challenge for europe. Environments - MDPI 8 (12). Available at https://doi.org/10.3390/environments8120136.

Guenka Scarcelli, P., Ruas, G., Lopez-Serna, R., Leite Serejo, M., Blanco, S., Árpád Boncz, M., et al., 2021. Integration of algae-based sewage treatment with anaerobic digestion of the bacterial-algal biomass and biogas upgrading. Bioresource Technology 340. Available at https://doi.org/10.1016/j.biortech.2021.125552.

Guo, B., Lu, X., Jiang, X., Shen, X.-L., Wei, Z., Zhang, Y., 2025. Artificial intelligence in advancing algal bioactive ingredients: production, characterization, and application. Foods 14 (10). Available at https://doi.org/10.3390/foods14101783.

Gurau, S., Imran, M., Ray, R.L., 2025. Algae: a cutting-edge solution for enhancing soil health and accelerating carbon sequestration – a review. Environmental Technology and Innovation 37. Available at https://doi.org/10.1016/j.eti.2024.103980.

Hamza, M.F., Guibal, E., Althumayri, K., Vincent, T., Yin, X., Wei, Y., et al., 2022. New process for the sulfonation of algal/PEI biosorbent for enhancing Sr(II) removal from aqueous solutions—application to seawater. Molecules (Basel, Switzerland) 27 (20). Available at https://doi.org/10.3390/molecules27207128.

He, J., Xi, L., Sun, X., Ge, B., Liu, D., Han, Z., et al., 2018. Enhanced hydrogen production through co-cultivation of Chlamydomonas reinhardtii CC-503 and a facultative autotrophic sulfide-oxidizing bacterium under sulfurated conditions. International Journal of Hydrogen Energy 43 (32), 15005–15013. Available at https://doi.org/10.1016/j.ijhydene.2018.06.081.

Hidayati, S., Zulferiyenni, Maulidia, U., Satyajaya, W., Hadi, S., 2021. Effect of glycerol concentration and carboxy methyl cellulose on biodegradable film characteristics of seaweed waste. Heliyon 7 (8). Available at https://doi.org/10.1016/j.heliyon.2021.e07799.

Hippler, M., Khosravitabar, F., 2024. Light-driven H2 production in Chlamydomonas reinhardtii: lessons from engineering of photosynthesis. Plants 13 (15). Available at https://doi.org/10.3390/plants13152114.

Hu, J., Meng, W., Su, Y., Qian, C., Fu, W., 2023. Emerging technologies for advancing microalgal photosynthesis and metabolism toward sustainable production. Frontiers in Marine Science 10. Available at https://doi.org/10.3389/fmars.2023.1260709.

Hu, L., Zhou, Y., Gong, Y., Luo, Y., Liu, X., Jiang, Z., 2025. Microalgal proteins for a circular bioeconomy: nutritional, material, and chemical valorization. Bioresource Technology 436. Available at https://doi.org/10.1016/j.biortech.2025.133049.

Imamoglu, E., 2024. Green extraction processes from renewable biomass to sustainable bioproducts. Bioresource Technology Reports 27. Available at https://doi.org/10.1016/j.biteb.2024.101952.

Irfan, F., Tasnim, N., Razzak, S.A., Uddin, S., 2025. Advances in capturing carbon and producing bioplastics through the microalgal approach towards a sustainable future: a review. Journal of Cleaner Production 486. Available at https://doi.org/10.1016/j.jclepro.2024.144576.

Janarthanan, Kumar, S., 2019. Extraction of alginate from brown seaweeds and evolution of bioactive alginate film coated textile fabrics for wound healing application. Journal of Industrial Textiles 49 (3), 328–351. Available at https://doi.org/10.1177/1528083718783331.

Jiang, X., Shan, X., Li, F., 2023. Improving the quality of reclaimed water via applying spirulina platensis to eliminate residual nitrate. International Journal of Environmental Research and Public Health 20 (3). Available at https://doi.org/10.3390/ijerph20032117.

Jui, T.J., Tasnim, A., Islam, S.M.R., Manjur, O.H.B., Hossain, M.S., Tasnim, N., et al., 2024. Optimal growth conditions to enhance Chlorella vulgaris biomass production in indoor phyto tank and quality assessment of feed and culture stock. Heliyon 10 (11). Available at https://doi.org/10.1016/j.heliyon.2024.e31900.

Kalita, N., Baruah, P.P., 2024. Copper removal efficacy and stress tolerance potential of Leptolyngbya sp. GUEco1015. Heliyon 10 (8). Available at https://doi.org/10.1016/j.heliyon.2024.e29131.

Kanno, M., Carroll, A.L., Atsumi, S., 2017. Global metabolic rewiring for improved CO2 fixation and chemical production in cyanobacteria. Nature Communications 8. Available at https://doi.org/10.1038/ncomms14724.

Kato, H., Suzuki, H., Wijffels, R.H., Schulze, P.S.C., Hulatt, C.J., 2024. Thermal responses of Tetradesmus obliquus for industrial outdoor cultivation. Bioresource Technology Reports 27. Available at https://doi.org/10.1016/j.biteb.2024.101909.

Kaur, M., Bhatia, S., Bagchi, D., Tak, Y., Kaur, G., Kaur, C., et al., 2025. Enhancing microalgal proteins for nutraceutical and functional food applications. Future Foods 11. Available at https://doi.org/10.1016/j.fufo.2025.100564.

Khan, N., Sudhakar, K., Mamat, R., 2021. Role of biofuels in energy transition, green economy and carbon neutrality. Sustainability (Switzerland) 13 (22). Available at https://doi.org/10.3390/su132212374.

Khanra, A., Vasistha, S., Rai, M.P., Cheah, W.Y., Khoo, K.S., Chew, K.W., et al., 2022. Green bioprocessing and applications of microalgae-derived biopolymers as a renewable feedstock: circular bioeconomy approach. Environmental Technology and Innovation 28. Available at https://doi.org/10.1016/j.eti.2022.102872.

Kishi, M., Yamada, Y., Katayama, T., Matsuyama, T., Toda, T., 2020. Carbon Mass Balance in Arthrospira platensis culture with medium recycle and high CO2 supply. Applied Sciences (Switzerland) 10 (1). Available at https://doi.org/10.3390/app10010228.

Koh, H.G., Cho, J.M., Jeon, S., Chang, Y.K., Lee, B., Kang, N.K., 2023. Transcriptional insights into Chlorella sp. ABC-001: a comparative study of carbon fixation and lipid synthesis under different CO2 conditions. Biotechnology for Biofuels and Bioproducts 16 (1). Available at https://doi.org/10.1186/s13068-023-02358-4.

Kumari, S., Dalania, K., Mogatra, S., 2025. Sustainable energy solutions: the role of biotechnology and algal biofuels in environmental preservation. Discover Biotechnology 2. Available at https://doi.org/10.1007/s44340-025-00023-0.

Kuppan, N., Padman, M., Mahadeva, M., Srinivasan, S., Devarajan, R., 2024. A comprehensive review of sustainable bioremediation techniques: eco friendly solutions for waste and pollution management. Waste Management Bulletin 2 (3), 154–171. Available at https://doi.org/10.1016/j.wmb.2024.07.005.

Lavado-Meza, C., De la Cruz-Cerrón, L., Lavado-Puente, C., Angeles-Suazo, J., Dávalos-Prado, J.Z., 2023. Efficient lead Pb(II) removal with chemically modified nostoc commune biomass. Molecules (Basel, Switzerland) 28 (1). Available at https://doi.org/10.3390/molecules28010268.

Le, V.V., Esterhuizen, M., Tran, Q.-G., Yun, J.-H., Jung, M.-Y., Lee, S.-A., 2025. Harnessing microalgae to mitigate the environmental impact of the cement industry: Emission reduction and bio-cement production. Carbon Capture Science and Technology 16. Available at https://doi.org/10.1016/j.ccst.2025.100448.

Li, S.F., Wang, C.C., Taidi, B., 2023. Effective CO2 capture by the fed-batch culture of Chlorella vulgaris. Journal of Environmental Chemical Engineering 11 (5). Available at https://doi.org/10.1016/j.jece.2023.110889.

Liao, K., Grandgeorge, P., Jimenez, A.M., Nguyen, B.H., Roumeli, E., 2023. Effects of mechanical cell disruption on the morphology and properties of spirulina-PLA biocomposites. Sustainable Materials and Technologies 36. Available at https://doi.org/10.1016/j.susmat.2023.e00591.

Liu, G., Zhang, T., 2025. Advances in hydrothermal carbonization for biomass wastewater valorization: optimizing nitrogen and phosphorus nutrient management to enhance agricultural and ecological outcomes. Water (Switzerland) 17 (6). Available at https://doi.org/10.3390/w17060800.

Liyanaarachchi, V.C., Nishshanka, G.K.S.H., Nimarshana, P.H.V., Ariyadasa, T.U., Attalage, R.A., 2020. Development of an artificial neural network model to simulate the growth of microalga Chlorella vulgaris incorporating the effect of micronutrients. Journal of Biotechnology 312, 44–55. Available at https://doi.org/10.1016/j.jbiotec.2020.02.010.

Lv, H., Wang, Q.-E., Wang, S., Qi, B., He, J., Jia, S., 2019. Enhancing biomass production of Dunaliella salina via optimized combinational application of phytohormones. Aquaculture (Amsterdam, Netherlands) 503, 146–155. Available at https://doi.org/10.1016/j.aquaculture.2018.12.077.

Malletzidou, L., Kyratzopoulou, E., Kyzaki, N., Nerantzis, E., Kazakis, N.A., 2025. Towards the sustainable removal of heavy metals from wastewater using Arthrospira platensis: a laboratory-scale approach in the context of a green circular economy. Applied Sciences (Switzerland) 15 (2). Available at https://doi.org/10.3390/app15020791.

Manochkumar, J., Jonnalagadda, A., Cherukuri, A.K., Vannier, B., Janjaroen, D., Chandrasekaran, R., et al., 2024. Machine learning-based prediction models unleash the enhanced production of fucoxanthin in Isochrysis galbana. Frontiers in Plant Science 15. Available at https://doi.org/10.3389/fpls.2024.1461610.

Manu, L., Mokolensang, J.F., Ben Gunawan, W., Setyawardani, A., Salindeho, N., Syahputra, R.A., et al., 2024. Photobioreactors are beneficial for mass cultivation of microalgae in terms of areal efficiency, climate implications, and metabolites content. Journal of Agriculture and Food Research 18. Available at https://doi.org/10.1016/j.jafr.2024.101282.

Metha, C., Pawar, S., Suvarna, V., 2024. Recent advancements in alginate-based films for active food packaging applications. Sustainable Food Technology 2 (5), 1246–1265. Available at https://doi.org/10.1039/D3FB00216K.

Michalak, I., Baśladyńska, S., Mokrzycki, J., Rutkowski, P., 2019. Biochar from a freshwater macroalga as a potential biosorbent for wastewater treatment. Water (Switzerland) 11 (7). Available at https://doi.org/10.3390/w11071390.

Minhas, L.A., Kaleem, M., Farooqi, H.M.U., Kausar, F., Waqar, R., Bhatti, T., et al., 2024. Algae-derived bioactive compounds as potential pharmaceuticals for cancer therapy: a comprehensive review. Algal Research 78. Available at https://doi.org/10.1016/j.algal.2024.103396.

Mohammed, A., Mohammed, C., Mautner, A., Kistow, M., Chaitram, P., Bismarck, A., et al., 2024. On the performance of Sargassum-derived calcium alginate ion exchange resins for Pb2+ adsorption: batch and packed bed applications. Environmental Science and Pollution Research 31 (21), 31224–31239. Available at https://doi.org/10.1007/s11356-024-33314-w.

Mohammed, A.H., Mohammed Shartooh, S., Trigui, M., 2025. Biosorption and isotherm modeling of heavy metals using Phragmites australis. Sustainability (Switzerland) 17 (12). Available at https://doi.org/10.3390/su17125366.

Mohan, A., Antony, A.R., Greeshma, K., Yun, J.-H., Ramanan, R., Kim, H.-S., 2022. Algal biopolymers as sustainable resources for a net-zero carbon bioeconomy. Bioresource Technology 344. Available at https://doi.org/10.1016/j.biortech.2021.126397.

Moreira, A.S.P., Gonçalves, J., Sousa, F., Maia, I., Pereira, H., Silva, J., et al., 2023. Potential of coccolithophore microalgae as fillers in starch-based films for active and sustainable food packaging. Foods 12 (3). Available at https://doi.org/10.3390/foods12030513.

Mosleh, Y., Nafea, E.-S., Heham, S., Mofeed, J., 2023. Heavy metals removal from industrial wastewater using dry green macro alga Ulva lactuca linnaeus. Egyptian Journal of Aquatic Biology and Fisheries 27 (2), 343–360. Available at https://doi.org/10.21608/ejabf.2023.292265.

Nair, L.G., Agrawal, K., Verma, P., 2022. An overview of sustainable approaches for bioenergy production from agro-industrial wastes. Energy Nexus 6. Available at https://doi.org/10.1016/j.nexus.2022.100086.

Nath, S., Das Astapati, A., Naha, A., Sharma, I., Shah, M.P., 2024. Algal-based membrane bioreactors for effective removal of hazardous and toxic contaminants: a comprehensive review. Current Research in Biotechnology 7. Available at https://doi.org/10.1016/j.crbiot.2024.100201.

Navarro López, E., Robles Medina, A., González Moreno, P.A., Jiménez Callejón, M.J., Esteban Cerdán, L., Martín Valverde, L., et al., 2015. Enzymatic production of biodiesel from nannochloropsis gaditana lipids: influence of operational variables and polar lipid content. Bioresource Technology 187, 346–353. Available at https://doi.org/10.1016/j.biortech.2015.03.126.

Nazloo, E.K., Danesh, M., Sarrafzadeh, M.-H., Moheimani, N.R., Ennaceri, H., 2024. Biomass and hydrocarbon production from Botryococcus braunii: a review focusing on cultivation methods. Science of the Total Environment 926. Available at https://doi.org/10.1016/j.scitotenv.2024.171734.

Nurlaila, I., Irawati, W., Purwandari, K., Pardamean, B., 2021. K-means clustering model to discriminate copper-resistant bacteria as bioremediation agents. Procedia Computer Science 179, 804–812. Available at https://doi.org/10.1016/j.procs.2021.01.068.

Nwoba, E.G., Parlevliet, D.A., Laird, D.W., Alameh, K., Moheimani, N.R., 2019. Light management technologies for increasing algal photobioreactor efficiency. Algal Research 39. Available at https://doi.org/10.1016/j.algal.2019.101433.

Obayomi, O.V., Olawoyin, D.C., Oguntimehin, O., Mustapha, L.S., Kolade, S.O., Oladoye, P.O., et al., 2024. Exploring emerging water treatment technologies for the removal of microbial pathogens. Current Research in Biotechnology 8. Available at https://doi.org/10.1016/j.crbiot.2024.100252.

Oladimeji, Oyedemi, Emetere, Agboola, Adeoye, Odunlami, 2024. Review on the impact of heavy metals from industrial wastewater effluent and removal technologies. Heliyon 10 (23). Available at https://doi.org/10.1016/j.heliyon.2024.e40370.

Omokaro, G.O., Nafula, Z.S., Iloabuchi, N.E., Chikukula, A.A., Osayogie, O.G., Nnoli, E.C., 2025. Microalgae as biofactories for sustainable applications: advancing carbon sequestration, bioenergy, and environmental remediation. Sustainable Chemistry for Climate Action 7. Available at https://doi.org/10.1016/j.scca.2025.100098.

Ooi, W.C., Dominic, D., Kassim, M.A., Baidurah, S., 2023. Biomass fuel production through cultivation of microalgae Coccomyxa dispar and Scenedesmus parvus in palm oil mill effluent and simultaneous phycoremediation. Agriculture (Switzerland) 13 (2). Available at https://doi.org/10.3390/agriculture13020336.

Oostlander, van Houcke, Wijffels, Barbosa, 2020. Microalgae production cost in aquaculture hatcheries. Aquaculture (Amsterdam, Netherlands) 525. Available at https://doi.org/10.1016/j.aquaculture.2020.735310.

Ordóñez, J.I., Cortés, S., Maluenda, P., Soto, I., 2023. Biosorption of heavy metals with algae: critical review of its application in real effluents. Sustainability (Switzerland) 15 (6). Available at https://doi.org/10.3390/su15065521.

Otálora, Guzmán, Acién, Berenguel, Reul, 2021. Microalgae classification based on machine learning techniques. Algal Research 55. Available at https://doi.org/10.1016/j.algal.2021.102256.

Parreidt, T.S., Müller, K., Schmid, M., 2018. Alginate-based edible films and coatings for food packaging applications. Foods 7 (10). Available at https://doi.org/10.3390/foods7100170.

Paul, S., Bera, S., Dasgupta, R., Mondal, S., Roy, S., 2021. Review on the recent structural advances in open and closed systems for carbon capture through algae. Energy Nexus 4. Available at https://doi.org/10.1016/j.nexus.2021.100032.

Pinto, G.C., Piazza, R.D., Mendes Soares, I.P., Anselmi, C., Coimbra, E.T.C., Tavares, P.M.H., et al., 2025. β-CPP bioceramics in alginate 3D scaffolds as a new material for mineralized tissue regeneration. Ceramics International 51 (18), 25125–25137. Available at https://doi.org/10.1016/j.ceramint.2025.03.194.

Pipliya, S., Kumar, S., Kumar Gupta, R., Das, R.S., Meena, D., Srivastav, P.P., et al., 2025. The future of algal proteins: innovations in extraction and modifications, functional properties, and sustainable food applications. Future Foods 11. Available at https://doi.org/10.1016/j.fufo.2025.100549.

Pohl, A., 2020. Removal of heavy metal ions from water and wastewaters by sulfur-containing precipitation agents. Water, Air, and Soil Pollution 231 (10). Available at https://doi.org/10.1007/s11270-020-04863-w.

Popescu, E.-R., Einbu, A., Israelsen, L., Eriksen, G., Hardy, N., Ingebrigtsen, R.A., et al., 2025. Web application for automated counting and size analysis of industrial microalgae monocultures using image processing and convolutional neural network based deep learning. Algal Research 90. Available at https://doi.org/10.1016/j.algal.2025.104173.

Pozzobon, V., Levasseur, W., Camarena-Bernard, C., 2024. Chlorella vulgaris cold preservation (4°C) as a means to stabilize biomass for bioreactor inoculation: a six-month study. Aquacultural Engineering 107. Available at https://doi.org/10.1016/j.aquaeng.2024.102449.

Pradhan, D., Sukla, L.B., Mishra, B.B., Devi, N., 2019. Biosorption for removal of hexavalent chromium using microalgae Scenedesmus sp. Journal of Cleaner Production 209, 617–629. Available at https://doi.org/10.1016/j.jclepro.2018.10.288.

Qiao, J., Chen, Y., Zhou, R., Chen, Y., Cai, R., Zhao, T., et al., 2025. Sodium alginate-based nanocomposite hydrogel membrane for removal of heavy metal ions and dyes in water. International Journal of Biological Macromolecules 307. Available at https://doi.org/10.1016/j.ijbiomac.2025.142109.

Rangel Pinto, J.D., Guerrero, J.L., Rivera, L., Parada-Pinilla, M.P., Cala, M.P., López, G., et al., 2024. Predicting the microalgae lipid profile obtained by supercritical fluid extraction using a machine learning model. Frontiers in Chemistry 12. Available at https://doi.org/10.3389/fchem.2024.1480887.

Riaz, A., Deng, F., Chen, G., Jiang, W., Zheng, Q., Riaz, B., et al., 2022. Molecular regulation and evolution of redox homeostasis in photosynthetic machinery. Antioxidants 11 (11). Available at https://doi.org/10.3390/antiox11112085.

Rossi, T.S., Francescato, L., Gupte, A.P., Favaro, L., Treu, L., Campanaro, S., 2025. Harnessing the potential of Cupriavidus necator for CO2 capture from alcoholic fermentation and its bioconversion into poly(3-hydroxybutyrate). Bioresource Technology 419. Available at https://doi.org/10.1016/j.biortech.2025.132060.

Rout, S., Yadav, S., Pansare, L.G., 2025. Simultaneous removal of heavy metals and Escherichia coli from water using silver-doped hydroxyapatite-coated activated carbon nano composite-alginate beads. Next Sustainability 5. Available at https://doi.org/10.1016/j.nxsust.2024.100076, https://doi.org/10.1016/j.nxsust.2024.100076.

Salinas-García, M., Calatrava-Arrizabalaga, P., Ciardi, M., Villaró-Cos, S., Lafarga, T., 2025. Development and reutilisation of a fertiliser-based culture medium for the commercial production of Chlorella sorokiniana. Scientific Reports 15 (1). Available at https://doi.org/10.1038/s41598-025-08320-8.

Samoraj, M., Çalış, D., Trzaska, K., Mironiuk, M., Chojnacka, K., 2024. Advancements in algal biorefineries for sustainable agriculture: biofuels, high-value products, and environmental solutions. Biocatalysis and Agricultural Biotechnology 58. Available at https://doi.org/10.1016/j.bcab.2024.103224.

Sanz Smachetti, M.E., Coronel, C.D., Salerno, G.L., Curatti, L., 2020. Sucrose-to-ethanol microalgae-based platform using seawater. Algal Research 45. Available at https://doi.org/10.1016/j.algal.2019.101733.

Sarangi, N.V., Rajkumar, R., Kumar, N.S., Arunkumar, P., Alromaeh, A.I., AlReshaidan, S.B., et al., 2025. Seaweed biosorption: a green solution for heavy metal remediation in aquatic and soil environments. Desalination and Water Treatment 321. Available at https://doi.org/10.1016/j.dwt.2025.101036.

Sarma, U., Hoque, M.E., Thekkangil, A., Venkatarayappa, N., Rajagopal, S., 2024. Microalgae in removing heavy metals from wastewater – an advanced green technology for urban wastewater treatment. Journal of Hazardous Materials Advances 15. Available at https://doi.org/10.1016/j.hazadv.2024.100444.

Sasounian, R., Martinez, R.M., Lopes, A.M., Giarolla, J., Rosado, C., Magalhães, W.V., et al., 2024. Innovative approaches to an eco-friendly cosmetic industry: a review of sustainable ingredients. Clean Technologies 6 (1), 176–198. Available at https://doi.org/10.3390/cleantechnol6010011.

Satchanska, G., Davidova, S., Petrov, P.D., 2024. Natural and synthetic polymers for biomedical and environmental applications. Polymers 16 (8). Available at https://doi.org/10.3390/polym16081159.

Satpati, G.G., Kundu, D., Rajak, R.C., Gupta, S., Kim, J.-W., Davoodbasha, M.A., 2024. Algal-based membrane reactor for the remediation of emerging contaminants from wastewater: mechanism, synthesis and technological advancement. Algal Research 79. Available at https://doi.org/10.1016/j.algal.2024.103465.

Sayadi, M.H., Rashki, O., Shahri, E., 2019. Application of modified Spirulina platensis and Chlorella vulgaris powder on the adsorption of heavy metals from aqueous solutions.

Journal of Environmental Chemical Engineering 7 (3). Available at https://doi.org/10.1016/j.jece.2019.103169.
Sharma, A.K., Jaryal, S., Sharma, S., Dhyani, A., Tewari, B.S., Mahato, N., 2025. Biofuels from microalgae: a review on microalgae cultivation, biodiesel production techniques and storage stability. Processes 13 (2). Available at https://doi.org/10.3390/pr13020488.
Shefer, S., Lebendiker, M., Finkelshtein, A., Chamovitz, D.A., Golberg, A., 2022. Ulvan crude extract's chemical and biophysical profile and its effect as a biostimulant on Arabidopsis thaliana. Algal Research 62. Available at https://doi.org/10.1016/j.algal.2021.102609.
Sheik, A.G., Kumar, A., Ansari, F.A., Raj, V., Peleato, N.M., Patan, A.K., et al., 2024. Reinvigorating algal cultivation for biomass production with digital twin technology - a smart sustainable infrastructure. Algal Research 84. Available at https://doi.org/10.1016/j.algal.2024.103779.
Shelake, R.M., Khalid, M.A., Kim, J.-Y., 2025. Microalgal bioengineering for futuristic applications in synthetic and space biology. Biotechnology Advances 84. Available at https://doi.org/10.1016/j.biotechadv.2025.108665.
Siddiqui, V.U., Ilyas, Sapuan, Hamid, N.H.A., Khoo, Chowdhury, A., et al., 2025. Alginate-based materials as adsorbent for sustainable water treatment. International Journal of Biological Macromolecules 298. Available at https://doi.org/10.1016/j.ijbiomac.2025.139946.
Silva, J.A., 2025. Sustainable water resource management to achieve net-zero carbon in the water industry: a systematic review of the literature. Water (Switzerland) 17 (14). Available at https://doi.org/10.3390/w17142136.
Singh, A., Sharma, A., Ahmed, A., Sundramoorthy, A.K., Furukawa, H., Arya, S., et al., 2021. Recent advances in electrochemical biosensors: applications, challenges, and future scope. Biosensors 11 (9). Available at https://doi.org/10.3390/bios11090336.
Singh, B.J., Chakraborty, A., Sehgal, R., 2023a. A systematic review of industrial wastewater management: evaluating challenges and enablers. Journal of Environmental Management 348. Available at https://doi.org/10.1016/j.jenvman.2023.119230.
Singh, V., Verma, M., Chivate, M.S., Mishra, V., 2023b. Machine learning-based optimisation of microalgae biomass production by using wastewater. Journal of Environmental Chemical Engineering 11 (6). Available at https://doi.org/10.1016/j.jece.2023.111387.
Six, Dauvillée, Lancelon-Pin, Dimitriades-Lemaire, Compadre, Dubreuil, et al., 2024. From raw microalgae to bioplastics: conversion of Chlorella vulgaris starch granules into thermoplastic starch. Carbohydrate Polymers 342. Available at https://doi.org/10.1016/j.carbpol.2024.122342.
Sousa, G., Ferreira-Dias, S., Tecelão, C., Alves, V.D., 2025. Potential of marine biomolecules: advances in extraction and applications of proteins, polysaccharides, and antioxidant compounds. Foods 14 (15). Available at https://doi.org/10.3390/foods14152555.
Srimongkol, P., Sangtanoo, P., Songserm, P., Watsuntorn, W., Karnchanatat, A., 2022. Microalgae-based wastewater treatment for developing economic and environmental sustainability: current status and future prospects. Frontiers in Bioengineering and Biotechnology 10. Available at https://doi.org/10.3389/fbioe.2022.904046.
Stoett, P., Scrich, V.M., Elliff, C.I., Andrade, M.M., de M Grilli, N., Turra, A., 2024. Global plastic pollution, sustainable development, and plastic justice. World Development 184. Available at https://doi.org/10.1016/j.worlddev.2024.106756.
Sun, N., Skaggs, R.L., Wigmosta, M.S., Coleman, A.M., Huesemann, M.H., Edmundson, S.J., 2020. Growth modeling to evaluate alternative cultivation strategies to enhance national microalgal biomass production. Algal Research 49. Available at https://doi.org/10.1016/j.algal.2020.101939.

Supuran, C.T., 2023. Carbonic anhydrase versatility: from pH regulation to CO2 sensing and metabolism. Frontiers in Molecular Biosciences 10. Available at https://doi.org/10.3389/fmolb.2023.1326633.

Syed, T., Krujatz, F., Ihadjadene, Y., Mühlstädt, G., Hamedi, H., Mädler, J., et al., 2024. A review on machine learning approaches for microalgae cultivation systems. Computers in Biology and Medicine 172. Available at https://doi.org/10.1016/j.compbiomed.2024.108248.

Tammina, S.K., Priyadarshi, R., Khan, A., Manzoor, A., Rahman, R.S.H.A., Banat, F., 2025. Recent developments in alginate-based nanocomposite coatings and films for biodegradable food packaging applications. International Journal of Biological Macromolecules 295. Available at https://doi.org/10.1016/j.ijbiomac.2025.139480.

Tarigan, M., Raji, S., Al-Fatesh, H., Czermak, P., Ebrahimi, M., 2025. The occurrence of micropollutants in the aquatic environment and technologies for their removal. Processes 13 (3). Available at https://doi.org/10.3390/pr13030843.

Teng, S.Y., Yew, G.Y., Sukačová, K., Show, P.L., Máša, V., Chang, J.-S., 2020. Microalgae with artificial intelligence: a digitalized perspective on genetics, systems and products. Biotechnology Advances 44. Available at https://doi.org/10.1016/j.biotechadv.2020.107631.

Tennakoon, P., Chandika, P., Yi, M., Jung, W.-K., 2023. Marine-derived biopolymers as potential bioplastics, an eco-friendly alternative. iScience 26 (4). Available at https://doi.org/10.1016/j.isci.2023.106404.

Thammasane, S., Kaosol, T., 2018. Impact of chemical coagulants for Oscillatoria sp. removal from raw water on chemical coagulation process. American Journal of Environmental Sciences 14 (6), 257–265. Available at https://doi.org/10.3844/ajessp.2018.257.265, https://doi.org/10.3844/ajessp.2018.257.265.

Tommasi, I.C., 2021. The mechanism of rubisco catalyzed carboxylation reaction: chemical aspects involving acid-base chemistry and functioning of the molecular machine. Catalysts 11 (7). Available at https://doi.org/10.3390/catal11070813.

Torres, E., 2020. Biosorption: a review of the latest advances. Processes 8 (12), 1–23. Available at https://doi.org/10.3390/pr8121584.

Touliabah, H.E.-S., Almutairi, A.W., 2021. Effect of phytohormones supplementation under nitrogen depletion on biomass and lipid production of nannochloropsis oceanica for integrated application in nutrition and biodiesel. Sustainability (Switzerland) 13 (2), 1–12. Available at https://doi.org/10.3390/su13020592.

Ugya, A.Y., Sheng, Y., Chen, H., Wang, Q., 2024. Microalgal bioengineering: a futuristic tool for carbon capture. Results in Engineering 24. Available at https://doi.org/10.1016/j.rineng.2024.102990.

Ugya, A.Y., Chen, H., Wang, Q., 2025. Enhancing the sustainability of microalgae cultivation through biosensing technology. Materials Today Sustainability 31. Available at https://doi.org/10.1016/j.mtsust.2025.101139.

Usman, H.M., Kamaroddin, M.F., Sani, M.H., Malek, N.A.N.N., Omoregie, A.I., Zainal, A., 2024. A comparative analysis assessing growth dynamics of locally isolated Chlorella sorokiniana and Chlorella vulgaris for biomass and lipid production with biodiesel potential. Bioresource Technology 403. Available at https://doi.org/10.1016/j.biortech.2024.130868.

Vavra, L., Gono, M., Klosok-Bazan, I., Svehlakova, H., 2025. Reuse of pretreated household wastewater for decentralized food production. Water (Switzerland) 17 (3). Available at https://doi.org/10.3390/w17030372.

Verma, S.K., Prasad, A., Sonika, Katiyar, V., 2024. State of art review on sustainable biodegradable polymers with a market overview for sustainability packaging. Materials Today Sustainability 26. Available at https://doi.org/10.1016/j.mtsust.2024.100776.

Wan Osman, W.N.A., Mat Nawi, N.I., Samsuri, S., Bilad, M.R., Khan, A.L., Hunaepi, H., et al., 2021. Ultra low-pressure filtration system for energy efficient microalgae filtration. Heliyon 7 (6). Available at https://doi.org/10.1016/j.heliyon.2021.e07367.

Waqar, R., Kaleem, M., Iqbal, J., Minhas, L.A., Haris, M., Chalgham, W., et al., 2023. Kinetic and equilibrium studies on the adsorption of lead and cadmium from aqueous solution using Scenedesmus sp. Sustainability (Switzerland) 15 (7). Available at https://doi.org/10.3390/su15076024.

Wei, L., You, W., Gong, Y., El Hajjami, M., Liang, W., Xu, J., et al., 2020. Transcriptomic and proteomic choreography in response to light quality variation reveals key adaption mechanisms in marine Nannochloropsis oceanica. Science of the Total Environment 720. Available at https://doi.org/10.1016/j.scitotenv.2020.137667.

Wei, Z., Alam, S., Verma, M., 2022. Integrating water quality data with a bayesian network model to improve spatial and temporal phosphorus attribution: application to the maumee river basin. ESS Open Archive. Available at https://doi.org/10.1002/essoar.10512828.1, https://doi.org/10.1002/essoar.10512828.1.

Wen, H., Soyekwo, F., Liu, C., 2022. Highly permeable forward osmosis membrane with selective layer "hooked" to a hydrophilic Cu-Alginate intermediate layer for efficient heavy metal rejection and sludge thickening. Journal of Membrane Science 647. Available at https://doi.org/10.1016/j.memsci.2021.120247.

Wirth, R., Lakatos, G., Böjti, T., Maróti, G., Bagi, Z., Kis, M., et al., 2015. Metagenome changes in the mesophilic biogas-producing community during fermentation of the green alga Scenedesmus obliquus. Journal of Biotechnology 215, 52–61. Available at https://doi.org/10.1016/j.jbiotec.2015.06.396.

Xu, Y., Zhang, X., Xiao, S., Peng, B.-Y., Chen, J., Yang, L., et al., 2024. Distinct exposure impact of non-degradable and biodegradable microplastics on freshwater microalgae (Chlorella pyrenoidosa): implications for polylactic acid as a sustainable plastic alternative. Journal of Hazardous Materials 480. Available at https://doi.org/10.1016/j.jhazmat.2024.136265.

Yang, Y., Qi, C., Zhang, R., Li, Y., Tian, G., Gao, L., et al., 2025. Upcycling algal biomass into biodegradable materials: enhancing structural properties and performance through hydrophobic modifications. Algal Research 90. Available at https://doi.org/10.1016/j.algal.2025.104175.

Yun, J.-H., Cho, D.-H., Lee, S., Heo, J., Tran, Q.-G., Chang, Y.K., et al., 2018. Hybrid operation of photobioreactor and wastewater-fed open raceway ponds enhances the dominance of target algal species and algal biomass production. Algal Research 29, 319–329. Available at https://doi.org/10.1016/j.algal.2017.11.037.

Zermeño-Pérez, D., Chouirfa, H., Rodriguez, B.J., Dürig, T., Duffy, P., Cróinín, T.Ó., 2024. Bioresorbable polyester coatings with antifouling and antimicrobial properties for prevention of biofilm formation in early stage infections on Ti6Al4V hard-tissue implants. ACS Applied Bio Materials 7 (8), 5728–5739. Available at https://doi.org/10.1021/acsabm.4c00832.

Zhang, M., Yao, L., Maleki, E., Liao, B.-Q., Lin, H., 2019. Membrane technologies for microalgal cultivation and dewatering: recent progress and challenges. Algal Research 44. Available at https://doi.org/10.1016/j.algal.2019.101686.

Zhang, H., Cheng, J., Ao, Q., 2021. Preparation of alginate-based biomaterials and their applications in biomedicine. Marine Drugs 19 (5). Available at https://doi.org/10.3390/md19050264.

Zhang, Z., Zhu, G., Peng, X., 2024. Photorespiration in plant adaptation to environmental changes. Crop and Environment 3 (4), 203–212. Available at https://doi.org/10.1016/j.crope.2024.07.001.

Zhao, L., Cai, Z., Li, Y., Zhang, Y., 2024. Engineering Rubisco to enhance CO2 utilization. Synthetic and Systems Biotechnology 9 (1), 55–68. Available at https://doi.org/10.1016/j.synbio.2023.12.006.

Zhao, Y., Wei, M., Wang, Y., Zhang, Y., Liu, J., Zhang, H., et al., 2025. Two-dimensional calcium alginate crosslinked Ti3C2Tx membranes with stable interlayer spacing for efficient water purification. Journal of Alloys and Compounds 1036. Available at https://doi.org/10.1016/j.jallcom.2025.181847.

CHAPTER 8

Algal solution for greener cities: Phycoremediation approaches and applications

Madhavi Tiwari[1] and Maulin P. Shah[2]
[1]Department of Applied Science, Shri Rawatpura Sarkar University, Raipur, C.G., India
[2]Department of Research Impact and Outcome, Research & Development Cell, Lovely Professional University, Phagwara, Punjab, India

8.1 Introduction to urban phycoremediation

Urbanization represents one of the most significant global transformations of the 21st century. Currently, more than 55% of the world's population resides in urban areas, a figure projected to increase to nearly 68% by 2050 (United Nations, 2018). The ongoing growth of cities is closely linked to greater energy demands, increased business activity, and more industrial and domestic activities in one place. Because of this, metropolises are frequently "hotspots" for pollution and resource use, which can place considerable stress on local ecosystems to handle. Metropolises are important for economic growth and new technologies, but they also produce too much air pollution, wastewater, and solid waste. Poor infrastructure, poor waste separation, and limited treatment capacity make urban environmental stress even worse, especially in areas that are growing quickly. Activated sludge systems, chemical treatment, and adsorption-based treatments are just a few examples of traditional remediation technologies that use a lot of energy and release carbon, which makes them less effective for long-term urban planning. Many traditional systems also only focus on getting rid of pollutants and not recovering resources, which makes the life cycle cost higher and the circularity weaker. Phycoremediation, which uses algae to clean up air, water, and waste in metropolises, has become an integrated and nature-based approach to protect the environment in response to these problems. Algae-based systems provide concurrent carbon sequestration, nutrient production, oxygen generation, and biomass generation within a circular bioeconomy framework (Christenson and Sims, 2011; Rawat et al., 2011).

Applied Phycology
DOI: https://doi.org/10.1016/B978-0-443-24758-3.00009-5

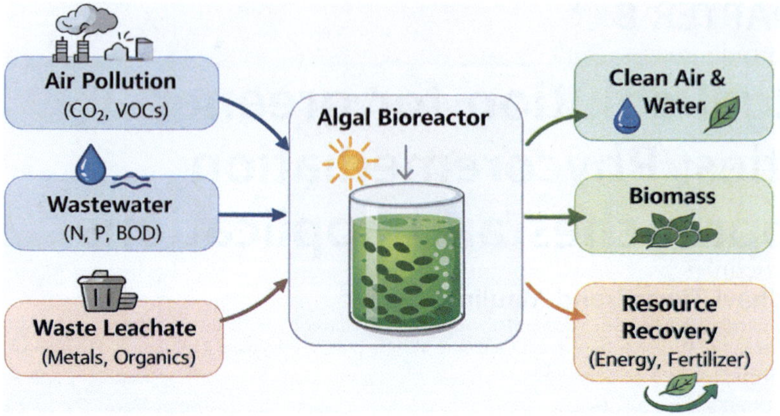

Figure 8.1 Conceptual model of urban phycoremediation. Integrated systems model of urban phycoremediation linking pollution streams (air, wastewater, landfill leachate) with algal bioprocessing and circular resource recovery pathways.

Phycoremediation can be designed as decentralized modular units, which is important because it makes it useful for premises, domestic clusters, commercial buildings, and settlements on the outskirts of metropolises. Fig. 8.1 shows how these functions work together in urban infrastructure.

8.1.1 Importance of algae for urban ecosystems

Microalgae and cyanobacteria are types of algae that can photosynthesize. They grow rapidlyand can alter their metabolism to adapt to different environmental conditions. Their functional importance for urban sustainability arises from their ability to absorb pollutants while simultaneously producing useful biomass. Algae can clean up the environment and generate resources at the same time, making them beneficial for circular urban metabolism. This distinguishes them from many other biological treatment agents. We can modify their effectiveness by controlling factors such as light, nutrients, aeration, and CO_2 supply, making it easy to integrate them into existing urban systems.

8.1.1.1 Fixing carbon and helping the climate

Microalgae have far higher photosynthetic efficiency than land plants when measured on an area basis (Chisti, 2007). They use the Calvin–Benson cycle to convert CO_2 from the atmosphere into organic biomass and release oxygen. In this way, they both capture carbon and enrich the surrounding environment with oxygen. This characteristic makes microalgae especially appealing in metropolitan areas with high levels of pollution, such as traffic corridors, industrial zones, and clusters of energy-intensive buildings. As a result, algae can be used into biofacades, rooftop photobioreactors, and roadside carbon capture modules. In addition to capturing CO_2, algal installations may also help regulate microclimates by shading building surfaces and reducing heat absorption. Fig. 8.1 shows that algal systems can be integrated into buildings and wastewater treatment units to act as decentralized carbon sinks while also generating biomass that can be harvested and utilized.

8.1.1.2 Taking in nutrients and cleansing water

Urban wastewater often contains high concentrations of nitrogen (NH_4^+, NO_3^-) and phosphorus (PO_4^{3-}), which contribute to eutrophication in receiving water bodies. Microalgae absorb these nutrients and transform them into cellular biomass. This decreases the nutrient content in the water and reduces the risk of downstream algal blooms (Pittman et al., 2011). This nutritional absorption is also significant in decentralized treatment settings, as small-scale units often face challenges with chemical dosage and operational performance. Additionally, algal cell walls feature functional groups including carboxyl, hydroxyl, and sulfate moieties that allow the uptake of heavy metals and organic contaminants (Wang and Chen, 2009). This implies that algae can effectively take up dissolved metal ions and certain types of organic matter, especially when they exist in mixed environments with bacteria that assist in these processes. Table 8.1 shows the different ways in which algae may assist cities in environmental management. It demonstrates how each mechanism works and how it affects the environment.

8.1.1.3 Combining a circular bioeconomy with spatial efficiency

Unlike traditional green spaces, algal farming systems do not require a lot of land and can be erected vertically in busy cities. This is very beneficial in regions where there is not enough room and different land-use considerations make it challenging to expand typical green belts. We can deploy algal systems on rooftops, building walls, and small tanks close to where wastewater originates. This makes point-of-treatment solutions feasible. Also, the collected biomass may be transformed into biofuels, biofertilizers, bioplastics, and charcoal, which helps the city's

Table 8.1 Functional roles of algae in urban ecosystems.

Urban challenge	Algal functional roles	Mechanism	Environmental outcome
CO_2 emissions	Carbon fixation	Photosynthesis	Reduced greenhouse gas emissions
Nutrient-rich wastewater	Nutrient uptake	Assimilation of nitrogen and phosphorus	Eutrophication control
Heavy metal contamination	Biosorption	Surface binding	Metal removal
Organic pollutants	Biodegradation	Enzymatic metabolism	Detoxification
Urban heat island	Oxygen release and shading	Biomass growth	Microclimate improvement

metabolism remain circular. This kind of valorization may transform "waste-derived nutrients" into products that can be sold, which helps fund initiatives and makes the process more sustainable in the long run. Fig. 8.1 depicts how these interrelated environmental services and circular flows work, and Table 8.1 offers a short description of how they function.

8.1.2 Summary of pollution in the air, water, and waste of metropolises

There are several pathways of urban pollution, including air, water, and land. The difficulty lies not only in the volume of pollutants but also in their interrelated transport between chambers, such as air deposition affecting water bodies, leachate contaminating groundwater, and runoff facilitating the transport of pollutants. Fig. 8.2 demonstrates how pollutants are redistributed via different media and pathways. Therefore, integrated approaches such as phycoremediation are increasingly essential, as they provide a multicompartment remediation strategy capable of simultaneously addressing cross-media pollutant transfer. By harnessing the metabolic versatility of microalgae, these systems enable concurrent carbon sequestration, nutrient recovery, and contaminant removal, thereby offering a sustainable and systems-level solution for complex urban pollution networks.

8.1.2.1 Air pollution in cities

Air pollution in cities is substantially caused by buses, factories, power plants, and construction work. These sources create intricate combinations of pollutants that vary based on location and time of day. When there is a lot of traffic or severe rainfall, they are generally at their worst. Some examples of primary pollutants include CO_2, nitrogen oxides (NO_x), sulfur dioxide (SO_2), volatile organic compound (VOC), and particulate matter ($PM_{2.5}$ and PM_{10}). Long-term exposure to these pollutants can cause cardiovascular and respiratory problems (WHO, 2021), as well as reduced productivity and increased healthcare costs. Air pollution in metropolises is harmful to human health, but it also reduces visibility, damages building structures, and affects the temperature in metropolises by trapping heat and influencing radiation balance. Consequently, urban air quality management is placing increasing emphasis on both measures that reduce emissions and those that capture pollutants. Fig. 8.2 depicts the primary ways that urban air pollutants spread and where they originate. This helps explain why algae-based mitigation measures are necessary in areas with high levels of exposure.

8.1.2.2 Water pollution in cities

Municipal wastewater contains high levels of biological oxygen demand (BOD), chemical oxygen demand (COD), nutrients, heavy metals, and

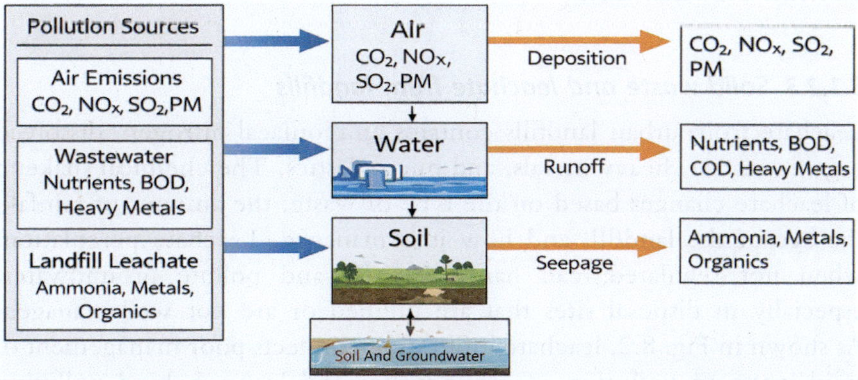

Figure 8.2 Simplified schematic of pollution sources and environmental pathways affecting urban air, water, and soil. The diagram illustrates major pollution sources, including air emissions, wastewater discharges, and landfill leachate, and their transport through environmental compartments via processes such as atmospheric deposition, surface runoff, and subsurface seepage, ultimately impacting soil and groundwater systems.

Table 8.2 Major urban pollutants and environmental impacts.

Pollution type	Major contaminants	Environmental impact	Public health risk
Air	CO_2, NO_x, $PM_{2.5}$	Climate change, smog	Respiratory disorders
Water	N, P, biological oxygen demand, metals	Eutrophication	Toxic exposure
Solid waste	Leachate, microplastics	Soil and groundwater contamination	Chronic toxicity

emerging pollutants, such as pharmaceuticals. In many cities, wastewater treatment capacity is either insufficient or not consistently available; consequently, some wastewater is only partially treated before being discharged directly into sewers, rivers, and ponds. Inadequate treatment leads to eutrophication, oxygen depletion, and ecological degradation (Rawat et al., 2011). These effects are generally accompanied by unpleasant odors and a loss of aquatic biodiversity. Wastewater is a good food source for algae since it has a lot of nutrients (N and P) in it. This creates an opportunityto treat wastewater and create more biomass at the same time. Table 8.2 lists the main types of pollutants, their usual effects, and the associated public health risks. This gives a reason for using algae to recover nutrients and clean them.

8.1.2.3 Solid waste and leachate from landfills

Leachate from urban landfills contains ammoniacal nitrogen, dissolved organic matter, heavy metals, and microplastics. The chemical makeup of leachate changes based on the type of waste, the amount of rainfall, the age of the landfill, and how it is managed. Leachate percolation, when not regulated, can harm the soil and pollute groundwater, especially in disposal sites that are unlined or are not well-managed. As shown in Fig. 8.2, leachate infiltration connects poor management of solid waste to pollution of groundwater and leaves behind polluting legacies that last for a long time. Algae-based systems may assist in leachate polishing (subsequent to primary conditioning) and in the treatment of runoff from landfills, particularly in scenarios where decentralized units are viable.

8.1.3 Nature-based and decentralized design strategies

New sustainability frameworks concentrate on design strategies that are decentralized and based in ecosystems. Decentralized systems, on the other hand, try to handle waste streams close to where they originate. This makes it easier to reclaim waste in the same area and cuts down on the need for transportation. Nature-based solutions (NBS) take this a step further by using natural processes that are beneficial to the environment and help ecosystems remain resilient. Urban phycoremediation is a blend of NBS and decentralization. It provides technological flexibility and offers clear environmental benefits.

8.1.3.1 Algal treatment systems that are not centralized

Rather than just having large wastewater treatment plants, homes, schools, and groups of businesses could have smaller algal units. These systems use less energy and piping, and they also encourage the recycling of nutrients and the reuse of water for purposes other than drinking. After primary treatment, decentralized units can also be used as "polishing" stages to improve the effluent and remove excess nutrients. Table 8.3 shows the differences between traditional systems and systems that use algae. It compares how much energy they use, how muchant carbon they emit, how many chemicals they require, and how well they recover resources. When choosing specialized options for long-term megacity planning, those in charge need to make these kinds of comparisons.

8.1.4 Nature-based solutions

Nature-based solutions utilize natural processes to address environmental challenges while enhancing ecosystem resilience (European Commission, 2015). Algal systems primarily rely on solar energy and require minimal

Table 8.3 Comparison between conventional treatment and algal phycoremediation.

Parameter	Conventional treatment	Algal phycoremediation
Energy requirement	High	Low-moderate
Chemical usage	Significant	Minimal
Carbon footprint	Positive	Neutral/negative
Resource recovery	Limited	High (biomass valorization)
Scalability	Centralized	Decentralized

chemical inputs, making them an environmentally sustainable option for urban infrastructure with a low carbon footprint. This makes them particularly suitable for cities aiming to reduce greenhouse gas emissions and promote climate-resilient development.

The incorporation of algae-based remediation systems into green walls and biofacades provides additional cobenefits. These systems not only improve air quality through carbon dioxide uptake and oxygen release but also contribute to indoor cooling and thermal regulation of buildings. Furthermore, algal installations enhance the aesthetic value of urban landscapes, promoting public acceptance of green technologies.

In addition, nature-based algal systems support biodiversity by creating microhabitats for beneficial microorganisms and contributing to ecological balance within urban environments. They also enable resource recovery by converting pollutants into valuable biomass, thereby supporting circular bioeconomy principles. Consequently, algae-based NBS represent a multifunctional approach that integrates environmental remediation, energy efficiency, and sustainable urban design.

8.1.5 Integration with smart urban systems

Adding IoT-enabled sensors to check pH, dissolved oxygen (DO), and CO_2 makes the system operate more reliably. Real-time monitoring makes it easier to make quick changes, such as adding nutrients, changing the inflow, or adjusting harvesting schedules. It also lowers the chance that the process will fail. Data-driven operation is especially important in decentralized units, where there might not be enough trained workers. Fig. 8.3 shows a theoretical model of decentralized algae-based remediation that uses digital monitoring technologies. It shows how advanced controls and natural remediation could work together. This kind of integration can also help with reporting compliance and assessing how well cities are managing the environment.

8.2 Choosing and growing different types of algae

The effectiveness of urban phycoremediation systems is significantly dependent upon the careful selection of algal species and the optimization of civilization methodologies. Urban wastewater, artificial water bodies, and atmospheric CO_2 emissions are all exemplifications of urban waste streams that have veritably different physical and chemical

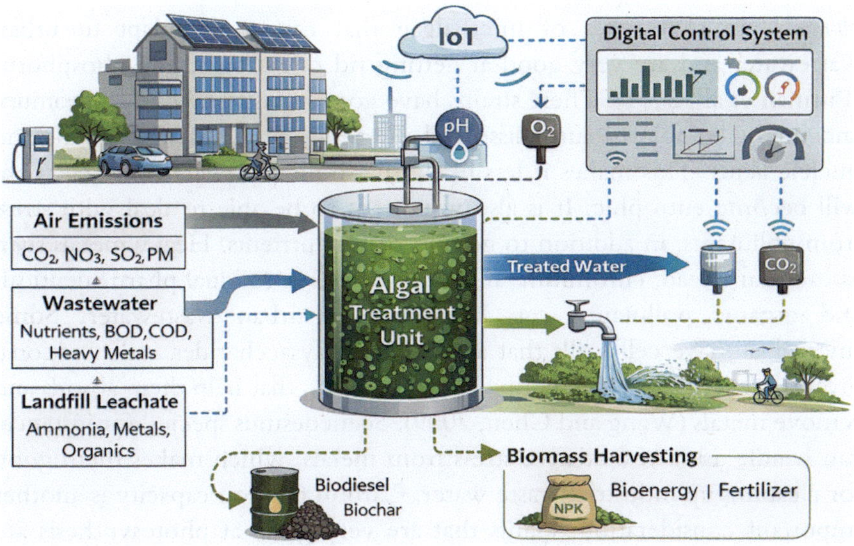

Figure 8.3 Smart decentralized algal remediation model for urban air and wastewater treatment. Decentralized algae-based remediation model integrated with smart urban infrastructure for localized treatment, monitoring, and resource recovery.

characteristics. These changes mean that we need strong and flexible algal strains that can handle changing environmental conditions while still producing a lot of biomass and removing pollutants. In addition, the civilization platform must facilitateeffective light utilization, gas exchange, and nutrient assimilation, while also being economically and operationally feasible within urban structure (Christenson and Sims, 2011; Rawat et al., 2011). Microalgae are especially appealing for remediation purposes due to their rapid growth rates, high photosynthetic efficiency, and metabolic flexibility. But not all species do well in wastewater conditions. So, when choosing species, you need to use specific performance criteria, similar as how well they can take up nutrients, how well they can handle toxic substances, resistance to contamination, and ease of harvesting.

8.2.1 Guidelines for selecting efficient algal strains

A good algal strain for cleaning up cities needs to grow quickly and produce a lot of biomass. Rapid growth facilitates wastewater treatment and speeds up the removal of nutrients. *Chlorella vulgaris* and *Scenedesmus obliquus* are two types of microalgae that can easily adapt to urban wastewater and are very good at getting rid of nitrogen and phosphorus (Pittman et al., 2011). These strains have good ways to take in ammonium and might be able to turn dissolved nutrients into cellular proteins and nucleic acids. This makes it less likely that bodies of water that get them will become eutrophic. It is also important to be able to deal with stress from pollutants, in addition to getting rid of nutrients. Heavy metals such as cadmium, lead, chromium, and zinc, as well as residual pharmaceuticals and organic pollutants, can be found in urban wastewater. Some microalgae have cell walls that are rich in polysaccharides and functional groups, such as carboxyl and hydroxyl moieties, that help them absorb and remove metals (Wang and Chen, 2009). Scenedesmus species, for instance, can handle a lot of oxidative stress from metals, which makes them good for cleaning up industrial waste water. Carbon fixation capacity is another important consideration. Plants that are very good at photosynthesis are very useful in systems that clean up both air and water. Chlorella and *Spirulina platensis* species have shown that they can absorb a lot of CO_2 when there is a lot of carbon in the air (Chisti, 2007). This function is helpful for decentralized systems located near roads or industrial areas. How well the system can handle biological contamination and protozoan grazing is also important for operational stability, especially in open systems such as raceway ponds. Strains that form thick cultures or have natural flocculation traits are less likely to get contaminated and are easier to harvest later.

Table 8.4 Representative microalgal strains for urban phycoremediation and their functional characteristics.

8.2.2 Cultivation systems

The cultivation system plays a major role in determining the productivity, contamination resistance, and scalability of a system. There are two main types of platforms for growing algae: open systems, including raceway ponds and high-rate algal ponds (HRAPs), and closed systems, such as photobioreactors. When it comes to cost, environmental control, and how hard it's to run, each system has its own benefits and downsides.

8.2.2.1 Raceway ponds

Waterway ponds are shallow, open-channel systems with paddle wheelsthat keep the water flowing continuously. They're useful for treating large

Table 8.4 Representative microalgal strains for urban phycoremediation and their functional characteristics.

Species	Key Characteristics	Primary Remediation Role	Reported Application
Chlorella vulgaris	High nutrient uptake, robust growth	N and P removal, CO_2 fixation	Municipal wastewater treatment
Scenedesmus obliquus	Metal tolerance, stress resistance	Heavy metal biosorption	Industrial effluent polishing
Spirulina platensis	High CO_2 assimilation	Carbon capture	Air remediation modules
Nannochloropsis sp.	High lipid accumulation	Biofuel integration	Integrated wastewater–bioenergy systems

volumes of urban wastewater since they're cheap to install and operate. The shallow depth permits greater light penetration, and mixing ensures that nutrients are distributed evenly and cells can circulate throughout the system. However, raceway systems may be subject to contamination, water loss through evaporation, and environmental fluctuations (Rawat et al., 2011). In open systems, biomass concentrations are generally lower than in closed systems. This may make harvesting more expensive.

8.2.2.2 High rates algal ponds

HRAPs are better than raceway systems because they facilitate nutrient removal and reduce costs. HRAPs have improved mixing dynamics and controlledhydraulic retention times, which help algae and bacteria function together. Algae produce oxygen through photosynthesis, which helps microorganisms break down organic matter. Bacteria, on the other hand, releaseCO_2, which helps algae grow (Pittman et al., 2011). This inter-actionbenefits both groups and makes remediation more effective while requiring lower aeration.

8.2.2.3 Photobioreactor systems

We can modify the temperature, CO_2 position, and peril of impurity in closed photobioreactors (PBRs), which gives you further control over the environment. In tubular photobioreactors, transparent tubes are arranged either horizontally or vertically. This provides them a high surface-to-volume ratio, which helps them capture more light. These systems are highly effective at removing CO_2 from industrial flue gas streams (Chisti, 2007). Flat- panel photobioreactors, on the other hand, disperse light more and can fit into into smaller spaces. This makes them suitable for urban biofacades and rooftop integration. A comparative evaluation of cultivation platforms is provided in Table 8.5. As shown in Table 8.5, closed photobioreactors offer greater operational control but require higher investment, whereas HRAPs provide a balanced compromise between cost and efficiency.

8.2.3 Settings for improvement

To get the most biomass and the best removal of contaminants, we need to optimize the operating parameters. pH, nutrient levels, CO_2 levels, and temperature are some of the most important factors. A pH level between 6.5 and 8.5 seems to be the best for growing microalgae. The pH can can

Table 8.5 Comparison of algal cultivation platforms for urban phycoremediation.

Parameter	Raceway pond	High-rate algal ponds	Photobioreactor
Capital cost	Low	Moderate	High
Environmental control	Limited	Improved	High
Contamination risk	High	Moderate	Low
Biomass density	Low	Moderate	High
Suitability for urban integration	Moderate	High	Very high

Table 8.6 Key operational parameters for optimized algal cultivation.

Parameter	Optimal range	Impact on system performance
pH	6.5–8.5	Influences nutrient solubility and enzyme activity
N:P ratio	10:1–16:1	Controls biomass productivity
CO_2 concentration	2%–10%	Enhances carbon fixation
Temperature	20°C–30°C	Regulates metabolic rate

increase duringphotosynthesis because bicarbonate is used up. If this is not kept under control, it could cause nutrient loss. We can keep the pH level just right and improve carbon fixation by controlling the amount of CO_2 (Chisti, 2007). Balanced nutrient availability is also important. The ratio of nitrogen to phosphorus has a big effect on how quickly things grow and what the biomass is made of. A molar ratio of NP between 10:1 and 16:1 is generally considered advantageous for long-term development (Pittman et al., 2011). Too much ammonia can be toxic, and not enough nutrients can cause lipids to accumulate. This could be useful in systems that make biofuels. Adding CO_2 speeds up photosynthesis and increases biomass production. People often use regulated systems with CO_2 levels that range from 2% to 10%. If the buffering capacity is not good enough, too much CO_2 can still make the water acidic and slow down growth. The speed of metabolism and the way enzymes work are both greatly affected by temperature. Microalgae that live in brackish water do best when the temperature is between 20°C and 30°C. Outdoor systems need to account for seasonal changes, which can affect productivity and nutrient removal. Table 8.6 shows the optimal operating ranges.

8.3 Using algae to clean water and wastewater

Microalgae-based treatment systems are getting more popular as sustainable alternatives to traditional wastewater treatment methods. These systems take advantage of algae's ability to take in nutrients, remove pollutants, and produce biomass that can be turned into useful products. Algal treatment systems are especially appealing in cities because they treat wastewater, capture carbon, and produce biomass all at the same time in a circular bioeconomy setting (Rawat et al., 2011; Christenson and Sims, 2011). Algae use solar energy and dissolved nutrients to drive natural remediation processes, which makes them suitable for both centralized wastewater treatment plants and decentralized urban water management systems. This is different from energy-intensive mechanical treatment methods. Photosynthesis, nutrient uptake, biosorption of heavy essence, and biodegradation of organic pollutants are the main ways that algae clean up wastewater. Algae release oxygen into the water through photosynthesis, which helps aerobic bacteria break down organic matter. This mutualistic relationship between algae and heterotrophic bacteria greatly speeds up the process of removing biological oxygen demand and COD from wastewater (Pittman et al., 2011). As a result, algal systems can be useful polishing tools in tertiary treatment processes.

8.3.1 Role of algae in municipal and decentralized water systems

Algae have been considerably studied for their use in municipal wastewater treatment systems, especially as components of tertiary treatment or polishing units. When municipal wastewater is released untreated into natural bodies of water, it can cause eutrophication because it contains high levels of nitrogen, phosphorus, organic matter, and suspended solids. Microalgae can use these nutrients as important growth substrates, which lowers nutrient levels while producing useful biomass (Rawat et al., 2011). In centralized municipal systems, algae-based technologies such as HRAPs are frequently used after primary and secondary treatment units. These systems make it easier to remove nutrients and reduce the costs of aeration and chemical dosing. In addition, the algal biomass collected from these systems can be converted into biofertilizers or bioenergy feedstock, which makes them even more economically feasible. In cities where centralized structure is inadequate or does not exist, decentralized wastewater

Table 8.7 Typical pollutant removal efficiencies in algae-based wastewater treatment systems.

Pollutant	Typical removal efficiency (%)	Mechanism
Nitrogen (NH_4^+, NO_3^-)	70%–95%	Assimilation into algal biomass
Phosphorus (PO_4^{3-})	60%–90%	Nutrient uptake and precipitation
Biological oxygen demand	60%–85%	Algal–bacterial symbiosis
Heavy metals	50%–90%	Biosorption and bioaccumulation

treatment systems are becoming increasingly popular. In these systems, small algal reactors can be installed at homes, schools, and businesses. These decentralized units treat greywater and partially treated sewage on-site, making it possible to use the water for landscaping, irrigation, and other nondrinking purposes. Decentralized algal treatment systems also support localized resource recovery and reduce the burden on municipal sewer networks. Table 8.7 shows the average removal efficiency that algal wastewater treatment systems can achieve.

8.3.2 Functional modes: batch and nonstop flow

Depending on the size, design, and objectives of the system, algal wastewater treatment systems can work in either batch or continuous flow modes. In batch systems, wastewater is put into the reactor and treated for a set period of time before being released. Batch operations work best for small, decentralized treatment units and laboratory-scale studies where wastewater is generated only occasionally. These systems let you keep a close eye on how quickly algae grows and how well it removes pollutants. However, batch reactors may need to be operated frequently, and they are not as suitable for large-scale applications. On the other hand, continuous flow systems maintaina steady flow of wastewater entering and treated effluent leaving at the same time. Municipal wastewater treatment plants use continuous systems extensively because they allow the plant to operate smoothly, treat water consistently, and remove nutrients efficiently over long periods of time (Christenson and Sims, 2011). Hydraulic retention time (HRT) is an important factor in continuous systems because it affects

Table 8.8 Comparison of batch and continuous algal wastewater treatment systems.

Parameter	Batch system	Continuous system
Operation mode	Fixed treatment cycles	Constant inflow and outflow
Process stability	Moderate	High
Operational complexity	Low	Moderate
Suitability	Small-scale/research	Large-scale municipal systems

both biomass production and pollutant removal. Table 8.8 shows how these functional modes are differ from each other.

8.3.3 Nutrient recovery (nitrogen and phosphorus)

Algae need nitrogen and phosphorus to grow and for their cells to function properly. In wastewater environments, these nutrients are mainly present as ammonium, nitrate, and phosphate ions. Microalgae take in these nutrients through active transport mechanisms and use them to make proteins, nucleic acids, and phospholipids (Pittman et al., 2011). The main ways to remove nitrogen are through ammonium assimilation and nitrate reduction pathways. Algae use ammonium directly to make amino acids during photosynthesis. Nitrate, on the other hand, has to be converted into ammonium through enzymatic reactions. Phosphorus is taken in as orthophosphate and used to make adenosine tri phosphate (ATP), nucleic acids, and phospholipids that make up cell membranes. Producing algal biomass to get nutrients back is a better option than using chemical precipitation methods. Harvested algal biomass contains high levels of nitrogen and phosphorus, which makes it a suitable biofertilizer or soil correction. This system not only gets rid of redundant nutrients from wastewater, but it also helps with long-term nutrient recycling in agricultural systems.

8.3.4 Metal biosorption and breakdown of organic pollutants

Microalgae have also shown significant potential for the removal of heavy metals and organic pollutants from wastewater, in addition to nutrient removal. The cell walls of microalgae are composed of polysaccharides, proteins, and lipids containing functional groups such as carboxyl, hydroxyl, and amino groups, which facilitate the binding of metal ions through electrostatic interactions and complexation mechanisms (Wang and Chen, 2009).

Heavy metal removal by algae primarily occurs through processes such as ion exchange, surface complexation, and intracellular accumulation, enabling the effective removal of metals such as cadmium, lead, chromium, and zinc from contaminated water.

In addition to metal removal, algal metabolic pathways can degrade organic pollutants such as dyes, phenols, pesticides, and pharmaceuticals. These degradation processes involve enzymatic oxidation, photodegradation, and microbial cometabolism within algal–bacterial consortia (Christenson and Sims, 2011), thereby transforming complex organic compounds into simpler and less toxic end products such as carbon dioxide and water.

As a result, algal systems are capable of treating a wide range of contaminants in wastewater, making them highly versatile and effective tools for integrated environmental remediation.

8.3.5 Protocols for sampling and performance testing

Algal wastewater treatment systems need dependable monitoring and performance evaluation to work well. Testing protocols must guarantee the representative of influent and effluent water samples for physicochemical analysis. Some of the most common parameters that are measured in algal treatment systems are BOD, COD, TN, or Total Nitrogen, Total Phosphorus (TP), Total Suspended Solids (TSS), heavy metal concentrations. Spectrophotometry, atomic absorption spectroscopy (AAS), and inductively coupled plasma optical emission spectroscopy (ICP-OES) are common methods used to measure metal and nutrient in samples in samples. Regular monitoring lets us check how well the treatment is working and make changes to the operating parameters to achieve the best results. Table 8.9 shows a summary of the most important monitoring parameters. A conceptual process diagram illustrating algal wastewater treatment mechanisms can be included as Fig. 8.4.

8.4 Enhancing air quality and carbon sequestration

One of the most critical environmental issues brought on by fast industrialization and urban growth is urban air pollution. Carbon dioxide (CO_2), sulfur dioxide (SO_2), nitrogen oxides (NO_x), VOCs, and particulate matter ($PM_2.^1$ and $PM_1{}^1$) are major urban pollutants. Transportation

Table 8.9 Key performance indicators in algal wastewater treatment systems.

Parameter	Analytical method	Significance
Biological oxygen demand	Biological oxygen demand incubation test	Organic load removal
Chemical oxygen demand	Dichromate digestion	Chemical oxygen demand
otal nitrogen	Kjeldahl method	Nutrient removal efficiency
Total phosphorus	Spectrophotometric assay	Phosphate removal
Heavy metals	ICP-OES/AAS	Toxic metal monitoring

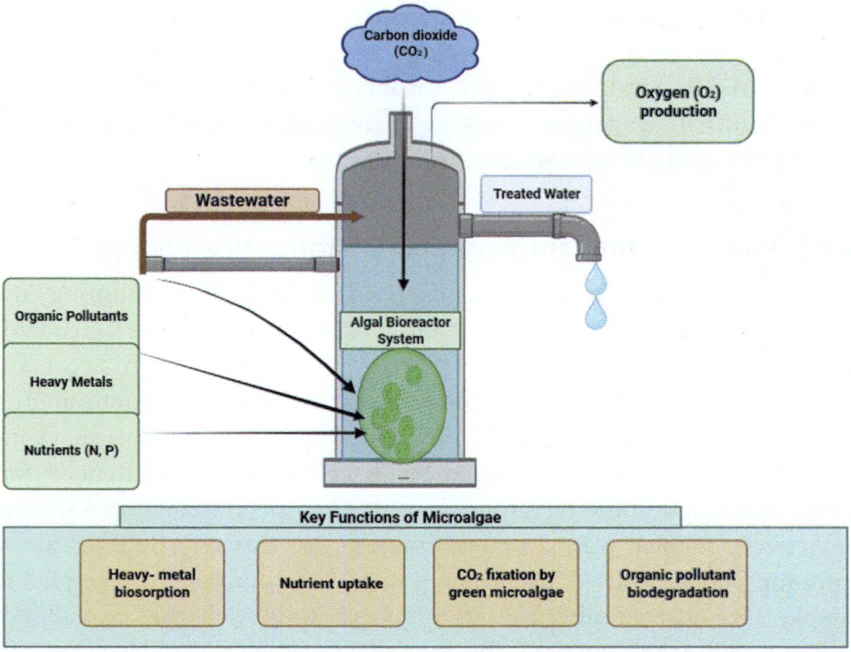

Figure 8.4 Algal wastewater treatment process illustrating nutrient uptake, metal biosorption, carbon dioxide fixation, and organic pollutant degradation within an algal bioreactor system, resulting in the production of oxygen and treated effluent.

networks, industrial processes, power generation, and construction activities are the main sources of these pollutants. Catalytic transformers and chemical scrubbers are examples of conventional air purification technologies that constantly demand large infrastructure and energy investments. Microalgae-based air remediation systems, on the other hand, provide a biologically driven and sustainable alternative that simultaneously produces biomass and oxygen while capturing atmospheric carbon dioxide (Chisti, 2007; Wang et al., 2008). Compared to terrestrial plants, microalgae have remarkably high photosynthetic efficiency, which allows for rapid carbon uptake and oxygen release. Microalgae act as natural carbon sinks by converting CO_2 and sunlight into organic biomass through photosynthesis. Algal technologies can be incorporated into roadside installations, air filtration systems, and architectural structures in urban settings to reduce pollution and promote the pretensions of the circular bioeconomy.

8.4.1 Algal green walls and biofacades at the urban scale

The creation of algal green walls and biofacades is one of the most promising uses of algae for urban air purification. These systems are made up of transparent panels or photobioreactor modules that are mounted on the outside of buildings and contain circulating microalgae cultures. Microalgae release oxygen into the surrounding environment while absorbing CO_2 and other gaseous pollutantsas ambient air flows through these systems (Pires et al., 2012). Beyond just purifying the air, biofacades provide a number of other environmental advantages. By lowering solar heat gain and improving thermal sequestration, algal panels can increase a building's energy efficiency. Algal growth in these systems also generates biomass that can be periodically harvested and used for the production of bioplastics, fertilizers, or biofuel. The BIQ House in Hamburg, Germany, is a well-known example of how microalgae-based photobioreactors have been incorporated into a building's facade to improve environmental sustainability and produce renewable energy (Berner et al., 2015). By lowering surface temperatures and enhancing microclimatic conditions, urban algal walls can also help alleviate the urban heat island effect. These systems show how buildings can be converted into active environmental remediation units using natural engineering methods.

8.4.2 Microalgal particulate matter and volatile unpredictable organic compound pollutants

Further research is being conductedon microalgae-based biofilters to remove particulate matter and VOCs from polluted air streams. Paints, detergents, fuel combustion, and industrial processes are common sources of VOCs, which are organic compounds that readily evaporate into the

Table 8.10 Major air pollutants and potential removal mechanisms by microalgae.

Pollutant	Major sources	Removal mechanism	Environmental benefit
CO_2	Vehicles, industries	Photosynthetic fixation	Carbon sequestration
Volatile organic compounds	Solvents, fuels, paints	Biodegradation	Reduced smog formation
$PM_{2.5}/PM_{10}$	Traffic, construction	Surface adsorption	Improved air quality
NO_x	Combustion engines	Biological assimilation	Reduced atmospheric toxicity

atmosphere. These substances may be harmful to human health and contribute to the formation of photochemical smog. VOCs can be removed by microalgae through absorption and metabolic breakdown. Through oxidation and biodegradation processes, some algal species possess enzymatic pathways that can change organic compounds into less harmful substances (Muñoz and Guieysse, 2006). In addition to removing VOCs, algal biofilters can capture airborne particulate matter through physical adsorption onto algal cells and associated biofilm matrices. Airflowrate, light availability, and the metabolic activity of the algal culture are some of the factors that affect how well microalgal air purification systems work. The effectiveness of contaminant declination is also improved by the interaction between algae and associated microbial communities. Algal biofilters have an advantage over traditional filtration technologies in that they degrade pollutants rather than simply capturing them. Table 8.10 lists the majorair pollutants targeted by algal remediation systems.

8.4.3 Capturing CO_2 from roads and industries

Cities emit large amounts of carbon dioxide because they have many vehicles and businesses. Microalgae have demonstrated significant potential for the direct absorption of CO_2 from flue gas streams and automotive exhaust streams. In controlled environments, numerousmicroalgal species can endure elevated CO_2 concentrations of 10% to 15%, making them suitable for integration with industrial emigration sources (Chisti, 2007). Roadside photobioreactors are an emerging approach for capturing CO_2

emissions from vehicles. These systems place transparent algal panels along roadways or urban thoroughfares. As vehicles pass by and release carbon dioxide, the algae continue to grow. There have also been proposals to install similar systems near power plants and industrial chimneys to capture concentrated CO_2 emissions before they enter the atmosphere. The captured carbon is converted into algal biomass. This biomass can then be used to produce biofuels, bioplastics, or supplementary animal feed. This technology has two benefits: it reduces greenhouse gas emissions and produces renewable resources that can be reused.

8.4.4 How to assess air quality and emissions

We need to use reliable methods to measure and monitor air quality to determine whether algae-based air-cleaning technologies are effective. The levelsof nitrogen oxides, VOCs, and other gaseous pollutants, as well as carbon dioxide and particulate matter, are all important indicators. We can compare pollution levels and evaluate how well remediation is working using air quality indicators (AQIs). Modern monitoring systems use advanced analytical tools to obtain highly accurate measurements of air pollution. For example, infrared gas analyzers are frequently used to monitor CO_2 levels, while gas chromatography–mass spectrometry (GC–MS) can find and measure VOCs. Laser-based optical particle counters are the most common tools for measuring particulate matter in the air. You can use sensor-based systems to monitor how well algal remediation systems perform in real time and change the settings as needed. IoT-enabled air quality sensors can automatically control the inflow of air, CO_2, and light in algal biofacade systems. Table 8.11 shows a list of the most common tools used to check air quality.

Table 8.11 Air quality monitoring parameters and analytical techniques.

Parameter	Analytical method	Environmental significance
CO_2 concentration	Infrared gas analyzer	Carbon emission monitoring
Volatile organic compounds	Gas chromatography–mass spectrometry	Organic pollutant detection
Particulate matter ($PM_{2.5}$/PM_{10})	Optical particle counter	Air quality index evaluation
NO_x and SO_2	Gas analyzers	Industrial emission monitoring

8.5 Collecting and using algal biomass

Microalgal biomass produced during phycoremediation and wastewater treatment processes constitutes a significant resource for various biotechnological applications. One of the biggest challenges with algal-based treatment systems, however, is recovering biomass efficiently from dilute cultures. Microalgal concentrations in wastewater treatment systems generally range from 0.2 to 1 g/L, which makes direct harvesting difficult and requires significant energy input (Christenson and Sims, 2011). So, for algal remediation technologies to be economically feasible, they need effective biomass recovery methods. People generally choose harvesting methods based on cell size, culture density, operating costs, and what they plan to do with the biomass. Flocculation, membrane-based filtration, and centrifugation are some of the most common methods for biomass recovery. Depending on the scale of operation and the setup of the system, each method has its own pros and cons.

8.5.1 Biomass recovery process

8.5.1.1 Flocculation

Flocculation is one of the most common and least expensive methods for collecting microalgal biomass. To make algal cells stick together into larger flocs that can be easily separated by sedimentation or flotation, chemical or natural agents are added to the algal culture to neutralize the negative face charge of the cells. Aluminum sulfate, ferric chloride, and natural bioflocculants such as chitosan are some of the most common flocculating agents (Uduman et al., 2010). Bioflocculation, which uses microbial polymers made by bacteria or algae, has become more popular because it is environmentally friendly. Bioflocculants are better than chemical flocculants because they reduce the risk of secondary contamination and improve the quality of biomass for use in downstream processes such as biofertilizer production. Flocculation is is best suited for large open-pond systems where you want to use as little energy as possible. But the effectiveness of flocculation may vary depending on the type of algae, the pH of the culture, and the ionic strength.

8.5.1.2 Membrane based filtration

Microfiltration and ultrafiltration are two types of membrane filtration technologies that are very effective at separating algal biomass from treated water. In membrane-based systems, the algal culture passes through membranes that allow water to pass but retain algal cells. This system allows for nonstop harvesting and makes biomass recovery more effective. Closed photobioreactor systems with high biomass concentrations are well suited for membrane filtration systems. However, membrane fouling and high operating costs are still major challenges for large-scale applications (Brennan and Owende, 2010). To keep the sludge working well, it needs to be gutted regularly and operated under optimal conditions.

8.5.1.3 Centrifugation

Centrifugation is one of the most effective methods for recovering algal biomass, especially for small microalgal species. This system uses centrifugal force to separate algal cells from the liquid medium based on how thick they are. In industrial microalgae processing, people frequently use disc-stack and tubular centrifuges. Centrifugation is a great way to rapidly collect biomass, but it generally uses a lot of energy and costs a lot to operate. So, it is frequently used as a secondary step in the harvesting process after flocculation or sedimentation to further concentrate algal biomass. Table 8.12 shows how different biomass harvesting styles compare with each other.

8.5.2 Valorization of algal biomass

Once collected, algal biomass can be used in numerous ways, turning wastewater treatment plants into full-fledged biorefineries. Valorization of algal biomass improves the economic viability of phycoremediation systems and supports circular bioeconomy models (Chisti, 2007).

Table 8.12 Comparison of major microalgal biomass harvesting techniques.

Method	Principle	Advantages	Limitations
Flocculation	Cell aggregation through charge neutralization	Low cost, suitable for large volumes	Chemical residues may affect biomass quality
Membrane filtration	Physical separation using porous membranes	High recovery efficiency	Membrane fouling and maintenance
Centrifugation	Density-based separation	Rapid and efficient	High energy consumption

8.5.3 Organic fertilizers

Algal biomass is a valuable source of organic fertilizer because it is rich in essential nutrients such as nitrogen, phosphorus, potassium, and trace elements. The use of algal biofertilizers has been shown to improve soil fertility, help plants grow, and boost crop yields (Mulbry et al., 2005). Algal biomass also contains compounds such as phytohormones and amino acids that promote plant growth. Using algal biomass as a biofertilizer also helps sustainable agriculture by recovering nutrients from wastewater systems.

8.5.4 Bioenergy production

Microalgal biomass can be used to make a number of bioenergy products, such as biodiesel, bioethanol, and biochar. Some types of microalgae accumulate large amounts of lipids when there are not enough nutrients, which makes them suitable for biodiesel production through transesterification (Chisti, 2007). Pyrolysis and other thermochemical conversion technologies can convert algal biomass into biochar and bio-oil. Algae-based biochar can be used for soil improvement and carbon sequestration, which is another way to help the environment.

8.5.5 Biodegradable plastics

Recent studies have explored the feasibility of using microalgae as a feedstock for the production of biodegradable polymers. We can make biopolymers such as polyhydroxyalkanoates (PHAs) and other biodegradable plastics can be produced from algal biomass. These materials are more environmentally friendly than plastics made from petroleum and help reduce plastic pollution (Reddy et al., 2003). Combining the use of algal biomass with wastewater treatment systems opens up the possibility for developing sustainable biorefineries that make numerous products that are worth further than what they bring. Table 8.13 shows a list of the most important ways to turn biomass into useful products.

Table 8.13 Major applications of harvested algal biomass.

Application	Product	Environmental benefit
Agriculture	Biofertilizers	Nutrient recycling
Bioenergy	Biodiesel, biochar	Renewable energy generation
Bioplastics	PHAs and biopolymers	Reduction of plastic pollution

8.6 Real-world applications and demonstration systems

The practical implementation of algae-based remediation technologies has progressed from laboratory trials to pilot-scale and full-scale demonstration systems in various parts of the world. These real-world applications highlight the potential of microalgal systems to address urbanenvironmental challenges such as wastewater treatment, carbon capture, and resource recovery. Unlike conventional treatment systems, algae-based technologies operate through natural processes that use sunlight, carbon dioxide, and nutrients, thereby reducing functional energy demand while generating valuable biomass (Rawat et al., 2011; Pittman et al., 2011). Several demonstration systems have been developed to evaluate the feasibility of integrating algal technologies into urban infrastructure. These systems demonstrate that microalgal technologies can be effectively implemented in decentralized wastewater treatment units, urban green infrastructure, and integrated food–energy–water systems. The integration of algal technologies with urbansustainability initiatives further enhances their applicability for climate-resilient city planning.

8.6.1 Use in decentralized water recycling units

Decentralized wastewater treatment systems are increasingly being adopted in urban areas where centralized sewage infrastructure is inadequate or overloaded. In such systems, algal bioreactors are used to treat greywater and partially treated municipal wastewater locally, allowing the recovered water to be reused for irrigation, landscaping, or nonpotable domestic applications. Microalgae remove redundant nutrients, organic matter, and certain pollutants from wastewater while producing biomass that can be harvested and used for biofertilizer or bioenergy production (Christenson and Sims, 2011). Decentralized algal treatment systems offer several advantages compared with conventional centralized plants. These systems require lower infrastructure investment, reduce the burden on networks, and enable localized resource recovery. In addition, small-scale algal reactors can be installed in residential complexes, universities, and industrial premises, creating localized circular water management systems. By recovering nutrients and producing reusable water, decentralized algal treatment units contribute significantly to urban water sustainability. Recent pilot studies have demonstrated that decentralized algal reactors can achieve nutrient removal efficiencies comparable to those of conventional tertiary treatment processes. The integration of such systems within smart city infrastructure also allows automated monitoring of water quality parameters, improving operational reliability and environmental performance.

8.6.2 Case studies from India, Europe, and Southeast Asia

Several countries have successfully implemented algae-based treatment and remediation systems in urban areas. In India, microalgae-based wastewater treatment has been explored in both municipal and industrial settings. Pilot-scale HRAPs have been used to treat domestic wastewater while producing biomass suitable for agricultural operations. These systems have demonstrated significant reductions in nitrogen and phosphorus concentrations, highlighting their potential for addressing eutrophication in freshwater bodies. In Europe, large-scale demonstration systems have integrated algae civilization with building architecture and wastewater treatment. One notable example is the BIQ House in Hamburg, Germany, which incorporates microalgal photobioreactors within its building facade. These reactors capture carbon dioxide and generate biomass that can be used for renewable energy production (Berner et al., 2015). Similar initiatives have been implemented in Spain and France to evaluate the feasibility of algae-based systems for industrial flue-gas treatment and wastewater polishing. Southeast Asian countries have also shown increasing interest in algal technologies due to the region's favorable climatic conditions for microalgal growth. Pilot systems in countries such as Thailand and Malaysia have focused on integrating algae-based wastewater treatment with aquaculture systems and agricultural irrigation. These systems not only improve water quality but also provide nutrient-rich biomass that supports sustainable agricultural practices. A summary of selected international demonstration systems is provided in Table 8.14.

8.6.3 Integration with vertical farming and rooftop systems

One of the new trends in urban sustainability is combining algae-based technologies with vertical farming and rooftop agriculture. These interconnected systems create a free-flowing circle of land where nutrients from wastewater are used to grow algae, and the biomass and treated water are used to support crop growth. Such systems improve food security while reducing water and nutrient wastage. Vertical farming systems usually

Table 8.14 Selected international demonstration projects of algae-based remediation systems.

Region	Project type	Key application	Outcome
India	High-rate algal ponds	Municipal wastewater treatment	Nutrient removal and biomass production
Germany	BIQ house biofacade	Carbon capture and energy generation	CO_2 reduction and renewable biomass
Thailand	Algae-based aquaculture systems	Wastewater recycling	Improved water quality and biomass utilization

operate in controlled environments where using resources efficiently is very important. Adding algal bioreactors to these systems allows them to use extra nutrients and carbon dioxide that are generated within the agricultural facilities to support algae grow. Algae also improve air in quality enclosed agricultural environments by releasing oxygen. Rooftop algal reactors are another new way to use microalgae in urban ecosystems. These systems grow microalgae using sunlight and carbon dioxide from the air, while also improving building sequestration and reducing heat immersion. The collected biomass can be utilized as biofertilizer for rooftop gardens, thereby creating a circular nutrient cycle within urban environments.

8.7 Innovations and regulatory perspectives

The rapid advancement of biotechnology, environmental engineering, and digital monitoring technologies has created new opportunities for enhancing the effectiveness and scalability of algae-based urban remediation systems. While microalgae offer significant potential for sustainable wastewater treatment, carbon insulation, and contaminant removal, successful implementation requires integration with ultramodern monitoring technologies, urban planning fabrics, and environmental regulations. Recent inventions have focused on combining biological remediation systems with smart monitoring tools, automated process control, and regulatory guidelines that ensure safe and effective deployment in urban environments (Pires et al., 2012; Rawat et al., 2011). The transition from pilot-scale demonstration systems to large-scale implementation also requires standardized operational protocols, environmental safety assessments, and supportive policy fabrics. Governments and environmental agencies are increasingly recognizing the role of NBS such as algal remediation systems, in achieving climate mitigation and sustainable urban development goals.

8.7.1 Smart monitoring with Internet of Things sensors

The integration of IoT technologies into algal remediation systems has significantly improved process monitoring and operational efficiency. IoT-enabled sensors allow nonstop real-time monitoring of key environmental parameters such as pH, DO, temperature, carbon dioxide concentration, and nutrient levels. These sensors transmit data to centralized control systems, enabling automated adaptations to optimize algal growth and contaminant removal efficiency. For example, real-time monitoring of DO and pH can help maintain optimal conditions for photosynthesis and nutrient assimilation within algal reactors. In addition, carbon dioxide sensors can regulate CO_2 supply in photobioreactors to maximize biomass productivity. Such automated monitoring systems reduce manual intervention and improve operational stability in large-scale installations. Recent developments in smart environmental monitoring have also incorporated artificial intelligence and machine learning algorithms to analyze sensor data and predict system performance. These predictive models help identify potential operational issues such as algal culture collapse or nutrient imbalance before they occur, thereby improving system reliability. The key parameters monitored in IoT-enabled algal systems are summarized in Table 8.15.

Table 8.15 Key parameters monitored in Internet of Things-enabled algal remediation systems.

Parameter	Sensor type	Importance
pH	Electrochemical sensor	Controls nutrient availability
Dissolved oxygen (DO)	Optical DO sensor	Indicates photosynthetic activity
CO_2 concentration	Infrared gas sensor	Carbon supply for algal growth
Temperature	Digital temperature probe	Influences metabolic activity
Nutrient levels (N, P)	Spectroscopic sensors	Determines nutrient removal efficiency

8.7.2 Guidelines for adding algae to city planning

For algae-based technologies to become essential components of urban-sustainability strategies, they must be integrated into the planning frameworks of cities. Urban planners and environmental experts are increasingly exploring how to incorporate algae systems into buildings, wastewater infrastructure, and green public spaces. Some examples include algal biofacades installed on buildings, roadside photobioreactors that capture carbon, and rooftop algae civilization systems. Urban planning guidelines for integrating algae generally focus on maximizing sunlight exposure, ensuring water safety, and preventing unforeseen environmental impacts. For instance, buildings designed to accommodate algal biofacades must include transparent panels, circulation channels, and automated monitoring system. In addition, algae-based decentralized wastewater treatment systems should be located close to wastewater generation sites to reduce transportation and infrastructure costs. Furthermore, when implementing visible algal technologies such as green walls or rooftop reactors, urban planners need to consider public acceptance and aesthetic appeal. With proper design and integration, these systems can be turned into beautiful green infrastructure that enhances the appearance of cities.

8.7.3 Conditions for standardization and environmental compliance

For large-scale implementation of algal remediation systems, consistent operational guidelines and regulatory frameworks need to be established. To ensure that algae-based technologies work well without harming the environment, environmental agencies need to set clear standards for water quality, biomass use, and system safety. Standardization efforts should concentrate on establishing acceptable thresholds for treated effluent quality, ensuring that water released from algal systems complies with environmental protection standards. In addition, standards need to include guidelines for the safe handling and utilization of algal biomass, especially when it is used as a feedstocik or raw material for industrial processes. Environmental compliance also involves monitoring potential ecological risks, including invasive algal species or unintended nutrient discharge. Establishing consistent monitoring protocols will help ensure that algae-based remediation systems remain environmentally safe and sustainable over the long term. International organizations and environmental regulatory authorities are increasingly recognizing the significance of nature-based solutions in environmental management. Developing regulatory frameworks that support algae-based technologies will be essential for their adoption in public infrastructure.

8.8 Conclusion

8.8.1 Technical summary and roadmap for implementation

Urban phycoremediation is a promising approach for addressing several environmental challenges associated with urbanization, such as water pollution, air contamination, and greenhousegas emissionsMicroalgae possess unique biological characteristics that enable them to absorb nutrients, sequester carbon dioxide, degrade pollutants, and generate valuable biomass. These traits make algae highly suitable for use in sustainable environmental applications. This chapter explores the possibility of algae-based systems in a number of important areas, including wastewater treatment, air purification, carbon sequestration, and biomass-to-energy conversion. Various cultivation platforms, such as raceway ponds, HRAPs, and photobioreactors, provide different approaches for implementing these technologies in urban environments. In addition, advances in biomass harvesting and resource recovery technologies have improved the economic possibility of algal remediation systems. The successful implementation of urban phycoremediation technologies requires a well-planned strategy that integrates biological processes, engineering design, monitoring technologies, and regulatory frameworks. Key steps in this strategy include selecting suitable algal strains, optimizing cultivation systems, developing efficient biomass harvesting methods, and incorporating monitoring technologies for system management.

8.8.2 Future potential of algae in sustainable city development

Algae-based technologies have considerable potential to contribute to environmentally sustainable and climate-resilient cities. Combining algal systems with smart city infrastructure, renewable energy sources, and circular bioeconomy models might turn citiess into into self-sustaining ecosystems that recycle resources while minimizing environmental impacts. Future developments in wastewater treatment are expected to be driven by innovative approaches such as genetically engineered microalgae with enhanced pollutant-removal efficiency, integrated algal–bacterial systems that exploit synergistic microbial interactions, and artificial intelligence-based platforms capable of automating process monitoring, predictive analysis, and treatment optimization for improved environmental performance. Progress in these areas might make algae remediation systems indeed more successful and easier to use on a larger scale. The adding worldwide focus on carbon impartiality and sustainable development is anticipated to quicken the abandonment of NBS, similar as algae-based remediation systems. Microalgae might be veritably important in making the coming generation of ecologically friendly urbanecosystems if technology keeps getting better, legislative fabrics keep getting better, and people from numerous fields work together.

AI disclosure

During the preparation of this manuscript, the authors used AI and AI-assisted technologies in image creation. The authors take full responsibility for the content, scientific accuracy, and integrity of all figures presented in this work.

References

Berner, F., Heimann, K., Sheehan, M., 2015. Microalgal biofilms for biomass production. Biotechnology Advances 33 (2), 277–287. Available at https://doi.org/10.1016/j.biotechadv.2014.12.001.

Brennan, L., Owende, P., 2010. Biofuels from microalgae—a review of technologies for production, processing, and extraction of biofuels and co-products. Renewable and Sustainable Energy Reviews 14 (2), 557–577. Available at https://doi.org/10.1016/j.rser.2009.10.009.

Chisti, Y., 2007. Biodiesel from microalgae. Biotechnology Advances 25 (3), 294–306. Available at https://doi.org/10.1016/j.biotechadv.2007.02.001.

Christenson, L., Sims, R., 2011. Production and harvesting of microalgae for wastewater treatment, biofuels, and bioproducts. Biotechnology Advances 29 (6), 686–702. Available at https://doi.org/10.1016/j.biotechadv.2011.05.015.

European Commission, 2015. Towards an EU Research and Innovation Policy Agenda for Nature-Based Solutions and Re-Naturing Cities. Publications Office of the European Union, Brussels.

Muñoz, R., Guieysse, B., 2006. Algal–bacterial processes for the treatment of hazardous contaminants: a review. Water Research 40 (15), 2799–2815. Available at https://doi.org/10.1016/j.watres.2006.06.011.

Mulbry, W., Kondrad, S., Pizarro, C., Kebede-Westhead, E., 2005. Treatment of dairy manure effluent using freshwater algae: algal productivity and recovery of manure nutrients using pilot-scale algal turf scrubbers. Bioresource Technology 96 (4), 159–166. Available at https://doi.org/10.1016/j.biortech.2004.05.007.

Pires, J.C.M., Alvim-Ferraz, M.C.M., Martins, F.G., Simões, M., 2012. Carbon dioxide capture from flue gases using microalgae: engineering aspects and bioremediation potential. Renewable and Sustainable Energy Reviews 16 (5), 3043–3053. Available at https://doi.org/10.1016/j.rser.2012.02.055.

Pittman, J.K., Dean, A.P., Osundeko, O., 2011. The potential of microalgae for nutrient removal from wastewater. Bioresource Technology 102 (1), 17–25. Available at https://doi.org/10.1016/j.biortech.2010.06.139.

Rawat, I., Kumar, R.R., Mutanda, T., Bux, F., 2011. Dual role of microalgae in wastewater treatment and biofuel production. Applied Energy 88 (10), 3411–3424. Available at https://doi.org/10.1016/j.apenergy.2010.11.025.

Reddy, C.S.K., Ghai, R., Rashmi, Kalia, V.C., 2003. Polyhydroxyalkanoates: an overview. Bioresource Technology 87 (2), 137–146 https://doi.org/10.1016/S0960-8524(02)00212-2.

Uduman, N., Qi, Y., Danquah, M.K., Forde, G.M., Hoadley, A., 2010. Dewatering of microalgal cultures: a major bottleneck to algae-based fuels. Journal of Renewable and Sustainable Energy 2 (1), 012701. Available at https://doi.org/10.1063/1.3294480.

Wang, B., Li, Y., Wu, N., Lan, C.Q., 2008. CO_2 bio-mitigation using microalgae. Applied Microbiology and Biotechnology 79 (5), 707–718. Available at https://doi.org/10.1007/s00253-008-1518-y.

Wang, J., Chen, C., 2009. Biosorbents for heavy metals removal and their future. Biotechnology Advances 27 (2), 195–226. Available at https://doi.org/10.1016/j.biotechadv.2008.11.002.

World Health Organization (WHO), 2021. Global Air Quality Guidelines: Particulate Matter (PM2.5 and PM10), Ozone, Nitrogen Dioxide, Sulfur Dioxide and Carbon Monoxide. WHO Press, Geneva.

United Nations, 2018. World Urbanization Prospects: The 2018 Revision. United Nations Department of Economic and Social Affairs, New York.

CHAPTER 9

Microalgae-based biofuels production

Laila Dina Amalia Purba and Devi Bentia Effendi
Research Center for Applied Microbiology, National Research and Innovation Agency (BRIN), Cibinong, West Java, Indonesia

9.1 Introduction

The International Energy Agency (IEA) predicted that by 2030, global energy consumption will surge by 50%, suggesting the need for alternative fuel sources (Abdullah et al., 2024). The pursuit of sustainable and low carbon energy alternatives has intensified in response to growing concerns over fossil fuel depletion, greenhouse gas emissions, and climate change. Among various bioenergy options, microalgae have garnered significant scientific interest as a next-generation feedstock for biofuel production (Wang et al., 2024). Microalgae exhibit superior attributes compared to terrestrial crops, including higher photosynthetic efficiency, faster biomass accumulation, and the capacity to produce substantial quantities of lipids suitable for conversion into biodiesel. Furthermore, their ability to grow on non arable land and utilize wastewater or flue gases as nutrient sources underscores their potential in integrated biorefinery systems that align with circular economy principles. Consequently, the development of biofuel from microalgal biomass represents a promising and strategically important avenue for achieving sustainable energy security and environmental resilience.

Microalgae are reported to be capable of absorbing and sequestering CO_2 from the environment at a rate of approximately 50% of their weight. During photosynthesis, microalgae absorb CO_2 and release oxygen during the day, and vice versa at night; these phenomena are similar to those of plants (Li et al., 2022b). Over the years, microalgae have been recognized as a potential raw material for producing large scale bio-based energy or biofuels and have been commercialized to a certain extent. In order to convert microalgae into biofuel, various technological approaches, from cultivation and harvesting to the conversion process of biomass into fuel, have been extensively developed. With their high CO_2 absorption

Applied Phycology
DOI: https://doi.org/10.1016/B978-0-443-24758-3.00011-3

capacity, rapid growth rate, and high proximate content such as carbohydrates, lipids, and proteins, microalgae are an environmentally friendly, renewable carbon-based energy source. This chapter will reveal the potential application of microalgae for biofuels as alternative renewable energy sources.

9.2 Current status of microalgae-based biofuel

Biofuel production is classified into different generations, as illustrated in Fig. 9.1 The first-generation biofuel was derived from oil-contained seeds and food crops through a series of fermentation and transesterification (Abdullah et al., 2024). This included utilization of starch, rapeseed, sunflower, palm, and sugarcane, while bioethanol may also be obtained from starch and sugar from crops such as sugarcane (Ganesan et al., 2020; Dey et al., 2021). Biodiesel may be extracted from oil-rich crops (e.g., soybean and palm oil) by transesterification of the lipids using alcohol. However, as a global food crisis occurred in 2007-2008, the debate over food versus fuel became a major issue. It was argued that the crops used in biofuel production should be reserved for food. The high demand for these crops also leads to their increasing market price, which is not desirable for the long-term biofuel production. As there will be an increasing demand for these crops, large agricultural areas would be required for biofuel production. For instance, the extended palm oil production might cause deforestation of tropical rainforests, leading to more CO_2 emissions, and the initial goal of sustainable fuel production would be eradicated. Due to these reasons, the first-generation biofuels were not recommended.

The biofuel production was then developed into the second generation through utilization of lignocellulosic biomass, agroindustrial residues, and waste from food sectors (e.g., animal fats, waste cooking oil, and wheat bran) (Kumar and Sharma, 2008). The *Jatropha curcas* also emerged as an alternative nonfood crop with promising biofuel productivity due to its ability to be grown in wasteland and nonproductive areas. Therefore, the production of second-generation would not compete with food production, resolving the issue found in first-generation biofuel. More than 95% of the second-generation biofuel has been commercially available, such as the cellulosic ethanol from straw residues, marketed as sunliquid from Clariant produced in Romania (Cavelius et al., 2023). Nevertheless, the complex characteristics of lignocellulosic biomass, especially its high lignin content, may hinder the extraction efficiency, requiring an additional

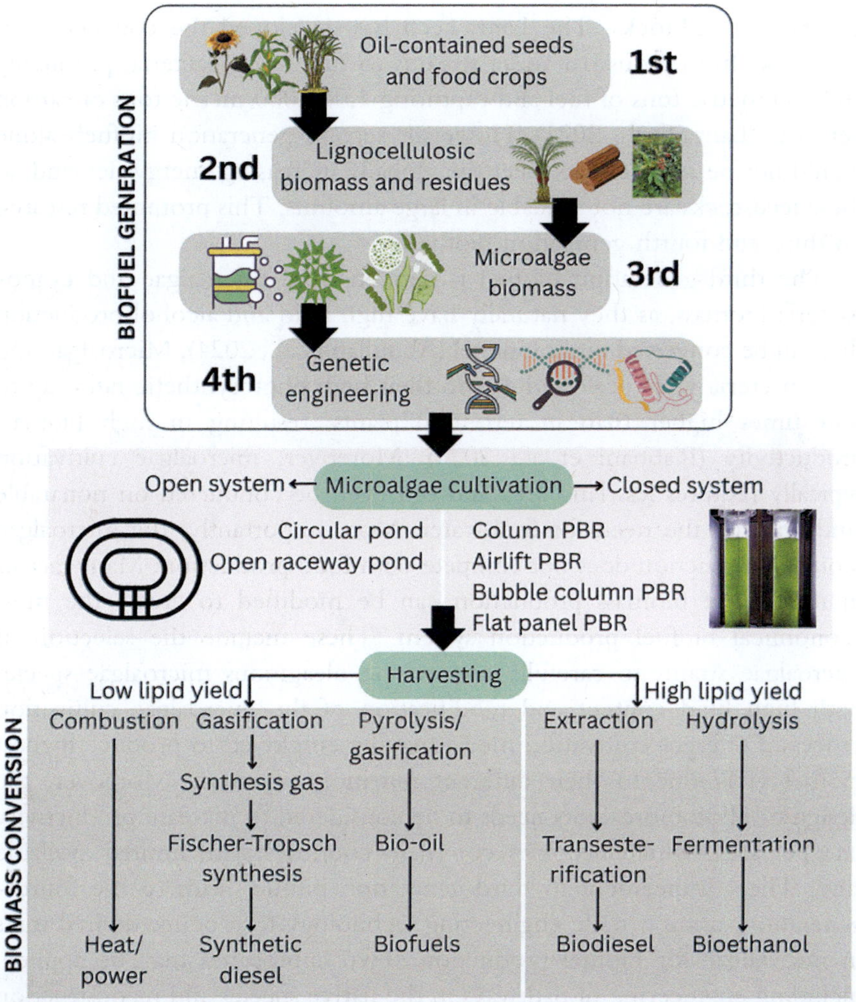

Figure 9.1 ***The advancement in biofuel generation.*** The advancement in biofuel generation, showing the feasibility of microalgae biomass in biofuel production through several strategies. Adapted from Wood (2021), Abdullah et al. (2024), Wang et al. (2024).

pretreatment process that may increase process time and costs. Lignocellulosic biomass can also be used to produce syngas by gasification process. Syngas, a mixture of carbon monoxide (CO), carbon dioxide (CO_2), and hydrogen gas (H_2), can be metabolized by anaerobic and methanogenic bacteria, such as *Acetobacterium* and *Clostridium,* to produce acetate, ethanol, and butanol. Thus, gasification can be used as a pretreatment for second-generation feedstocks. The LanzaTech has developed the conversion of feedstocks from industrial waste streams to fuel using bacteria, producing 600,000 metric tons of fuel and capturing 1,000,000 metric tons of carbon per year (LanzaTech, 2021). However, second-generation biofuels alone would not be sufficient to meet the globally increasing energy demand, as these feedstocks are not available in large amounts. This prompted research on third and fourth generation biofuels.

The third-generation biofuel is derived from microalgae and cyanobacteria biomass, as they naturally have high lipid and alcohol production that can be converted into biodiesel (Abdullah et al., 2024). Microalgae and cyanobacteria were preferred due to their high photosynthetic rates, up to four times higher than in terrestrial plants, resulting in high biomass productivity (Rabbani et al., 2024). Moreover, microalgae cultivation typically requires less land area and can even be conducted on nonarable land without the need for freshwater. More importantly, the microalgae biomass production does not compete with food production. Many factors in microalgae biomass production can be modified to create the most economical biofuel production system. These include the selection of microalgae strains to carefully choose the oleaginous microalgae species with high lipid content and modification of the microalgae cultivation process. Different cultivation media may be employed to produce highest biofuel yield due to their different nutritional content. Moreover, the design of photobioreactors needs to be considered to increase productivity rate per area, which becomes crucial in countries with limited available land. The advancement in third-generation biofuel leads to the fourth-generation, using genetic engineering technology to produce desired traits in microalgae for biofuel production. Two approaches may be applied, including engineering of pathways in the native species and reconstruction of pathways in accessible model organisms (Cavelius et al., 2023). The engineering of native species is used to optimize biomass and lipid yield while creating a more efficient carbon source utilization. Model microorganisms, including bacteria, yeast, and algae, can also be used as the hosts for biofuel production by introducing certain pathway genes. For example, butanol pathway genes were isolated from *Clostridia* and reintroduced into *Escherichia coli*, *Pseudomonas putida,* and *Bacillus subtilis* for production of bioethanol (Yeong et al., 2018). However, genetic engineering for native biofuel producers in microalgae is still limited to date.

During the production of microalgae-based biofuel, it is crucial to select the most suitable microalgae strain with high biomass production to ensure the economic feasibility of the process. Naturally, all microalgae species have the ability to accumulate lipids, though some species are capable of producing high lipid yields. Various literature has reported the numerous microalgae species along with their lipid production rates (Yang et al., 2023). Some genera, including *Chlorella, Scenedesmus,* and *Nannochloropsis,* were found to have superior lipid production rates compared to the others, along with high biomass production and strong adaptability in extreme environments (i.e., varied nutrient composition and high or low temperatures) (Yang et al., 2020). Microalgae species have varied lipid production rates, for instance, *Botryococcus braunii* may exhibit 25 – 75% lipid accumulation, while *Dunaliella tertiolecta* has up to 70% lipid content, making them the promising candidates for biofuel production from microalgae biomass (Jung et al., 2019). In addition, *Dunaliella* has excellent adaptability in saline environments, although its low biomolecule yield remains a major challenge (Goswami et al., 2022). *Chlorella* and *Scenedesmus* are the two most widely studied microalgae species due to their abundance in the environment, straightforward cultivation process, and high biomass production. *Chlorella vulgaris* and *Chlorella sorokiniana* may contain up to 53% and 28% lipid content, similar to *Scenedesmus obliquus* and *Scenedesmus acutus* (Qiu et al., 2017; Liu et al., 2020). Therefore, the biomass and lipid production from microalgae are highly dependent on species and culture conditions.

In the production of microalgae-based biofuel, the fatty acid composition also needs to be considered along with its biomass and lipid production (Yadav et al., 2021). Microalgae store triacylglycerols (TAG), a nonpolar lipid that can be converted into biodiesel, in their chloroplasts. The chain length and degree of unsaturation are important factors that need to be considered for biofuel production from microalgae (Arif et al., 2020). The profile of fatty acid methyl esters (FAME) from microalgae is categorized into saturated fatty acids (SFA), monounsaturated fatty acids (MUFA), and polyunsaturated fatty acids (PUFA). Higher SFA and MUFA are typically desirable for biodiesel production, while high PUFA content may lead to unstable biodiesel properties due to its high oxidation (Mandal et al., 2020). SFA, such as palmitic acid (C16:0) and stearic acid (C18:0), are abundant in FAME composition of microalgae, usually followed by oleic acid (C18:1). MUFA composition plays important roles in regulating the melting point of biodiesel, which is crucial in cold climate countries or regions (Qiu et al., 2017). However, some microalgae may naturally produce high PUFA content, including linoleic acid (C18:2) and linolenic acid (C18:3), such as

Desmodesmus sp., *Monoraphidium* sp., and *Tetradesmus obliquus* (Kabir et al., 2020). On the other hand, FAME from *C. vulgaris*, *Scenedesmus dimorphus*, and *Nannochloropsis* sp. show excellent biodiesel characteristics (Yang et al., 2023). The FAME profile in microalgae is also influenced by cultivation conditions, whereby mixotrophic and autotrophic cultivations have been reported to shift production of PUFA to MUFA. Nitrogen depletion in the medium may also increase the SFA production (C16:0 and C18:0) in *T. obliquus* and *Desmodesmus abundans* (Rugnini et al., 2020).

Modification of environmental conditions could improve the lipid production and FAME composition from microalgae. The modifications are classified into changes in physical environmental conditions, modification of medium composition, and regulation of microalgae metabolism. The physical environmental conditions, including temperature, light/illumination, and aeration strategy, can be altered to induce lipid accumulation. Different temperatures during microalgae cultivation process may result in different lipid yields, whereby the enzyme activities related to carbon fixation (ribulose diphosphate carboxylase or Rubisco) and acetyl-CoA carboxylase are highly dependent on surrounding temperature. Optimum temperature may vary among different species; however, temperatures above 30°C have been shown to inhibit lipid accumulation. Lipid productivity of mixed cultures containing *Scenedesmus* sp., *Chlorella* sp., and *Spyrogyra* sp. increased at 30°C compared to 25°C, but decreased significantly at 35°C (Subhash et al., 2022). Similar results were reported in *Nannochloropsis* sp. and *C. vulgaris* (Saranya and Shanthakumar, 2020). Light is also essential in photosynthesis process that can directly affect both biomass production and lipid accumulation. Increasing light intensity commonly benefits the lipid production, although excessive light can lead to photoinhibition. The maximum and optimal light intensity need to be identified should a species is directed for biofuel production since high light intensity will result in higher initial and operational costs of microalgae cultivation. Blue light (465 nm) may also be utilized to induce lipid production in microalgae, as the Rubisco enzyme shows elevated activities under blue light illumination (Li et al., 2021). Green light (520 nm) illumination results in enhanced lipid production due to stress response because chlorophyll cannot absorb green light (Jung et al., 2019). In addition to that, optimized CO_2 addition in aeration process could enhance both biomass and lipid yield in microalgae. Up to 5% CO_2 concentration can be tolerated by various microalgae species, while concentrations higher than 5% lead to decreased enzyme cell activity and affect lipid synthesis process (Yang et al., 2023). Recent reports also showed that application of low-strength ultrasound and magnetic fields may also improve the biomass and lipid production in microalgae (Feng et al., 2020; Ren et al., 2020).

The modification of medium composition, such as nutrient depletion, has been widely studied in microalgae. Due to limited nutrient availability, photosynthetic ability significantly decreased, and the excess energy is stored as lipids. Common modifications include varying total nitrogen (TN) and total phosphorus (TP) concentrations. Limited TN and TP availabilities showed elevated lipid content regardless of the species and cultivation conditions; however, biomass productivity would also be affected in a negative manner (Chu et al., 2022). Under same TP and decreasing TN concentrations, lipid content showed significant elevation (Liu et al., 2020). Nevertheless, when TN concentration was maintained and TP was limited, no significant effect was observed. This indicated that lipid production in microalgae was more affected by nitrogen availability compared to phosphorus. Moreover, some microalgae have the ability for P luxury uptake, allowing for polyphosphate formation and P storage in the cells, even during P starvation. The optimal TN and TP concentrations have been derived for most microalgae species, whereby the optimal TN concentration was at 15–35 mg/L and TP concentration at 4–16 mg/L to maintain both biomass production and high lipid content (Yang et al., 2023). A two-stage cultivation process may also be applied, whereby high biomass production was achieved in the first stage under optimal conditions, followed by stress conditions to stimulate lipid production in the second stage.

The modification of medium composition may also be achieved through addition of trace elements and specific chemicals. Trace elements have been reported to significantly affect microalgae growth and lipid production. The addition of Fe, Zn, and Mo could produce biolipid with high biodiesel quality due to different FAME composition (Mandal et al., 2019). However, excessive additions of Zn^{2+}, Mn^{2+}, and Mo^{2+} may result in decreased growth and lipid production, demonstrating that excessive trace elements would cause cell toxicity (Dou et al., 2013). The addition of specific chemicals, including phytohormones and nanoparticles, has also shown a positive correlation to biomass and lipid yield (Dehghanipour and Zamani, 2024; Zhang et al., 2024). Phytohormones reduce the oxidative damage and can upregulate the lipogenic gene expression, thus enhancing lipid metabolism in microalgae cells (Yahya et al., 2018). The utilization of industrial wastewater containing high chemical organic matter could also enhance lipid production. However, should additional chemicals are used, their environmental impact needs to be considered to ensure sustainability of the system. Recent studies have also reported strategies to increase biolipid production in microalgae through genetic engineering and cocultivation processes that will be discussed in the following subsection. Several modifications are more feasible to implement on a full scale than others; these include nutrient depletion, heterotrophic cultivation strategies, and temperature control. While addition of chemicals and salinity stress may cause biological toxicity and lead to higher operational costs in large-scale cultivation.

9.3 Microalgae cultivation system for biofuel production

Microalgae cultivation can be conducted in open and closed systems, each having its own advantages and disadvantages. Optimal cultivation process should allow maximum illumination penetration to promote photosynthesis process, effective mass transfer, low contamination risk, minimal land requirement, and an easy operational procedure. The open pond system, which includes open ponds and raceway ponds, has been shown to have high feasibility and direct operation in microalgal biomass production (Costa and de Morais, 2014). Moreover, it has minimal energy requirements and can be applied in massive production. The open cultivation system can utilize natural water bodies, such as lakes and ponds, along with nonnatural water bodies, for example, the raceway ponds or tanks. The earliest application of open system for microalgae cultivation utilized a circular pond with a width of 45 m and a depth of 30–70 cm (Abdullah et al., 2024). The circulation was provided using an agitator at the center of the pond. The system has also been applied for cultivation of *Chlorella* in Japan and Taiwan, where several obstacles were observed (Tan et al., 2020). The wide diameter of the pond causes strain on the stirrer due to its high-water resistance. The agitation process also requires high energy to prevent the biomass sedimentation. Therefore, open raceway pond is now commonly utilized in mass cultivation of microalgae biomass. The circulation in raceway pond was provided using a paddlewheel attached to a recirculation channel with a depth of 15–20 cm.

During cultivation in an open pond system, several parameters need to be considered, such as variation of light and temperature of the system, as well as the rainfall discharge in the area (Thomas Kiran and Tiwari, 2018).

The light could only penetrate from the surface of the pond, whereby its intensity will decrease at the middle and bottom layers. Thus, depth of the pond becomes a crucial factor, showing an optimal depth of 30 cm (Li et al., 2021). The length to wide ratio (L/W) has also been optimized, whereby an L/W ratio of 10 can minimize the shear stress. Additionally, the recommended distance between the tip of paddlewheel and the bottom of the pond is 5–8 cm. In an open raceway pond, dead zones may occur in low-velocity areas, causing cell sedimentation and anaerobic conditions that may lead to growth inhibition and bacterial contamination. Paddlewheel configuration shall be optimized to minimize the dead zones, such as the semicircular baffle configuration and application of a deflector baffle. Nevertheless, some parameters, such as heavy rainfall and temperature, are the factors that cannot be controlled. Contamination in an open system is also unavoidable; however, this can be overcome by applying a severe alkaline or saline environment and utilizing multiple microalgae strains to promote resilience of the system and eventually produce higher biomass yield. Despite its straightforward operational process, the open pond system typically produces lower biomass yield; thus only a handful of suitable microalgae species can be cultivated in this system. *Chlorella* spp. is viable for cultivation in an open pond system because of its fast growth rate that can outgrow potential contaminants (Molina-Grima et al., 2022). In addition, *Spirulina* sp. can survive in extreme conditions, making it a suitable species to be cultivated in an open pond system (Assunção and Xavier Malcata, 2020). The system has also been applied for cultivation of *Dunaliella salina*.

The closed cultivation system called photobioreactors (PBRs) was developed to overcome the barriers in open systems. Various PBR configurations would meet the requirements for microalgae cultivation, including controlled temperature, light intensity, and optimized agitation strategies (Barboza-Rodríguez et al., 2024). Artificial lights are typically used in closed PBRs; thus, the intensity, duration, and wavelength can be strictly controlled (Maltsev et al., 2021). Agitation can be provided with different strategies, whereby aeration with O_2 and CO_2 is the most commonly used circulation mechanism (Kirnev et al., 2020). The aeration or gas injection that involves formation of bubbles will also facilitate the mass transfer and enhance biomass production. At least three components are required in PBR operations, namely the culture tank, illumination equipment, and control units. The PBR can be operated in batch, semicontinuous, and continuous operational modes, which can be combined with phototrophic, mixotrophic, and photoheterotrophic systems. Based on their geometries, PBRs can be classified into tubular, bubble column, and flat-panel, while based on the liquid flow, PBRs are categorized into airlift, stirred-tank, and bubble column.

The tubular PBRs are most widely used in large-scale production due to their high versatility. The length of the tube in tubular PBRs shall be optimized to maximize biomass yield, light penetration, and oxygen transfer. A larger size of tubular PBR will cause certain challenges in removing the oxygen that will decrease biomass productivity (Kirnev et al., 2020). Moreover, dimension of the tube will affect the ratio of illuminated surface to reactor volume. Thus, the tube length should be kept at a minimal level while its width could be increased (Sirohi et al., 2022). Due to its design, the tubular PBR might have less photosynthetic efficiency as the oxygen accumulates in the system. The tubular PBR system can be divided into two sections: the photo stage loop where photosynthesis occurs, and retention and mixing tank to extract the accumulated oxygen (Barboza-Rodríguez et al., 2024). On the other hand, bubble column PBR offers a straightforward design, whereby the column typically has a 2:1 height to diameter (H:D) ratio. Column PBR could be constructed using glass, plastic, and polyethylene bags, depending on its configuration. In bubble column PBR, a sparger is present in the internal structure to provide aeration in the system. The uniform bubble size of 4–6 mm was recommended for optimal mass transfer, although the velocity needs to be further optimized (Mcclure et al., 2013). The mechanical simplicity of bubble column PBR can significantly reduce the capital and operational costs while maintaining optimal cultivation conditions and mass transfer properties; thus it is suitable to be used in biofuel production.

The flat-panel PBR is a vertical rectangular frame consisting of two parallel transparent plates that form a rectangular channel, containing microalgae biomass (Chanquia et al., 2022). The agitation in flat-panel PBR is provided through aeration, while illumination can be more flexible due to its compact design. Flat-panel PBR can utilize direct sunlight for illumination, whereby its position can be tilted or adjusted following the sun's direction. In extreme sun exposure, the position of flat-panel PBR can also be adjusted to avoid excessive heat and illumination. The width of flat-panel PBR may reach 10 cm with height varying between 1 and 2 m (Nasr, 2019). However, since flat-panel PBR has more complex control variables, its application is mainly limited in laboratory scale (Carolin Christopher et al., 2021). Moreover, in large-scale cultivation, efficient illumination is challenging since a series of flat-panel PBRs will hinder the light penetration. The areal requirement also needs to be carefully considered. Introduction of environmental stressors focusing on biofuel production is also challenging in flat-panel PBR. In addition to that, the airlift PBR was introduced, by having physical separation of the upflow and downflow streams in the column. Three main components of airlift PBR include a sparger in the internal loop, an internal loop, and an external loop vessel (Barboza-Rodríguez et al., 2024). The gas was introduced through the sparger, generating gas holdup and subsequently lowering the fluid's density. The liquid will move upward, reaching the top of the vessel. The difference between upward and downward sections drives the liquid circulation in airlift reactors (Gupta et al., 2015). Draft tubes may be added as internal loop airlifts, enhancing circulation of the culture. These characteristics in airlift PBR create a more stable and well-defined flow pattern, as opposed to the abstract movement seen in bubble column PBR (Öncel et al., 2016).

In the production of biofuel from microalgae biomass, the selection of microalgae cultivation needs to be carefully considered. A suitable system should allow for maximized production of microalgae biomass and be highly adaptable for production of lipid or biofuel, as summarized in Table 9.1 If oleaginous species are used, the system will require less modification. However, environmental stressors might be necessary for certain species to promote the lipid production in microalgae. After successful cultivation process, the biomass needs to be separated from the cultivation media. The main mechanisms in microalgae harvesting methods are centrifugation, filtration, flocculation, flotation, and sedimentation (Barros et al., 2015). Some harvesting methods, such as flocculation, flotation, and sedimentation, are inexpensive and noninvasive. However, they require longer operation times and have lower separation efficiency. Centrifugation and filtration are commonly used in laboratory-scale production, although they might require large capital and operational costs in full-scale production, along with high energy requirements. Especially in filtration process, the filter needs to be replaced regularly to maintain its high efficiency. Biofuel production system needs to consider the most efficient biomass harvesting method to maximize the yield.

Table 9.1 Various cultivation systems for production of microalgae biomass and bio-oil for potential biofuel application.

Cultivation system	Species	Biomass production	Lipid content	Application	Reference
Open raceway pond	*Marvania coccoides*	1205 mg/L	19.46%	Biodiesel and biochar	Gawali et al. (2025)
	Dunaliella tertiolecta, *Chlorella vulgaris*, *Anabaena* sp., *Phormidium* sp.	12.5 g/m^2.day 12.1 g/m^2.day 8.9 g/m^2.day 15.1 g/m^2.day	8.1%–13.5% 5.2%–13.2% 5.0%–7.2% 10.5%–11.9%	Bio-oil and biodiesel	Kannan et al. (2025)
	Picochlorum costavermella, *Nephroselmis pyriformis*	0.41 g/L 0.42 g/L	18.5% 7.3%	Bio-oil	Dritsas et al. (2025)
Airlift PBR	*Chlorella* sp.	104 mg/L.d	36.22%	Biodiesel	Daniela Rios Ramirez et al. (2024)
	Chlorella sorokiniana	1.3 g/L	38.2%	Biocrude oil	Mahata et al. (2023)
	Chlorella sp.	0.718 g/L	23.94%	Biocrude oil	Condori et al. (2024)
Bubble column PBR	*Chlorella sorokiniana*	0.097 g/L.d	Not reported	-	Uyar et al. (2024)
Stirred tank PBR	*Chlorella sorokiniana*	0.064 g/L.d	Not reported	-	Uyar et al. (2024)
Bubble column PBR	*Chlorella sorokiniana*	Up to 2.75 g/L	Up to 24.14%	Biocrude oil	Ghosh et al. (2024)
Flat-panel PBR	*Scenedesmus obliquus*	Up to 3.0 g/L	40%	Biodiesel	Abdel-Baset et al. (2024)
Flat-panel PBR	*Chlorella sorokiniana*	1.89 g/L	19%	Biodiesel	Banerjee et al. (2023)

9.4 Conversion of microalgae biomass into biofuels

The conversion process of microalgae biomass into various forms of biofuels is a key aspect of utilizing carbon-based energy sources to replace fossil-based energy systems. Generally, this conversion technology can be divided into two major groups: the direct (thermochemical) and the indirect (chemical or biochemical) pathways. In essence, each method has its own advantages and limitations. Factors influencing this conversion include moisture content during the process, the composition of the biomass produced, energy requirements, technological capabilities, and the production scale. With advancements in science and technology, several methods have been developed to address the limitations of the conversion process, making it more energy efficient and environmentally friendly both before (pretreatment) and during the actual process. The details of each process are illustrated in Fig. 9.1 and will be explained in the following subsections. (Fig. 9.2)

9.4.1 Direct process

9.4.1.1 Pyrolysis

Pyrolysis works by thermally decomposing biomass using heat without oxygen, producing noncondensable gases (syngas), liquids (bio-oil), and solids (biochar). Microalgae are excellent candidates for this process due to their high levels of carbohydrates and proteins, because they have low lignin content and a more homogeneous cell structure compared to other cellulose-based biomass sources (Suali and Sarbatly, 2012). The thermal decomposition process of algae can be divided into three main stages: (1) at temperatures below 200°C, dehydration occurs in the algae; (2) at temperatures between 200°C and 500°C, the main pyrolysis process of devolatilization occurs; (3) at temperatures above 550°C, solid decomposition occurs. In the pyrolysis process, the biofuel products obtained will depend on the thermal conditions applied, such as: (1) low temperature and low heating rate will produce biochar, (2) low temperature and high heating rate will produce liquid bio-oil, and (3) high temperature, low heating rate, and prolonged vapor phase time will produce syngas (Maguyon and Capareda, 2013).

As the dried microalgae biomass can be easily converted using a thermal process on the dried raw material, pyrolysis is a common alternative conversion process. By adjusting relatively simple parameter settings,

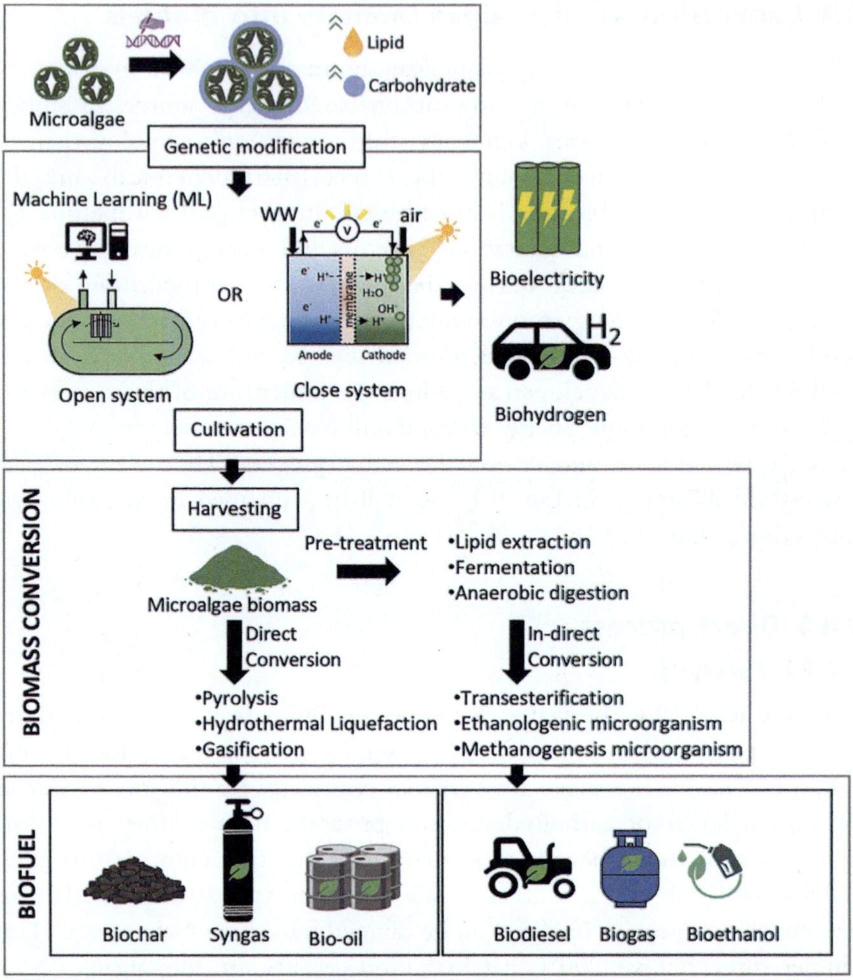

Figure 9.2 ***Conversion of microalgae.*** The emerging modifications in microalgae cultivation techniques and conversion of microalgae biomass into different types of biofuel.

particularly regarding heating temperature and rate flow, the desired products can be tailored to our needs, whether in the form of gas (syngas), liquid (bio-oil), or solid (biochar). However, the primary challenge of pyrolysis remains in its economic feasibility, so this technology is still primarily in the developmental phase, though continous technological advancements are gradually paving the way toward commercialization.

9.4.1.2 Hydrothermal liquefaction

The combination of thermal processes and elevated pressure allows hydrothermal liquefaction, or HTL toconvert high-moisture biomass into biocrude with relatively moderate parameters (200°C–400°C and 100–200 bar). Specifically, biochar could be produced in the temperature range of 250°C–300°C, biooil could be produced in the temperature range of 300°C–350°C, and syngas could be produced at temperatures above 350°C (Jaiswal et al., 2020). The HTL process can be operated immediately after the washing process, meaning that wet biomass or, at least, biomass with above 10% moisture content can be fed directly into the biofuel conversion process. Generally, the conversion process from HTL consists of three stages: depolymerization, decomposition, and recombination. For the conversion process using HTL, microalgae with high lipid content are more commonly used to produce high-quality biofuel. However, microalgae with low lipid content can also be used as raw material for conversion processes using HTL, given that they have a fast growth rate (Alba et al., 2012).

The use of the HTL conversion process can cut the drying process from raw biomass to biofuel, making it more energy efficient and potentially reducing greenhouse gas emissions. Furthermore, the resulting aqueous phase after the HTL conversion contains phosphorus, nitrogen, and other nutrients that can be recycled as cultivation material for algae or as material for anaerobic digestion (AD) to produce biomethane. However, high energy input, dependence on the impurities of the biomass and catalyst, and environmental impact of the industrial scale of this process are still quite a challenge for now. Additionally, the bio-oil produced from the HTL process still does not meet the requirements for direct use as fuel due to its low C-H ratio and high acidity level. Therefore, the development of increasing the scale of production from this conversion is still being studied (Zhang et al., 2021; Shan et al., 2021).

9.4.1.3 Gasification

The gasification conversion uses the principle of converting organic material or biomass into a gas mixture at high temperatures (700°C–1000°C) through partial oxidation. There are four general stages in conversion using gasification, namely drying, pyrolysis, oxidation, and reduction. The primary gas products are CO (7%–52%), methane (CH_4), and H_2 (10%–50%), alongside byproducts such as char and tar, which influence the quality of the gas produced during the conversion process (Ramzan et al., 2011). The organic material or biomass used in gasification requires a relatively low moisture content of around 15% (McKendry, 2002). However, previous research has shown that microalgae with higher moisture content can still be tolerated and produce good biofuel yields (Raheem et al., 2015).

The conversion of microalgal biomass using gasification can be divided into two major processes. The first is conventional gasification, which involves the reaction between oxidizers such as oxygen, air, and steam with dry algal biomass under heating conditions of approximately 800°C–1000°C and a moderate pressure of approximately 1–10 bar (Khoo et al., 2013). Conventional gasification has been widely used on a large or commercial scale; however, the challenge with this process lies in its low-temperature efficiency, especially for biomass with high moisture content. The second type of gasification is supercritical water gasification (SCWG), which combines moderate heating temperatures (375°C–550°C) with high pressure (221–360 bar) to exceed the supercritical point of water, enabling microalgal biomass to be directly converted into gas without undergoing a drying process (Amin, 2009). Currently, gasification using SCWG is still in its early developmental stages, and requires further optimization for commercial applications.

9.4.2 Indirect process

9.4.2.1 Lipid extraction and transesterification

The conversion of biomass into biofuel on a commercial scale involves a lipid extraction stage followed by a transesterification process. The development of lipid extraction methods can generally be divided into three categories. First, it involves extracting lipids using organic solvents. In lipid extraction using organic solvents, the triglyceride component in algae oil reacts chemically with the aid of enzymatic catalysts, acids, or bases to form FAMEs, a process known as transesterification. Second, biodiesel from biomass is extracted under acidic conditions at atmospheric pressure and room temperature (Wahlen et al., 2011). Third, the supercritical conditions of alcohol are utilized, that is, under high temperature and pressure without the use of a catalyst (Patil et al., 2011).

The microalgae used in this process are typically those with high lipid content, including both saturated and unsaturated fatty acids. Rizwanul Fattah et al. (2020) demonstrated an optimization approach for lipid extraction, noting that ultrasonication produces significantly better oil yields, namely 1.38 g for *Spirulina* sp. and 2.82 g for *Chlorella* sp., compared to conventional soxhlet extraction (1.05 g and 1.25 g) and solvent extraction methods (0.68 g and 0.77 g). The lipid extraction and transesterification conversion method is widely used due to its economic feasibility, but the environmental impact of solvent waste remains a challenge. Additionally, the use of enzymatic catalysts presents scalability and cost limitations at an industrial level (Rizwanul Fattah et al., 2020).

9.4.2.2 Fermentation

The fermentation pathway generally consists of two stages. The first stage involves converting carbohydrates into simple sugars, and the second stage involves ethanologenic microorganisms using these simple sugars as a growth medium and fermenting ethanol (Kee et al., 2018). Microalgae are a source of carbohydrates that can be used to produce bioethanol or biobutanol. In the process, microalgal biomass is enzymatically or acidically hydrolyzed to break down into simple sugars or fermentable sugars, such as mannose, galactose, and glucose. At this stage, enzymatic hydrolysis is more commonly used because it produces more specific products and fewer inhibitors. After hydrolysis, the simple sugars are fermented using microorganisms such as *Saccharomyces cerevisiae, Dekkera bruxellensis, Pichia stipitis,* and *Spathaspora passalidarum* (Uchida et al., 2014; Takagi et al., 2015; Hwang et al., 2016).

Because microalgae contain a minimum lignin content, making the fermentation process simpler than that of lignocellulosic biomass. At this stage of fermentation, genetically engineered *Pseudomonas* sp. or *E. coli* can ferment specific simple sugars to increase ethanol yield. However, this technology is not yet commercially available and requires further study.

9.4.2.3 Anaerobic digestion

Anaerobic digestion (AD) is a conversion process that uses microorganisms to convert organic compounds in biomass into biogas, producing methane and CO_2. AD can also be used to process residues from lipid extraction or fermentation, thereby maximizing the commercial utilization of biomass. The conversion process in AD is divided into four main stages: hydrolysis, acidogenesis, acetogenesis, and methanogenesis. During the hydrolysis stage, bacterial genera such as *Bacteroides* and *Clostridium* can secrete hydrolytic enzymes that break down organic molecules in microalgae, making them easier for microorganisms to utilize in subsequent stages (Menzel et al., 2020). During the acidogenesis stage, bacteria like *Clostridium*, *Lactobacillus*, and *Streptococcus*, can break down the hydrolyzed organic material into organic acids (volatile fatty acids) used in subsequent processes (Li et al., 2021). Acetogenesis converts organic acids into acetate, hydrogen, and CO_2. The microorganisms involved in this stage belong to *Syntrophomonas* and *Syntrophobacter* (Raza et al., 2021). The final stage is methanogenesis by archaeal genera such as *Methanobacterium* and *Methanosarcina*. In this stage, materials such as acetate, hydrogen, and CO_2 from the acetogenesis stage are synthesized into methane and CO_2 (Holmes and Smith, 2016).

For conversion using AD, the cell wall of microalgae presents the main challenge because most microalgae consist of carbohydrates in the formation of their cell walls and can survive in digesters for more than 30 days, thus complicating the conversion process. However, if microalgae lack cell walls or are composed of protein components, they can be easily converted in the digester and produce biomethane (Menzel et al., 2020). Various pretreatment methods to optimize this conversion have been widely implemented, such as steam heating before AD, ultrasonic treatment, and alkali (NaOH) treatment. The primary objective is to maintain substrate availability and improve cell lysis. AD also heavily depends on how we maintain the balance of the microbial ecosystem.

9.4.3 Types of biofuels

As previously discussed, The conversion of microalgae biomass has been explained in terms of both direct methods such as pyrolysis, hydrothermal liquefaction, and gasification, as well as indirect methods such as fermentation, transesterification, and AD. Therefore, an in-depth examination of the types of biofuel products produced by these processes is necessary, as each conversion process can yield different forms of energy and physical states: solid, liquid, gas, and electricity depending on the reaction conditions set and the organic content within the microalgal biomass. The subsequent sections will detail specific energy outputs such as biodiesel, bioelectricity, bioethanol, biohydrogen, biomethane, and biooil. Analyzing their unique characteristics, advantages, and associated challenges is crucial for establishing a robust foundation for the future development of microalgae-based bioenergy technology.

9.4.3.1 Biodiesel

Biodiesel is a renewable and biodegradable biofuel produced through the transesterification of lipids, generationg glycerol as a byproduct. Microalgae, which can easily accumulate lipids, are an ideal feedstock for making biodiesel. The species and growth conditions of microalgae significantly influence the lipids produced, such as SFAs, MUFAs, PUFAs, glycolipids, and triacylglycerols (TAGs). Remarkably, the lipids produced from microalgae can reach up to 50% of their dry weight (Chisti, 2007; Hu et al., 2008).

Microalgal biodiesel can be utilized directly in diesel engines requiring onlyminor structural modifications, or it can be blended with conventional petroleum diesel. Biodiesel from microalgae is also a promising alternative for the aviation industry due to its high energy density, low freezing point (Giwa et al., 2018), and compliance with the EN14214 standard (Arguelles, 2021). Since not all species have the ability to synthesize triacylglycerols (TAGs), future research should focus on producing biodiesel from different species based on their carbon sources and genetic modifications.

9.4.3.2 Bioelectricity

Bioelectricity is a form of energy generated from a bioelectrochemical system, which is a biocatalytic reaction of electroactive microorganisms (Sharma and Kundu, 2010; Angelaalincy et al., 2018), so this system is better known as a microbial fuel cell (MFC). In general, an MFC consists of three parts: an anode, a cathode, and an ion-exchange membrane. The operating principle of the MFC involves the anode, which generates electrons (e^-) and protons (H^+) from the breakdown of organic material. The electrons are then transported to the cathode through an external circuit, thereby generating an electric current, while the protons pass through the membrane toward the cathode and react with O_2 to produce water.

The use of microalgae in this MFC system utilizes photosynthesis, which acts as an in situ O_2 supplier that can facilitate reactions in the cathode chamber. This system is also able to effectively use nitrogen and phosphorus sources from wastewater for growth media, which can increase the efficiency of the MFC system (Greenman et al., 2019) and create an ecofriendly and sustainable bioelectricity system. However, challenges include not only the lack of mature technology for industrial-scale applications, but also energy consumption and costs associated with cultivation and harvesting. Additionally, the integration within the MFC often leads to operational issues, such as pH differences between the anode (acidic) and cathode (alkaline), resulting in suboptimal voltage within the cell and power output.

9.4.3.3 Bioethanol

Bioethanol is a liquid biofuel produced from the fermentation of sugars derived from biomass. Microalgae contain complex carbohydrates such as cellulose, starch, and other polysaccharides, depending on the type and growing conditions. These complex carbohydrates are then hydrolyzed into simple sugars, which are subsequently fermented into ethanol by other microorganisms. The process of converting microalgal biomass into bioethanol involves several stages, including biomass harvesting, pretreatment (enzymatic, chemical, and mechanical), carbohydrate simplification through hydrolysis, and fermentation (Lakatos et al., 2019). The bioethanol produced must still be separated from its mixture through a distillation process. The residue from the use of microalgae biomass in this conversion still contains other bioactive compounds and proteins, so it can be used as raw material for bioproducts or other animal feed. However, the cost of using enzymes remains a challenge, and the efficiency of bioethanol production still needs to be further studied (Beigbeder and Lavoie, 2022).

9.4.3.4 Biohydrogen

Biohydrogen is a gaseous biofuel with great potential as an environmentally friendly alternative to fossil fuels, as it produces no emissions. Microalgae posses unique biological mechanisms that enable themto produce hydrogen by utilizing solar energy and organic compounds from the environment (Li et al., 2022b). Solar energy is required for photosynthesis, but under certain conditions, such as sulfur deprivation (Haosagul et al., 2021), the enzymes used for hydrogen production (hydrogenase enzymes) become active and reduce protons into H_2 gas (Khandekar et al., 2020). Organic compounds such as acetate, glucose, and glycerol are required for dark fermentation reactions by heterotrophic or mixotrophic microalgae (Gao et al., 2021). During this dark fermentation, the enzyme formate hydrogen lyase plays a role in converting the substrate into hydrogen and carbon dioxide. The biohydrogen system from microalgae is one of the cleanest sources of bioenergy compared to other biomass uses. However, low product efficiency and enzyme sensitivity to oxygen make closed photobioreactor systems costly for large-scale production.

9.4.3.5 Biomethane

Biomethane is a high purity gaseous biofuel produced through the anaerobic digestion (AD) of biomass by methanogenic microorganisms. To obtain pure and significant biogas production, microalgae biomass usually undergoes a pretreatment process to degrade rigid and recalcitrantmicroalgae cell walls. However, some types of microalgae have a high protein content, resulting in a low C/N ratio, which necessitates the addition of other substrates, such as agricultural waste, to optimize the AD process (Mussgnug et al., 2010). To achieve biomethane purity exceeding 95%, a biogas upgrading process is necessary to make it suitable for use as vehicle fuel (Venkiteshwaran et al., 2022).

The production of biomethane using microalgae certainly has an environmentally friendly system because it can be integrated with liquid waste, such as industrial, domestic, and livestock waste as its substrate, making it a candidate for low-carbon biofuel (Bose et al., 2020). However, the stability of the anaerobic process, the potential for ammonia formation from protein deamination, and pH control remain challenges for future research.

9.4.3.6 Biooil

Biooil is a form of liquid biofuel that has the potential to replace fossil-derived oil as a feedstock for fuel or chemicals. The high organic compound content of microalgae biomass makes it an ideal feedstock for conversion into volatile biofuel compounds. Specifically, fast pyrolysis technology has been identified as a highly effective method for producing biooil from microalgae , it efficiently accomodates the small size and high moisture content of microalgae, resulting in more optimal bio-oil production and fewer byproducts in the form of char (Miao and Wu, 2004).

The challenge with microalgae bio-oil is its composition, which can contains nitrogen that generates harmful NO_x emissions during combustion. Various methods for upgrading biooil, such as hydrodeoxygenation (HDO), are still being developed to obtain pure biooil that can be used as fuel (Ramos-Tercero et al., 2015).

9.5 Next generation of microalgae-based biofuel

The conventional process of converting microalgal biomass into biofuel has indeed shown progress in terms of efficiency and scale. However, to fully optimize production for fussil fuel replacement, ensure economic viability, and enhance overall sustainability, next generation conversion must focus on cocultivation techniques, integration with wastewater treatment, optimization with the aid of machine learning (ML), and genetic engineering. The following section will discuss the potential and approaches of technological innovations for biofuel conversion from microalgae.

9.5.1 Genetic modification

One of the main factors that needs to be improved in the technology of converting microalgae biomass into biofuel is the selection of microalgae strains that have high lipid content. Several methods have been employed to increase lipid content in microalgae, such as inducing phosphate or nitrogen starvation conditions (James et al., 2011, 2013); however, these methods have negative effects, such as inhibited growth rates, leading to reduced biomass production (Chisti, 2007), which poses a challenge for industrial-scale production. Genetic engineering technology in microalgae has continued to advance and now offers solutions for improving strains (in this case, strains capable of producing excess lipids) without negative impacts on growth and biomass production (Mata et al., 2010). One of the technologies used to create these strains is CRISPR/Cas9, a technology capable of inducing mutations or overexpression in microalgal strains by deleting or inserting target genes. One example of a gene that can enhance lipid production is diacylglycerol acyltransferase(DGAT), which plays a role in the mechanism of triacylglycerol (TAG) formation (Fan et al., 2014), which serves as the raw material for biodiesel or biooil. The challenges in genetic modification technology currently lie in regulations and public concerns about genetically modified organisms (GMOs), as genetically modified strains still need to be evaluated, especially in terms of their potential ecological impacts.

9.5.2 Cocultivation strategy

Cocultivation systems mimic natural ecosystems by establishing mutualistic symbiosis between microorganisms, specifically the cultivation of microalgae with bacteria or fungi. Microalgae provide organic matter and oxygen, while bacteria or fungi produce CO_2 and vitamin B12. Thus, these microorganisms create a mutually beneficial system (Perera et al., 2022). Research by Makut et al. (2019) on the use of cocultivation of microalgae and bacterial consortia demonstrated an increase in microalgae biomass and optimized overal growth parameters. Additionally, the use of wastewater in the study showed a reduction in cultivation costs.

9.5.3 Immobilization technique

Immobilization is the cultivation of microalgae on solid surfaces such as sodium alginate, cellulose-based hydrogels, and polyvinyl alcohol (PVA). With this system, simplifying biomass separation , and biomass density and product concentration are higher (Lam and Lee, 2012). Srinuanpan et al. (2019) explain that immobilizing microalgae can purify methane to exceed 95% by capturing CO_2 from the biogas and effectively removing phosphorus and nitrogen from the effluent. Additionaly, this method streamlines biomass harvesting easier without requiring significant energy for processing.

9.5.4 Integration with wastewater treatment system

Agricultural, domestic, and industrial wastewaters are rich in nitrogen and phosphorus, which are needed by microalgae as a growth medium. Thus, biomass from microalgae that utilize this waste can reduce operational costs and lessen environmental impact (Chisti, 2007). A study by Geremia et al. (2021) explains that microalgae can remove nitrogen pollutants by 24%–100% and phosphorus by 25%–100%, and also reduce chemical oxygen demand (COD) levels by 23%–93%, depending on the specific wastewater characteristics. Integrating wastewater systems with conventional processes can support the concept of a circular bioindustry and is economically viable.

9.5.5 Machine learning and advances in cultivation systems

With advances in technology, particularly in the fields of computing and automation, the production of biofuel from microalgae has become more sophisticated through integration with ML. Research on the conversion of microalgae into biofuel has been ongoing for some time, with raw data from databases serving as a source for ML to predict solutions and automate processes at both small and large scales, as well as improve energy efficiency. In recent years, biofuel conversion using microalgae has been applied across various stages, including strain identification, culture monitoring, growth modeling, cultivation, and conversion of microalgae, biofuel production, and environmental implications (Zewdie and Ali, 2020; Araya et al., 2021).

The basic concepts of ML used in microalgae research consist of four parts, namely artificial neural networks (ANN), support vector machines (SVM), decision trees (DT), and random forests (RF). These four concepts can be used to create models from environmental factor data for microalgae growth, enabling predictions of biomass growth, production of organic compounds (carbohydrates, proteins, and lipids), and nutrient consumption (Haro et al., 2022; Fu et al., 2023). One example of ML mechanisms in harvesting systems is the development of algorithms that predict culture saturation levels and can automatically schedule harvesting (e.g., membrane cleaning in filtration processes) to mitigate fouling (Wang et al., 2022).

9.6 Challenges and opportunities

A number of economic, technical, and policy challenges continue to obstruct to the production of microalgae-based biofuels, even though microalgae have great potential to mitigate the global energy crisis and environmental pollution. Technological developments are opening up promising opportunities to improve efficiency and environmental impact. This section will explain the challenges and future perspectives.

9.6.1 Economic viability: techno-economic analysis and life cycle assessment

Technoeconomic analyses (TEA) provide detailed evaluations of energy and production costs, highlighting that economic feasibility remains a critical bottleneck for pilot-scale microalgal biomass conversion. The estimated price for biodiesel from microalgae ranges from 4 to 10 USD per liter, depending on technology, location, and scale. Therefore, it is still unable to replace fossil diesel, which is priced at around 1 USD per liter (Chisti, 2007). Conversely, Life cycle assessment (LCA) also explains that the production of biofuel from microalgae can reduce greenhouse gas emissions, especially when integrated with the ability of microalgae to capture CO_2 and use wastewater as a growth medium (Adesanya et al., 2014).

9.6.2 Harvesting process and biomass recovery

The harvesting stage represents a major economic barrie, accounting for approximately 20%–30% of the total of biofuel production cost. Due to the small size and density of microalgae, conventional harvesting methods still require a large amount of energy input or the use of expensive reagents.

Some harvesting methods, such as centrifugation, have limitations, such as high energy requirements. Flocculation has challenges, such as the use of chemical additives that can affect downstream processes. Membrane filtration has challenges related to membrane maintenance costs and requires regular cleaning. To address these challenges, integrating hybrid or multistage harvesting methods can enhance overall efficiency (Huy et al., 2022).

9.6.3 Scalability

For scaling up production from laboratory to industrial or commercial levels, closed systems or photobioreactors and open raceway ponds are typically used for open systems. Photobioreactor systems remain challenging to operate due to issues such as biofouling, oxygen accumulation, and overheating. Conversely, For open raceway ponds, biological contamination is the primary challenge, and performance is highly dependent on environmental conditions. Moving forward, innovations for scaling up, such as modular bioreactors, floating photobioreactors, and vertical cultivation systems, have the potential to overcome these structural limitations (Quiroz Arita et al., 2015).

9.6.4 Biological and technical limitations

Technical bottlenecks that occur during microalgae cultivation in general include contamination from bacteria, fungi, and even competing algae strains, decreased light intensity due to algae growth processes that increase density, cell wall rigidity that hinders extraction.Furthermore, conventional stress induction methods used to trigger lipid accumulation often severely compromise cell growth. Biological solutions such as mutagenesis and genetic engineering of future strains can reduce energy input and improve yield. Additionally, ML technology integrated with bioreactors can be an option for regulating production conditions from upstream to downstream purification (Piyatilleke et al., 2025).

9.6.5 Opportunities of microalgae-based biofuel in developed countries and its prospects to support sustainability

In terms of research and technology investment, developed countries such as the United States, the Netherlands, Japan, and Germany are pioneers in the field of microalgae-based biofuel research. The transition to clean energy from fossil fuels and the significant reduction of carbon emissions are the reasons why developed countries are pursuing this research. In the United States, research on algae is directly supported by the Department of Energy (DOE) through collaborative studies to test the economic integration and efficiency of pilot-scale biofuel production from microalgae (Sheehan et al., 2012). In the Netherlands, research on developing modular photobioreactor systems to enhance biomass production efficiency and its conversion into biofuel has been conducted at AlgaePARC. In Japan, one company is collaborating with Malaysia on the Chitose Carbon Capture Central (C4) program to produce biofuel from microalgae biomass on approximately 5 hectares of land. Meanwhile, Germany stands as a leading producer of microalgae and *Spirulina*, serving as a hub for advanced cultivation methodologies;; however, its application use for biofuel conversion remains limited compared to high-value food production and nutraceutical production (Araújo et al., 2021).

The focus in these developed countries also plays a role in the development of integrated energy systems, the utilization of CO_2 emissions from industry as a source of carbon, and the application of TEA and LCA so that the sustainability impact of this technology can be measured comprehensively. With the increasing efficiency of technology and the significant carbon emissions in the environment, the development of biofuel conversion from microalgae will present tremendous potential in both developed and developing countries. The summary of biofuel production in various countries is presented in Table 9.2

9.7 Conclusion

Microalgae-based biofuel production emerged as the solution to first- and second-generation biofuels, which interrupted the food supply and required large land areas. The process began with selection of suitable microalgae strain with desired metabolite composition, ideally one that is locally available. This is followed by a cultivation process that may utilize open or closed systems, whereby each system has its own advantages and disadvantages. Various parameters (e.g., nutrient composition, light intensity, and aeration strategy) could be adjusted during the cultivation process to maximize the biomass productivity and biofuel yield. The harvested microalgae biomass is then subjected to conversion into biofuels through direct or indirect methods. Different processes will yield varying biofuel types and properties; thus, they need to be selected appropriately.

Table 9.2 Reports on various biofuel productions in different countries.

Country (year)	Biofuel production	Types of biofuel	Cost of biofuel	Reference
India (2024)	Pyrolysis and condensation	Biochar Bio-oil	$ 0.42/Kg	Mohit et al. (2024)
United Kingdom (2023)	Transesterification Fermentation	Biodiesel Biobutanol	$ 2.21/L $ 8.57/L	Figueroa-Torres and Theodoropoulos (2023)
China (2021)	Pyrolysis Codigestion with activated sludge from municipal waste	Biochar Bio-oil Biomethane	$ 3.22/Kg	Li et al. (2022a)
Canada (2021)	Hydrothermal liquefaction (HTL) in a methanol-water system with reutilization of hydrochar by-product as catalyst	Bio-oil	$ 2.2/L	Masoumi and Dalai (2021)
India (2021)	Reactive flash volatilization with low-temperature shift and H_2 purification Reactive flash volatilization with COmethanation and low-temperature shift	Biohydrogen Biomethane	$ 10/Kg	Gholkar et al. (2021)
United States (2017)	HTL Microvace-assisted pyrolysis Lipid extraction anaerobic digestion of lipid-extracted residue	Bio-oil Biogas Biochar Algal lipids Biogas	- - - -	Roostaei and Zhang (2017)
United States (2016)	Pyrolysis	Bio-oil	$ 2.23/ gallon	Xin et al. (2016)

Moreover, the cultivation and extraction processes account for majority of energy requirements. Optimization of these processes is necessary to minimize the operational costs, supporting the commercialization of biofuel to compete with fossil fuel. Recent strategies have been developed, such as genetic modification of the microalgae strain, cocultivation with other species or microorganisms to enhance its applicability, and ML to support the cultivation and harvesting process. Although some challenges remain, substantial efforts are being made for real application of microalgae-based biofuel.

References

Abdel-Baset, A., Matter, I.A., Ali, M.A., 2024. Enhanced scenedesmus obliquus cultivation in plastic-type flat panel photobioreactor for biodiesel production. Sustainability 16. Available at https://doi.org/10.3390/su16083148.

Abdullah, M., Ali, Z., Yasin, M.T., Amanat, K., Sarwar, F., Khan, J., et al., 2024. Advancements in sustainable production of biofuel by microalgae: recent insights and future directions. Environmental Research 262 (Pt 2), 119902. Available at https://doi.org/10.1016/j.envres.2024.119902.

Adesanya, V.O., Cadena, E., Scott, S.A., Smith, A.G., 2014. Life cycle assessment on microalgal biodiesel production using a hybrid cultivation system. Bioresource Technology 163, 343–355. Available at https://doi.org/10.1016/j.biortech.2014.04.051, www.elsevier.com/locate/biortech.

Alba, L.G., Torri, C., Samorì, C., et al., 2012. Hydrothermal Treatment (HTT) of Microalgae : Evaluation of the Process As Conversion Method in an Algae Biorefinery Concept 642–657.

Amin, S., 2009. Review on biofuel oil and gas production processes from microalgae. Energy Conversion and Management 50 (7), 1834–1840. Available at https://doi.org/10.1016/j.enconman.2009.03.001.

Angelaalincy, M.J., Navanietha Krishnaraj, R., Shakambari, G., Ashokkumar, B., Kathiresan, S., Varalakshmi, P., 2018. Biofilm engineering approaches for improving the performance of microbial fuel cells and bioelectrochemical systems. Frontiers in Energy Research 6 Available from. Available at https://doi.org/10.3389/fenrg.2018.00063, http://www.frontiersin.org/Energy_Research/.

Araya, M., García, S., Rengel, J., Pizarro, S., Álvarez, G., 2021. Determination of free and protein amino acid content in microalgae by HPLC-DAD with pre-column derivatization and pressure hydrolysis. Marine Chemistry 234, 103999. Available at https://doi.org/10.1016/j.marchem.2021.103999.

Araújo, R., Vázquez Calderón, F., Sánchez López, J., Azevedo, I.C., Bruhn, A., Fluch, S., et al., 2021. Current status of the algae production industry in Europe: an emerging sector of the blue bioeconomy. Frontiers in Marine Science 7. Available at https://doi.org/10.3389/fmars.2020.626389.

Arguelles, E.D.L.R., 2021. Biochemical composition and bioactive properties of chlorella minutissima (Chm1) as a potential source of chemical compounds for nutritional feed supplement and disease control in aquaculture. Current Applied Science and Technology 21 (1), 65–77. Available at https://li01.tci-thaijo.org/index.php/cast/article/download/245911/168108.

Arif, M., Bai, Y., Usman, M., Jalalah, M., Harraz, F.A., Al-Assiri, M.S., et al., 2020. Highest accumulated microalgal lipids (polar and non-polar) for biodiesel production with advanced wastewater treatment: Role of lipidomics. Bioresource Technology 298, 122299. Available at https://doi.org/10.1016/j.biortech.2019.122299.

Assunção, J., Xavier Malcata, F., 2020. Enclosed "non-conventional" photobioreactors for microalga production: a review. Algal Research 52, 102107. Available at https://doi.org/10.1016/j.algal.2020.102107.

Banerjee, S., Dasgupta, S., Atta, A., Das, D., Dayal, D., Malik, S., et al., 2023. Flow rate optimization in a flat-panel photobioreactor for the cultivation of microalgae for mitigating waste gas. Water (Switzerland) 15 (15). Available at https://doi.org/10.3390/w15152824, http://www.mdpi.com/journal/water.

Barboza-Rodríguez, R., Rodríguez-Jasso, R.M., Rosero-Chasoy, G., Rosales Aguado, M.L., Ruiz, H.A., 2024. Photobioreactor configurations in cultivating microalgae biomass for biorefinery. Bioresource Technology 394, 130208. Available at https://doi.org/10.1016/j.biortech.2023.130208.

Barros, A.I., Gonçalves, A.L., Simões, M., Pires, J.C.M., 2015. Harvesting techniques applied to microalgae: a review. Renewable and Sustainable Energy Reviews 41, 1489–1500. Available at https://doi.org/10.1016/j.rser.2014.09.037, https://www.journals.elsevier.com/renewable-and-sustainable-energy-reviews.

Beigbeder, J.-B., Lavoie, J.-M., 2022. Effect of photoperiods and CO2 concentrations on the cultivation of carbohydrate-rich P. kessleri microalgae for the sustainable production of bioethanol. Journal of CO2 Utilization 58, 101934. Available at https://doi.org/10.1016/j.jcou.2022.101934.

Bose, A., O'Shea, R., Lin, R., Murphy, J.D., 2020. A perspective on novel cascading algal biomethane biorefinery systems. Bioresource Technology 304, 123027. Available at https://doi.org/10.1016/j.biortech.2020.123027.

Carolin Christopher, F., Kumar, P.S., Vo, D.V.N., Joshiba, G.J., 2021. A review on critical assessment of advanced bioreactor options for sustainable hydrogen production. International Journal of Hydrogen Energy 46 (10), 7113–7136. Available at https://doi.org/10.1016/j.ijhydene.2020.11.244, http://www.journals.elsevier.com/international-journal-of-hydrogen-energy/.

Cavelius, P., Engelhart-Straub, S., Mehlmer, N., Lercher, J., Awad, D., Brück, T., 2023. The potential of biofuels from first to fourth generation. PLoS Biology 21 (3), e3002063. Available at https://doi.org/10.1371/journal.pbio.3002063.

Chanquia, S.N., Vernet, G., Kara, S., 2022. Photobioreactors for cultivation and synthesis: specifications, challenges, and perspectives. Engineering in Life Sciences 22, 712–724. Available at https://doi.org/10.1002/elsc.202100070.

Chisti, Y., 2007. Biodiesel from microalgae. Biotechnology Advances 25 (3), 294–306. Available at https://doi.org/10.1016/j.biotechadv.2007.02.001.

Chu, R., Hu, D., Zhu, L., Li, S., Yin, Z., Yu, Y., 2022. Recycling spent water from microalgae harvesting by fungal pellets to re-cultivate Chlorella vulgaris under different nutrient loads for biodiesel production. Bioresource Technology 344 (Pt B), 126227. Available at https://doi.org/10.1016/j.biortech.2021.126227.

Condori, M.A.M., Condori, M.M., Gutierrez, M.E.V., Choix, F.J., García-Camacho, F., 2024. Bioremediation potential of the Chlorella and Scenedesmus microalgae in explosives production effluents. Science of the Total Environment 920. Available at https://doi.org/10.1016/j.scitotenv.2024.171004, https://www.sciencedirect.com/science/journal/00489697.

Costa, J.A.V., de Morais, M.G., 2014. An Open Pond System for Microalgal Cultivation. Elsevier BV, pp. 1–22. Available at 10.1016/b978-0-444-59558-4.00001-2.

Daniela Rios Ramirez, K., Botero Ñañez, K., Leonardo Gonzalez Gomez, C., Thiago Andrade Moreira, Í., 2024. Efficient PAHs removal and CO2 fixation by marine microalgae in wastewater using an airlift photobioreactor for biofuel production.

Environmental Research 261, 119672. Available at https://doi.org/10.1016/j.envres.2024.119672.

Dehghanipour, A., Zamani, H., 2024. Interaction of Fe2O3 nanoparticles with marine microalga Chlorella sorokiniana: Analysis of growth, morphological changes and biochemical composition. Plant Physiology and Biochemistry 207, 108385. Available at https://doi.org/10.1016/j.plaphy.2024.108385.

Dey, S., Reang, N.M., Das, P.K., Deb, M., 2021. A comprehensive study on prospects of economy, environment, and efficiency of palm oil biodiesel as a renewable fuel. Journal of Cleaner Production 286, 124981. Available at https://doi.org/10.1016/j.jclepro.2020.124981.

Dou, X., Lu, X.-H., Lu, M.-Z., Yu, L.-S., Xue, R., Ji, J.-B., 2013. The effects of trace elements on the lipid productivity and fatty acid composition of Nannochloropis oculata. Journal of Renewable Energy 2013, 1–6. Available at https://doi.org/10.1155/2013/671545.

Dritsas, P., Patsialou, S., Kampantais, D., Roussos, E., Kotzamanis, Y., Tekerlekopoulou, A., et al., 2025. Investigating the potential of newly isolated microalgae strains from the Ionian Sea (Greece) cultured in an open raceway pond. Applied Sciences 15 (12), 6680. Available at https://doi.org/10.3390/app15126680.

Fan, J., Cui, Y., Wan, M., Wang, W., Li, Y., 2014. Lipid accumulation and biosynthesis genes response of the oleaginous Chlorella pyrenoidosaunder three nutrition stressors. Biotechnology for Biofuels 7 (1). Available at https://doi.org/10.1186/1754-6834-7-17.

Feng, X., Chen, Y., Lv, J., Han, S., Tu, R., Zhou, X., et al., 2020. Enhanced lipid production by Chlorella pyrenoidosa through magnetic field pretreatment of wastewater and treatment of microalgae-wastewater culture solution: magnetic field treatment modes and conditions. Bioresource Technology 306, 123102. Available at https://doi.org/10.1016/j.biortech.2020.123102.

Figueroa-Torres, G.M., Theodoropoulos, C., 2023. Techno-economic analysis of a microalgae-based biorefinery network for biofuels and value-added products. Bioresource Technology Reports 23, 101524. Available at https://doi.org/10.1016/j.biteb.2023.101524.

Fu, W., Li, X.'e, Yang, Y., Song, D., 2023. Enhanced degradation of bisphenol A: Influence of optimization of removal, kinetic model studies, application of machine learning and microalgae-bacteria consortia. Science of The Total Environment 858 (Pt 2), 159876. Available at https://doi.org/10.1016/j.scitotenv.2022.159876.

Ganesan, R., Manigandan, S., Samuel, M.S., Shanmuganathan, R., Brindhadevi, K., Lan Chi, N.T., et al., 2020. A review on prospective production of biofuel from microalgae. Biotechnology Reports 27, e00509. Available at https://doi.org/10.1016/j.btre.2020.e00509.

Gao, F., Yang, Z.-Y., Zhao, Q.-L., Chen, D.-Z., Li, C., Liu, M., et al., 2021. Mixotrophic cultivation of microalgae coupled with anaerobic hydrolysis for sustainable treatment of municipal wastewater in a hybrid system of anaerobic membrane bioreactor and membrane photobioreactor. Bioresource Technology 337, 125457. Available at https://doi.org/10.1016/j.biortech.2021.125457.

Gawali, S., Alam, S.N., Tilva, J., Jagadale, K., Singh, B., Guldhe, A., 2025. Development of scalable cultivation strategy for microalgal biomass generation using wastewater and integrated biorefinery approach for biofuel production. Journal of Environmental Chemical Engineering 13 (2). Available at https://doi.org/10.1016/j.jece.2025.115441, https://www.sciencedirect.com/science/journal/22133437.

Geremia, E., Ripa, M., Catone, C.M., Ulgiati, S., 2021. A review about microalgae wastewater treatment for bioremediation and biomass production—a new challenge for europe. Environments - MDPI 8 (12). Available at https://doi.org/10.3390/environments8120136, www.mdpi.com/journal/environments.

Gholkar, P., Shastri, Y., Tanksale, A., 2021. Renewable hydrogen and methane production from microalgae: a techno-economic and life cycle assessment study. Journal of Cleaner Production 279, 123726. Available at https://doi.org/10.1016/j.jclepro.2020.123726.

Ghosh, G., Bhimrao Daile, S., Chakraborty, S., Atta, A., 2024. Influence of super-optimal light intensity on the acetic acid uptake and microalgal growth in mixotrophic culture of Chlorella sorokiniana in bubble-column photobioreactors. Bioresource Technology 393, 130152. Available at https://doi.org/10.1016/j.biortech.2023.130152.

Giwa, A., Adeyemi, I., Dindi, A., Lopez, C.G.-B., Lopresto, C.G., Curcio, S., et al., 2018. Techno-economic assessment of the sustainability of an integrated biorefinery from microalgae and Jatropha: a review and case study. Renewable and Sustainable Energy Reviews 88, 239–257. Available at https://doi.org/10.1016/j.rser.2018.02.032.

Goswami, R.K., Agrawal, K., Verma, P., 2022. Microalgae Dunaliella as biofuel feedstock and β-carotene production: an influential step towards environmental sustainability. Energy Conversion and Management: X 13, 100154. Available at https://doi.org/10.1016/j.ecmx.2021.100154.

Greenman, J., Gajda, I., Ieropoulos, I., 2019. Microbial fuel cells (MFC) and microalgae; photo microbial fuel cell (PMFC) as complete recycling machines. Sustainable Energy & Fuels 3, 2546–2560. Available at https://doi.org/10.1039/C9SE00354A.

Gupta, P.L., Lee, S.-M., Choi, H.-J., 2015. A mini review: photobioreactors for large scale algal cultivation. World Journal of Microbiology and Biotechnology 31, 1409–1417. Available at https://doi.org/10.1007/s11274-015-1892-4.

Haosagul, S., Oaew, S., Prommeenate, P., Sawasdee, V., Pisutpaisal, N., 2021. DNA microarray for detection and identification of sulfur oxidizing bacteria in biogas clean-up system. Energy Reports 7, 559–568. Available at https://doi.org/10.1016/j.egyr.2021.07.097.

Haro, S., Jesus, B., Oiry, S., Papaspyrou, S., Lara, M., González, C.J., et al., 2022. Microphytobenthos spatio-temporal dynamics across an intertidal gradient using Random Forest classification and Sentinel-2 imagery. Science of The Total Environment 804, 149983. Available at https://doi.org/10.1016/j.scitotenv.2021.149983.

Holmes, D.E., Smith, J.A., 2016. Biologically Produced Methane as a Renewable Energy Source. 97, Elsevier BV, pp. 1–61. Available at 10.1016/bs.aambs.2016.09.001.

Hu, Q., Sommerfeld, M., Jarvis, E., Ghirardi, M., Posewitz, M., Seibert, M., et al., 2008. Microalgal triacylglycerols as feedstocks for biofuel production: perspectives and advances. The Plant Journal 54 (4), 621–639. Available at https://doi.org/10.1111/j.1365-313x.2008.03492.x.

Huy, M., Kumar, G., Sharma, P., Sirohi, R., Pandey, A., Kim, S.-H., 2022. Effective recovery of microalgal biomass using various types of emulsion polymers. Journal of Biotechnology 358, 25–32. Available at https://doi.org/10.1016/j.jbiotec.2022.08.010.

Hwang, J.H., Kabra, A.N., Ji, M.K., Choi, J., El-Dalatony, M.M., Jeon, B.H., 2016. Enhancement of continuous fermentative bioethanol production using combined treatment of mixed microalgal biomass. Algal Research 17, 14–20. Available at https://doi.org/10.1016/j.algal.2016.03.029, http://www.sciencedirect.com/science/journal/aip/22119264.

Jaiswal, K.K., Banerjee, I., Singh, D., et al., 2020. Ecological stress stimulus to improve microalgae biofuel generation: a review. Octa Journal of Bioscinces 8, 48–54.

James, G.O., Hocart, C.H., Hillier, W., Chen, H., Kordbacheh, F., Price, G.D., et al., 2011. Fatty acid profiling of Chlamydomonas reinhardtii under nitrogen deprivation. Bioresource Technology 102 (3), 3343–3351. Available at https://doi.org/10.1016/j.biortech.2010.11.051.

James, G.O., Hocart, C.H., Hillier, W., et al., 2013. Bioresource technology temperature modulation of fatty acid profiles for biofuel production in nitrogen deprived Chlamydomonas reinhardtii. Bioresource Technology 127, 441–447. Available at https://doi.org/10.1016/j.biortech.2012.09.090.

Jung, J.H., Sirisuk, P., Ra, C.H., Kim, J.M., Jeong, G.T., Kim, S.K., 2019. Effects of green LED light and three stresses on biomass and lipid accumulation with two-phase culture of microalgae. Process Biochemistry 77, 93–99. Available at https://doi.org/10.1016/j.procbio.2018.11.014, www.elsevier.com/inca/publications/store/4/2/2/8/5/7.

Kabir, F., Gulfraz, M., Raja, G.K., Inam-ul-Haq, M., Awais, M., Mustafa, M.S., et al., 2020. Screening of native hyper-lipid producing microalgae strains for biomass and lipid production. Renewable Energy 160, 1295–1307. Available at https://doi.org/10.1016/j.renene.2020.07.004, http://www.journals.elsevier.com/renewable-and-sustainable-energy-reviews/.

Kannan, D.C., Magar, C.S., Bhattacharya, S., Bajpai, N., Maurya, R., Dhamanse, N., et al., 2025. Integrated production of algal biofuels and biocommodities. Algal Research 89, 104063. Available at https://doi.org/10.1016/j.algal.2025.104063.

Kee, C., Chyuan, H., Chen, W., et al., 2018. Overview: comparison of pretreatment technologies and fermentation processes of bioethanol from microalgae. Energy Conversion Management 173, 81–94. Available at https://doi.org/10.1016/j.enconman.2018.07.054.

Khandekar, N.S., Jagtap, R.S., Sajane, S.J., Jagtap, S.R., 2020. Sporopollenin: the ground discussion. Research Journal of Pharmacy and Technology 13 (3), 1555–1564. Available at https://doi.org/10.5958/0974-360X.2020.00282.6, http://www.rjptonline.org/.

Khoo, H.H., Koh, C.Y., Shaik, M.S., Sharratt, P.N., 2013. Bioenergy co-products derived from microalgae biomass via thermochemical conversion – life cycle energy balances and CO2 emissions. Bioresource Technology 143, 298–307. Available at https://doi.org/10.1016/j.biortech.2013.06.004.

Kirnev, P.C.S., Carvalho, J.C., Vandenberghe, L.P.S., Karp, S.G., Soccol, C.R., 2020. Technological mapping and trends in photobioreactors for the production of microalgae. World Journal of Microbiology and Biotechnology 36 (3). Available at https://doi.org/10.1007/s11274-020-02819-0, www.wkap.nl/journalhome.htm/0959-3993.

Kumar, A., Sharma, S., 2008. An evaluation of multipurpose oil seed crop for industrial uses (Jatropha curcas L.): a review. Industrial Crops and Products 28 (1), 1–10. Available at https://doi.org/10.1016/j.indcrop.2008.01.001.

Lakatos, G.E., Ranglová, K., Manoel, J.C., Grivalský, T., Kopecký, J., Masojídek, J., 2019. Bioethanol production from microalgae polysaccharides. Folia Microbiologica 64 (5), 627–644. Available at https://doi.org/10.1007/s12223-019-00732-0, www.biomed.cas.cz/mbu/folia/index.html.

Lam, M.K., Lee, K.T., 2012. Immobilization as a feasible method to simplify the separation of microalgae from water for biodiesel production. Chemical Engineering Journal 191, 263–268. Available at https://doi.org/10.1016/j.cej.2012.03.013.

LanzaTech. 2021 Annual Report. 2021.

Li, P., Luo, Y., Yuan, X., 2022. Life cycle and techno-economic assessment of source-separated wastewater-integrated microalgae biofuel production plant: a nutrient organization approach. Bioresource Technology 344, 126230. Available at https://doi.org/10.1016/j.biortech.2021.126230.

Li, S., Li, F., Zhu, X., Liao, Q., Chang, J.-S., Ho, S.-H., 2022. Biohydrogen production from microalgae for environmental sustainability. Chemosphere 291 (Pt 1), 132717. Available at https://doi.org/10.1016/j.chemosphere.2021.132717.

Li, Y., Xu, H., Yi, X., Zhao, Y., Jin, F., Chen, L., et al., 2021. Study of two-phase anaerobic digestion of corn stover: focusing on the conversion of volatile fatty acids and microbial characteristics in UASB reactor. Industrial Crops and Products 160, 113097. Available at https://doi.org/10.1016/j.indcrop.2020.113097.

Liu, X., Chen, G., Tao, Y., Wang, J., 2020. Application of effluent from WWTP in cultivation of four microalgae for nutrients removal and lipid production under the supply of CO2. Renewable Energy 149, 708–715. Available at https://doi.org/10.1016/j.renene.2019.12.092.

Maguyon, M.C.C., Capareda, S.C., 2013. Evaluating the effects of temperature on pressurized pyrolysis of Nannochloropsis oculata based on products yields and characteristics. Energy Conversion and Management 76, 764–773. Available at https://doi.org/10.1016/j.enconman.2013.08.033.

Mahata, C., Mishra, S., Dhar, S., Ray, S., Mohanty, K., Das, D., 2023. Utilization of dark fermentation effluent for algal cultivation in a modified airlift photobioreactor for biomass and biocrude production. Journal of Environmental Management 330, 117121. Available at https://doi.org/10.1016/j.jenvman.2022.117121.

Makut, B.B., Das, D., Goswami, G., 2019. Production of microbial biomass feedstock via co-cultivation of microalgae-bacteria consortium coupled with effective wastewater treatment: a sustainable approach. Algal Research 37, 228–239. Available at https://doi.org/10.1016/j.algal.2018.11.020.

Maltsev, Y., Maltseva, K., Kulikovskiy, M., Maltseva, S., 2021. Influence of light conditions on microalgae growth and content of lipids, carotenoids, and fatty acid composition. Biology 10 (10), 1060. Available at https://doi.org/10.3390/biology10101060.

Mandal, M.K., Saikia, P., Chanu, N.K., Chaurasia, N., 2019. Modulation of lipid content and lipid profile by supplementation of iron, zinc, and molybdenum in indigenous microalgae. Environmental Science and Pollution Research 26 (20), 20815–20828. Available at https://doi.org/10.1007/s11356-019-05065-6, http://www.springerlink.com/content/0944-1344.

Mandal, M.K., Chanu, N.K., Chaurasia, N., 2020. Exogenous addition of indole acetic acid and kinetin under nitrogen-limited medium enhances lipid yield and expression of glycerol-3-phosphate acyltransferase & diacylglycerol acyltransferase genes in indigenous microalgae: a potential approach for biodie. Bioresource Technology 297, 122439. Available at https://doi.org/10.1016/j.biortech.2019.122439.

Masoumi, S., Dalai, A.K., 2021. Techno-economic and life cycle analysis of biofuel production via hydrothermal liquefaction of microalgae in a methanol-water system and catalytic hydrotreatment using hydrochar as a catalyst support. Biomass and Bioenergy 151, 106168. Available at https://doi.org/10.1016/j.biombioe.2021.106168.

Mata, T.M., Martins, A.A., Caetano, N.S., 2010. Microalgae for biodiesel production and other applications: a review. Renewable and Sustainable Energy Reviews 14, 217–232. Available at https://doi.org/10.1016/j.rser.2009.07.020.

McKendry, P., 2002. Energy production from biomass (part 1): overview of biomass. Bioresource Technology 83 (1), 37–46. Available at https://doi.org/10.1016/s0960-8524(01)00118-3.

Mcclure, D.D., Kavanagh, J.M., Fletcher, D.F., Barton, G.W., 2013. Development of a CFD model of bubble column bioreactors: part one - a detailed experimental study. Chemical Engineering and Technology 36 (12), 2065–2070. Available at https://doi.org/10.1002/ceat.201300544.

Menzel T., Neubauer P., Junne S. (2020) Role of Microbial Hydrolysis in Anaerobic Digestion.

Miao, X., Wu, Q., 2004. High yield bio-oil production from fast pyrolysis by metabolic controlling of Chlorella protothecoides. Journal of Biotechnology 110 (1), 85–93. Available at https://doi.org/10.1016/j.jbiotec.2004.01.013.

Mohit, A., Mishel, V.K., Remya, N., 2024. Life cycle assessment and technoeconomic analysis of biofuels produced from polyculture microalgae cultivated in greywater.

Journal of Environmental Management 356, 120711. Available at https://doi.org/10.1016/j.jenvman.2024.120711.

Molina-Grima, E., García-Camacho, F., Acién-Fernández, F.G., Sánchez-Mirón, A., Plouviez, M., Shene, C., et al., 2022. Pathogens and predators impacting commercial production of microalgae and cyanobacteria. Biotechnology Advances 55, 107884. Available at https://doi.org/10.1016/j.biotechadv.2021.107884.

Mussgnug, J.H., Klassen, V., Schlüter, A., Kruse, O., 2010. Microalgae as substrates for fermentative biogas production in a combined biorefinery concept. Journal of Biotechnology 150 (1), 51–56. Available at https://doi.org/10.1016/j.jbiotec.2010.07.030.

Nasr, M., 2019. Design Considerations of Algal Systems for Wastewater Treatment. Springer Science and Business Media LLC, pp. 411–426. Available at 10.1007/978-3-030-13913-1_19.

Öncel, S., Köse, A., Öncel, D., 2016. In: Pacheco-Torgal, F., Rasmussen, E., Granqvist, C.-G. (Eds.), 11 - Façade Integrated Photobioreactors for Building Energy Efficiency. Woodhead Publishing, pp. 237–299.

Patil, P.D., Gude, V.G., Mannarswamy, A., Deng, S., Cooke, P., Munson-McGee, S., et al., 2011. Optimization of direct conversion of wet algae to biodiesel under supercritical methanol conditions. Bioresource Technology 102 (1), 118–122. Available at https://doi.org/10.1016/j.biortech.2010.06.031.

Perera, I.A., Abinandan, S., Panneerselvan, L., Subashchandrabose, S.R., Venkateswarlu, K., Naidu, R., et al., 2022. Co-culturing of microalgae and bacteria in real wastewaters alters indigenous bacterial communities enhancing effluent bioremediation. Algal Research 64, 102705. Available at https://doi.org/10.1016/j.algal.2022.102705.

Piyatilleke, S., Thevarajah, B., Nimarshana, P.H.V., Ariyadasa, T.U., 2025. Microalgal biofuels: challenges and prospective in the framework of circular bioeconomy. Energy Nexus 17, 100338. Available at https://doi.org/10.1016/j.nexus.2024.100338.

Qiu, R., Gao, S., Lopez, P.A., Ogden, K.L., 2017. Effects of pH on cell growth, lipid production and CO2 addition of microalgae Chlorella sorokiniana. Algal Research 28, 192–199. Available at https://doi.org/10.1016/j.algal.2017.11.004, http://www.sciencedirect.com/science/journal/aip/22119264.

Quiroz Arita, C.E., Peebles, C., Bradley, T.H., 2015. Scalability of combining microalgae-based biofuels with wastewater facilities: a review. Algal Research 9, 160–169. Available at https://doi.org/10.1016/j.algal.2015.03.001.

Rabbani, M., Hosseini, A., Karim, M.A., Fahimi, A., Karimi, K., Vahidi, E., 2024. Environmental impact assessment of a novel third-generation biorefinery approach for astaxanthin and biofuel production. Science of The Total Environment 912, 168733. Available at https://doi.org/10.1016/j.scitotenv.2023.168733.

Raheem, A., Wan Azlina, W.A.K.G., Taufiq Yap, Y.H., Danquah, M.K., Harun, R., 2015. Thermochemical conversion of microalgal biomass for biofuel production. Renewable and Sustainable Energy Reviews 49, 990–999. Available at https://doi.org/10.1016/j.rser.2015.04.186.

Ramos-Tercero, E.A., Bertucco, A., Brilman, D.W.F.(W.), 2015. Process water recycle in hydrothermal liquefaction of microalgae to enhance bio-oil yield. Energy & Fuels 29, 2422–2430. Available at https://doi.org/10.1021/ef502773w.

Ramzan, N., Ashraf, A., Naveed, S., Malik, A., 2011. Simulation of hybrid biomass gasification using Aspen plus: a comparative performance analysis for food, municipal solid and poultry waste. Biomass and Bioenergy 35 (9), 3962–3969. Available at https://doi.org/10.1016/j.biombioe.2011.06.005.

Raza, F., Khalid, H., El-mashad, H.M., et al., 2021. Science of the total environment Functions of bacteria and archaea participating in the bioconversion of organic waste for methane production. The Science of the Total Environment 763, 143007. Available at https://doi.org/10.1016/j.scitotenv.2020.143007.

Ren, H.-Y., Dai, Y.-Q., Kong, F., Xing, D., Zhao, L., Ren, N.-Q., et al., 2020. Enhanced microalgal growth and lipid accumulation by addition of different nanoparticles under xenon lamp illumination. Bioresource Technology 297, 122409. Available at https://doi.org/10.1016/j.biortech.2019.122409.

Rizwanul Fattah, I.M., Ong, H.C., Mahlia, T.M.I., Mofijur, M., Silitonga, A.S., Rahman, S.M.A., et al., 2020. State of the art of catalysts for biodiesel production. Frontiers in Energy Research 8. Available at https://doi.org/10.3389/fenrg.2020.00101, http://www.frontiersin.org/Energy_Research/.

Roostaei, J., Zhang, Y., 2017. Spatially explicit life cycle assessment: opportunities and challenges of wastewater-based algal biofuels in the United States. Algal Research 24, 395–402. Available at https://doi.org/10.1016/j.algal.2016.08.008.

Rugnini, L., Rossi, C., Antonaroli, S., Rakaj, A., Bruno, L., 2020. The influence of light and nutrient starvation on morphology, biomass and lipid content in seven strains of green microalgae as a source of biodiesel. Microorganisms 8 (8), 1254. Available at https://doi.org/10.3390/microorganisms8081254.

Saranya, D., Shanthakumar, S., 2020. Effect of culture conditions on biomass yield of acclimatized microalgae in ozone pre-treated tannery effluent: a simultaneous exploration of bioremediation and lipid accumulation potential. Journal of Environmental Management 273, 111129. Available at https://doi.org/10.1016/j.jenvman.2020.111129.

Shan T., Anto S., Mathimani T., et al. 2021. Upgrading of bio-oil from thermochemical conversion of various biomass – mechanism, challenges and opportunities. 287.

Sharma, V., Kundu, P.P., 2010. Biocatalysts in microbial fuel cells. Enzyme and Microbial Technology 47 (5), 179–188. Available at https://doi.org/10.1016/j.enzmictec.2010.07.001.

Sheehan J.D., Dunahay T., Benemann J.R., Roessler P., 2012. A Look Back at the U.S. Department of Energy's Aquatic Species

Sirohi, R., Kumar Pandey, A., Ranganathan, P., Singh, S., Udayan, A., Kumar Awasthi, M., et al., 2022. Design and applications of photobioreactors- a review. Bioresource Technology 349, 126858. Available at https://doi.org/10.1016/j.biortech.2022.126858.

Srinuanpan, S., Cheirsilp, B., Boonsawang, P., Prasertsan, P., 2019. Immobilized oleaginous microalgae as effective two-phase purify unit for biogas and anaerobic digester effluent coupling with lipid production. Bioresource Technology 281, 149–157. Available at https://doi.org/10.1016/j.biortech.2019.02.085.

Suali, E., Sarbatly, R., 2012. Conversion of microalgae to biofuel. Renewable and Sustainable Energy Reviews 16 (6), 4316–4342. Available at https://doi.org/10.1016/j.rser.2012.03.047.

Subhash, G.V., Rajvanshi, M., Kumar, G.R.K., et al., 2022. Challenges in microalgal biofuel production: a perspective on techno economic feasibility under biorefinery stratagem. Bioresource Technology 343, 126155.

Takagi, T., Uchida, M., Matsushima, R., Kodama, H., Takeda, T., Ishida, M., et al., 2015. Comparison of ethanol productivity among yeast strains using three different seaweeds. Fisheries Science 81 (4), 763–770. Available at https://doi.org/10.1007/s12562-015-0875-6.

Tan, J.S., Lee, S.Y., Chew, K.W., Lam, M.K., Lim, J.W., Ho, S.H., et al., 2020. A review on microalgae cultivation and harvesting, and their biomass extraction processing using ionic liquids. Bioengineered 11 (1), 116–129. Available at https://doi.org/10.1080/21655979.2020.1711626, http://www.tandfonline.com/loi/kbie20.

Thomas Kiran, M., Tiwari, A., 2018. In: Nageswara-Rao, M., Soneji, J. (Eds.), Biofuels from Microalgae. IntechOpen, Rijeka.

Uchida, M., Miyoshi, T., Kaneniwa, M., Ishihara, K., Nakashimada, Y., Urano, N., 2014. Production of 16.5% v/v ethanol from seagrass seeds. Journal of Bioscience and Bioengineering 118 (6), 646–650. Available at https://doi.org/10.1016/j.jbiosc.2014.05.017.

Uyar, B., Ali, M.D., Uyar, G.E.O., 2024. Design parameters comparison of bubble column, airlift and stirred tank photobioreactors for microalgae production. Bioprocess and Biosystems Engineering 47, 195–209. Available at https://doi.org/10.1007/s00449-023-02952-8.

Venkiteshwaran, K., Xie, T., Seib, M., Tale, V.P., Zitomer, D., 2022. Anaerobic digester biogas upgrading using microalgae. Integrated Wastewater Management and Valorization using Algal Cultures. Elsevier, United States, pp. 183–214. Available at https://www.sciencedirect.com/book/9780323858595, 10.1016/B978-0-323-85859-5.00004-X.

Wahlen, B.D., Willis, R.M., Seefeldt, L.C., 2011. Bioresource technology biodiesel production by simultaneous extraction and conversion of total lipids from microalgae, cyanobacteria, and wild mixed-cultures. Bioresource Technology 102, 2724–2730. Available at https://doi.org/10.1016/j.biortech.2010.11.026.

Wang, M., Ye, X., Bi, H., Shen, Z., 2024. Microalgae biofuels: illuminating the path to a sustainable future amidst challenges and opportunities. Biotechnology for Biofuels and Bioproducts 17 (1). Available at https://doi.org/10.1186/s13068-024-02461-0.

Wang, Q., Yang, Y., Shen, Q., Chen, X., Li, F., Wang, J., et al., 2022. Energy saving and rapid establishment of granular microalgae system from tiny microalgae cells: effect of decrease in upflow air velocity under intermittent aeration condition. Bioresource Technology 363, 127860. Available at https://doi.org/10.1016/j.biortech.2022.127860.

Wood, D.A., 2021. Microalgae to biodiesel - review of recent progress. Bioresource Technology Reports 14, 100665. Available at https://doi.org/10.1016/j.biteb.2021.100665.

Xin, C., Addy, M.M., Zhao, J., Cheng, Y., Cheng, S., Mu, D., et al., 2016. Comprehensive techno-economic analysis of wastewater-based algal biofuel production: a case study. Bioresource Technology 211, 584–593. Available at https://doi.org/10.1016/j.biortech.2016.03.102, www.elsevier.com/locate/biortech.

Yadav, G., Sekar, M., Kim, S.-H., Geo, V.E., Bhatia, S.K., Sabir, J.S.M., et al., 2021. Lipid content, biomass density, fatty acid as selection markers for evaluating the suitability of four fast growing cyanobacterial strains for biodiesel production. Bioresource Technology 325, 124654. Available at https://doi.org/10.1016/j.biortech.2020.124654.

Yahya, N.A., Suhaimi, N., Kaha, M., Hara, H., Zakaria, Z., Sugiura, N., et al., 2018. Lipid production enhancement in tropically isolated microalgae by azide and its effect on fatty acid composition. Journal of Applied Phycology 30 (6), 3063–3073. Available at https://doi.org/10.1007/s10811-018-1609-9, www.wkap.nl/journalhome.htm/0921-8971.

Yang, L., Ren, L., Tan, X., Chu, H., Chen, J., Zhang, Y., et al., 2020. Removal of ofloxacin with biofuel production by oleaginous microalgae Scenedesmus obliquus. Bioresource Technology 315, 123738. Available at https://doi.org/10.1016/j.biortech.2020.123738.

Yang, Y., Ge, S., Pan, Y., Qian, W., Wang, S., Zhang, J., et al., 2023. Screening of microalgae species and evaluation of algal-lipid stimulation strategies for biodiesel production. Science of The Total Environment 857 (Pt 1), 159281. Available at https://doi.org/10.1016/j.scitotenv.2022.159281.

Yeong, T.K., Jiao, K., Zeng, X., Lin, L., Pan, S., Danquah, M.K., 2018. Microalgae for biobutanol production – technology evaluation and value proposition. Algal Research 31, 367–376. Available at https://doi.org/10.1016/j.algal.2018.02.029, http://www.sciencedirect.com/science/journal/aip/22119264.

Zewdie, D.T., Ali, A.Y., 2020. Cultivation of microalgae for biofuel production: coupling with sugarcane-processing factories. Energy, Sustainability and Society 10 (1). Available at https://doi.org/10.1186/s13705-020-00262-5, https://energsustainsoc.biomedcentral.com/.

Zhang, M., Hu, Y., Wang, H., Li, H., Han, X., Zeng, Y., et al., 2021. A review of bio-oil upgrading by catalytic hydrotreatment: advances, challenges, and prospects. Molecular Catalysis 504, 111438. Available at https://doi.org/10.1016/j.mcat.2021.111438.

Zhang, X., Zhang, Z., Peng, Y., Zhang, Y., Li, Q., Sun, D., 2024. Salicylic acid enhances cell growth, fatty acid and astaxanthin production in heterotrophic Chromochloris zofingiensis without reactive oxygen species elevation. Biotechnology for Biofuels and Bioproducts 17 (1). Available at https://doi.org/10.1186/s13068-023-02449-2.

CHAPTER 10

Microalgae with artificial intelligence

Tengku Nilam Baizura Tengku Ibrahim[1,2], Aida Soraya Shamsuddin[3], Muhamad Syahril Khamis[1], Siti Nurshahida Nazli[1,4], Nurhidayah Sabri[1], and Nur Azalina Suzianti Feisal[5,6]

[1]Department of Environmental Health, Faculty of Health Sciences, Universiti Teknologi MARA Cawangan Pulau Pinang, Bertam, Pulau Pinang, Malaysia
[2]Occupational Health and Safety Risk Management (OHSeRM) Research Initiative Group, Faculty of Health Sciences, Universiti Teknologi MARA, Puncak Alam, Selangor, Malaysia
[3]Southeast Asia Disaster Prevention Research Initiative, Institute for Environment and Development (LESTARI), Universiti Kebangsaan Malaysia (UKM), Bangi, Selangor, Malaysia
[4]Child Health Research Centre, The University of Queensland, South Brisbane, QLD, Australia
[5]MSU Centre for Climate Resilience and Strategy (m-CREST), Management and Science University, Shah Alam, Selangor, Malaysia
[6]Department of Diagnostic and Allied Health Science, Faculty of Health and Life Sciences, Management and Science University, Shah Alam, Selangor, Malaysia

10.1 Introduction

Microalgae, including prokaryotic cyanobacteria and eukaryotic photosynthetic protists, are the most abundant primary producers, playing a fundamental role in aquatic food webs. They are responsible for nearly 50% of the Earth's oxygen production and play a key role in primary productivity (Hu et al., 2023). Although microalgae are unicellular and lack complex structure compared to plants, they are able to perform photosynthesis using sunlight, carbon dioxide, and water owing to the presence of photosynthetic pigments such as chlorophylls in their cells (Tan et al., 2020). Microalgae have a long evolutionary history spanning billions of years and have undergone various endosymbiosis events, leading to the emergence of different groups, including green algae, red algae, heterokont algae, dinoflagellates, cryptophytes, and euglenids. About 80,000 species of microalgae were estimated to exist; however, only half of them have been isolated, described, and explored at a large scale for commercial and industrial purposes (Hu et al., 2023; Olabi et al., 2023). They have adapted to various environmental conditions such as light, temperature, salinity, pH, nutrient availability, and stress factors. This adaptation has led to remarkable diversity in morphological, physiological, biochemical, and genetic traits (Aydin, 2024; Grama et al., 2022). These allow different

Applied Phycology
DOI: https://doi.org/10.1016/B978-0-443-24758-3.00003-4

species to thrive in extreme and variable habitats ranging from freshwater ponds to hypersaline lakes and wastewater lagoons.

Microalgae play a crucial role in maintaining the carbon-oxygen balance in the atmosphere by helping to stabilize oxygen (O_2) and carbon dioxide (CO_2) levels. Phytoplankton exhibit significantly higher biomass productivity and CO_2 fixation efficiency than other photosynthetic organisms, contributing to the fixation of approximately 50 gigatons of CO_2 annually, about half of the global primary productivity. Due to their ability to assimilate CO_2 10–50 times more efficiently than vascular plants without competing with food sources, microalgae have attracted global interest for carbon sequestration applications (Prasad et al., 2021).

Microalgae effectively uptake excess nutrients from wastewater. They assimilate ammonium, nitrate, and phosphate into biomass, achieving very high removal efficiencies for preventing eutrophication and improving water quality (Srimongkol et al., 2022). They can assimilate ammoniacal nitrogen (NH^{4+}) and nitrate (NO^{3-}) as their primary nitrogen sources, and orthophosphates ($PO_4{}^{3-}$) for phosphorus. Studies have shown that high-rate algal ponds (HRAPs) can achieve substantial removal percentages, with ammonia nitrogen removal reaching up to 90% and phosphate removal around 50% (Velásquez-Orta et al., 2024). This nutrient assimilation process not only cleans wastewater but also produces valuable biomass, offering a dual benefit for sustainable water management. Microalgae store starch as a metabolite when they experience a lack of nutrients, such as nitrogen, phosphorus, and sulfur. The starch produced by microalgae can be utilized for generating biofuels, including third-generation bioethanol and biohydrogen, and can also act as a raw material for bio-based chemicals (Maia et al., 2020; Hu et al., 2023). In short, microalgae combine environmental resilience with metabolic versatility, making them promising for sustainable applications.

Heavy metals commonly found in wastewater include arsenic (As), cadmium (Cd), copper (Cu), chromium (Cr), lead (Pb), manganese (Mn), mercury (Hg), nickel (Ni), and zinc (Zn) (Dhokane et al., 2023). Microalgae have demonstrated a high capacity for biosorbing these metals due to their unique cellular structures and surface chemistry. The advantages of using microalgae for heavy metal removal include rapid metal uptake, environmental friendliness, energy and time efficiency, low cost, reusability, fast growth rates compared to terrestrial plants, and the production of nonhazardous waste (Srimongkol et al., 2022). Biosorption, also known as passive adsorption, is a metabolism-independent process that occurs in both living and dead algal cells. In this process, heavy metal ions are captured and bound to functional groups on the cell surface through mechanisms such as ion exchange, complexation, chelation, and micro-precipitation. The microalgal cell walls contain functional groups such as amino (-NH_2), carboxyl (-COOH), hydroxyl (-OH), and sulfhydryl (-SH), which provide binding sites for cationic metal ions (Ayele and Godeto, 2021). These interactions enable efficient sequestration of heavy metals from contaminated water.

Artificial intelligence (AI) and machine learning (ML) have recently begun to transform microalgae biotechnology by discovering optimal growth conditions and automating cultivation. Microalgae are used in wastewater treatment, nutrient recycling, biofuel production, and carbon capture, but their cultivation involves many interacting factors such as light, nutrients, temperature, pH, and CO_2 (Wu et al., 2025b). Modern AI tools can analyze large datasets to predict ideal conditions, often with very high accuracy. For example, AI models have been shown to boost algal biomass yields by 15%–57% and improve CO_2 fixation and lipid production by 20%–43% compared to traditional methods (Wu et al., 2025b). Coupled with sensors and the Internet of Things (IoT), AI enables real-time monitoring and control of photobioreactors, automatically adjusting parameters as conditions change and reducing manual oversight (Imamoglu, 2024).

This emerging AI-driven use is evident in both research and industrial approaches. In experimental systems, ML models have accurately predicted nutrient removal and biomass output in sewage-treatment algae bioreac tors, allowing multicriteria optimization of flow rates and nutrient loading. More generally, AI decision tools can continuously fine-tune light, temperature, and feed rates in large-scale reactors, maximizing yield while minimizing energy and nutrient use (Oruganti et al., 2023). This integration promises more efficient and sustainable microalgae production for diverse applications.

Integration of AI, microalgal biotechnology, and environmental systems is increasingly explored as a framework for sustainable environmental management. The environmental systems will include essential inputs in optimum range for microalgal growth, such as carbon dioxide (CO_2), nutrients, light, and water. These elements are not only fundamental to photosynthesis but also influence the metabolic activity and productivity of microalgae. Parallel to this, environmental monitoring captures real-time parameters like pH, temperature, CO_2 concentration, and nutrient availability, which are critical for maintaining optimal cultivation conditions

(Wei et al., 2023). These environmental and biological data streams are processed by AI systems, which are central to this integration. Through ML, computer vision, optimization algorithms, and data analytics, AI can analyze large datasets to predict microalgal growth trends, detect stress symptoms, and optimize operational parameters.

On the biological side, the microalgal processes are categorized into cultivation and monitoring, and harvesting and processing. AI systems enhance cultivation by regulating growth rate, biomass production, and health indicators, thus ensuring high efficiency and minimal waste (Pearson, 2025). Postharvesting, the biomass can be processed for diverse applications, from biofuel production to pharmaceuticals and carbon capture, offering high economic and ecological value. The AI-generated control signals form a feedback loop, automatically adjusting environmental conditions based on real-time assessments to improve yield and sustainability. This conceptual model represents an innovative step toward intelligent, automated, and scalable environmental biotechnologies (Fig. 10.1).

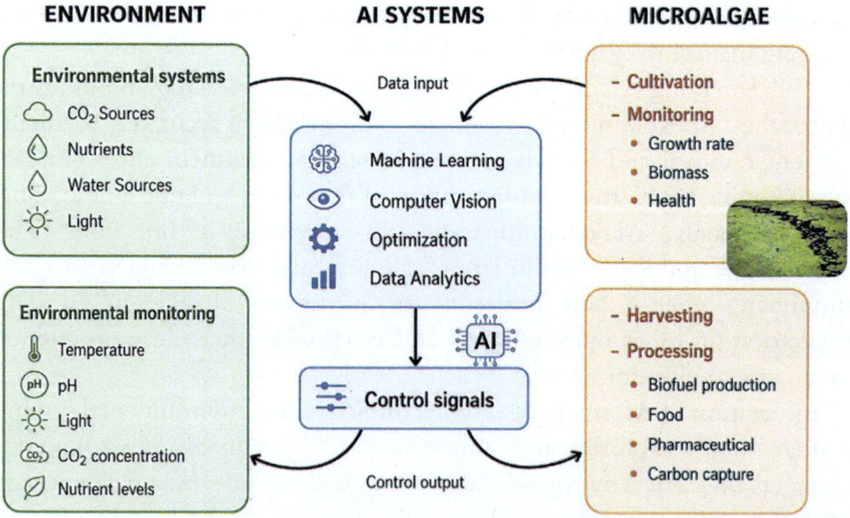

Figure 10.1 ***Conceptual framework of artificial intelligence (AI)–microalgae–environmental systems integration.*** Conceptual diagram illustrating the integration of environmental signals (e.g., nutrients, CO_2, and light) with AI systems (e.g., machine learning, computer vision, and optimization) to support microalgae cultivation and harvesting applications.

10.2 Modeling approaches in microalgal systems

Modeling microalgal systems is important for understanding, predicting, and improving their biological and technical performance in a wide range of applications, from producing biomass to cleaning up wastewater (Sumanasekara et al., 2025; Imamoglu, 2024). Modeling methods have evolved over time from traditional, rule-based frameworks based on microbiological kinetics and fluid mechanics to new, data-driven paradigms that utilize AI and ML. There are pros and cons to each method, depending on the complexity of the system, data availability, and the desired application (Yang et al., 2024). This section compares traditional modeling methods with AI-based modeling methods, focusing on their theoretical foundations, real-world applications, and significance in modern environmental biotechnology.

10.2.1 Conventional models

Classical microbial kinetics and reactor engineering are the basis for the conventional approach to modeling microalgal systems. The Monod kinetics model is an example that is widely known. It describes how the growth rate of microbes changes based on the concentration of a limiting substrate in a saturating (Michaelis–Menten-type) manner. Because the concepts are simple and based on enzyme kinetics, Monod-type equations have been widely used to predict the growth of microalgae biomass and the uptake of nutrients (Imamoglu, 2024).

In applied microalgal research, Monod-based models, logistic models, along with first-order kinetics, are all commonly used to predict biomass growth and improve cultivation because they are easy to understand mathematically and do not require a lot of data (Kasbawati et al., 2022). These models, on the other hand, are based on limiting assumptions, which include the idea that growth is controlled by a single limiting substrate while other factors remain constant, and that parameters like μ_max and K_s are static and usually estimated from steady-state lab conditions (Galinha et al., 2018). For example, kinetic modeling in open ponds and photobioreactors does not work as well when environmental conditions change, because Monod-type equations cannot account for colimitation effects or changing light and nutrient levels. Extended or multiplicative models that include more than one resource (like light, N, or P) fit better, but they still have trouble with time variability, storage effects, or physiological acclimation (Romagnoli et al., 2021). These weaknesses are especially evident in dynamic, real-world systems like outdoor raceway ponds or wastewater-fed reactors, where factors such as irradiance, temperature, nutrient loads, and algal physiology fluctuate over time. As a result, models that are calibrated in one specific context may struggle to apply effectively in different situations, leading to diminished predictive reliability (Darvehei et al., 2018). Furthermore, conventional models exhibit a lack of adaptability, as their rigid framework prevents them from learning from new real-time data, which restricts their usefulness in monitoring or control systems. The limitations encountered have driven the advancement and integration of AI-driven and hybrid modeling approaches, which provide greater flexibility and predictive capability in understanding intricate, multivariate environmental dynamics (Romagnoli et al., 2021).

In regard to the constraints associated with Monod kinetics, mechanistic models have developed into more accurate representations for understanding the growth of microalgae. The models are designed to reflect the biochemical and cellular processes that form the basis of algal metabolism, providing a flexible framework for simulating internal nutrient regulation, photoacclimation, and biomass composition (Solimeno et al., 2015). An illustrative case is the Droop quota model, which separates the availability of external substrates from immediate growth by incorporating an internal nutrient quota. This mechanism facilitates growth to continue even in the absence of extracellular nutrients, as long as internal reserves are adequate. This dynamic more accurately mirrors the behavior observed in real-world batch or semicontinuous systems (Schediwy et al., 2019; Saccardo et al., 2023). Mechanistic models also incorporate modules for photosynthesis and respiration kinetics, pigment accumulation, and light attenuation, making them particularly useful in modeling systems exposed to diurnal light cycles or varying irradiance. Such physiological depth enables greater extrapolative power, particularly in environmental or large-scale production scenarios where light, temperature, and nutrient regimes fluctuate (Ye et al., 2013). Unlike empirical models, which rely purely on observed correlations, mechanistic approaches are grounded in first principles, allowing them to maintain predictive integrity under changing conditions. However, this enhanced realism comes at a cost: mechanistic models often require a large number of parameters, many of which are species-specific or difficult to measure directly (Bernardi et al., 2017). The trade-off is that mechanistic models require extensive knowledge of cellular processes and many parameters, which can make them complex to construct and validate. In contrast, some practitioners opt for empirical models, for example, statistical regressions or simple polynomial fits that forego mechanistic interpretation in favor of directly fitting observed data. Empirical models (such as logistic growth curves or response surface methodology [RSM] designs) are relatively easy to develop and can capture system behavior under specific conditions. However, their validity is narrow, as they may fail to extrapolate beyond the conditions under which they were calibrated. In general, a purely empirical model might lack predictive power if any key variable strays outside the originally observed range or if the underlying system changes (e.g., different microalgal species or light regimen) (Belany et al., 2024).

Another pillar of conventional modeling in microalgal cultivation involves reactor engineering models, including the use of computational fluid dynamics (CFD). CFD simulations are employed to model the hydrodynamics of photobioreactors or open ponds—tracking fluid flow patterns, mixing intensity, gas–liquid mass transfer, light distribution, and heat transfer. These simulations, when coupled with kinetic equations for growth, allow engineers to analyze how reactor design and operating conditions affect microalgae productivity (Imamoglu, 2024). For example, integrated models have the potential to combine Monod-type growth kinetics with CFD predictions concerning light gradients and nutrient transport. This integration would enable the estimation of spatially resolved growth rates within a photobioreactor. Models based on CFD are essential for improving reactor designs and aeration techniques, providing important insights into aspects like shear stress on cells and the distribution of carbon dioxide (Park and Li, 2014). Although models based on CFD demonstrate a strong capability in precisely capturing physical dynamics, they are limited in their ability to forecast biological results in microbalgal cultivation systems. While CFD excels in simulating aspects such as fluid flow, light distribution, and the transfer of mass and heat, it falls short in its ability to accurately represent cellular responses. These responses, including photosynthetic acclimation and nutrient uptake dynamics, are influenced by changing environmental conditions. This limitation occurs because CFD is inherently based on physical principles and necessitates that all biological interactions, such as the impact of turbulence on nutrient uptake or the influence of light exposure on growth rates, be clearly and precisely defined. This task is complicated by the highly nonlinear and context-dependent characteristics of these processes, as well as existing gaps in our mechanistic understanding (Akenteng et al., 2025; Luzi and McHardy, 2022). Another major limitation is computational expense: high-fidelity CFD coupled with detailed kinetics becomes impractical for large-scale or system-wide optimization. Exhaustive design sweeps evaluating various reactor geometries, mixing strategies, or illumination schemes quickly become infeasible due to time and resource constraints (Amanna et al., 2024; Imamoglu, 2024). Recognizing these challenges, researchers have increasingly turned to hybrid modeling approaches. One prominent strategy is the development of surrogate models using ML, trained on a limited set of CFD–kinetic simulation outputs. For instance, deep neural networks have been used to approximate complex hydrodynamic and biological relationships, reducing computation time from months to days while maintaining comparable predictive accuracy (Gargalo et al., 2024). Such data-driven surrogates can facilitate multiobjective optimization, generating Pareto frontiers for design decisions, and are proving invaluable for real-time control and scale-up applications in algal bioprocess engineering.

Conventional modeling approaches are increasingly inadequate when confronted with the complex, dynamic nature of real-world microalgal cultivation systems. These systems are governed by a web of interacting variables such as light intensity and photoperiod, nutrient concentrations (e.g., nitrogen, phosphorus, CO_2), temperature, pH, mixing regimes, and potential biotic interactions in cocultures. These factors not only fluctuate temporally due to day/night cycles, nutrient depletion, or growth phase transitions but also vary spatially within the reactor, creating steep gradients in light and nutrient availability. Capturing such multivariate, nonstatic behavior within a rule-based framework presents a significant challenge (Schediwy et al., 2019). To address this complexity, traditional models often layer multiple submodels or introduce correction factors (e.g., stacked Monod kinetics, light attenuation functions). However, this increases mathematical complexity, heightens parameter uncertainty, and rarely results in robust, generalizable predictions. More fundamentally, these models typically fail to account for the adaptive nature of microalgae themselves (Bekirogullari et al., 2020; Romagnoli et al., 2021). Cellular responses such as pigment regulation, enzyme activity shifts, or energy storage adjustments occur on varying time scales from seconds to days, rendering static formulations inherently limited. Another critical flaw lies in the mismatch between laboratory-derived parameters and field-scale conditions. Photosynthetic performance measured under controlled light may not reflect behavior under fluctuating outdoor environments. In open raceway ponds, for example, environmental conditions can shift rapidly due to weather or diurnal cycles (Igou et al., 2023). Without continuous recalibration, conventional models lose relevance and predictive accuracy under such variability. Furthermore, spatial heterogeneity, such as self-shading in dense cultures, compromises the applicability of uniform kinetic assumptions across reactor volumes. Compounding this, physiological heterogeneity within algal populations (e.g., differences in cell size, nutrient status) introduces a degree of biological variability that conventional deterministic models are not equipped to handle. In essence, these models lack adaptability. Once established, their fixed structure cannot accommodate deviations from initial assumptions without significant manual tuning. This rigidity represents a fundamental shortcoming in using traditional modeling approaches for dynamic, real-world microalgal systems, particularly those subject to complex environmental fluctuations and biological feedback mechanisms (Imamoglu, 2024).

Statistical tools like RSM have long been employed to optimize multivariate processes in microalgal research, particularly for fine-tuning nutrient concentrations or extraction conditions. By fitting a second-order polynomial to experimental data, RSM can reduce the number of required trials while predicting optimal conditions within a defined range. However, despite its structured approach and ease of implementation, RSM is fundamentally limited by its underlying assumptions (Reza et al., 2024; Imamoglu, 2024). The method is inherently constrained to quadratic approximations, which are often too simplistic to capture the complex, nonlinear interactions typical of biological systems. Microalgal responses are rarely smooth or symmetrical; instead, they can exhibit abrupt thresholds, multiphase behavior, or interaction effects that defy second-order modeling. This limitation is not just theoretical. Numerous studies have reported that RSM-predicted "optimal" conditions often fail under validation either because the true optimum lies just outside the experimental region or because the system's actual response surface is more complex than assumed (Veza et al., 2023). In such cases, RSM not only provides misleading guidance but may also result in resource waste or flawed conclusions. These issues underscore a broader problem in relying on RSM—although it offers convenience and efficiency, it lacks the flexibility and precision needed to model high-dimensional, biologically adaptive systems. Its reliance on predefined model structures makes it ill-suited for capturing emergent behaviors or nonlinear dynamics—both of

Table 10.1 Comparative overview of modeling approaches in algal biotechnology, including Monod-type kinetics, mechanistic models, empirical models, and computational fluid dynamics (CFD)-based reactor models. The table outlines their descriptions, strengths, limitations, and representative studies, providing a framework for evaluating their applicability in different cultivation and optimization scenarios.

Model approach	Description/examples	Strengths	Limitations	Reference
Monod-type kinetics	Saturation-growth models relating growth rate to a single limiting substrate (e.g., nutrient concentration). Originally developed for bacteria, often applied to algal growth.	Simple form with a few parameters, grounded in enzyme kinetics, making it somewhat interpretable. Widely used for nutrient-limited growth predictions.	Assumes one limiting factor at a time, cannot directly handle multiple nutrient limitations or light limitation without extensions. Parameters are static—model must be recalibrated for different strains or conditions. Fails to capture acclimation or time-varying dynamics in the system.	Kasbawati et al. (2022)
Mechanistic models	Detailed models incorporating physiological processes (e.g., Droop internal nutrient quota model, photosynthesis-irradiance models, etc.) may involve multiple equations for nutrients, light, and cell composition.	Built on first principles and biological understanding, it can generalize to new conditions if the mechanisms remain the same. Can yield reliable predictions across scenarios and scales when well-validated.	High complexity with many parameters requiring extensive experimental data to fit. Demands deep domain knowledge to construct. Computationally heavier to solve if many differential equations are involved. Risk of overparameterization or identifiability issues.	Schediwy et al. (2019)

Empirical models	Black-box fits to data (e.g., logistic growth curves, polynomial regressions, or response surface methodology designs). No explicit biological basis; purely data-driven within observed range.	Easy and quick to develop if sufficient data are available. Useful for interpolation and local optimization. Does not require detailed mechanistic insight.	Poor extrapolation beyond the data range—lacks predictive power if conditions change. Provides limited insight into causal factors. Typically needs to be re-built for each new scenario. Can miss true optima or complex interactions if the chosen empirical form is oversimplified.	Imamoglu (2024)
CFD-based reactor models	CFD simulations of the cultivation system, often coupled with basic growth kinetics. Simulates fluid flow, mixing, light distribution, etc., in photobioreactors or ponds.	Captures spatial and physical effects (hydrodynamics, light attenuation) with high resolution. Useful for design and scale-up—can identify zones of poor mixing or illumination. Allows virtual testing of design modifications (aeration rate, paddle wheel speed, etc.) on culture performance.	Heavy computational requirements, especially for 3D and multiphase simulations. Need to mesh with biological models—CFD alone does not predict biological outcomes. Complex models may become unstable or impractically slow for large scales. Usually requires simplifications (e.g., assuming uniform cells, average kinetics) to be tractable.	Imamoglu (2024)

which are central to real-world microalgal cultivation (Imamoglu, 2024). As discussed in Table 10.1, RSM, like other conventional models, carries trade-offs that must be carefully weighed when applied to complex bioprocesses.

Table 10.1 shows that conventional models are fundamentally rule-based, depending on established equations and parameters to characterize the behavior of microalgae. This approach may be effective for systems that are relatively stable and well-defined; however, it tends to become challenging as the complexity of the system increases. When several factors change at the same time, such as light, nutrients, and hydrodynamics, the number of required rules or interconnected equations increases significantly. This complexity can be compounded by our limited understanding of the underlying biological processes, which may hinder progress. The limitations encountered in modeling dynamic, multivariate systems have opened avenues for exploring alternative methodologies. In recent years, scientists have turned the focus toward AI and ML as valuable tools for addressing the complexities that conventional models often find challenging, as will be elaborated in the following section.

10.3 Artificial intelligence algorithms, tools, and applications

10.3.1 Artificial neural networks

ANNs have emerged as one of the most widely adopted AI approaches in algal biotechnology due to their ability to capture complex, nonlinear interactions between cultivation parameters and biological responses. Inspired by the architecture of the human brain, ANNs consist of interconnected layers of neurons that transform input data (e.g., light, temperature, nutrient levels, CO_2 supply) into predictive outputs such as biomass yield, lipid accumulation, or photosynthetic efficiency. Unlike classical regression or RSMs that require assumptions of linearity or polynomial form, ANNs are data-driven and can approximate highly nonlinear functional relationships (Liyanaarachchi et al., 2021; Wu et al., 2025b). This flexibility is particularly important in algal systems, where feedback loops, threshold effects, and synergistic interactions dominate.

In practice, ANNs have been applied to optimize both indoor photobioreactor systems and outdoor open ponds. Liyanaarachchi et al. (2021), for instance, developed ANN models to optimize *Chlorella vulgaris* cultivation for biodiesel production, achieving predictive accuracies above

R^2=0.92 and outperforming RSM. Their ANN framework successfully identified cultivation conditions that simultaneously maximized biomass (2663 mg/L) and lipid yield (1266 mg/L), underscoring ANN's strength in multiobjective optimization. Similarly, Onay (2023) integrated ANN with RSM to predict lipid content in *Chlorella minutissima*, achieving near-perfect correlation (R=0.9961), further demonstrating ANN's superiority in capturing nonlinear dynamics that traditional regression oversimplifies.

Beyond yield optimization, ANNs have also been deployed for process control and forecasting. They have been used to simulate growth under variable light–dark cycles, to predict CO_2 sequestration rates, and to model nutrient uptake efficiency under fluctuating conditions. Such predictive capabilities are invaluable for scale-up, where environmental variability is difficult to replicate in controlled laboratory experiments. When coupled with online sensor data streams, ANNs can form the predictive backbone of "smart bioreactor" systems, enabling real-time control of aeration, mixing, and nutrient dosing.

Despite these strengths, ANNs are often criticized as "black-box" models: while they can achieve high predictive accuracy, they do not yield explicit mechanistic insight into biological processes. For example, while an ANN may predict that increasing CO_2 delivery enhances lipid content, it cannot elucidate whether this effect is mediated through altered pH, enzyme regulation, or metabolic fluxes. Furthermore, overfitting is a persistent risk when training datasets are small, a common limitation in algal studies where large experimental datasets are expensive and time-consuming to generate. Regularization, cross-validation, and hybrid approaches (e.g., ANN-RSM or ANN-GA models) are, therefore, frequently used to mitigate these issues.

In sum, ANNs provide a powerful predictive framework for algal biotechnology, excelling in scenarios where system dynamics are too complex for mechanistic modeling. Their strength lies in optimizing cultivation conditions and forecasting biomass or metabolite yields under diverse inputs. However, to move from predictive accuracy to biological insight, ANNs are most effective when paired with interpretable methods or mechanistic models, ensuring that operational guidance does not come at the expense of understanding underlying causal mechanisms.

10.3.2 Random forests

Random forests (RF) represent one of the most versatile ensemble learning methods in algal biotechnology, particularly valued for their ability to manage large, heterogeneous datasets where ecological and cultivation parameters interact in nonlinear and sometimes counterintuitive ways. Unlike linear regression models or generalized additive models (GAMs), which assume additive or smooth relationships, RF constructs multiple decision trees from bootstrapped datasets and averages their predictions, thereby reducing overfitting and capturing higher-order interactions that are common in algal systems (Liu et al., 2023).

In algal cultivation studies, RF has been employed to disentangle the relative importance of diverse variables, including meteorological drivers (solar radiation, wind speed), optical indicators (chlorophyll-a, turbidity), and nutrient chemistry (nitrate, phosphate) within a unified predictive framework. This capacity is particularly useful in outdoor ponds and lakes where system dynamics are driven by multiple, often collinear drivers. For example, Liu et al. (2023) demonstrated the ability of RF to predict phytoplankton biomass under fluctuating meteorological and nutrient regimes, highlighting the model's robustness against noisy field datasets. Similarly, Zhang et al. (2021) applied RF in Lake Okeechobee, showing that nutrient and light factors influence phytoplankton biomass at different spatial scales, with RF successfully capturing both lake-wide and local heterogeneity, something conventional regression models struggled to resolve.

One of the distinctive strengths of RF is its capacity to generate variable importance rankings, which provide biological insight beyond mere prediction. For instance, in bloom forecasting models, RF frequently identifies nutrient ratios or underwater light availability as dominant predictors, which can guide targeted management interventions. However, this interpretability is relative: while RF offers more transparency than "black-box" ANNs, it does not provide mechanistic equations that can be directly translated into bioprocess design, limiting its explanatory power compared to mechanistic or hybrid models. Despite these strengths, RF is not without drawbacks. Overfitting remains a risk when tree depth is not carefully controlled, particularly in small experimental datasets common in algal biotechnology. Moreover, RF struggles with extrapolation outside the training domain; predictions tend to regress toward the mean when exposed to unseen extreme conditions, which is problematic for climate change scenarios or novel cultivation settings. Finally, while RF can highlight influential predictors, it does not establish causal mechanisms, meaning it should ideally be used in tandem with mechanistic models or experimental validation. Collectively, these features position RF as a pragmatic "middle ground" between black-box accuracy and ecological interpretability. Its utility lies not only in forecasting biomass and bloom dynamics but also in screening influential variables for further mechanistic exploration. When integrated into IoT-enabled monitoring frameworks, RF can act as a robust decision-support tool for algal cultivation, particularly in operational contexts where real-time adaptability is critical.

10.3.3 Support vector machines

Support vector machines (SVMs) have established themselves as powerful classification and regression tools in algal biotechnology, particularly in applications requiring precise discrimination among species, cell states, or environmental conditions. Unlike ANNs, which rely on multilayered architectures to capture complex nonlinearities, SVMs work by constructing an optimal hyperplane in high-dimensional feature space that maximally separates data classes. This property makes SVMs especially effective when dealing with small to medium-sized datasets with high-dimensional input conditions typical of algal biotechnology, where spectral, morphological, or flow cytometry profiles often consist of hundreds of features but relatively few labeled samples.

One of the most impactful uses of SVMs is in species identification and contamination detection, where subtle spectral or morphological differences must be captured with high reliability. For example, Izadi et al. (2021) applied SVMs, along with other ML models, to forecast harmful algal blooms (HAB) of *Karenia brevis* in Florida, leveraging Moderate Resolution Imaging Spectroradiometer (MODIS) satellite data. Their models achieved up to eight-day prediction accuracy, with SVMs showing particular strength in handling high-dimensional remote sensing inputs and outperforming traditional threshold-based approaches. Similarly, in ecological monitoring, SVMs have been used to classify algal cell health from flow cytometry profiles, enabling early detection of stress in reef-associated *Symbiodiniaceae* before declines in coral productivity occur (Ariza Rangel, 2025). These examples underscore the versatility of SVMs across both remote sensing and cellular-scale applications.

From a theoretical perspective, the use of kernel functions allows SVMs to transform input data into higher-dimensional spaces where nonlinear boundaries become linearly separable. This flexibility is crucial in algal systems, where growth, physiology, and stress responses rarely follow simple linear patterns. Kernelized SVMs have been applied to hyperspectral datasets to distinguish morphologically similar microalgae species with accuracies exceeding 95%, outperforming conventional methods such as linear discriminant analysis. However, SVMs are not without limitations. Their performance is highly dependent on the choice of kernel and hyperparameter tuning, which can be computationally intensive and nonintuitive. Moreover, while SVMs provide strong classification boundaries, they are less interpretable than RF, which can rank predictor importance, and less flexible than ANNs for modeling continuous nonlinear processes such as biomass yield forecasting. In noisy datasets, a frequent occurrence in outdoor algal cultivation, SVMs can also be sensitive to outliers, potentially skewing classification boundaries.

Despite these challenges, SVMs remain an indispensable tool in algal biotechnology when precision and robustness are prioritized over interpretability. Their strengths are most evident in classification-based tasks distinguishing harmful from nonharmful algal species, identifying viable versus nonviable cells, and detecting contamination in large-scale cultures. When integrated into automated monitoring systems, SVMs can significantly reduce reliance on manual microscopy, thereby accelerating response times in both industrial and ecological contexts. As such, SVMs occupy a critical niche within the AI toolkit for algal biotechnology, complementing the predictive strengths of ANNs and the ensemble interpretability of RF.

10.3.4 Genetic algorithms

Genetic algorithms (GA) belong to the family of evolutionary computation techniques inspired by Darwinian natural selection. They operate through iterative processes of selection, crossover, and mutation to explore large, multidimensional search spaces. This feature makes GA particularly attractive in algal biotechnology, where cultivation outcomes such as biomass productivity, lipid yield, or metabolite accumulation are influenced by numerous interdependent parameters (e.g., light intensity, nutrient ratios, CO_2 supply, and aeration regimes). Unlike gradient-based optimization methods that often converge prematurely to local optima, GA is capable of escaping such traps, allowing researchers to identify global solutions in highly nonlinear and rugged landscapes.

Applications of GA in algal systems range from nutrient medium optimization to process control. Camacho-Rodriguez et al. (2021)

demonstrated the potential of GA by optimizing the nutrient formulation for *Isochrysis galbana*, simultaneously enhancing polyunsaturated fatty acid (PUFA) productivity by 15% and carotenoid content by 50%, while reducing micronutrient requirements by up to 60%. Such results highlight GA's ability not only to maximize yield but also to improve sustainability by lowering input costs and resource consumption. Similarly, Purnomo et al. (2015) integrated GA with an extreme learning machine (ELM) to forecast the growth of *Chlamydomonas* sp. under varying pH conditions. Their GA-ELM framework significantly reduced regression errors compared to standalone ELM, illustrating how GA can act as a metaoptimizer to refine ML models for improved predictive reliability. Beyond laboratory optimization, GA has also been explored for operational scheduling in large-scale systems. For instance, GA has been applied to fine-tune aeration cycles in *Haematococcus pluvialis* cultivation, cutting energy costs without compromising astaxanthin accumulation. In such contexts, GA is especially valuable because aeration interacts dynamically with light penetration, mixing efficiency, and nutrient availability relationships that are often too complex for manual calibration.

Despite its strengths, GA is computationally demanding and may require a large number of evaluations before converging on an optimal solution. This can be a limiting factor when experiments are costly or time-consuming, as in large photobioreactors or outdoor ponds. Moreover, while GA excels at finding optimal parameter sets, it does not inherently provide mechanistic insight into why a particular solution works, meaning it is best used as an optimization layer rather than a standalone explanatory model. Hybrid frameworks, such as genetic algorithm–adaptive neuro-fuzzy inference system (GA–ANFIS), have therefore been developed to combine GA's optimization power with the interpretability of fuzzy systems, delivering both accuracy and transparency in process control (Kushwaha et al., 2023).

In summary, GA offers a robust evolutionary approach to parameter optimization in algal biotechnology, enabling improvements in productivity, cost efficiency, and sustainability. Its ability to explore complex, multidimensional parameter spaces distinguishes it from traditional trial-and-error or one-factor-at-a-time methods, making it particularly suited for modern, data-driven cultivation frameworks. When integrated into hybrid AI systems, GA provides a powerful optimization engine that can accelerate the transition from laboratory success to industrial-scale feasibility.

10.3.5 Fuzzy logic

Fuzzy logic represents a rule-based computational approach that excels in handling uncertainty, imprecision, and variability conditions that are ubiquitous in algal cultivation systems, particularly in outdoor environments. Unlike conventional control methods such as proportional–integral–derivative (PID) controllers, which rely on strict numerical thresholds, fuzzy logic uses linguistic rules and membership functions to mimic human reasoning. This allows for more adaptive and nuanced control, making it especially useful in bioprocesses where environmental factors (light, temperature, pH, dissolved oxygen [DO], and salinity) fluctuate unpredictably and interact in nonlinear ways.

One of the primary applications of fuzzy logic in algal biotechnology is real-time process control. Rauf (2018) highlighted how conventional Proportional–Integral (PI) controllers were used to regulate pond temperature in outdoor *H. pluvialis* cultivation, improving biomass productivity. Building upon this, fuzzy controllers have been proposed as more robust alternatives because they can adapt to diurnal and seasonal variability without requiring precise mathematical models of the system. By encoding expert knowledge in the form of "if–then" rules (e.g., *if DO is low and temperature is high, then increase aeration*), fuzzy systems can dynamically adjust aerators, pumps, or CO_2 injection rates to maintain near-optimal growth conditions.

The flexibility of fuzzy logic has also been demonstrated in hybrid models, where it serves as the interpretive layer for more computationally intensive optimization techniques. For example, Kushwaha et al. (2023) developed a GA–ANFIS model to optimize pH and temperature for CO_2 fixation in microalgae. By combining GA's global search ability with fuzzy logic's adaptive rule-based reasoning, the hybrid approach achieved superior predictive accuracy compared to standalone models, underscoring fuzzy logic's value in managing environmental uncertainty. In aquaculture-integrated systems, fuzzy logic has further been applied to IoT-enabled water-quality management. Recent studies have shown how fuzzy inference systems can process continuous sensor data streams (temperature, pH, DO, salinity) to regulate pond aeration and water exchange in real time, achieving up to 98% accuracy in maintaining optimal conditions (Kushwaha et al., 2023). Such frameworks demonstrate how fuzzy logic can form the decision-making "brain" of smart aquaculture and algal cultivation systems.

Nevertheless, fuzzy logic has its limitations. The accuracy of a fuzzy model depends heavily on the quality of its rule base, which is often derived from expert opinion rather than systematic data analysis. This subjectivity can lead to inconsistencies when scaling across different species, cultivation systems, or climates. Furthermore, fuzzy systems are less effective when confronted with highly complex datasets involving hundreds of variables, where ML models such as RFs or deep learning architectures may outperform them. To address these shortcomings, fuzzy logic is increasingly being integrated into hybrid AI frameworks, where it complements data-driven models by providing interpretability and robust decision-making in noisy, real-world environments.

In summary, fuzzy logic provides a pragmatic, adaptive, and interpretable control strategy for algal biotechnology, particularly well suited for open systems where uncertainty and variability dominate. Its ability to formalize expert knowledge into automated decision rules makes it a valuable bridge between traditional process engineering and modern AI-driven bioprocess control. When combined with evolutionary algorithms or neuro-fuzzy systems, fuzzy logic not only improves stability and productivity but also enhances scalability for industrial and aquaculture-linked applications.

10.3.6 Deep learning: convolutional neural networks and recurrent neural networks

Deep learning, encompassing convolutional neural networks (CNNs) and recurrent neural networks (RNNs), has rapidly expanded the scope of AI in algal biotechnology. Unlike traditional ML algorithms that depend heavily on feature engineering, deep learning architectures are capable of automatically extracting hierarchical representations from raw data, making them highly effective for both image-based classification and time-series forecasting. CNNs are particularly transformative for algal species classification and contamination monitoring. By applying convolutional filters across microscopy or hyperspectral images, CNNs can detect morphological and spectral patterns that distinguish species, growth stages, or cell viability. This eliminates the need for manual preprocessing or handcrafted features, which are both time-consuming and prone to error.

For example, Soar et al. (2023) applied a transfer learning CNN (EfficientNetV2-B3) with Grad-CAM explainability to classify *Nannochloropsis* and *Spirulina*, achieving >98% accuracy. Their model not only delivered high predictive performance but also highlighted image regions most relevant to classification, making the results interpretable for biologists. Similarly, Memmolo et al. (2020) used CNNs with holographic microscopy to classify diatoms on dry slides, demonstrating how CNNs can accelerate species identification while reducing reliance on skilled taxonomists. These applications show CNNs' potential as the foundation for automated quality control systems in industrial algae production facilities. Yet, CNNs are not without limitations. Their "black-box" nature can obscure biological meaning, and their performance often depends on large, well-annotated datasets, which remain scarce in algal biotechnology. Data augmentation, transfer learning, and explainable AI (e.g., Grad-CAM) have been employed to address these barriers, but systematic datasets for broader species coverage are still needed.

Whereas CNNs excel in image-based tasks, RNNs—and particularly long short-term memory (LSTM) models—are better suited for temporal forecasting in dynamic cultivation environments. RNNs capture sequential dependencies in time-series data, allowing them to model the effects of past environmental conditions on current algal performance. This is crucial in open ponds and photobioreactors, where diurnal cycles, weather fluctuations, and nutrient dynamics drive growth patterns.

Fournier et al. (2024) demonstrated the potential of LSTM models to forecast algal bloom dynamics by integrating meteorological and water-quality data, outperforming traditional time-series approaches such as Autoregressive Integrated Moving Average (ARIMA). Similarly, Yeh et al. (2023) showed that incorporating light history data into LSTM models improved predictions of outdoor algal growth, capturing diurnal photosynthetic rhythms that are often missed by static models. Such capabilities make LSTMs particularly valuable for operational decision-making in raceway ponds, where predicting photosynthetic efficiency and biomass yield in advance can optimize aeration, harvesting, and nutrient dosing schedules. Nevertheless, RNNs face challenges in terms of computational intensity and interpretability. Training deep temporal models requires extensive datasets with consistent temporal resolution, which can be difficult to obtain in outdoor cultivation systems. Overfitting to local conditions also limits their generalizability. Hybrid frameworks that combine LSTM forecasting with mechanistic models or ensemble methods (e.g., RF-LSTM hybrids) are emerging as promising solutions.

Together, CNNs and RNNs extend the AI toolkit in algal biotechnology by addressing two critical domains: species classification (image-based) and biomass forecasting (time series). When integrated into IoT-enabled monitoring systems, CNNs can provide real-time feedback on species composition and contamination, while LSTMs can predict near-future growth trajectories and stress responses. This dual application offers a comprehensive platform for "smart cultivation," enabling precision management of algal systems that is both proactive and adaptive. In summary, deep learning represents the frontier of AI in algal biotechnology. CNNs excel in automating species identification with high accuracy, while RNNs/LSTMs capture temporal complexity to forecast growth and photosynthetic activity in fluctuating environments. Despite challenges related to data requirements and interpretability, their integration into hybrid and IoT-driven frameworks positions them as central to the next generation of intelligent algal cultivation systems.

Taken together, the integration of AI into algal biotechnology illustrates how different algorithms occupy distinct niches within the cultivation and monitoring process. ANNs excel at modeling nonlinear input–output relationships, enabling optimization of biomass and metabolite yields across diverse cultivation conditions, though they suffer from black-box limitations. SVMs, by contrast, thrive in classification tasks where precision is critical, such as distinguishing harmful from nonharmful taxa or viable from nonviable cells, though they require careful kernel tuning and are less interpretable. RF offers a balance between predictive accuracy and interpretability, with the added ability to rank variable importance, making it particularly useful for ecological studies and heterogeneous datasets. GA provides an evolutionary optimization framework that can navigate vast parameter spaces, offering sustainability gains in nutrient and energy management, but at the cost of computational demand and limited mechanistic insight. Fuzzy logic addresses uncertainty and imprecision, excelling in real-time process control in outdoor ponds, though its reliance on expert-derived rules constrains scalability. Finally, deep learning models (CNNs and RNNs/LSTMs) extend AI capabilities to high-dimensional image and temporal datasets, automating species classification and forecasting growth under fluctuating environments, though their effectiveness remains contingent on large, high-quality datasets. Collectively, these approaches underscore that no single method is universally superior; rather, their combined use through hybrid and ensemble frameworks offers the greatest promise for advancing smart, resilient, and sustainable algal biotechnologies (Table 10.2).

10.4 Artificial intelligence in algal image analysis, classification, and genetic engineering

Accurate identification and quantification of algal cells is a cornerstone of both ecological monitoring and industrial biotechnology. Traditional

Table 10.2 Comparative summary of AI techniques applied in algal biotechnology (Section 3). The table highlights the primary applications, methodological strengths, and limitations of each approach, along with representative studies, providing a concise reference point for evaluating their suitability in biomass optimization, bloom forecasting, species classification, and cultivation system control.

Artificial intelligence techniques	Primary applications	Key strengths	Limitations	Example studies
Artificial neural network	Biomass and lipid yield optimization; CO_2 sequestration modeling	Captures nonlinear dynamics; high predictive accuracy	Black box; prone to overfitting; large data needs	Liyanaarachchi et al. (2021), Onay (2023)
Support vector machine	Species identification; harmful algal bloom forecasting; cell viability classification	High precision; robust with small, high-dimensional datasets	Sensitive to kernel/hyperparameters; limited interpretability	Izadi et al. (2021), Ariza Rangel (2025)
Random forest	Phytoplankton biomass prediction; bloom dynamics; environmental factor ranking	Handles heterogeneous data; interpretable (variable importance)	Limited extrapolation; risk of overfitting small datasets	Liu et al. (2023), Zhang et al. (2021)
Genetic algorithms (GA)	Nutrient medium optimization; aeration scheduling; hybrid model tuning	Explores complex parameter spaces; avoids local optima	Computationally intensive; lacks causal insight	Camacho-Rodiguez et al. (2021), Purnomo et al. (2015)

Fuzzy logic	Pond water-quality control; temperature and dissolved oxygen regulation; hybrid genetic algorithm–adaptive neuro-fuzzy inference system frameworks	Manages uncertainty; interpretable rule-based decisions	Rule-based subjective; less scalable for high-dimensional data	Rauf (2018), Kushwaha et al. (2023)
Convolutional neural networks (CNNs) and recurrent neural networks	Species classification (CNN); growth and photosynthetic activity forecasting long short-term memory	Automates feature extraction; excels in image/time-series tasks	Data-hungry; computationally heavy; black-box	Soar et al. (2023), Fournier et al. (2024)

microscopy-based approaches, while effective at small scales, suffer from subjectivity, labor demands, and an inability to scale with the demands of modern cultivation systems (Deore et al., 2020). Subtle morphological differences between taxa or viability states often go undetected, particularly when dealing with morphologically similar genera such as *Chlorella*, *Scenedesmus*, and *Nannochloropsis* (Madkour et al., 2023; Imamoglu, 2024). AI-based image analysis, particularly with CNNs, addresses these bottlenecks by automating feature extraction and enabling high-throughput classification. Unlike classical statistical models or even earlier ML methods that rely on handcrafted features, CNNs autonomously learn discriminative spatial features such as chloroplast patterns, pigmentation, or cell wall thickness directly from raw image data. This approach has demonstrated classification accuracies exceeding 98%, even under varying optical and environmental conditions (Soar et al., 2023; Sonmez et al., 2023). Importantly, CNNs are not limited to species identification but also allow for health status assessment, capturing subtle morphological shifts that signal stress before productivity declines occur. When combined with flow cytometry and imaging flow cytometry, CNNs can process tens of thousands of cells in minutes, providing culture managers with real-time feedback. Despite these advances, challenges remain in generalizing models across laboratories, imaging platforms, and environmental contexts, underscoring the need for larger, standardized datasets to improve transferability (Carloto et al., 2021).

HABs are one of the most visible and costly manifestations of algal overproliferation, leading to fish kills, shellfish contamination, and widespread ecological damage (Gobler et al., 2024). Forecasting these events is notoriously difficult due to the nonlinear interplay of nutrient enrichment, hydrodynamics, light availability, and climate forcing. Conventional approaches, including threshold-based chlorophyll-a monitoring and regression models, often underperform because they cannot accommodate complex spatiotemporal feedbacks (Khan et al., 2021; Izadi et al., 2021). AI-based methods, particularly deep learning, have demonstrated superior predictive power by learning directly from diverse, high-dimensional datasets. CNNs applied to satellite remote sensing can detect bloom initiation with finer spatial granularity, while RNNs and LSTM networks excel in capturing sequential dependencies, enabling forecasts that extend beyond simple "yes/no" bloom occurrence to include bloom trajectory and persistence. Hybrid models that combine CNN spatial filters with LSTM temporal forecasting have shown predictive accuracies above 95%, with lead times of up to a week (Liu et al., 2025; Tang et al., 2022). These advances translate into practical early warning systems for fisheries and coastal managers. However, the "black-box" nature of deep models still raises concerns about interpretability, and overfitting remains a risk when training on limited bloom datasets. Thus, the integration of explainable AI (e.g., Grad-CAM heatmaps for image data) and ensemble modeling strategies is increasingly emphasized to balance predictive accuracy with scientific credibility (Samek and Müller, 2019).

The convergence of AI with genome editing represents a frontier that goes beyond monitoring algal physiology to actively reshape cellular metabolism. CRISPR/Cas systems offer unprecedented precision in gene editing, yet their practical success hinges on the efficient selection of guide RNAs and the minimization of off-target effects (Abbasi et al., 2025). AI, through supervised learning models and deep neural networks, is now used to predict gRNA efficiency, specificity, and likely editing outcomes, significantly improving experimental success rates (Khoshandam et al., 2024). At a higher systems level, AI contributes to metabolic pathway optimization by integrating omics datasets and simulating flux distributions. For example, reinforcement learning and evolutionary algorithms can identify optimal gene knockouts or overexpression targets to maximize lipid accumulation, astaxanthin synthesis, or PUFA production under realistic cultivation conditions (Cheng et al., 2023). This AI-guided optimization accelerates the "design–build–test–learn" (DBTL) cycle, reducing the experimental burden and cost of strain development. Importantly, AI's ability to handle multiobjective optimization means it can balance productivity with robustness, identifying solutions that maximize yield while maintaining growth stability under fluctuating environmental conditions. Looking forward, the integration of AI with laboratory automation and high-throughput omics pipelines could enable semiautonomous or fully autonomous strain design, bringing the vision of "intelligent biofoundries" for algal biotechnology closer to reality. Nevertheless, ethical considerations, biosecurity, and regulatory oversight must keep pace with these technological capabilities to ensure responsible deployment (Khoshandam et al., 2024; Dhokane et al., 2023).

In algal biotechnology, AI applications can be broadly divided into image-based monitoring/classification and genome-level design and optimization. As illustrated in Fig. 10.2, image analysis pipelines leverage

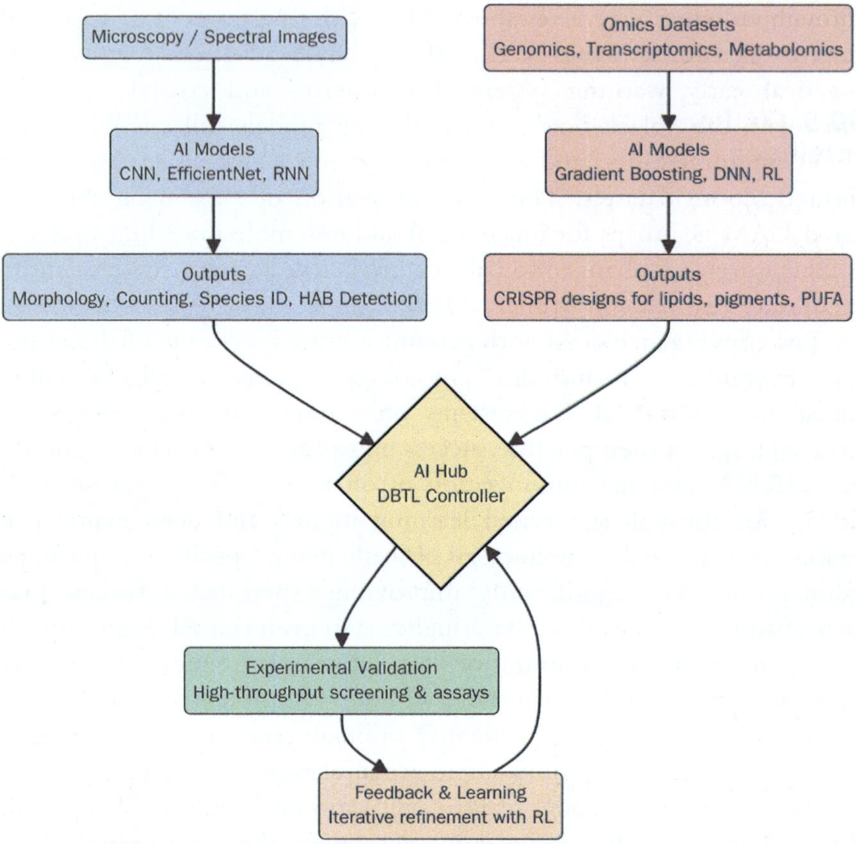

Figure 10.2 ***Artificial intelligence (AI)-assisted pipelines in algal biotechnology.*** AI-assisted pipelines in algal biotechnology integrate image-based classification (morphology, counting, species identification, harmful algal bloom [HAB] detection) with omics-driven genome editing (CRISPR/Cas designs for lipids, pigments, and polyunsaturated fatty acid [PUFA] pathways). Both streams are coupled through an AI-driven design–build–test–learn (DBTL) controller that enables iterative refinement via experimental validation and reinforcement learning.

CNNs, EfficientNet, and RNNs to automate tasks such as cell morphology assessment, counting, and HAB detection. In parallel, omics-driven approaches use models such as gradient boosting, Deep Neural Network (DNNs), and reinforcement learning to support CRISPR/Cas design for enhancing lipid, pigment, and PUFA pathways. Both tracks converge in an AI-driven DBTL cycle, where predictions are validated through high-throughput assays and iteratively refined with reinforcement learning to improve performance over successive cycles.

10.5 Challenges, ethics, and the future of artificial intelligence in microalgae

The central risk in AI-for-microalgae is epistemic: models are only as strong as the data regimes that birthed them. High-quality data are fundamental for developing effective AI models; however, in the field of microalgae research, the quality and availability of data are often limited (Wu et al., 2025b). Most reported gains, whether in growth prediction, bloom detection, or phenotype mapping, are trained on compact, locally idiosyncratic datasets. This creates brittle generalization across strains, photobioreactor geometries, media chemistries, and light fields (Ning et al., 2022). The field's "quiet assumption" is that accuracy demonstrated in one reactor or lake will transfer seamlessly to another; experience in HAB forecasting demonstrates otherwise, with performance degrading under domain shift and seasonality drift (Baydaroglu, 2025; Imamoglu, 2024). Stronger data governance, findability, accessibility, interoperability, and reusability (FAIR)-aligned curation, rich metadata, and versioned preprocessing can reduce hidden confounders and make models inspectable and reusable rather than one-off artifacts. Importantly, FAIR principles should extend beyond datasets to the code, workflows, and model cards that accompany them; without this broader scope, transfer learning and reproducibility remain largely aspirational (Ali and Dahlhaus, 2022).

Yet even when data volume appears sufficient, the challenge of integration is formidable. Multiomics signals include transcriptomic, proteomic, and metabolomic data that arrive with batch effects, missing values, and high dimensionality; imaging and process sensor data add asynchronous sampling and heterogeneous noise. Physics-informed learning and probabilistic modeling can mitigate some of these issues, but they demand rigorous validation under realistic deployment conditions, such as leave-location-out cross-validation or forward-chaining splits, coupled with explicit uncertainty quantification before any real-world actuation. The contrast with computer vision is telling: public benchmarks there have driven progress through shared, well-curated datasets, whereas in applied phycology we still lack comparable resources for (1) high-frequency photobioreactor time series with synchronized actuation and (2) HAB detection and forecasting that spans basins, sensor types, and multiple years (Hill et al., 2019). Without such shared infrastructure, the community risks an insular research cycle in which each laboratory solves its own problem in isolation, with little cumulative progress. A coordinated community challenge with transparent leaderboards would provide a foundation for comparability and reproducibility. Recent explainable HAB-modeling studies reinforce this point—deep learning architectures can perform impressively, but only when the ecological drivers they identify are interpretable and grounded in rigorous analysis (Chen et al., 2025).

As these computational capabilities expand, so too do the ethical, legal, and biosafety considerations. The acceleration of microalgal biotechnology through AI, whether for target prioritization in metabolic engineering, closed-loop cultivation control, or semiautonomous field operations, means that governance cannot remain an afterthought (Wu et al., 2025a). Biosecurity literature warns that advanced AI systems and biodesign tools alter the risk calculus for dual-use, even in ostensibly "low-risk" chassis such as microalgae. The prudent posture is anticipatory: implementing red-teaming to simulate misuse, enforcing tiered access to powerful design models, and maintaining auditable pipelines for any AI component that materially influences biological state (Pannu et al., 2024).

It is tempting to assume that microalgae, generally considered low-risk organisms in biosafety terms, automatically imply low regulatory scrutiny when combined with AI. This assumption, however, is misleading. Under the EU AI Act (Imamoglu, 2024), risk classification is determined not solely by the biological hazard profile but by the operational context, the criticality of the system, and the downstream consequences of AI-driven decisions (Mennella et al., 2024). This means that even an innocuous species such as *Chlorella* or *Nannochloropsis* could form part of a high-risk AI system if the model's outputs influence public health, environmental protection, or the management of critical infrastructure (Kusche, 2024).

One example is the AI-controlled wastewater bioremediation system deployed by a municipality to remove nitrogen and phosphorus from effluent using large-scale photobioreactors inoculated with *C. vulgaris*. Here, the AI autonomously manages CO_2 injection, nutrient dosing, harvesting schedules, and hydraulic retention times in real time, drawing on continuous sensor feedback and predictive modeling. While the microalgae themselves pose minimal biological risk, the system falls under the AI Act's category of "management and operation of critical infrastructure" (Annex III) because any significant error could result in untreated discharges, eutrophication in downstream waters, and consequent public health implications. As a result, this application demands a risk management plan, comprehensive technical documentation, continuous monitoring, human oversight, and robust post-market incident reporting (Kusche, 2024).

A second case illustrates how even purely analytical AI models can reach the high-risk threshold. In coastal aquaculture, an AI system trained on multispectral satellite imagery and historical toxin records forecasts HABs, such as those caused by Alexandrium or Karenia. Its outputs directly feed automated alert systems that trigger shellfish closures and adjust drinking water treatment protocols. A false negative could release toxin-contaminated seafood into markets, while a false positive could cause unnecessary closures with major economic repercussions. Although the AI does not manipulate algae, it performs a safety-related decision-making function, thereby falling within the EU AI Act's classification of high-risk uses in environmental monitoring and public health (Panteli et al., 2025; Pagallo, 2025). This case resonates with recent molecular surveys in the Gulf of Thailand, which uncovered unexpectedly diverse toxic dinoflagellates and diatoms overlooked by microscopy, including *Alexandrium tamiyavanichii*, *Gymnodinium catenatum*, and new Pseudo-nitzschia records (Fu et al., 2021). Metabarcoding targeting Large Subunit ribosomal RNA (LSU) and Internal Transcribed Spacer (ITS) regions revealed hidden ribotypes and rare harmful taxa, leading to calls for routine molecular monitoring. Together, these insights highlight why AI-enhanced predictive systems are vital: metabarcoding refines taxonomic resolution, but AI's integration of genomic, remote sensing, and environmental datasets in real time elevates monitoring into proactive, high-stakes decision-making.

Similarly, AI-guided production of nutraceutical microalgae may fall within the EU AI Act's high-risk category when the AI functions as a safety component in regulated production equipment, a concern underscored by documented safety issues, hazardous substances, heavy-metal bioaccumulation, and labeling transparency problems in microalgae products (EU Art 6; Imamoglu, 2024). A biotechnology firm, for instance, might employ AI to optimize *Spirulina* cultivation for the extraction of phycocyanin, a pigment marketed as an antioxidant dietary supplement (Prates, 2025). The AI controls light cycles, nutrient regimes, and harvesting intervals to maximize pigment yield. Because the biomass is destined for human consumption, any AI-induced deviation, whether through contamination, toxin build-up, or failure to meet compositional standards, could directly compromise food safety (Prates, 2025; Guo et al., 2025). The AI Act classifies such systems under "AI in safety components of products" regulated by food safety legislation, obligating adherence to stringent safety and traceability requirements.

AI has become a transformative force across biotechnology, revolutionizing drug discovery, genomics, personalized medicine, diagnostics, and bioprocess optimization by accelerating processes that were once costly and time-consuming. In this scenario, a genetically engineered microalgal strain is cultivated in closed photobioreactors to produce a precursor for an anticancer drug (Pattanaik et al., 2025). The cultivation process is optimized in real time by a digital twin, a dynamic, AI-powered simulation that adjusts environmental parameters to maximize both yield and purity. The harvested biomass is then processed and purified before incorporation into medicinal formulations. Because the AI functions within a Good Manufacturing Practice (GMP)-regulated production chain for pharmaceuticals, it is considered a safety-critical component (Niazi, 2025). Accordingly, it must comply with the AI Act's highest safety standards, alongside the regulatory frameworks governing medicinal products, ensuring full validation, traceability, and postmarket surveillance.

For AI-enabled genetic engineering of microalgae more broadly, biosafety regulations apply regardless of AI's role. Under the Cartagena Protocol on Biosafety, genetically modified microalgal strains fall within the category of living modified organisms (LMOs), defined as organisms with a novel combination of genetic material obtained through modern biotechnology and capable of replication, and are therefore subject to risk assessment, advanced informed agreement procedures, and monitoring requirements before transboundary movement or release into the environment. Where genetic resources are involved, the Nagoya Protocol mandates fair and equitable benefit-sharing through mutually agreed terms (MATs), including both monetary and nonmonetary returns—critical when AI models are trained on omics data derived from biodiversity-rich regions (Park and Park, 2025). In the digital era, this obligation extends into digital sequence information (DSI), with growing calls for coordinated global governance to ensure that as genomics data drives AI in biotechnology, benefit-sharing obligations remain enforceable and inclusive (Aubry et al., 2022). While AI promises breakthroughs in genomic insight and experimental design, it also lowers the technical threshold for potentially hazardous biodesign, raising pressing biosafety and biosecurity concerns. In Malaysia, these risks intersect with the existing legal landscape: the Biosafety Act 2007 governs genetically modified organism (GMO) development, import, and environmental release, while the Environmental Quality Act 1974, which is currently focused on pollution control and water quality, could be adapted to oversee AI-assisted environmental monitoring, ensuring technological advances align with environmental protection (Idris et al., 2013; Andrew et al., 2018).

On the technical frontier, digital twins offer a pragmatic bridge between idealized simulation and the complex variability of operational systems. These continuously updated, physics-informed surrogates of raceways and photobioreactors allow safe policy testing before physical implementation. Recent studies have shown how "algal digital twins" can retrofit legacy raceways, integrating light-field and hydrodynamic simulation to guide design and operational strategy (Sheik et al., 2024). Yet fidelity is not guaranteed; sensor calibration drift, fouling, and latency can erode model accuracy unless the twin is periodically re-identified and its uncertainty explicitly communicated. Adopting a safety-envelope approach, akin to that in aviation, would allow operators to recognize when the twin's predictions are out-of-distribution and mandate human review before acting on high-impact recommendations (Ward et al., 2021), National Academies of Sciences, Engineering, and Medicine (Imamoglu, 2024).

When coupled with autonomous sensing platforms, unmanned aerial vehicles (UAVs) for high-resolution chlorophyll mapping and remotely operated vehicles (ROVs) for biomass and water-quality transects, digital twins can close the loop from environmental perception to operational action (Ngwenya et al., 2025). However, autonomy carries its own fragility: edge-deployed models may encounter untrained optics (glint, turbidity, atmospheric haze) or unexpected pigment profiles that compromise inference. Here, explainable AI (XAI) is not a cosmetic add-on but an operational safeguard (Rane and Paramesha, 2024). Techniques such as SHapley Additive exPlanations (SHAP) attributions, spectral-band saliency, and counterfactual testing can reveal which variables drive a given forecast and whether the model is extrapolating beyond its knowledge base. Studies in HAB risk prediction show that explainability can materially alter intervention priorities, shifting sampling effort to the most influential ecological drivers (Demiray et al., 2025). Nonetheless, XAI must be used with caution: it is an interpretive tool, not definitive proof of causality, and should be complemented by mechanistic checks and targeted ablations to assess robustness under domain shift. This dual approach, combining AI-derived insights with ecological theory, reduces the risk of overfitting to transient patterns or sensor artifacts, and aligns with WHO guidance on AI governance, which emphasizes accountability, transparency, and human control over automated decision-making (Imamoglu, 2024).

All of these developments point back to the underlying sustainability agenda. The circular-economy vision for microalgae converting CO_2 and nutrient-rich waste streams into fuels, feeds, nutraceuticals, pigments, and biopolymers while delivering ecosystem services such as wastewater polishing is compelling, but it is vulnerable to overstatement. AI can strengthen this vision by optimizing multiobjective trade-offs in productivity, quality, energy use, and environmental impact, as well as by orchestrating downstream scheduling to reduce losses in storage or processing. Yet life cycle assessment (LCA) repeatedly shows that energy-intensive steps such as harvesting, dewatering, drying, and extraction can erode or even negate climate benefits unless these processes are codesigned with energy recovery, waste-heat utilization, and market realities in mind. Wastewater-based cultivation is particularly promising when it replaces conventional treatment methods and when downstream markets can accommodate the specific quality profile of algal-derived products. Embedding AI directly into techno-economic assessment and LCA pipelines, with explicit uncertainty quantification, can surface Pareto frontiers rather than single-point optima, allowing policymakers and investors to make transparent trade-offs between carbon mitigation, nutrient recovery, profitability, and social acceptance (Oduque de Jesus et al., 2021; Mahmud et al., 2021). Such integration also facilitates compliance with emerging sustainability and traceability standards, such as ISO 59014 (Imamoglu, 2024) for secondary materials, which can provide a credible verification framework for circular bioeconomy claims in microalgal production systems.

10.6 Conclusion

The convergence of AI and microalgal biotechnology is redefining the boundaries of environmental engineering, bioresource production, and sustainable process optimization. Microalgae possess remarkable metabolic flexibility, rapid growth rates, and the ability to thrive in diverse aquatic environments, enabling applications that range from carbon sequestration and nutrient recovery to biofuel generation and high-value compound synthesis. Yet, despite decades of research, translating microalgae's biological potential into consistent, scalable industrial outcomes has remained challenging. The fundamental issue lies in the complexity of their cultivation systems, which are multivariate, dynamic, and influenced by biological, chemical, and physical interactions that are not easily captured by traditional control methods. AI is shifting this paradigm by enabling the integration of large, heterogeneous datasets from genomics and metabolomics profiles to real-time environmental sensor streams into predictive and adaptive operational models. Algorithms such as ANNs, SVMs, RFs, GA, and fuzzy logic, along with deep learning models like CNNs and RNNs, have demonstrated the capacity to optimize light regimes, CO_2 dosing, nutrient supplementation, and harvesting schedules with unprecedented precision. When coupled with IoT-enabled biosensors, AI–bioreactor feedback loops, and digital twin frameworks, these tools facilitate real-time decision-making, anomaly detection, and scenario testing that far surpass conventional manual monitoring or static modeling approaches. However, these advancements raise critical operational and governance questions. AI systems are only as robust as the data they are trained on, and limitations such as sensor calibration drift, environmental variability, and data sparsity can introduce uncertainty into model predictions. Moreover, the "black box" nature of many AI models complicates transparency and interpretability issues that become especially significant when such systems interface with public health, water treatment, or environmental protection. Under the EU AI Act (Imamoglu, 2024), risk classification is determined not solely by the biological hazard profile of the organism but by the operational context and downstream impacts of AI-driven decisions. This regulatory framing means that even "low-risk" species such as Chlorella or Nannochloropsis can be considered high-risk if their AI control systems govern critical infrastructure, where a malfunction could trigger untreated discharges, eutrophication, or other ecological and societal consequences. The future of AI-enabled microalgal systems will hinge on three interconnected priorities. First, technological integration must advance toward hybrid frameworks that merge mechanistic understanding with data-driven adaptability, supported by open and standardized datasets to improve reproducibility. Second, regulatory alignment requires active dialog between researchers, policymakers, and industry stakeholders to ensure compliance while fostering innovation. Third, sustainability optimization demands that AI systems be designed not just for maximum productivity, but for minimal energy consumption, closed-loop nutrient cycling, and alignment with circular-economy principles. Ultimately, the question is not whether AI can enhance microalgal biotechnology it already has but whether the scientific, regulatory, and industrial communities can collectively ensure that this enhancement is transparent, equitable, and ecologically sound. If approached with foresight, interdisciplinary collaboration, and rigorous governance, AI-enabled microalgal systems have the potential to move from niche research applications to mainstream, scalable solutions that bridge computational intelligence with biological productivity, delivering tangible gains in climate resilience, resource efficiency, and environmental protection.

AI disclosure

During the preparation of this work, the author(s) used ChatGPT in order to assist with language refinement, summarization, and improving readability of the chapter. After using this tool/service, the author(s) reviewed and edited the content as needed and take(s) full responsibility for the content of the publication.

References

Abbasi, A.F., Asim, M.N., Dengel, A., 2025. Transitioning from wet lab to artificial intelligence: a systematic review of AI predictors in CRISPR. Journal of Translational Medicine 23 (1). Available at https://doi.org/10.1186/s12967-024-06013-w, https://translational-medicine.biomedcentral.com.

Akenteng, Y.D., Chen, H., Opoku, K.N., Ullah, F., Wang, S., Kumar, S., 2025. The role of computational fluid dynamics (CFD) in phytohormone-regulated microalgae-based carbon dioxide capture technology. Sustainability 17 (3), 860. Available at https://doi.org/10.3390/su17030860.

Ali, B., Dahlhaus, P., 2022. The role of FAIR data towards sustainable agricultural performance: a systematic literature review. Agriculture 12 (2), 309. Available at https://doi.org/10.3390/agriculture12020309.

Amanna, B., Bahri, P.A., Moheimani, N.R., 2024. Application of computational fluid dynamics in optimizing microalgal photobioreactors. Algal Research 83. Available at https://doi.org/10.1016/j.algal.2024.103718, https://www.sciencedirect.com/science/journal/22119264.

Ariza Rangel, M.A. 2025. ZooxClassML: Evaluation of Machine Learning Models for Classification of Live and Dead Symbiodiniaceae algae.

Aubry, S., Frison, C., Medaglia, J.C., Frison, E., Jaspars, M., Rabone, M., et al., 2022. Bringing access and benefit sharing into the digital age. Plants People Planet 4 (1), 5–12. Available at https://doi.org/10.1002/ppp3.10186, nph.onlinelibrary.wiley.com/journal/25722611.

Aydin, S., 2024. Introductory chapter: microalgae – a versatile and valuable resource for sustainable development. In: Aydin, S. (Ed.), Microalgae - Current and Potential Applications. IntechOpen. Available at https://doi.org/10.5772/intechopen.1004110.

Ayele, A., Godeto, Y.G., 2021. Bioremediation of chromium by microorganisms and its mechanisms related to functional groups. Journal of Chemistry 2021. Available at https://doi.org/10.1155/2021/7694157, http://www.hindawi.com/journals/chem/contents/.

Baydaroğlu, Ö., 2025. Harmful algal bloom prediction using empirical dynamic modeling. Science of the Total Environment 959. Available at https://doi.org/10.1016/j.scitotenv.2024.178185, https://www.sciencedirect.com/science/journal/00489697.

Bekirogullari, M., Figueroa-Torres, G.M., Pittman, J.K., Theodoropoulos, C., 2020. Models of microalgal cultivation for added-value products - a review. Biotechnology

Advances 44. Available at https://doi.org/10.1016/j.biotechadv.2020.107609, www.elsevier.com/inca/publications/store/5/2/5/4/5/5/index.htt.

Belany, P., Hrabovsky, P., Sedivy, S., Cajova Kantova, N., Florkova, Z., 2024. A comparative analysis of polynomial regression and artificial neural networks for prediction of lighting consumption. Buildings 14 (6), 1712. Available at https://doi.org/10.3390/buildings14061712.

Bernardi, A., Nikolaou, A., Meneghesso, A., Chachuat, B., Morosinotto, T., Bezzo, F., 2017. Semi-empirical modeling of microalgae photosynthesis in different acclimation states – application to N. gaditana. Journal of Biotechnology 259, 63–72. Available at https://doi.org/10.1016/j.jbiotec.2017.08.002.

Camacho-Rodríguez, J., Gallardo-Rodríguez, J.J., Cerón-García, M.C., García-Camacho, F., Molina-Grima, E., 2021. A new culture medium based on genetic algorithms for Isochrysis galbana production relevant to hatcheries. Journal of Applied Phycology 33 (6), 3647–3656. Available at https://doi.org/10.1007/s10811-021-02564-3.

Carloto, I., Johnston, P., Pestana, C.J., Lawton, L.A., 2021. Detection of morphological changes caused by chemical stress in the cyanobacterium Planktothrix agardhii using convolutional neural networks. Science of the Total Environment 784. Available at https://doi.org/10.1016/j.scitotenv.2021.146956, www.elsevier.com/locate/scitotenv.

Chen, S., Huang, J., Huang, J., Wang, P., Sun, C., Zhang, Z., et al., 2025. Explainable deep learning identifies patterns and drivers of freshwater harmful algal blooms. Environmental Science and Ecotechnology 23, 100522. Available at https://doi.org/10.1016/j.ese.2024.100522.

Cheng, Y., Bi, X., Xu, Y., Liu, Y., Li, J., Du, G., et al., 2023. Machine learning for metabolic pathway optimization: a review. Computational and Structural Biotechnology Journal 21, 2381–2393. Available at https://doi.org/10.1016/j.csbj.2023.03.045.

Demiray, B.Z., Mermer, O., Baydaroğlu, Ö., Demir, I., 2025. Predicting harmful algal blooms using explainable deep learning models: a comparative study. Water 17 (5), 676. Available at https://doi.org/10.3390/w17050676.

Deore, P., Beardall, J., Noronha, S., 2020. A perspective on the current status of approaches for early detection of microalgal grazing. Journal of Applied Phycology 32 (6), 3723–3733. Available at https://doi.org/10.1007/s10811-020-02241-x.

Dhokane, D., Shaikh, A., Yadav, A., Giri, N., Bandyopadhyay, A., Dasgupta, S., et al., 2023. CRISPR-based bioengineering in microalgae for production of industrially important biomolecules. Frontiers in Bioengineering and Biotechnology 11 Available at https://doi.org/10.3389/fbioe.2023.1267826, http://journal.frontiersin.org/journal/bioengineering-and-biotechnology#archive.

Fournier, C., Fernandez-Fernandez, R., Cirés, S., López-Orozco, J.A., Besada-Portas, E., Quesada, A., 2024. LSTM networks provide efficient cyanobacterial blooms forecasting even with incomplete spatio-temporal data. Water Research 267, 122553. Available at https://doi.org/10.1016/j.watres.2024.122553.

Fu, Z., Piumsomboon, A., Punnarak, P., Uttayarnmanee, P., Leaw, C.P., Lim, P.T., et al., 2021. Diversity and distribution of harmful microalgae in the Gulf of Thailand assessed by DNA metabarcoding. Harmful Algae 106. Available at https://doi.org/10.1016/j.hal.2021.102063, https://www.journals.elsevier.com/harmful-algae.

Galinha, C.F., Sanches, S., Crespo, J.G., 2018. Chapter 6 - Membrane bioreactors. In: Luis, P. (Ed.), Fundamental Modelling of Membrane Systems. Elsevier, pp. 209–249. Available at https://doi.org/10.1016/B978-0-12-813483-2.00006-X.

Gargalo, C.L., Malanca, A.A., Aouichaoui, A.R.N., Huusom, J.K., Gernaey, K.V., 2024. Navigating industry 4.0 and 5.0: the role of hybrid modelling in (bio)chemical engineering's digital transition. Frontiers in Chemical Engineering 6. Available at https://doi.org/10.3389/fceng.2024.1494244, https://www.frontiersin.org/journals/chemical-engineering.

Gobler, C.J., Drinkwater, R.W., Anthony, A., Goleski, J.A., Famularo-Pecora, A.M.E., Wallace, M.K., et al., 2024. Sewage-and fertilizer-derived nutrients alter the intensity, diversity, and toxicity of harmful cyanobacterial blooms in eutrophic lakes. Frontiers in Microbiology 15. Available at https://doi.org/10.3389/fmicb.2024.1464686, https://www.frontiersin.org/journals/microbiology#.

Grama, S.B., Liu, Z., Li, J., 2022. Emerging Trends in Genetic Engineering of Microalgae for Commercial Applications. Marine Drugs 20 (5), 285. Available at https://doi.org/10.3390/md20050285.

Guo, B., Lu, X., Jiang, X., Shen, X.L., Wei, Z., Zhang, Y., 2025. Artificial Intelligence in Advancing Algal Bioactive Ingredients: Production, Characterization, and Application. Foods 14 (10). Available at https://doi.org/10.3390/foods14101783.

Hill, P.R., Kumar, A., Temimi, M., Bull, D.R., 2019. HABNet: machine learning, remote sensing based detection and prediction of harmful algal blooms. Arxiv. Available at https://doi.org/10.48550/arxiv.1912.02305, https://arxiv.org.

Hu, J., Meng, W., Su, Y., Qian, C., Fu, W., 2023. Emerging technologies for advancing microalgal photosynthesis and metabolism toward sustainable production. Frontiers in Marine Science 10. Available at https://doi.org/10.3389/fmars.2023.1260709, https://www.frontiersin.org/journals/marine-science#.

Igou, T., Zhong, S., Reid, E., Chen, Y., 2023. Real-time sensor data profile-based deep learning method applied to open raceway pond microalgal productivity prediction. Environmental Science & Technology 57 (46), 17981–17989. Available at https://doi.org/10.1021/acs.est.2c07578.

Izadi, M., Sultan, M., Kadiri, R.E., Ghannadi, A., Abdelmohsen, K., 2021. A Remote Sensing and Machine Learning-Based Approach to Forecast the Onset of Harmful Algal Bloom. Remote Sensing 13 (19).

Imamoglu, E., 2024. Artificial intelligence and/or machine learning algorithms in microalgae bioprocesses. Bioengineering 11 (11), 1143. Available at https://doi.org/10.3390/bioengineering11111143.

Khan, R.M., Salehi, B., Mahdianpari, M., Mohammadimanesh, F., Mountrakis, G., Quackenbush, L.J., 2021. A Meta-Analysis on Harmful Algal Bloom (HAB) Detection and Monitoring: A Remote Sensing Perspective. Remote Sensing 13 (21).

Kasbawati, K., Mardawati, E., Samsir, R., Suhartini, S., Kalondeng, A., 2022. An appropriate unstructured kinetic model describing the batch fermentation growth of Debaryomyces hansenii for xylitol production using hydrolysis of oil palm empty fruit bunch. Biotechnology and Biotechnological Equipment 36 (1), 462–471. Available at https://doi.org/10.1080/13102818.2022.2098816, http://www.tandfonline.com/loi/tbeq20#.Vx8TFE1f3cs.

Khoshandam, M., Soltaninejad, H., Hosseinkhani, S., Saltanatpour, Z., Goudarzi, M.T.H., Hamidieh, A.A., 2024. CRISPR/Cas and artificial intelligence to improve precision medicine: future perspectives and potential limitations. Human Gene 42. Available at https://doi.org/10.1016/j.humgen.2024.201356, https://www.sciencedirect.com/science/journal/27730441.

Kusche, I., 2024. Possible harms of artificial intelligence and the EU AI act: fundamental rights and risk. Journal of Risk Research 1–14. Available at https://doi.org/10.1080/13669877.2024.2350720.

Kushwaha, O.S., Uthayakumar, H., Kumaresan, K., 2023. Modeling of carbon dioxide fixation by microalgae using hybrid artificial intelligence (AI) and fuzzy logic (FL) methods and optimization by genetic algorithm (GA). Environmental Science and Pollution Research 30 (10), 24927–24948. Available at https://doi.org/10.1007/s11356-022-19683-0.

Liu, M., Huang, Y., Hu, J., He, J., Xiao, X., 2023. Algal community structure prediction by machine learning. Environmental Science and Ecotechnology 14, 100233. Available at https://doi.org/10.1016/j.ese.2022.100233.

Liu, Y., Yang, B., Xie, K., Sun, J., Zhu, S., 2025. Dongting Lake algal bloom forecasting: Robustness and accuracy analysis of deep learning models. Journal of Hazardous Materials 485, 136804. Available at https://doi.org/10.1016/j.jhazmat.2024.136804.

Liyanaarachchi, V.C., Nishshanka, G.K.S.H., Sakarika, M., Nimarshana, P.H.V., Ariyadasa, T.U., Kornaros, M., 2021. Artificial neural network (ANN) approach to optimize cultivation conditions of microalga Chlorella vulgaris in view of biodiesel production. Biochemical Engineering Journal 173, 108072. Available at https://doi.org/10.1016/j.bej.2021.108072.

Luzi, G., McHardy, C., 2022. Modeling and Simulation of Photobioreactors with Computational Fluid Dynamics—A Comprehensive Review. Energies 15 (11).

Maia, J.L.d, Cardoso, J.S., Mastrantonio, D.J.d S., Bierhals, C.K., Moreira, J.B., Costa, J.A.V., et al., 2020. Microalgae starch: A promising raw material for the bioethanol production. International Journal of Biological Macromolecules 165, 2739–2749. Available at https://doi.org/10.1016/j.ijbiomac.2020.10.159.

Memmolo, P., Carcagnì, P., Bianco, V., Merola, F., Goncalves da Silva Junior, A., Garcia Goncalves, L.M., et al., 2020. Learning diatoms classification from a dry test slide by holographic microscopy. Sensors 20 (21), 6353. Available at https://doi.org/10.3390/s20216353.

Madkour, D.M., Shapiai, M.I., Mohamad, S.E., Aly, H.H., Ismail, Z.H., Ibrahim, M.Z., 2023. A Systematic Review of Deep Learning Microalgae Classification and Detection. IEEE Access 11, 57529–57555. Available at https://doi.org/10.1109/ACCESS.2023.3280410.

Mennella, C., Maniscalco, U., De Pietro, G., Esposito, M., 2024. Ethical and regulatory challenges of AI technologies in healthcare: a narrative review. Heliyon 10 (4), e26297. Available at https://doi.org/10.1016/j.heliyon.2024.e26297.

Mahmud, R., Moni, S., High, K., Carbajales-Dale, M., 2021. Integration of techno-economic analysis and life cycle assessment for sustainable process design – A review. Journal of Cleaner Production 317, 128247. Available at https://doi.org/10.1016/j.jclepro.2021.128247.

Ngwenya, N., Bangira, T., Sibanda, M., Kebede Gurmessa, S., Mabhaudhi, T., 2025. UAV-based remote sensing of chlorophyll-a concentrations in inland water bodies: a systematic review. Geocarto International 40 (1). Available at https://doi.org/10.1080/10106049.2025.2452246, http://www.tandfonline.com/toc/tgei20/current.

Ning, H., Li, R., Zhou, T., 2022. Machine learning for microalgae detection and utilization [Review]. Frontiers in Marine Science 9, 2022. Available at https://doi.org/10.3389/fmars.2022.947394.

Niazi, S.K., 2025. Regulatory perspectives for AI/ML implementation in pharmaceutical GMP environments. Pharmaceuticals 18 (6). Available at https://doi.org/10.3390/ph18060901, https://www.mdpi.com/journal/pharmaceuticals.

Oduque de Jesus, J., Oliveira-Esquerre, K., Lima Medeiros, D., 2021. Integration of Artificial Intelligence and Life Cycle Assessment Methods. IOP Conference Series: Materials Science and Engineering 1196 (1), 012028. Available at https://doi.org/10.1088/1757-899X/1196/1/012028.

Olabi, A., Shehata, N., Sayed, E.T., Rodriguez, C., Anyanwu, R.C., Russell, C., et al., 2022. Role of microalgae in achieving sustainable development goals and circular economy. The Science of the Total Environment 854, 158689. Available at https://doi.org/10.1016/j.scitotenv.2022.158689.

Onay, A., 2023. Theoretical Models Constructed by Artificial Intelligence Algorithms for Enhanced Lipid Production: Decision Support Tools. Bitlis Eren Üniversitesi Fen

Bilimleri Dergisi 12 (4), 1195–1211. Available at https://doi.org/10.17798/bitlisfen.1362136.

Oruganti, R.K., Biji, A.P., Lanuyanger, T., Show, P.L., Sriariyanun, M., Upadhyayula, V.K.K., et al., 2023. Artificial intelligence and machine learning tools for high-performance microalgal wastewater treatment and algal biorefinery: a critical review. Science of the Total Environment 876. Available at https://doi.org/10.1016/j.scitotenv.2023.162797, www.elsevier.com/locate/scitotenv.

Pagallo, U., 2025. On Twelve Shades of Green: Assessing the Levels of Environmental Protection in the Artificial Intelligence Act. Minds and Machines 35 (1), 10. Available at https://doi.org/10.1007/s11023-025-09713-4.

Panteli, D., Adib, K., Buttigieg, S., Goiana-da-Silva, F., Ladewig, K., Azzopardi-Muscat, N., et al., 2025. Artificial intelligence in public health: promises, challenges, and an agenda for policy makers and public health institutions. Lancet Public Health 10 (5), e428–e432. Available at https://doi.org/10.1016/s2468-2667(25)00036-2.

Pannu, J., Bloomfield, D., Zhu, A., MacKnight, R., Gomes, G., Cicero, A., et al., 2024. Prioritizing high-consequence biological capabilities in evaluations of artificial intelligence models. arXiv. Available at https://doi.org/10.48550/arXiv.2407.13059, https://arxiv.org.

Park, S., Li, Y., 2014. Integration of Biological Kinetics and Computational Fluid Dynamics to Model the Growth of Nannochloropsis salina in an Open Channel Raceway. Biotechnology and Bioengineering 112. Available at https://doi.org/10.1002/bit.25509.

Park, T.J., Park, S.P., 2025. Legal and economic perspectives on fair and equitable benefit sharing in the Nagoya Protocol. Conservation Biology 39 (2). Available at https://doi.org/10.1111/cobi.14410, http://onlinelibrary.wiley.com/journal/10.1111/(ISSN)1523-1739.

Pattanaik, A., Sukla, L.B., Pradhan, D., 2025. Artificial intelligence and virtual environment for microalgal source for production of nutraceuticals. Biomedical Journal 1 (5). Available at https://doi.org/10.26717/BJSTR.2019.13.002459.

Pearson, S., 2025. Artificial intelligence in horticulture. The Journal of Horticultural Science and Biotechnology 100 (5), 564–570. Available at https://doi.org/10.1080/14620316.2025.2475401.

Prasad, R., Gupta, S.K., Shabnam, N., Oliveira, C.Y.B., Nema, A.K., Ansari, F.A., et al., 2021. Role of microalgae in global CO2 sequestration: physiological mechanism, recent development, challenges, and future prospective. Sustainability (Switzerland) 13 (23). Available at https://doi.org/10.3390/su132313061, https://www.mdpi.com/2071-1050/13/23/13061/pdf.

Prates, J.A.M., 2025. Applications of bioactive compounds from marine microalgae in health, cosmetics, and functional foods. Applied Sciences 15 (11), 6144. Available at https://doi.org/10.3390/app15116144.

Purnomo, D.M. J., Purbarani, S.C., Wibisono, A., Hendrayanti, D., Bowolaksono, A., Mursanto, P., et al. (2015). Genetic algorithm optimization for extreme learning machine based microalgal growth forecasting of Chlamydomonas sp. 2015 International Conference on Advanced Computer Science and Information Systems (ICACSIS).

Rane, N.L., Paramesha, M., 2024. Explainable Artificial Intelligence (XAI) as a Foundation for Trustworthy Artificial Intelligence. Deep Science Publishing. Available at http://doi.org/10.70593/978-81-981367-4-9_1.

Reza, A., Chen, L., Mao, X., 2024. Response surface methodology for process optimization in livestock wastewater treatment: A review. Heliyon 10 (9), e30326. Available at https://doi.org/10.1016/j.heliyon.2024.e30326.

Romagnoli, F., Weerasuriya-Arachchige, A.R.P.P., Paoli, R., Feofilovs, M., Ievina, B., 2021. Growth kinetic model for microalgae cultivation in open raceway ponds: a system dynamics tool. Environmental and Climate Technologies 25 (1), 1317–1336.

Available at https://doi.org/10.2478/rtuect-2021-0100, https://sciendo.com/pl/journal/RTUECT.
Rauf, S. (2018). *Design and implementation of control system for optimal growth of microalgae in outdoor ponds.*
Saccardo, A., Felices-Rando, B., Sforza, E., Bezzo, F., 2023. Droop model identification via model-based design of experiments to describe microalgae nitrogen uptake in continuous photobioreactors. Chemical Engineering Journal 468, 143577. Available at https://doi.org/10.1016/j.cej.2023.143577.
Samek, W., Müller, K.-R., 2019. Towards Explainable Artificial Intelligence. In: Samek, In. W., Montavon, G., Vedaldi, A., Hansen, L.K., Müller, K.-R. (Eds.), Explainable AI: Interpreting, Explaining and Visualizing Deep Learning. Springer International Publishing, pp. 5–22. Available at https://doi.org/10.1007/978-3-030-28954-6_1.
Schediwy, K., Trautmann, A., Steinweg, C., Posten, C., 2019. Microalgal kinetics — a guideline for photobioreactor design and process development. Engineering in Life Sciences 19 (12), 830–843. Available at https://doi.org/10.1002/elsc.201900107.
Sheik, A.G., Kumar, A., Ansari, F.A., Raj, V., Peleato, N.M., Patan, A.K., et al., 2024. Reinvigorating algal cultivation for biomass production with digital twin technology - a smart sustainable infrastructure. Algal Research 84. Available at https://doi.org/10.1016/j.algal.2024.103779, https://www.sciencedirect.com/science/journal/22119264.
Solimeno, A., Samsó, R., Uggetti, E., Sialve, B., Steyer, J.P., Gabarró, A., et al., 2015. New mechanistic model to simulate microalgae growth. Algal Research 12, 350–358. Available at https://doi.org/10.1016/j.algal.2015.09.008, http://www.sciencedirect.com/science/journal/aip/22119264.
Soar, J., Oh, S.L., Wen, L., Ward, A., Sharma, E., Deo, R., et al., 2023. Deep Image Analysis Methods for Microalgae Identification. Available at https://doi.org/10.20944/preprints202305.1323.v1.
Sonmez, M.E., Altinsoy, B., Ozturk, B.Y., Gumus, N.E., Eczacioglu, N., 2023. Deep learning-based classification of microalgae using light and scanning electron microscopy images. Micron (Oxford, England: 1993) 172, 103506. Available at https://doi.org/10.1016/j.micron.2023.103506.
Srimongkol, P., Sangtanoo, P., Songserm, P., Watsuntorn, W., Karnchanatat, A., 2022. Microalgae-based wastewater treatment for developing economic and environmental sustainability: current status and future prospects. Frontiers in Bioengineering and Biotechnology 10. Available at https://doi.org/10.3389/fbioe.2022.904046, http://journal.frontiersin.org/journal/bioengineering-and-biotechnology#archive.
Sumanasekara, H., Jayasingha, H., Amarasooriya, G., Dayarathne, N., Mainali, B., Senevirathna, L., et al., 2025. Hybrid Machine Learning Models for Predicting the Impact of Light Wavelengths on Algal Growth in Freshwater Ecosystems. Phycology 5 (2).
Tan, J.S., Lee, S.Y., Chew, K.W., Lam, M.K., Lim, J.W., Ho, S.H., et al., 2020. A review on microalgae cultivation and harvesting, and their biomass extraction processing using ionic liquids. Bioengineered 11 (1), 116–129. Available at https://doi.org/10.1080/21655979.2020.1711626, http://www.tandfonline.com/loi/kbie20.
Tang, Y., Feng, Y., Fung, S., Xomchuk, V.R., Jiang, M., Moore, T., et al., 2022. Spatiotemporal Deep-Learning-Based Algal Bloom Prediction for Lake Okeechobee Using Multisource Data Fusion. IEEE Journal of Selected Topics in Applied Earth Observations and Remote Sensing 15, 8318–8331. Available at https://doi.org/10.1109/JSTARS.2022.3208620.
Velásquez-Orta, S.B., Yáñez-Noguez, I., Ramírez, I.M., Ledesma, M.T.O., 2024. Pilot-scale microalgae cultivation and wastewater treatment using high-rate ponds: a meta-analysis. Environmental Science and Pollution Research 31 (34), 46994–47021.

Available at https://doi.org/10.1007/s11356-024-34000-7, https://www.springer.com/journal/11356.

Veza, I., Spraggon, M., Fattah, I.M.R., Idris, M., 2023. Response surface methodology (RSM) for optimizing engine performance and emissions fueled with biofuel: review of RSM for sustainability energy transition. Results in Engineering 18. Available at https://doi.org/10.1016/j.rineng.2023.101213, https://www.journals.elsevier.com/results-in-engineering.

Ward, R., Choudhary, R., Gregory, A., Jans-Singh, M., Girolami, M., 2021. Continuous calibration of a digital twin: comparison of particle filter and Bayesian calibration approaches. Data-Centric Engineering 2 (3). Available at https://doi.org/10.1017/dce.2021.12.

Wei, Y., Han, C., Yu, Z., 2023. An environment safety monitoring system for agricultural production based on artificial intelligence, cloud computing and big data networks. Journal of Cloud Computing 12 (1). Available at https://doi.org/10.1186/s13677-023-00463-1.

Wu, Y., Shan, L., Zhao, W., Lu, X., 2025a. Harnessing Artificial Intelligence to Revolutionize Microalgae Biotechnology: Unlocking Sustainable Solutions for High-Value Compounds and Carbon Neutrality. Marine Drugs 23 (5). Available at https://doi.org/10.3390/md23050184.

Wu, Y., Shan, L., Zhao, W., Lu, X., 2025b. Harnessing Artificial Intelligence to Revolutionize Microalgae Biotechnology: Unlocking Sustainable Solutions for High-Value Compounds and Carbon Neutrality. Marine Drugs 23 (5). Available at https://doi.org/10.3390/md23050184.

Yang, C.T., Kristiani, E., Leong, Y.K., Chang, J.S., 2024. Machine learning in microalgae biotechnology for sustainable biofuel production: advancements, applications, and prospects. Bioresource Technology 413. Available at https://doi.org/10.1016/j.biortech.2024.131549, https://www.sciencedirect.com/science/journal/09608524.

Ye, Z.P., Suggett, D.J., Robakowski, P., Kang, H.J., 2013. A mechanistic model for the photosynthesis-light response based on the photosynthetic electron transport of photosystem II in C3 and C4 species. New Phytologist 199 (1), 110–120. Available at https://doi.org/10.1111/nph.12242.

Yeh, Y.-C., Syed, T., Brinitzer, G., Frick, K., Schmid-Staiger, U., Haasdonk, B., et al., 2023. Improving Microalgae Growth Modeling of Outdoor Cultivation with Light History Data using Machine Learning Models: A Comparative Study. Bioresource Technology 390, 129882. Available at https://doi.org/10.1016/j.biortech.2023.129882.

Zhang, J., Zhi, M., Zhang, Y., 2021. Combined Generalized Additive model and Random Forest to evaluate the influence of environmental factors on phytoplankton biomass in a large eutrophic lake. Ecological Indicators 130, 108082. Available at https://doi.org/10.1016/j.ecolind.2021.108082.

CHAPTER 11

Genetic modification in algae: understanding meiosis for breeding and selection

Pablo Bolaños-Villegas[1,2,3] and Arnaud Ronceret[4]
[1]School of Agronomy, Faculty of Food and Agricultural Sciences, University of Costa Rica, San Jose, Costa Rica
[2]Fabio Baudrit Agricultural Research Station, University of Costa Rica, Alajuela, Costa Rica
[3]Lankester Botanical Gardens, University of Costa Rica, Cartago, Costa Rica
[4]Instituto de Biotecnología, National Autonomous University of Mexico, Cuernavaca, Mexico

11.1 Introduction

Sexual reproduction has a deep evolutionary origin in eukaryotes and is instrumental in the creation of genetic diversity by allowing chromosomal recombination and proper chromosomal disjunction during meiosis (Hasan and Ness, 2020). A remarkable assortment of sexual traits has arisen across algal lineages, including differences in the determination of clear sexual and mating types, hermaphroditism, and differences in the dominant life cycle as haploids or diploids (Coelho and Umen, 2021). Algae are extremely valuable organisms for their industrial and agricultural use as high-density sources of proteins and vitamins for human consumption, as in the red algae (Rhodophyta) polyextremophile genus *Galdiera* (Hirooka et al., 2022); as feed for fisheries, as in the red alga *Glacilariopsis lemaneiformis* (Zhu et al., 2023); as a sweetener and thickener, as in the brown alga (Phaeophyceae), *Saccharina latissima* (the sugar kelp) (Pearson et al., 2019); as a biofuel feedstock, as in the green alga *Chlorella sorokiana* (Hovde et al., 2018); or as a CO_2-biomitigating agent within bioreactors (Saravanan et al., 2022). A better understanding of reproductive biology and meiosis, often through genomics, may allow for introducing and selecting novel traits in algal crops such as *Galdiera* and *Saccharina* by facilitating the development of breeding tools such as high-density genetic maps and the identification of quantitative trait loci (QTL) (Coelho and Cock, 2020; Hirooka et al., 2022) (Fig. 11.1. In at least the green alga *Chlamydomonas reinhardtii*, both the effective population recombination rate and crossover rate have been determined in detail (Hasan and Ness, 2020).

Applied Phycology
DOI: https://doi.org/10.1016/B978-0-443-24758-3.00010-1

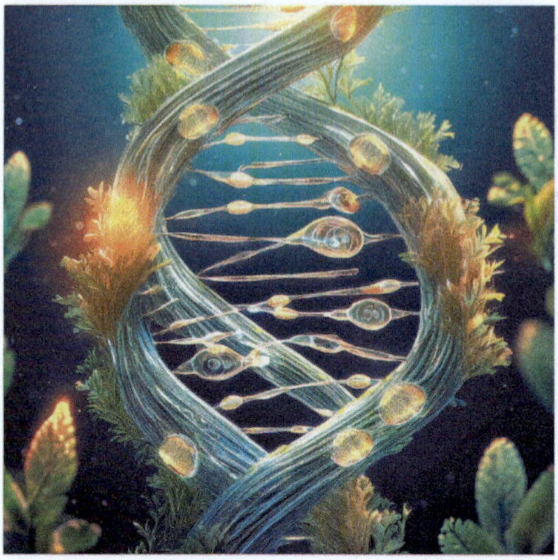

Figure 11.1 ***Opportunities for applied and basic research in algae.*** Algae are complex organisms. The study of their reproductive biology may hold clues on how to breed superior lines for human consumption and industrial use. Original image file. Permission granted by Elsevier.

Algae are autotrophic photosynthetic organisms distributed broadly, with representatives in most branches of the eukaryotic tree of life (Bourdareau et al., 2021). This broad distribution can be attributed to the mechanisms that enabled photosynthesis in each group. For instance, in green algae (Chlorophyta), red algae (Rhodophytes), and glaucophytes (archaeplastids or freshwater algae), photosynthesis may have resulted from engulfment and enslavement of a putative cyanobacterium that over time became a plastid, whereas in the rest of algae (brown algae, diatoms, dinoflagellates, and foraminiferans), photosynthesis may have resulted from several rounds of capture and engulfment, this time of green and red algae by others (Bourdareau et al., 2021; Vigneau and Borg, 2021).

In algae, two different free-living developmental life forms, sporophyte and gametophyte, may alternate (Vigneau and Borg, 2021). The sporophyte is often diploid and, through meiosis, gives rise to haploid spores that further differentiate into gametophytes (Vigneau and Borg, 2021). Then, the eventual fusion of gametes at fertilization creates a diploid sporophyte, and a haplo-diplontic life cycle loop is completed (Vigneau and Borg, 2021), at least in most multicellular algae (Bourdareau et al., 2021).

Nonetheless, algal lineages are notorious for several cytological, developmental, and genomic quirks (Lin et al., 2022). For example, in dinoflagellates, a group that contributes to coral reef growth and harmful algal blooms, the life cycle is haploid-dominant (haplontic) and features an asexual haploid vegetative stage, plus a diploid transient stage that develops into a resting cyst that undergoes meiosis once the environmental conditions are favorable (Lin et al., 2022). Dinoflagellates possess some of the largest eukaryotic nuclear genomes, up to 250 Gb (Wisecaver and Hackett, 2011), and feature some remarkable traits such as: (1) chromosomes attached to the nuclear membrane that remain condensed during the cell cycle; (2) chromosomes remain confined within a modified eukaryotic nucleus (e.g., dinokaryon nucleus) and divide by closed mitosis (dinomitosis), and (3) chromosomes segregate using an extranuclear spindle, without membrane breakdown (Gavelis et al., 2019). Chromosomes in dinoflagellates also show histone depletion, and if present, regular histones are replaced instead by special histone-like proteins known as dinoflagellate viral nucleoproteins and dinoflagellate histone-like proteins (Irwin et al., 2018).

In unicellular green algae such as *Chlamydomonas*, the life cycle is haplontic (Fig. 11.1), with a sporophytic generation limited to the diploid zygote (Vigneau and Borg, 2021). In contrast, in multicellular brown algae, evolution has favored an irreversible diploid-dominant (diplontic) life cycle in most clades (Heesch et al., 2021), probably because frequent sex tends to favor a dimorphic (e.g., dioecious) diploid condition: mean fitness increases due to the masking of deleterious alleles, and selfing is less likely to occur when male and female gametes are produced by separate individuals (Heesch et al., 2021).

The red alga (Rhodophyta), *Galdierapartita* (Glaucophycaceae/order Cyanidiales), lives in acidic hot springs and is one of the few red algae known to reproduce sexually (Hirooka et al., 2022). This alga exhibits alternation between a cell-walled diploid and a motile haploid without a cell wall, dubbed a "tadpole" (Fig. 11.1). Both can proliferate asexually, but remarkably, the haploid may also self-diploidize by endoreduplication and mate to create heterozygotes by cell fusion (Hirooka et al., 2022). This life cycle is called triphasic haplo-diplontic (Vigneau and Borg, 2021).

Finally, some haptophyte algae (formerly known as golden algae), such as *Phaeocystis globosa*, which form floating harmful blooms associated with marine foams (Smith and Trimborn, 2024), exhibit alternation between meiotically derived solitary cells that proliferate rapidly during unfavorable conditions and thick colonies embedded in an elastic, transparent polysaccharide matrix as a strategy to deter predation by zooplankton (Liang et al., 2023).

The common view is that algae species mostly reproduce by asexual fission or vegetatively. Sexual reproduction involving meiosis occurs only under stressful environmental conditions such as nutrient limitation (Ji et al., 2020), including but not limited to phosphorus and nitrogen deficiency, salinity stress, parasite infection, ocean warming, and acidification (Kalinina et al., 2023; Liang et al., 2023). Moreover, in *C. reinhardtii*, diploid zygotes will initiate meiosis in the presence of blue (465 nm) or red light (635 nm), a process mediated by an animal-like cryptochrome (Zou et al., 2017). However, in such a complex group as algae, reproductive strategies may defy commonly held beliefs. For instance, active meiosis during the bloom of unrelated dinoflagellates such as *Prorocentrum shikokuense* and *Karenia mikimotoi* may serve to simultaneously promote cell proliferation and genetic diversity to boost or extend a bloom (Lin et al., 2022). This strategy is called sex for proliferation (or SXP) (Lin et al., 2022) and involves overexpression of key meiotic genes coupled with two consecutive rounds of DNA halving (Lin et al., 2022). This complex behavior illustrates the need to better understand algal meiosis as a tool to exploit genetic diversity for breeding or to predict harmful blooms (Hovde et al., 2018; Ji et al., 2020) (Fig. 11.1).

11.2 Meiosis in algae

Meiosis is often described as a type of specialized cell division that promotes genetic diversity and comprises a series of interlocked mechanisms highly conserved across eukaryotes (Thangavel et al., 2023), but with some variability across species (Loidl, 2016). These processes include: (1) sister chromatid cohesion and formation of the axial elements, (2) DNA double-stranded break (DSB) formation and processing, (3) homologous chromosome pairing, (4) synapsis (i.e., the polymerization of the central and transversal elements of the synaptonemal complex [SC] allowing the formation of a bivalent between the two parental homologous chromosomes), (5) crossover (CO) establishment and resolution, and (6) segregation of chromosomes (Thangavel et al., 2023). Segregation of chromosomes during meiosis is different from that during mitosis because, in meiosis, a single round of DNA replication is followed by two rounds of chromosome segregation (meiosis I and meiosis II), as required to produce haploid gametes from a diploid progenitor cell (Zickler and Kleckner, 2023). Additionally, the segregation of homologous chromosomes during meiosis I requires a physical connection in the form of one or a few reciprocal crossovers, each involving one chromatid of each homolog, combined with cohesion between sister chromatids that is maintained until meiosis II (Zickler and Kleckner, 2023). Thus, meiosis I segregates homologous chromosomes, whereas meiosis II segregates sister chromatids (Chen and Weir, 2024).

Meiotic chromosomes are also architecturally distinct from those in mitosis (Chen and Weir, 2024). Meiotic homologous chromosomes feature a protein axis from which loops of chromatin protrude (Chen and Weir, 2024). Throughout meiotic prophase, each homolog is organized as a cooriented sister linear loop array with a single sister-chromatid axis. Along this axis, loops are tightly packed with an evolutionarily conserved density of 20 per micron (Zickler and Kleckner, 2023). Then DSBs, which are essential for recombination, are inflicted upon these DNA loops, with the break-forming machinery simultaneously loaded onto the axis, kick-starting the process of pairing that is then followed by synapsis. Synapsis is the close alignment of homologous chromosomes to induce repair of breaks between them as the template, instead of the sister chromatid (Chen and Weir, 2024), and involves the assembly of the SC, a tripartite protein scaffold with elements running in parallel, namely a central element and two outer lateral elements with transversal filaments holding the elements together (Shah et al., 2020). Loops of sister chromatids are anchored to each lateral element. Eventually, the process of meiotic synapsis will zip down the scaffolding, with crossovers occurring at recombination nodules; nonsynapsed segments of the lateral elements are called the axial elements (Shah et al., 2020).

How meiosis evolved in eukaryotes is unclear. Some compelling hypotheses suggest that meiosis may have evolved from chromosome reduction in diploid cells, from haploid cell-to-cell fusion, or that perhaps molecular regulators of meiosis evolved from archaean DNA repair machinery deployed to tackle the oxidative stress caused by primitive mitochondrial endosymbionts (Rizos et al., 2024). Alternatively, sexual recombination might have arisen as a response to unwanted DNA replication events. In any case, meiosis offers protection against deleterious mutations and the possibility to deftly adjust to environmental alterations (Rizos et al., 2024).

Although the expression of meiotic genes is often used to determine how sexual reproduction is triggered in algae (Zhu et al., 2023), not all meiotic genes are conserved across the green and freshwater algae (the charophytes) or the dinoflagellates (Berdieva et al., 2021; Thangavel et al., 2023), so comparisons across different algal clades must be approached carefully. The current molecular approach to meiosis involves high-throughput Illumina/Solexa sequencing of the transcriptome, followed by differential gene expression and Gene Ontology analyses (Zhu et al., 2023). This method is called the *meiosis detection toolkit* and was first proposed by Schurko and Logsdon to distinguish asexual lineages in eukaryotes with six meiotic genes (Schurko and Logsdon, 2008). Then the toolkit expanded to 51 genes to study budding yeast, ciliates, and dinoflagellates (Morse, 2019), and it was finally refined by testing the presence of specific conserved domains and motifs in selected homologs (Berdieva et al., 2021) (Table 11.1.). Other parallel methods involve the use of flow cytometry to detect the halving of DNA that occurs during meiosis (Kalinina et al., 2023) and molecular analyses of meiotic tetrads to detect variation (Yu et al., 2024). A current (and simplified) list of meiotic toolkit genes reported for meiosis in eukaryotes and algae contains: (1) for double strand formation during prophase I, the core topoisomerase *SPO11*; (2) for the MRE11, RAD50, and XRS2 (MRX) complex that catalyzes *SPO11* activity, *MRE11* and *RAD50*, which may tether the broken ends of DNA; (3) for crossovers, *MSH4* and *MSH5*, which may keep homologous chromosomes together; (4) the chromosomal meiotic axis regulator *ASY1/HOP1*; (5) for strand invasion and recombination, *DMC1*, *RAD51*, *MND1*, *HOP2*, and *MEI5*, which assemble on DSBs with the help of ASY1; and (6) for resolution of double Holliday junctions as crossovers, *EXO1*, *MUS81*, *SGS1*, *MLH1*, and *MLH3* (Berdieva et al., 2021; Chen and Weir, 2024; Lin et al., 2022; Sayers et al., 2022; Thangavel et al., 2023) (Table 11.1).

Two distinct pathways regulate crossover formation and positioning in eukaryotes (Kim et al., 2024). The class I pathway mediates the formation of most crossovers and depends on the activity of the ZMM protein complex (corresponding to Zip1/4, MER3, and MSH4/5 in baker's yeast) and the MLH1/3 endonucleases (Kim et al., 2024). Class I crossovers show mutual interference and thus are wider than expected at random. In contrast, class II crossovers are noninterfering and depend on the endonuclease MUS81 (Kim et al., 2024). In plants, the formation of noninterfering crossovers is repressed by antirecombination factors such as

Table 11.1 List of algae of interest in which the presence or the expression profile of meiotic genes has been determined.

Function	Genes	Sources
Cohesion and condensation of meiotic chromosomes	*SMC1, SMC2; SMC3, SMC4; PDS5, REC8*	Hovde et al. (2018); Pearson et al. (2019); Shah et al. (2020); Zhu et al. (2023); Liang et al. (2023); Thangavel et al. (2023).
Double-strand break formation on DNA	*SPO11*	Hofstatter et al. (2020); Berdieva et al. (2021); Lin et al. (2022); Thangavel et al. (2023); Zhu et al. (2023).
Tethering of broken ends of DNA	*MRE11, RAD50, RAD52*	Hofstatter et al. (2020); Berdieva et al. (2021); Lin et al. (2022); Thangavel et al. (2023); Del Mondo et al. (2019).
Synapsis	*MSH4, MSH5*	Hofstatter et al. (2020); Berdieva et al. (2021); Lin et al. (2022); Thangavel et al. (2023); Shah et al. (2020).
Maintenance of axis structure	*ASY1/HOP1, ASY3*	Hofstatter et al. (2020); Berdieva et al. (2021); Lin et al. (2022); Thangavel et al. (2023).
Strand invasion and homologous recombination	*DMC1, RAD50, RAD51B/C/D, RAD54, MND1, HOP2, MEI5, ATM, ATR, BRCA2, XRCC2/3*	Hofstatter et al. (2020); Berdieva et al. (2021); Lin et al. (2022); Thangavel et al. (2023); Hovde et al. (2018); Pearson et al. (2019); Shah et al. (2020); Liang et al. (2023); He et al. (2021); Ota et al. (2019), Fauser et al. (2022).
Crossover formation	*EXO1, MUS81, SGS1, RECQ, MLH1, MLH3, MUS81, MER3, MSH2/3/6, HEI10*	Hofstatter et al. (2020); Berdieva et al. (2021); Lin et al. (2022); Thangavel et al. (2023); Hovde et al. (2018); Shah et al. (2020); Liang et al. (2023); He et al. (2021), Fauser et al. (2022); Thangavel et al. (2023).

(*Continued*)

Table 11.1 (Continued)

Function	Genes	Sources
Meiotic DNA replication	*MCM2/3/4/5/6/7/8/9, POLA, DNA polymerases* $\alpha/\delta/\varepsilon/\theta$, *PCNA*	Hovde et al. (2018); Pearson et al. (2019); Liang et al. (2023); He et al. (2021), Fauser et al. (2022).
Meiotic DNA methylation	*DNMT1*	Ge et al. (2024).
Chromosomal segregation	*AURKX*	Pearson et al. (2019).

This list shows the most relevant meiotic genes reported in economically important algae.

the FANCM and RECQ4A/4B helicases. Moreover, among the ZMM proteins, the ZIP3/HEI10 family of meiotic E3 ligases plays a key role in dosage-dependently controlling crossover number and position (Kim et al., 2024). These genes for meiotic sex are generally shared across all domains of life and are proposed to be already present in the hypothetical universal ancestor, so this machinery is truly common to all lineages and consistently regulates each subprocess everywhere (Hofstatter et al., 2020) (Table 11.1).

In the case of the dinoflagellate *P. globosa*, genes involved in DNA repair and meiosis that are overexpressed during explosive toxic algal blooms are the cohesins *SMC1, SMC2,* and *SMC3*; the MLH1-interacting and DNA mismatch repair factors *MSH2* and *MSH6*; the DNA polymerases $\alpha/\delta/\varepsilon$; the DNA replication and repair factor *PCNA*; the DNA replication licensing factor *MCM6*; the RAD51-associated recombination factor *XRCC3*; and the endoplasmic reticulum (ER) factor *MUG157* (Liang et al., 2023; Sayers et al., 2022). In *P. globosa*, reproduction is believed to be triggered by heat and acidification, two processes that are stressful because they may inflict DNA damage and cell death; sexual reproduction may increase genetic diversity and enhance its adaptive capacity to global warming (Liang et al., 2023) (Table 11.1).

In the red alga crop *G. lemaneiformis*, two diploid sporophytic phases exist: carposporophyte and tetrasporophyte (Zhu et al., 2023). The tetrasporophyte forms branches that contain a fertile cortical layer called the tetrasporangium, in which meiosis takes place. These tissues can be sampled at the tip to analyze transcriptional patterns with a modified meiosis detection toolkit. Results suggest that these tetrasporangia feature significant upregulation of the cohesion subunit gene *PDS5* (Zhu et al., 2023), which in *Arabidopsis thaliana* affects chromosome segregation at anaphase I and telophase I, the formation of chiasmata, and regulation of somatic homologous recombination after radiation exposure (Pradillo et al., 2015). Other genes showing significant changes in expression are meiotic genes *SPO11*, *MND1*, *DMC1*, and *MSH4*, as well as genes related to ribosomal activity, RNA transport, and protein processing at the ER, which together may facilitate cytoplasmic growth during meiosis (Table 11.1). Taken together, these findings suggest that in this valuable crop, truly sexual spores might be recovered for breeding and selection (Zhu et al., 2023) (Table 11.2.).

Nonetheless, a meiotic toolkit alone may not provide a full picture of meiotic division unless it is combined with confocal microscopy and flow cytometry, as shown in *Cladocopium latusorum* (*Symbiodinaceae*), an obligate endosymbiont of corals (Figueroa, Howe-Kerr, and Correa, 2021). Coral reefs are increasingly collapsing because of coral bleaching, a phenomenon brought about by ocean acidification and warming (Figueroa, Howe-Kerr, and Correa, 2021; Shah et al., 2020). Thus, a better understanding of meiosis in *C. latusorum* may facilitate breeding and selection for thermal resilience (Table 11.2). Confocal fluorescent microscopy and propidium iodide staining revealed that conjugation and syngamy of vegetative cells lead to the initial formation of diploid zygotes (2 N, 2 C) with very large nuclei. DNA replicates once, giving rise to a 4 C DNA content cell, which then divides into two cells (a dyad) during meiosis I. Further rounds of asynchronous division during meiosis II lead to the formation of a triad and then to four haploid tetrads (1 C) linearly arranged (Figueroa, Howe-Kerr and Correa, 2021). Chromosome segregation is also different from what is observed in flowering plants . For instance, chromosome segregation occurs by binding of chromosomes to nuclear envelope areas that surround cytoplasmic channels and to microtubule bundles. A main channel is placed at the center of the cell and can be seen by microscopy at meiosis I (Figueroa, Howe-Kerr, and Correa, 2021). However, there is no genomic evidence of factors that may code for an SC in the *Symbiodinaceae* (Shah et al., 2020). Hence, how crossovers form, if any, is unclear, but they might be processed and resolved as in *Tetrahymena thermophila* and *Saccharomyces pombe* (Loidl, 2016), two organisms that did away with the SC and only form class II noninterfering crossovers by means of the meiotic endonuclease MUS81 and the helicase SGS1 (Shah et al., 2020). This situation is

Table 11.2 List of algae of commercial or environmental interest in which research opportunities may exist regarding sexual reproduction, meiosis, and breeding.

Scientific name	Description	Use and opportunities for breeding and transformation
Cladocopium latusorum	Dinoflagellate. Endosymbiont of coral reefs. Its successful association with coral organisms may stabilize marine organisms under global warming.	Breeding and selection for enhanced thermotolerance and general tolerance to coral bleaching. Enhanced CO_2 capture Figueroa, Howe-Kerr, and Correa (2021).
Chlamydomonas reinhardtii	Green microalga, considered a living factory of lipids, proteins, and pigments. Model organism that lacks the central element of the meiotic synaptonemal complex.	Study of homologous recombination may facilitate genome editing Fauser et al. (2022). Recombination rates across the genome have been estimated Hasan and Ness (2020).
Chlorella sorokiniana	Green microalga that is extremely efficient at the accumulation of carbohydrates (60%). Considered the best feedstock and bioethanol source.	Breeding might allow for selecting desirable traits, such as accumulation of secondary metabolites and increased photosynthetic rate Hovde et al. (2018).
Galdiera partita	Unicellular red alga that tolerates extreme temperatures and low pH. Can grow in the absence of light on a large number of carbon sources, including glucose and mannitol. A good source of polyunsaturated fatty acids used in the manufacture of pharmaceutical products.	Meiosis apparently occurs at low pH. The resulting haploid has no cell wall. The study of meiosis in *Galdiera sp.* may allow the introduction and selection of novel traits in the haploid Hirooka et al. (2022).
Glacilariopsis lemaneiformis	Multicellular red alga farmed for agar production or as feedstock for fisheries.	The identification of meiotic transcriptional patterns in the upper branches where tetra-sporophytes are located suggests the possible recovery of recombinant haploid spores for breeding Zhu et al. (2023).

(*Continued*)

Table 11.2 (Continued)

Scientific name	Description	Use and opportunities for breeding and transformation
Pyropia yezoensis	Multicellular red alga. Resilient algal crop farmed for Nori that populates shallow intertidal zones. F_1 hybrids have been developed in China.	Tetrads can be recovered in this alga, and segregation of markers is possible. Genomics and conventional breeding tools may allow selection of novel traits. Nonetheless, meiosis is suspected to be highly mutagenic in this species Yu et al. (2024).
Saccharina latissima, Saccharina japonica	Multicellular brown algae. Used as sweeteners and thickeners. Right after meiosis, the mature sporophytes form zoospores that differentiate into haploid female and male gametophytes. Under laboratory conditions, the gametophytes enter vegetative reproduction and form multicellular gametophytes useful for breeding apomictic parental homozygous lines.	The study of meiosis through genomics may allow the development of genetic maps and the identification of quantitative trait loci (QTL) for breeding. Agricultural traits for selection have been identified in China. Elite lines already exist Shan and Pang (2021).
Undaria pinnatifida	Multicellular brown alga with similar life and agricultural traits to those of *S. latissima*. Meiosis allows the segregation of haploid gametophyte clonal lines and the isolation of double haploid lines.	The haploid gametophyte may be used to map sex-linked loci. In the future, immortalized F_2 populations could be created from F_1 lines. Linkage maps exist. The study of meiosis may allow identification of QTL for marker-assisted selection Shan and Pang (2021).

This table describes in detail the agricultural or industrial features of a few selected algae.

likely because the core meiotic machinery is in place with genes such as *SMC1*, *SMC3*, *RAD51*, *DMC1*, *HOP2*, *MND1*, *MLH1*, *MER3*, *MSH4/MSH5*, the MRX kinase gene *ATR*, as well as *MUS81* and *SGS1* (Shah et al., 2020; Cooper et al., 2014). In any case, direct immunostaining of crossovers may be needed to clarify the issue (Shah et al., 2020). As a note of caution, not all cultured cells of *C. latusorum* undergo meiosis when cultured; only 1.5% do, so what really triggers sexual reproduction in this environmentally important symbiont remains for investigation (Figueroa, Howe-Kerr, and Correa, 2021).

Although in algae the relationship between sexual reproduction and the creation of genetic diversity is widely accepted, a new hypothesis was recently put forth: whether meiosis could be mutagenic as well. The team of Yunxiang Mao at the Ocean University of China, in Qingdao city, analyzed the mutation rate on tetrads from the red alga *Pyropia yezoensis* (Yu et al., 2024). This is a resilient algal crop farmed for Nori (dry seaweed sheets). It populates shallow intertidal zones where it is brutally exposed to drought, UV radiation, and heat. Its life cycle is complex and alternates between multicellular gametophytes (haploid) and sporophytes (diploid) that propagate mitotically (Yu et al., 2024) (Table 11.2). Nonetheless, the switch between stages requires meiosis. Fortunately, its genome has been fully sequenced, so high-throughput molecular analyses of meiosis are possible (Yu et al., 2024).

To analyze meiosis in F_1 progenies, two lines of *P. yezoensis* were crossed: a maternal line, RZ/PYL201306-440, collected from a Nori farm in Shandong province, and a paternal laboratory (mutant) line, HT/PYL-349 (Yu et al., 2024). These progenies were also compared to wild gametophytes collected across China (Yu et al., 2024). In each case, single-nucleotide polymorphisms and indels from monospores and regenerated thalli were compared. Initially, 44 mutations were identified, but only 18 were confirmed by polymerase chain reaction. The estimated mutation rate was 2.97×10^{-8} bp per site per generation (Yu et al., 2024). The authors speculated that in this alga, homologous recombination during meiosis might induce mutations by two mechanisms: faulty resection of 3'-ends of single-stranded DNA and faulty repair of heteroduplexes at crossover sites by translesion polymerases such as polymerase θ (Sayers et al., 2022; Yu et al., 2024). The authors suggested that this high rate of mutagenesis might compensate for low genetic diversity and for a small population size; nonetheless, they also suggested not drawing any conclusions before further outdoor selection experiments are performed to evaluate fitness

(Yu et al., 2024). Genes for homologous recombination and DNA repair that are overexpressed in meiotic mother cells of this species (the diploid conchospores) are those that code for the meiotic cohesin SMC6, which regulates the resolution of chromosome linkages during the premeiotic S phase (Farmer et al., 2011); the mismatch repair proteins MSH3/6; the MRX complex factor RAD50; recombinases RAD54B and MND1; the crossover helicase RECQ; and the DNA replication factors MCM8/9) (He et al., 2021) (Table 11.1).

Galdiera is a genus of thermo-acidophilic microalgae that belongs to the order Cyanidiales and represents the only eukaryotes known to science that can handle environments with pH <3.0, high temperatures up to 56°C, and high concentrations of salts and heavy metals (López et al., 2019). Natural habitats are thermal mineral springs. These algae can grow both photo-autotrophically and chemo-heterotrophically in the absence of light in more than 50 carbon sources, including glucose, galactose, mannitol, xylose, and sucrose (López et al., 2019; Hirooka et al., 2022). The molecules harvested from *Galderia* sp. are usually high-value polyunsaturated fatty acids used in the manufacture of cosmeceuticals, nutraceuticals, fine chemicals, pharmaceutics, and biodiesel (López et al., 2019). Remarkably, the culture of *G. partita* at pH 1.0 leads to the formation of haploid cells without a cell wall, or "tadpoles." A combination of electron microscopy, flow cytometry, and transcriptional analysis suggested that these tadpoles are actually isogamous haploid gametes, also suggesting that *G. partita* is able to undergo meiosis to switch from a diploid sporophyte to haploid gametes (Hirooka et al., 2022). The related species *G. sulphuraria* possesses a copy of the meiotic recombinase *RAD52*, which may cooperate with the MRX complex (Del Mondo et al., 2019; Chen and Weir, 2024). It also shows evidence of linkage disequilibrium, which decays with pairwise distance across intrachromosomal polymorphisms and is considered a trademark of meiotic recombination (Downing et al., 2022). The gametes of *G. partita* are amenable to stable transformation via homologous recombination with an *mTP-mVenus* construct and a blasticidin S deaminase selectable marker, allowing for stepwise transformation at multiple loci. Hence, this approach may allow for increased biotechnological use of *G. partita* via the haploid (Hirooka et al., 2022) (Table 11.2).

Currently, the most promising biofuel source organism for its heterotrophic versatility and high lipid accumulation performance is the *C. sorokiana* strain UTEX 1230 (Hovde et al., 2018). This is a freshwater green alga (class Chlorophyceae, order Chlorellales) that shows high genetic diversity across strains. Hence, strategic breeding might allow for the selection of desirable traits such as the accumulation of secondary metabolites and a high photosynthetic rate (Hovde et al., 2018). Green algae form extraordinarily complex thalli and have been extensively studied cytologically. Detailed chromosome counts exist for approximately 250 taxa, even from meiocytes in which chiasmata have been estimated (Singh, Sahu, 2024). However, until recently, the genus *Chlorella* was presumed to be mostly asexual because of the lack of flagella (Ota et al., 2019). Analysis and comparison of the genome of strain UTEX 1230 indicated the presence of 39 meiosis genes, including those that code for SMC1/SMC3, SPO11, MRE11, RAD51B/C/D, DMC1, and BRCA2, which may interact with DMC1 (Li and Engebrecht, 2021); the kinase ATM, which interacts with the MRX complex (Cooper et al., 2014); crossover endonucleases MUS81 and EXO1; and a large set of mini chromosome complex proteins (MCM2/3/4/5/6/7). Therefore, *C. sorokiana* may be capable of sexual reproduction (Fig. 11.1). Flow cytometry may be needed to track changes in ploidy associated with meiosis (Hovde et al., 2018). A similar study in the *Parachlorella kessleri* strain NIES-2152 concluded that its meiotic genes may not be pseudogenes, their expression varies according to culture conditions, and their expression seems to be correlated with that of presumed flagellar genes that indicate haploidy (Ota et al., 2019). Hence, sexual reproduction may truly operate, but with a twist. The constant expression of *RAD50*, an MRX gene for mitotic DNA repair, may imply that in the absence of meiosis, the genome of green algae may continue to rearrange and evolve (Ota et al., 2019).

C. reinhardtii is a green microalga with unparalleled potential as a living factory that is highly capable of producing valuable molecules with a direct impact on human health. Such molecules include lipids, functional proteins, pigments, hormones, vaccines, and even antibodies (Masi et al., 2023). *C. reinhardtii* is a model algal organism with a completely sequenced genome (Merchant et al., 2007; Payne et al., 2023; Craig et al., 2023). Because of its high carbohydrate content (60%), it also has exciting potential as a feedstock to produce ethanol via fermentation with engineered bacterial strains (Kaloudas, Pavlova, and Penchovsky, 2021) (Table 11.2). The cytology of meiosis in *C. reinhardtii* has been described using SYBR Green stain on mature zygotes first exposed to light for 16–18 hours, then treated with a solution of 3:1 methanol and acetic acid as fixative (Aoyama, Kuroiwa, and Nakamura, 2008). The number of chromosomes at meiosis has been estimated at 17–18 bivalents, and pairing is observed at pachytene. However, the small size of chromosomes has not allowed visualization of chiasmata at diplotene or diakinesis (Aoyama, Kuroiwa, and Nakamura, 2008). The recombination landscape of *C. reinhardtii* has been analyzed by tools that measure linkage disequilibrium. Results suggest the existence of hotspot and coldspot regions across the genome (Hasan and Ness, 2020). Additionally, meiotic recombination and tetrad viability are influenced by cytosine methylation enforced by the DNA methyltransferase 1 (Table 11.1), which operates as a repressor of recombination at the sex determining region (the mating type region) on chromosome 6 (Ge et al., 2024). Nonetheless, in *C. reinhardtii*, the existence of separate mating types or sexes (*mt*+/*mt*−) is considered a positive evolutionary trait that may increase the efficiency of recombination and breeding by ensuring that matings only take place between non-identical individuals in a population (Ge et al., 2024).

The assembly of the SC seems to differ in some algae and lower plants as compared with budding yeast and higher plants (Shah et al., 2020; Thangavel et al., 2023). In the case of *C. reinhardtii*, even though at pachytene, chromosomes show two lateral elements measuring 30–40 nm in diameter, the core of the SC is not identifiable by electron microscopy (Storms and Hastings, 1977). Accordingly, the *C. reinhardtii* genome contains only the meiotic cohesin REC8 and the SC axial element factors ASY1 and ASY3 (Thangavel et al., 2023) (Table 11.1). Other proteins for the SC , such as ASY4 (Chambon et al., 2018), ZYP1, SCEP1, and SCEP2, have not been identified within the *Chlamydomonas* genome (Vrielynck et al., 2023; Thangavel et al., 2023). Although dimorphic *mt*/+ and *mt*/− strains exist in *C. reinhardtii*, meiosis occurs only after fusion of the zygote nuclei (karyogamy), with the parental chromosomes combined. Thus, unlike in flowering plants, in *C. reinhardtii*, there is no evidence of recombinational bias in one sex (e.g., heterochiasmy) (Capilla-Pérez et al., 2021). In *Arabidopsis*, the central region of the SC shows the presence of proteins ZYP1, SCEP1, and SCEP2, known to regulate key aspects of the recombinational landscape at the bivalent level (Rafiei and Ronceret, 2023; Vrielynck et al., 2023), such as obligate CO formation, CO interference, and heterochiasmy (Capilla-Pérez et al., 2021; France et al., 2021; Vrielynck et al., 2023). The role of the different elements of the SC is still poorly understood, even in yeast. The analysis of a possible bipartite SC in algae, lacking the central and transversal elements, could constitute a new approach to dissect the specific role of different SC subunits in the control of distinct phenomena related to recombination. Such phenomena include the formation of hotspots versus coldspots, the formation of obligate COs, and perhaps heterochiasmy. Nonetheless, the absence of central elements in the SC of *Chlamydomonas* (Storms and Hastings, 1977) and perhaps other green algae suggests a mode of recombination fundamentally different.

For instance, the formation of recombination nodules of different sizes (from 120 to 250 nm), suggesting CO interference, has been observed between two axial elements set apart by 110 nm (Storms and Hastings, 1977). The *C. reinhardtii* genome encodes the protein HEI10 and proteins of the ZMM complex that are required for the formation of class I COs (Thangavel et al., 2023) (Table 11.1). In *Arabidopsis*, HEI10 may control the size and spacing of recombination nodules, which in turn may enable it to regulate CO interference in the absence of elements of the SC, such as ZYP1 (Fozard, Morgan and Howard, 2023). However, a key partner of HEI10 called HEIP1 is missing in most genomes of chlorophyte and rhodophyte algae (Thangavel et al., 2023). This absence may suggest that even though in these organisms the machinery for the formation of interfering class I crossover is partially present, this pathway may not be fully functional, if at all. The *C. reinhardtii* genome contains key regulators for homologous recombination and DNA repair, such as genes for the translesion repair factor DNA polymerase θ, MRX kinases ATM and ATR, and the crossover junction endonuclease MUS81, which marks noninterfering class II COs in *Arabidopsis* (Fauser et al., 2022) (Table 11.1). However, results from nanopore sequencing of different *mt*/+ and *mt*/− mutagenized laboratory strains suggest genetic instability due to the deletion of the RECQ3 crossover helicase gene (Craig et al., 2023) (Table 11.2).

Two important crops for which breeding is a top priority are *Saccharina japonica* and *Undaria pinnatifida*, the world's most important cultivated brown algae (Shan and Pang, 2021). These two multicellular species are mostly characterized by a life cycle that features alternation of heteromorphic generations, whereby the sporophytes are diploid and macroscopic, forming a monoecious thallus. In contrast, the gametophytes are haploid and microscopic, forming a dioecious filament (Yue et al., 2024). Right after meiosis, the mature sporophytes form zoospores that differentiate into haploid female and male gametophytes. These, in turn, produce eggs and spermatozoids that eventually fuse and give rise to diploid zygotes. Finally, the zygotes enter mitosis and grow into multicellular sporophytes, a process that completes the life cycle (Table 11.2). Under laboratory conditions, the gametophytes can enter vegetative reproduction to form multicellular gametophytes useful for breeding presumed apomictic parental homozygous lines (Shan & Pang, 2021; Yue et al., 2024). Apomixis is the generation of offspring without meiosis and fertilization (Yue et al., 2024). In the case of *U. pinnatifida*, the F_1 hybrid can be directly used as a commercial cultivar with high heterosis. However, alternatively, several cultivated maternal gametophyte lines are both monoecious and apomictic, so they can be selfed to harvest double haploid clonal sporophytes that show high fecundity and thus allow for eliminating the time-consuming cultivation of haploid gametophytic filaments (Table 11.2). The potential breeding of double haploid lines in both species may facilitate the identification of genetic linkage in traits, analyses of QTL, and association studies with DNA polymorphisms, all to conduct marker-assisted selection and boost production (Shan and Pang, 2021) (Table 11.2). Regardless, two problems remain: lines may need to be selected by flow cytometry because ploidy mosaics might be formed, and new meiosis toolkits would be required to rule out partial or leaky meiosis (Yue et al., 2024). Clonal male haploid gametophytes of the related species *S. latissima* (strains 3300 and 3123) may not show obvious overexpression of key meiotic genes (Table 11.1), but indicate continuous DNA replication and repair via expression of the mitotic and meiotic recombinase *XRCC2* (Da Ines et al., 2013), the mitotic and meiotic condensin *SMC4* (Hagstrom et al., 2002), *MCM3*, the DNA polymerase alpha subunit A (*POLA1*), and the Aurora kinase (*AURKX*) (Pearson et al., 2019).

11.3 Conclusions and prospective lines of future work

Sexual reproduction and meiosis are often easy to observe in large multicellular taxa such as flowering plants and mammals, as opposed to algae, in which asexual reproduction is widespread and reproductive systems are vastly different (Krueger-Hadfield, 2024). Additionally, confusing algal jargon has been identified as a barrier that must be approached with care when studying meiotic sex in algae (Krueger-Hadfield, 2024). Nonetheless, the widespread presence of molecular meiotic machinery is evolutionarily meaningful, so efforts should focus on characterizing the regulation of meiotic processes (Hofstatter et al., 2020). One way to do this is to develop immunocytochemistry probes for crossover formation in organisms in which cytological procedures are well established (Shah et al., 2020; Wang et al., 2024) and combine this with transcriptional analysis of selected genes (Table 11.1) and flow cytometry (Lin et al., 2022). Eventually, a better understanding of the life cycle in algae might lead to high-density cultivation, industrialization, and genetic transformation (Hirooka et al., 2022). Genome editing via CRISPR/Cas9 has also been suggested for brown algae and *Chlamydomonas* (Coelho and Cock, 2020; Fauser et al., 2022), opening the way for precision breeding via homologous recombination. Thus far, successful transformation *via* electroporation has been reported in *C. vulgaris* (strain ARS-A1) and in red alga *P. yezoensis* (strain TU-1) (Robben et al., 2026; Izumi et al., 2025). Therefore, very soon, algal biomass could play a role in resolving the conflict between food production and biofuel production (Mahmood et al., 2023), breeding for thermotolerance in *Symbiodinaceae* may stabilize coral reef growth and CO_2 fixation (Liu et al., 2018), and monitoring for meiosis may allow early forecasting of toxic blooms (Lin et al., 2022), simultaneously allowing for farmers' well-being, adaptation to climate change, and industrial innovation (Krueger-Hadfield, 2024; Mahmood et al., 2023) (Fig. 11.1). Further functional genome characterization will be required. The foreseeable rewards in terms of food security and pharmaceutical product development may be immense.

Conflict of interest statement

The authors declare that they have no conflicts of interest.

Acknowledgments

Bolaños-Villegas thanks the Vicerrectoría de Investigación of the University of Costa Rica for funding projects C3022, C3020, C2060, and C0244. Ronceret thanks the Mexican National Council on Humanities, Sciences, and Technologies (CONAHCYT) for awarding the supporting grant for basic science (no. A1-S-8496). Neither the University of Costa Rica nor the CONAHCYT of Mexico had any direct involvement in the writing of this manuscript and in the decision to submit the chapter for publication. Special thanks go to Ms. Laura Smales for carefully editing this manuscript.

Declaration of competing interests

The authors declare that they do not have any financial or personal relationships with other people or organizations that could inappropriately influence this work, such as employment, consultancies, stock ownership, patent applications, or grants.

Declaration of generative AI in scientific writing

The authors declare that AI tools were not used during the writing process and that the authorship and all responsibility are entirely human.

References

Aoyama, H., Kuroiwa, T., Nakamura, S., 2008. Observations of chromosomal behaviour in living meiotic zygotes of *Chlamydomonas reinhardtii* (Chlorophyceae). European Journal of Phycology 43 (4), 389–394. Available at https://doi.org/10.1080/09670260701853566.

Berdieva, M.A., Pozdnyakov, I.A., Kalinina, V.O., Skarlato, S.O., 2021. Putative meiotic toolkit in the dinoflagellate *Prorocentrum cordatum*: additional evidence for sexual process from transcriptome. Journal of Eukaryotic Microbiology 68 (3). Available at https://doi.org/10.1111/jeu.12845, https://doi.org/10.1111/jeu.12845.

Bourdareau, S., Mignerot, L., Heesch, S., Peters, A.F., Coelho, S.M., Cock, J.M., 2021. Alternation of generations in plants and algae. In: Nuño de la Rosa, L., Müller, G.B. (Eds.), Evolutionary Developmental Biology. Springer International Publishing, Cham, Switzerland, pp. 631–644. Available at https://doi.org/10.1007/978-3-319-32979-6_58.

Capilla-Pérez, L., Durand, S., Hurel, A., Lian, Q., Chambon, A., Taochy, C., et al., 2021. The synaptonemal complex imposes crossover interference and heterochiasmy in Arabidopsis. Proceedings of the National Academy of Sciences 118 (12). Available at https://doi.org/10.1073/pnas.2023613118.

Chambon, A., West, A., Vezon, D., Horlow, C., De Muyt, A., Chelysheva, L., et al., 2018. Identification of ASYNAPTIC4, a component of the meiotic chromosome axis. Plant Physiology 178 (1), 233–246. Available at https://doi.org/10.1104/pp.17.01725, http://www.plantphysiol.org/content/plantphysiol/178/1/233.full.pdf.

Chen, L., Weir, J.R., 2024. The molecular machinery of meiotic recombination. Biochemical Society Transactions 52 (1), 379–393. Available at https://doi.org/10.1042/BST20230712, https://portlandpress.com/biochemsoctrans/issue.

Coelho, S.M., Cock, J.M., 2020. Brown algal model organisms. Annual Review of Genetics 54, 71–92. Available at https://doi.org/10.1146/annurev-genet-030620-093031, http://www.annualreviews.org/journal/genet.

Coelho, S.M., Umen, J., 2021. Switching it up: algal insights into sexual transitions. Plant Reproduction 34 (4), 287–296. Available at https://doi.org/10.1007/s00497-021-00417-0, http://www.springer.com/life+sciences/cell+biology/journal/497.

Cooper, T.J., Wardell, K., Garcia, V., Neale, M.J., 2014. Homeostatic regulation of meiotic DSB formation by ATM/ATR. Experimental Cell Research 329 (1), 124–131. Available at https://doi.org/10.1016/j.yexcr.2014.07.016, http://www.journals.elsevier.com/experimental-cell-research/.

Craig, R.J., Gallaher, S.D., Shu, S., Salomé, P.A., Jenkins, J.W., Blaby-Haas, C.E., et al., 2023. The Chlamydomonas Genome Project, version 6: reference assemblies for mating-type plus and minus strains reveal extensive structural mutation in the laboratory. The Plant Cell 35 (2), 644–672. Available at https://doi.org/10.1093/plcell/koac347, https://academic.oup.com/plcell/issue.

Downing, J.M., Lock, S.C.L., Iovinella, M., Davey, J., Mackinder, L.C.M., Chong, J.P.J., et al., 2022. Comparisons between complete genomes of the eukaryotic extremophile *Galdieria sulphuraria* reveals complex nuclear chromosomal structures. bioRxiv. United Kingdom bioRxiv. Available at https://doi.org/10.1101/2022.10.04.510839, https://www.biorxiv.org.

Farmer, S., San-Segundo, P.A., Aragón, L., Lichten, M., 2011. The Smc5–Smc6 complex is required to remove chromosome junctions in meiosis. PLoS One 6 (6), e20948. Available at https://doi.org/10.1371/journal.pone.0020948.

Fauser, F., Vilarrasa-Blasi, J., Onishi, M., Ramundo, S., Patena, W., Millican, M., et al., 2022. Systematic characterization of gene function in the photosynthetic alga *Chlamydomonas reinhardtii*. Nature Genetics 54 (5), 705–714. Available at https://doi.org/10.1038/s41588-022-01052-9, http://www.nature.com/ng/index.html.

Figueroa, R.I., Howe-Kerr, L.I., Correa, A.M.S., 2021. Direct evidence of sex and a hypothesis about meiosis in Symbiodiniaceae. Scientific Reports 11 (1). Available at https://doi.org/10.1038/s41598-021-98148-9.

Fozard, J.A., Morgan, C., Howard, M., 2023. Coarsening dynamics can explain meiotic crossover patterning in both the presence and absence of the synaptonemal complex. eLife 12. Available at https://doi.org/10.7554/eLife.79408.

France, M.G., Enderle, J., Röhrig, S., Puchta, H., Franklin, F.C.H., Higgins, J.D., 2021. ZYP1 is required for obligate cross-over formation and cross-over interference in Arabidopsis. Proceedings of the National Academy of Sciences 118 (14). Available at https://doi.org/10.1073/pnas.2021671118.

Gavelis, G.S., Herranz, M., Wakeman, K.C., Ripken, C., Mitarai, S., Gile, G.H., et al., 2019. Dinoflagellate nucleus contains an extensive endomembrane network, the nuclear net. Scientific Reports 9 (1). Available at https://doi.org/10.1038/s41598-018-37065-w, www.nature.com/srep/index.html.

Ge, T., Gui, X., Xu, J.X., Xia, W., Wang, C.H., Yang, W., et al., 2024. DNA cytosine methylation suppresses meiotic recombination at the sex-determining region. Science Advances 10 (41). Available at https://doi.org/10.1126/sciadv.adr2345, https://www.science.org/doi/epdf/10.1126/sciadv.adr2345.

Hagstrom, K.A., Holmes, V.F., Cozzarelli, N.R., Meyer, B.J., 2002. *C. elegans* condensin promotes mitotic chromosome architecture, centromere organization, and sister chromatid segregation during mitosis and meiosis. Genes and Development 16 (6), 729–742. Available at https://doi.org/10.1101/gad.968302.

Hasan, A.R., Ness, R.W., 2020. Recombination rate variation and infrequent sex influence genetic diversity in *Chlamydomonas reinhardtii*. Genome Biology and Evolution 12 (4), 370–380. Available at https://doi.org/10.1093/GBE/EVAA057, http://gbe.oxfordjournals.org/.

Heesch, S., Serrano-Serrano, M., Barrera-Redondo, J., Luthringer, R., Peters, A.F., Destombe, C., et al., 2021. Evolution of life cycles and reproductive traits: insights from the brown algae. Journal of Evolutionary Biology 34 (7), 992–1009. Available at https://doi.org/10.1111/jeb.13880, http://onlinelibrary.wiley.com/journal/10.1111/(ISSN)1420-9101.

He, B., Gu, W., Wang, L., Zheng, Z., Shao, Z., Huan, L., et al., 2021. RNA-seq between asexual archeospores and meiosis-related conchospores in *Neopyropia yezoensis* using Smart-seq2. Journal of Phycology 57 (5), 1648–1658. Available at https://doi.org/10.1111/jpy.13197.

Hirooka, S., Itabashi, T., Ichinose, T.M., Onuma, R., Fujiwara, T., Yamashita, S., et al., 2022. Life cycle and functional genomics of the unicellular red alga Galdieria for elucidating algal and plant evolution and industrial use. Proceedings of the National Academy of Sciences 119 (41). Available at https://doi.org/10.1073/pnas.2210665119.

Hofstatter, P.G., Ribeiro, G.M., Porfirio-Sousa, A.L., Lahr, D.J.G., 2020. The sexual ancestor of all eukaryotes: a defense of the "meiosis toolkit": a rigorous survey supports the obligate link between meiosis machinery and sexual recombination. BioEssays 42 (9). Available at https://doi.org/10.1002/bies.202000037, http://onlinelibrary.wiley.com/journal/10.1002/(ISSN)1521-1878.

Hovde, B.T., Hanschen, E.R., Steadman Tyler, C.R., Lo, C.C., Kunde, Y., Davenport, K., et al., 2018. Genomic characterization reveals significant divergence within *Chlorella sorokiniana* (Chlorellales, Trebouxiophyceae). Algal Research. 35, 449–461. Available at https://doi.org/10.1016/j.algal.2018.09.012, http://www.sciencedirect.com/science/journal/aip/22119264.

Irwin, N.A. T., Martin, B.J. E., Young, B.P., Browne, M.J. G., Flaus, A., Loewen, C.J. R., Keeling, P.J., Howe, L.J., 2018. Viral proteins as a potential driver of histone depletion in dinoflagellates. Nature Communications 9 (1). Available at 10.1038/s41467-018-03993-4.

Da Ines, O., Degroote, F., Amiard, S., Goubely, C., Gallego, M.E., White, C.I., 2013. Effects of XRCC2 and RAD51B mutations on somatic and meiotic recombination in *Arabidopsis thaliana*. Plant Journal 74 (6), 959–970. Available at https://doi.org/10.1111/tpj.12182.

Izumi, H., Uji, T., Matsumoto, K., Nagaosa, K., Fukuda, S., 2025. Electroporation-mediated gene delivery in *Pyropia yezoensis* (Rhodophyta) without cell wall removal. Marine Biotechnology 27 (5). Available at https://doi.org/10.1007/s10126-025-10506-0, https://doi.org/10.1007/s10126-025-10506-0.

Ji, N., Huang, J., Zhang, Z., Zhou, L., Shen, X., Lin, S., 2020. Identification and expression analysis of meiosis-related genes in the harmful alga *Heterosigma akashiwo* (Raphidophyceae). Harmful Algae 92, 101736. Available at https://doi.org/10.1016/j.hal.2019.101736.

Kalinina, V., Berdieva, M., Aksenov, N., Skarlato, S., 2023. Phosphorus deficiency induces sexual reproduction in the dinoflagellate *Prorocentrum cordatum*. Scientific Reports 13 (1). Available at https://doi.org/10.1038/s41598-023-41339-3.

Kaloudas, D., Pavlova, N., Penchovsky, R., 2021. Lignocellulose, algal biomass, biofuels and biohydrogen: a review. Environmental Chemistry Letters 19 (4), 2809–2824. Available at https://doi.org/10.1007/s10311-021-01213-y.

Kim, H., Kim, J., Son, N., Kuo, P., Morgan, C., Chambon, A., et al., 2024. Control of meiotic crossover interference by a proteolytic chaperone network. Nature Plants 10 (3), 453–468. Available at https://doi.org/10.1038/s41477-024-01633-y.

Krueger-Hadfield, S.A., 2024. Let's talk about sex: why reproductive systems matter for understanding algae. Journal of Phycology 60 (3), 581–597. Available at https://doi.org/10.1111/jpy.13462, http://onlinelibrary.wiley.com/journal/10.1111/(ISSN)1529-8817.

Liang, D., Xiang, H., Jin, P., Xia, J., 2023. Response mechanism of harmful algae *Phaeocystis globosa* to ocean warming and acidification. Environmental Pollution 320, 121008. Available at https://doi.org/10.1016/j.envpol.2023.121008.

Lin, S., Yu, L., Wu, X., Li, M., Zhang, Y., Luo, H., et al., 2022. Active meiosis during dinoflagellate blooms: A 'sex for proliferation' hypothesis. Harmful Algae 118, 102307. Available at https://doi.org/10.1016/j.hal.2022.102307.

Liu, H., Stephens, T.G., González-Pech, R.A., Beltran, V.H., Lapeyre, B., Bongaerts, P., et al., 2018. Symbiodinium genomes reveal adaptive evolution of functions related to coral-dinoflagellate symbiosis. Communications Biology 1, 95. Available at https://doi.org/10.1038/s42003-018-0098-3.

Li, Q., Engebrecht, J.A., 2021. BRCA1 and BRCA2 tumor suppressor function in meiosis. Frontiers in Cell and Developmental Biology 9. Available at https://doi.org/10.3389/fcell.2021.668309.

Loidl, J., 2016. Conservation and variability of meiosis across the Eukaryotes. Annual Review of Genetics 50 (1), 293–316. Available at https://doi.org/10.1146/annurev-genet-120215-035100.

López, G., Yate, C., Ramos, F.A., Cala, M.P., Restrepo, S., Baena, S., 2019. Production of polyunsaturated fatty acids and lipids from autotrophic, mixotrophic and heterotrophic cultivation of Galdieria sp. strain USBA-GBX-832. Scientific Reports 9 (1). Available at https://doi.org/10.1038/s41598-019-46645-3, http://www.nature.com/srep/index.html.

Mahmood, T., Hussain, N., Shahbaz, A., Mulla, S.I., Iqbal, H.M.N., Bilal, M., 2023. Sustainable production of biofuels from the algae-derived biomass. Bioprocess and Biosystems Engineering 46 (8), 1077–1097. Available at https://doi.org/10.1007/s00449-022-02796-8, https://www.springer.com/journal/449.

Masi, A., Leonelli, F., Scognamiglio, V., Gasperuzzo, G., Antonacci, A., Terzidis, M.A., 2023. *Chlamydomonas reinhardtii*: a factory of nutraceutical and food supplements for human health. Molecules 28 (3). Available at https://doi.org/10.3390/molecules28031185, http://www.mdpi.com/journal/molecules.

Merchant, S.S., Prochnik, S.E., Vallon, O., Harris, E.H., Karpowicz, S.J., Witman, G.B., et al., 2007. The *Chlamydomonas* genome reveals the evolution of key animal and plant functions. Science 318 (5848), 245–251. Available at https://doi.org/10.1126/science.1143609.

Del Mondo, A., Iovinella, M., Petriccione, M., Nunziata, A., Davis, S.J., Cioppa, D., et al., 2019. A Spotlight on Rad52 in Cyanidiophytina (Rhodophyta): a relic in algal heritage. Plants 8 (2), 46. Available at https://doi.org/10.3390/plants8020046.

Morse, D., 2019. A transcriptome-based perspective of meiosis in dinoflagellates. Protist 170 (4), 397–403. Available at https://doi.org/10.1016/j.protis.2019.06.003.

Ota, S., Oshima, K., Yamazaki, T., Takeshita, T., Bišová, K., Zachleder, V., et al., 2019. The *Parachlorella* genome and transcriptome endorse active RWP-RK, meiosis and flagellar genes in Trebouxiophycean algae. Cytologia 84 (4), 323–330. Available at https://doi.org/10.1508/cytologia.84.323.

Payne, Z.L., Penny, G.M., Turner, T.N., Dutcher, S.K., 2023. A gap-free genome assembly of *Chlamydomonas reinhardtii* and detection of translocations induced by CRISPR-mediated mutagenesis. Plant Communications 4 (2). Available at https://doi.org/10.1016/j.xplc.2022.100493, https://doi.org/10.1016/j.xplc.2022.100493.

Pearson, G.A., Martins, N., Madeira, P., Serrão, E.A., Bartsch, I., 2019. Sex-dependent and -independent transcriptional changes during haploid phase gametogenesis in the sugar kelp *Saccharina latissima*. PLoS One 14 (9). Available at https://doi.org/10.1371/journal.pone.0219723, https://journals.plos.org/plosone/article/file?id=10.1371/journal.pone.0219723&type=printable.

Pradillo, M., Knoll, A., Oliver, C., Varas, J., Corredor, E., Puchta, H., et al., 2015. Involvement of the cohesin cofactor PDS5 (SPO76) during meiosis and DNA repair in *Arabidopsis thaliana*. Frontiers in Plant Science 6. Available at https://doi.org/10.3389/fpls.2015.01034, http://journal.frontiersin.org/article/10.3389/fpls.2015.01034/pdf.

Rafiei, N., Ronceret, A., 2023. Crossover interference mechanism: new lessons from plants. Frontiers in Cell and Developmental Biology 11. Available at https://doi.org/10.3389/fcell.2023.1156766.

Rizos, I., Frada, M.J., Bittner, L., Not, F., 2024. Life cycle strategies in free-living unicellular eukaryotes: diversity, evolution, and current molecular tools to unravel the private life of microorganisms. Journal of Eukaryotic Microbiology 71 (6). Available at https://doi.org/10.1111/jeu.13052, http://onlinelibrary.wiley.com/journal/10.1111/(ISSN)1550-7408.

Robben, D.M., Amin, Z., Budiman, C., Kumar, V.S., 2026. A step-by-step approach to establishing an efficient genetic transformation protocol for *Chlorella vulgaris* using electroporation. Molecular Biology Reports 53 (1). Available at https://doi.org/10.1007/s11033-025-11167-x, https://doi.org/10.1007/s11033-025-11167-x.

Saravanan, A., Deivayanai, V.C., Senthil Kumar, P., Rangasamy, G., Varjani, S., 2022. CO2 bio-mitigation using genetically modified algae and biofuel production towards a carbon net-zero society. Bioresource Technology 363, 127982. Available at https://doi.org/10.1016/j.biortech.2022.127982.

Sayers, E.W., Bolton, E.E., Brister, J.R., Canese, K., Chan, J., Comeau, D.C., et al., 2022. Database resources of the national center for biotechnology information. Nucleic Acids Research 50 (1), D20. Available at https://doi.org/10.1093/nar/gkab1112, https://academic.oup.com/nar/issue.

Schurko, A.M., Logsdon, J.M., 2008. Using a meiosis detection toolkit to investigate ancient asexual "scandals" and the evolution of sex. BioEssays 30 (6), 579–589. Available at https://doi.org/10.1002/bies.20764.

Shah, S., Chen, Y., Bhattacharya, D., Chan, C.X., 2020. Sex in Symbiodiniaceae dinoflagellates: genomic evidence for independent loss of the canonical synaptonemal complex. Scientific Reports 10 (1). Available at https://doi.org/10.1038/s41598-020-66429-4, www.nature.com/srep/index.html.

Shan, T., Pang, S., 2021. Breeding in the economically important brown alga *Undaria pinnatifida*: a concise review and future prospects. Frontiers in Genetics 12. Available at https://doi.org/10.3389/fgene.2021.801937.

Singh, I.P., Sahu, N., 2024. Nuclear characteristics and chromosome of green algae: a review. In: Sahu, N., Sridhar, S. (Eds.), Algal Biotechnology Current Trends, Challenges and Future Prospects for a Sustainable Environment. CRC Press, Boca Raton, pp. 196–211. Available at https://doi.org/10.1201/9781003219156 (2024).

Smith, W.O., Trimborn, S., 2024. Phaeocystis: a global enigma. Annual Review of Marine Science 16, 417–441. Available at https://doi.org/10.1146/annurev-marine-022223-025031, http://www.annualreviews.org/journal/marine.

Storms, R., Hastings, P.J., 1977. A fine structure analysis of meiotic pairing in *Chlamydomonas reinhardtii*. Experimental Cell Research 104 (1), 39–46. Available at https://doi.org/10.1016/0014-4827(77)90066-0.

Thangavel, G., Hofstatter, P.G., Mercier, R., Marques, A., 2023. Tracing the evolution of the plant meiotic molecular machinery. Plant Reproduction 36 (1), 73–95. Available at https://doi.org/10.1007/s00497-022-00456-1, https://www.springer.com/journal/497.

Vigneau, J., Borg, M., 2021. The epigenetic origin of life history transitions in plants and algae. Plant Reproduction 34 (4), 267–285. Available at https://doi.org/10.1007/s00497-021-00422-3.

Vrielynck, N., Peuch, M., Durand, S., Lian, Q., Chambon, A., Hurel, A., et al., 2023. SCEP1 and SCEP2 are two new components of the synaptonemal complex central element. Nature Plants 9 (12), 2016–2030. Available at https://doi.org/10.1038/s41477-023-01558-y.

Wang, B., Jia, Y., Dang, N., Yu, J., Bush, S.J., Gao, S., et al., 2024. Near telomere-to-telomere genome assemblies of two Chlorella species unveil the composition and evolution of centromeres in green algae. BMC Genomics 25 (1). Available at https://doi.org/10.1186/s12864-024-10280-8, https://bmcgenomics.biomedcentral.com/.

Wisecaver, J.H., Hackett, J.D., 2011. Dinoflagellate genome evolution. Annual Review of Microbiology 65, 369–387. Available at https://doi.org/10.1146/annurev-micro-090110-102841.

Yue, S., Xue, N., Yi, C., Sun, J., Li, X., Chen, S., et al., 2024. Apomixis in *Saccharina japonica*: parthenogenesis in male and apogamy in female gametophytes. Aquaculture 591. Available at https://doi.org/10.1016/j.aquaculture.2024.741142, https://doi.org/10.1016/j.aquaculture.2024.741142.

Yu, X., Tang, X., Xu, K., Wang, L., Mao, Y., 2024. Meiosis I causes a high spontaneous mutation rate in a multicellular red alga (*Pyropia yezoensis*) with a complex life cycle. Algal Research 83, 103694. Available at https://doi.org/10.1016/j.algal.2024.103694.

Zhu, W., Yang, Z., Xu, D., Yang, J., Zhang, Y., Pan, X., 2023. Transcriptome analysis during tetrasporogenesis of *Gracilariopsis lemaneiformis* and preliminary study of the expressions of its meiotic genes. Journal of Ocean University of China 22 (2), 541–554. Available at https://doi.org/10.1007/s11802-023-5289-y.

Zickler, D., Kleckner, N., 2023. Meiosis: dances between homologs. Annual Review of Genetics 57 (1), 1–63. Available at https://doi.org/10.1146/annurev-genet-061323-044915.

Zou, Y., Wenzel, S., Müller, N., Prager, K., Jung, E.M., Kothe, E., et al., 2017. An animal-like cryptochrome1 controls the *Chlamydomonas* sexual cycle. Plant Physiology 174 (3), 1334–1347. Available at https://doi.org/10.1104/pp.17.00493, http://www.plantphysiol.org/content/plantphysiol/174/3/1334.full.pdf.

CHAPTER 12

Low carbon initiative: industrial flue gas utilization as source of CO_2 for microalgal biomass production

Joko Prayitno and Khairul Anam
Research Center for Applied Microbiology, National Research and Innovation Agency (BRIN), Cibinong, Indonesia

12.1 Introduction: industrial flue gas utilization

Rapid growth of the global economy and population stimulates the expansion of new plants from a wide range of industries as well as the emergence of new industries to accommodate human needs. As a result, CO_2 gas emissions from industrial flue gas are steadily increasing, particularly since 2000 (Bashmakov et al., 2022). Industrial flue gas derived from burning of fossil fuels, such as coal, oil, and natural gas, makes a significant contribution to global CO_2 gas emissions and thus climate change. According to IPCC reports (Bashmakov et al., 2022), direct emissions from industries accounted for 24% of global CO_2 emissions in 2019 (or 14.1 $GtCO_2$-eq), of which half (7.1 $GtCO_2$-eq) is from combustion alone. Global actions to mitigate CO_2 emissions from industries were adopted by many countries since the ratification of the Kyoto Protocol in 1997. Later, a low carbon initiative has been introduced by the IPCC in 2007 (Solomon et al., 2007) and adopted by many countries to reduce CO_2 emissions from the industrial sector (Song et al., 2022).

The low carbon initiative is a global effort to address the issues of climate change by reducing greenhouse gas emissions through the use of less fossil fuel and more sustainable energy production, consumption, and general practices. The initiative is a further step to combat climate change since the ratification of the Kyoto Protocol in 1997 and prior to the Paris Agreement in 2015. The Kyoto Protocol established binding emissions reduction targets for developed countries (UNFCCC, 1998), while the Paris Agreement set the goal of limiting global warming well below 2°C above preindustrial levels and pursuing efforts to limit it to 1.5°C

Applied Phycology
DOI: https://doi.org/10.1016/B978-0-443-24758-3.00002-2

(UNFCCC, 2015). The low carbon initiatives are interconnected with circular economy programs that utilize and recycle waste and emissions for other production chains. In this context, actions to reduce CO_2 emissions from industrial flue gas through carbon capture and utilization have become important mitigation options to address climate change.

Flue gas plays a significant role in low carbon initiatives as it is a primary source of carbon emissions. Flue gas is the gas that is emitted from combustion processes, such as those in power plants, industrial facilities, and residential heating systems (Speight, 2018). It is produced when fossil fuels or biomass are burned, and it consists of a mixture of gases, including carbon dioxide (CO_2), nitrogen oxides (NOx), sulfur dioxide (SO_2), carbon monoxide (CO), and traces of other pollutants (Van Den Hende et al., 2012). In recent years, there has been growing interest in using flue gas as a resource rather than simply releasing it into the atmosphere. Flue gas utilization offers the potential to reduce greenhouse gas emissions and create economic opportunities. Capturing and utilizing CO_2 from flue gas helps mitigate climate change, reduce the environmental impact of combustion processes, and contribute to developing a more sustainable and low-carbon economy.

Microalgae naturally use atmospheric CO_2 as the carbon source for their growth. In microalgae cultivation systems, an additional supply of CO_2 is often required and incorporated to improve biomass production and valuable products. During the past two decades, there have been increasing efforts to utilize CO_2 from industrial flue gas for microalgae growth due to its advantages as an environmentally friendly method for mitigation and low cost for biomass production (Cheah et al., 2015; Ighalo et al., 2022; Van Den Hende et al., 2012). The degree of CO_2 mitigation in the microalgal cultivation system ranged between 10% and 50% (Doucha et al., 2005), with the conversion value of 1.83 kg CO_2 to produce 1 kg of microalgae biomass (Li et al., 2020). Accordingly, the utilization of CO_2 from industrial flue gas for microalgae biomass production has emerged as one of low carbon initiatives that need further consideration for development and application (Hunt et al., 2010; Cheah et al., 2015).

In the perspective of microalgae production, industrial flue gas can be utilized as a substitute for more costly pure CO_2 as a carbon source for microalgae growth. In this scenario, CO_2 released from the flue gas is transmitted and introduced into bioreactors or ponds to induce microalgal growth. Although there has been growing attention on the use of flue gas to reduce CO_2 in the atmosphere as part of a low carbon initiative in industrial processes, its application for microalgae biomass production is still limited. Therefore, consistent development of its utilization and application is required to generate benefits for society. In this chapter, the prospects of flue gas utilization for microalgal biomass production and strategies to improve its utilization for maximum growth are discussed. First, the key components of industrial flue gas and their characteristics that affect the growth of microalgae, as well as the requirements and tolerance of microalgae to various types of gases in the flue gas, are examined. Second, the technology employed for the utilization of industrial flue gas for microalgal production is reviewed, and in the last section, future challenges of its utilization are discussed.

12.2 Conditions for flue gas utilization in microalgae production

Flue gas generated from burning of fossil fuels typically has a high temperature and contains various types of gases and particulate matter. The main constituents of flue gas are CO_2, N_2, water vapor, and O_2 (Speight, 2018). CO_2 from the flue gas becomes the major concern in mitigation initiatives because of its significant contribution to greenhouse gas emissions. CO_2 is not classified as a pollutant or toxic substance, although its high concentration in the air can cause disorders in humans, such as dizziness, nausea, and unconsciousness. Depending on the fuel type and combustion process, flue gas typically contains 5%–15% CO_2 but can be as high as 28% in the steel industry (Kao et al., 2014; Nagappan et al., 2020). Other constituents of the flue gas are in minor amounts and depend on the fuel feedstock, including CO, NOx, SOx, halogen acids, and particulate matter such as unburned carbon and heavy metals (Larki et al., 2023). Many of these gases and particulate matter at specific concentrations are known to have negative effects when exposed to organisms, including microalgae. Therefore, it is common practice to consider some factors before incorporating flue gas into microalgal cultivation system.

Nevertheless, many studies have shown that microalgae are able to utilize CO_2 from flue gas and incorporate it into their metabolic pathways. Based on a wide range of studies, it is known that the success of flue gas utilization for microalgae growth depends on the composition of the flue gas, particularly components such as CO_2, NOx, SOx, and particulates, and the tolerance of microalgae to the flue gas being applied. Flue gas compositions can vary depending on the type of fuel burned and the combustion process in the plants. Coal is the most widely used fuel for combustion in the stoke, as it is inexpensive, abundant, widely available across the globe, and the most secure energy source compared to other types of fuels. Currently, coal is mainly used in power generation industries and steel production. The flue gas from coal-fire combustion generally contains around 10%–28% CO_2. It also contains approximately 120–200 ppm SO_2 and 150–200 ppm NO_x, but may have SO_2 and NO_x content as high as 2000 and 1000 ppm, respectively (Table 12.1), that may contribute to the phenomenon of acid rain. Burning of the coal also generates 1–10 mg/m^3 of dust (Larki et al., 2023).

The key components of the flue gas critical to the growth of microalgae are discussed as follows.

12.2.1 Temperature

Typical flue gas temperatures released from the stacks are 120°C–180°C (Table 12.1), which are far above the tolerance limit of most microorganisms to survive, including microalgae. Although some thermophilic microalgae can tolerate high temperatures of 40°C during their growth (Kumar et al., 2011; Varshney et al., 2015), most microalgal species typically prefer mesothermal conditions at around 20°C–25°C (Ma et al., 2019) and cannot tolerate temperatures higher than 30°C. Temperature has a positive effect on CO_2 solubility in the liquid medium (Aslam and Tahira, 2016; Ahmad and Gersen, 2014), which may also have a positive effect on the intake of CO_2 by microalgal cells. It has been demonstrated that CO_2 solubility in water may increase by 50 ppm with an increase in temperature of 1°C within temperature range of 5°C–40°C (Ahmad and Gersen, 2014). For flue gas utilization, temperature adjustment to mesothermal conditions is considered to be more critical for biomass production than maximizing CO_2 solubility in the growth medium.

12.2.2 Carbon dioxide

CO_2 is the main target of flue gas utilization for the formation of microalgal biomass. An increase in biomass production is obtained when higher CO_2 concentrations are applied to microalgae cultivation system up to certain levels. Most microalgal species can tolerate CO_2 concentrations of up to 30%, although it was reported that some species were able to grow at

Table 12.1 Typical flue gas components from coal and waste incinerator plants.

Source	Temperature (°C)	H_2O (%)	CO_2 (%)	O_2 (%)	SO_2 (ppm)	NO_x (ppm)	dust (mg/m^3)	Reference
Coal	160–180	20–23	10–11	4–5	120–200	150–250	NA[a]	Aouini et al. (2014)
Coal-fired boiler	135–180	5–12	10–16	4–6	100–2000	100–1000[b]	NA	Nagappan et al. (2020)
Cement kiln	150–230	5–35	14–33	2–4	10–2500	475–1900[c]	NA	Nagappan et al. (2020)
Coal-fired coke oven	160–180	NA	23–27	6–8	80–90	70–80	1–8	Kao et al. (2014)
Coal-fired hot stove	120–160	NA	24–28	2–3	15–20	8–10	1–5	Kao et al. (2014)
Coal-fired power generator	150–190	NA	22–26	3–6	15–20	25–30	1–10	Kao et al. (2014)
Waste incinerator	180–230	10–18	5–14	6–15	100–300	100–1000[d]	NA	Nagappan et al. (2020)
Municipal waste incinerator	NA	NA	NA	NA	84.7[e]	194.5[f]	2×10^{-3}	Wen et al. (2018)

[a] NA, data not available
[b] as NO.
[c] as NO.
[d] as NO.
[e] from the value conversion of mg/Nm^3.
[f] from the value conversion of mg/Nm^3.

concentrations of 80% (Kumar et al., 2011). High CO_2 concentrations can inhibit microalgal growth due to the direct induction of environmental stress by CO_2 and indirect effect of lowering the pH of the growth medium (Hosseini et al., 2018). When injected into the growth medium, part of CO_2 is dissolved in the form of an equilibrium between bicarbonate (HCO_3^-) and carbonate (CO_3^{2-}) ions, while the other part remains in the gas phase. Microalgae can directly utilize CO_2 in solution, which is the most available in the medium, although they preferentially uptake bicarbonate over CO_2 gas (Ighalo et al., 2022). Depending on the microalgal species, CO_2 concentrations in flue gas that induce optimum microalgal growth are in the range of 5%–10%. In fact, maximum growth of *Chlorella* is achieved when the concentration of CO_2 is 5% (Yadav et al., 2019). However, continuous supply of CO_2 from flue gas may reduce pH of the medium, which subsequently may affect microalgal growth.

12.2.3 Sulfur dioxide and nitrogen oxides

The most common type of fuel for combustion in the stoke is coal, since it is low-cost and abundant compared to other types of fuel. Flue gas from coal-fired power plants typically contains higher levels of pollutants like SO_2, NOx, and particulate matter than natural gas combustion (Speight, 2018; Table 12.1). For that reason, these two gases received particular attention when decision was made to use the flue gas from coal burning for microalgal culture. SO_2 and NOx in the flue gas may cause acidification of the growth medium and may have a direct toxic effect on microalgal cells (Van Den Hende et al., 2012), and thus may limit the growth of microalgae. The effect of pH change on microalgae growth due to SO_2 is stronger than the minor effect of pH change due to CO_2. SO_2 may also interfere with the CO_2-HCO equilibrium in the algal growth medium that lowers the pH (Ronda et al., 2014). For example, the growth rate of *Chlorella* sp. was reduced to 25% (or to 1.24 g/L/day) when supplied with flue gas containing 60 ppm SO_2, and increase in SO_2 concentration delayed lag phase of microalgae culture (Lee et al., 2002). Moreover, a study reported that 50 ppm of SO_2 in the flue gas caused complete growth inhibition of *Chlorella* sp. (Yanagi et al., 1995). Another study showed that 50 ppm SO_2 did not inhibit the growth of *Scenedesmus obliquus* PF3, but increase of SO_2 to 100 ppm significantly reduced its growth (Ma et al., 2019). Therefore, it is suggested that the allowed concentration of SO_2 in microalgae culture should be less than 50 ppm. Control of the medium pH by using a buffered solution is also a common method to prevent pH drop in the medium due to SO_2.

In the case of NOx, previous studies showed that up to 45 mg NOx per m^3 in the flue gas had no negative effect on microalgal growth (Doucha et al., 2005). Similarly, application of coal-fired flue gas containing 400 ppm SO_2, NO, and ash to microalgae culture did not have any negative effects on CO_2 fixation by *Chlorella fusca* (Duarte et al., 2016). Likewise, *Scenedesmus* sp. did not show growth inhibition when 100 ppm of NO_2, as in the flue gas, was applied (Ma et al., 2019). Hence, unlike SO_2, microalgae can tolerate higher concentrations of NOx in the flue gas. Microalgae may also use NO_x as nitrogen source for their growth, as previously suggested (Van Den Hende et al., 2012). The growth of other microalgal species when supplied with flue gas or simulated flue gas is shown in Table 12.2.

12.2.4 Particulates

Particulates in coal-fired flue gas known as fly ashes may contain unburned carbon, calcite, oxides such as alumina (Al_2O_3), calcium oxide (CaO), silicate (SiO_2), hematite (Fe_2O_3), and heavy metals (Aslam and Tahira, 2016; Cheng et al., 2019). Particulates in the combustion gas may affect microalgae growth due to an increase in medium turbidity and heavy metal content, such as Zn, Ni, Cr, Cu, As, Co, Pb, Cd, Se, and Hg. Although several trace elements may be utilized by microalgae cells for metabolism in very low amounts, they can be toxic when accumulated in high concentrations in cells, causing reduction in CO_2 fixation rate, biomass production, and lipid synthesis. Particulates from diesel engine flue gas with a size of less than 1 μm have been shown to be toxic due to their agglomeration with microalgal cells that causes physical damage to the cells (Pikula et al., 2019). Therefore, removal of particulates in the flue gas is considered important for reducing its toxic effect on microalgal growth.

12.3 Flue gas treatments for microalgae

Flue gas conditions should be favorable for microalgae growth and free from harmful pollutants before sparging it into the cultivation system. Accordingly, pretreatment of the flue gas released from the chimney is considered important before it is used as a CO_2 source for microalgae cultivation. Critical components of the flue gas that need to be adjusted for microalgae growth include temperature, CO_2 concentration, and the

Table 12.2 Microalgae growth when supplied with the flue gas.

Species	Flue gas source	CO_2 (%)	SO_2 (ppm)	NOx (ppm)	Cultivation methods	Results	Reference
Chlorella sp. MTF-15	Coke oven	1/2 dilution of 25	1/2 dilution of 80–90	1/2 dilution of 70–80	24 × 50 L tubular photobioreactors (PBRs)	Increase biomass yield to 1.56 g/L	Kao et al. (2014)
Chlorella sp. MTF-15	Hot stove	1/4 dilution of 26	1/4 dilution of 15–20	1/4 dilution of 8–10	24 × 50 L tubular PBRs	Increase biomass yield to 1.27 g/L	Kao et al. (2014)
Chlorella sp. MTF-15	Power plant	1/4 dilution of 24	1/4 dilution of 15–20	1/4 dilution of 25–30	24 × 50 L tubular PBRs	Increase biomass yield to 1.25 g/L	Kao et al. (2014)
Chlorella sp.	Coal-fired power plant	1–10	30a	61	500mL columnar PBRs	Maximum biomass productivity 0.21 g/L/day using 5% CO_2 + wastewater	Yadav et al. (2019)
Chlorella LAMB 31	Gas cylinders	40	-	-	800 ml PBRs	Increase growth rate to 0.19/day	Li et al. (2020)
Chlorella LAMB 31	Coal-fired power plant	13	30	129	3 L indoor PBRs	Growth rate 0.15/day; biomass production 0.36 g/L	Li et al. (2024)
Chlorella fusca LEB 111	Gas cylinders	10	400	400	1.8 L tubular PBRs	SO_2 and NO did not affect CO_2 biofixation	Duarte et al. (2016)
Chlorella pyrenoidosa XQ-20044	Gas cylinders	15	300	300	8 L indoor column PBRs	Lipid content of 38%, fatty acid composition suitable for biodiesel feedstocks	Du et al. (2019)

Chlorella pyrenoidosa	Simulated flue gas	10	18	149	3.6 L oval tube PBRs	Biomass productivity 0.13 g/L/day	Nad et al. (2023)
Chlorella vulgaris	Coal-burning boiler	6	180b	250	350 mL flasks, mixotrophic culture with sewage wastewater	Biomass production was 1.0 g/L	Kumar et al. (2018)
Chlorococcum sp.	Coal-fired power plant	1–10	30c	61	500 mL columnar PBRs	Maximum biomass productivity 0.11 g/L/day using 5% CO_2 + wastewater	Yadav et al. (2019)
Scenedesmus acutus UTEX B72	Coal-fired power plant	10–13	0–30d	50–80	1200 L PET PBRs	biomass productivity of 0.1–0.2 g/L/day	Mohler et al. (2019)
Scenedesmus obliquus PF3	Gas cylinders	10	50, 100	100	600 mL bubble column reactors	No toxicity to microalgae	Ma et al. (2019)
Scenedesmus dimorphus	Gas cylinders	15	400	300	Indoor PBRs	Biomass production 3.2 g/L	
Scenedesmus dimorphus	Gas cylinders	10	35e	121.7 f	3-L airlift PBRs, 0.2 L/min	No effect of SO_2 and NOx on microalgal growth and carbohydrate, lipid, and protein content	Wang et al. (2022)
Scenedesmus sp.	Diesel-fueled heating boiler	10.6	0.0	38.3	100 m^2 raceway ponds	Transfer rate of 10 gC/min; biomass productivity of 17 $g/m^2/d$	de Godos et al. (2014)

(*Continued*)

Table 12.2 (Continued)

Species	Flue gas source	CO_2 (%)	SO_2 (ppm)	NOx (ppm)	Cultivation methods	Results	Reference
Nannochloropsis oculata	Coal-fired power plant	12	50	120	0.26 m depth and 310 m^3 raceway ponds	Flue gas flow rate of 150 m^3/h increased biomass productivity to 14.1 $g/m^2/d$	Cheng et al. (2018)
Nannochloropsis oceanica KA2	Coal-fired power plants	13	30	129	2000-L, 4000-L, or 8000-L indoor raceway ponds	Biomass production 0.32–0.34 g/L	Zhu et al. (2014)
Mixed microalgal cultures	Coal-fired boilers	Sequential increase from 1.1% to 3.3%, 5.5%, and 11%	-	-	12 × 30 L PE PBRs	No growth at 3.3%; at 11% (undiluted flue gas), pH dropped to 2.3	Aslam and Tahira (2016)

[a] =as SOx.
[b] =as SOx.
[c] =as SOx.
[d] =as SOx.
[e] =from the value conversion of mg/Nm^3.
[f] =from the value conversion of mg/Nm^3.

presence of SO_2, NOx, particulates, and heavy metals. Currently, pollution control technology is implemented by industries worldwide as part of strict regulations to limit the release of pollutants from emissions into the environment. A brief explanation of each of these factors is described as follows.

12.3.1 Temperature

As mentioned earlier, the temperature of flue gases released from the boiler stacks is 120°C–180°C (Table 12.1); thus reduction of flue gas temperature to the critical range for microalgae growth is essential. The most commonly used method to directly reduce flue gas temperature is by using heat exchanger. Moreover, the installation of a condensing scrubber after the heat exchanger provides an additional benefit for achieving further temperature reduction of the flue gas while effectively removeing pollutants from the flue gas. The main principle of heat exchanger is to transfer heat from the flue gas to liquid medium such as water. By using this system, the temperature of the flue gas can be lowered to 35°C–40°C (Yadav et al., 2019).

12.3.2 Carbon dioxide concentration

The flue gas composition, particularly the CO_2 concentration, is crucial for microalgae cultivation. Flue gas with a higher CO_2 concentration may be preferred, as it provides a more abundant and readily available carbon source for microalgae. However, high CO_2 concentration may cause acidification of microalgal culture (Li et al., 2020) and may limit microalgal growth. In this case, pH of the growth medium needs to be monitored and maintained at the optimal range, that is, pH 7–9. Adjustment of pH by adding alkaline substance such as NaOH or sodium bicarbonate is a common practice when the pH medium falls below 7.

12.3.3 Sulfur dioxide and nitrogen oxides

A wide range of methods, mostly physical and chemical approaches, have been available to remove SO_2 and NOx from the flue gas. Those technologies can eliminate SO_2, NOx, and particulates effectively, while leaving CO_2 largely unchanged. Examples of physical treatments commonly employed in industrial stacks include cyclone, electrostatic precipitator, and baghouse filter to trap particulate matter from the flue gas;

and water scrubber, gas absorber, and gas adsorber to capture gases and particulates (Larki et al., 2023). SO_2 may be removed by desulfurization processes using relatively cheap and abundant sorbents such as CaO and limestone before injecting the flue gas into the cultivation system (Hosseini et al., 2018). Moreover, NO_x can be reduced by using denitrification methods or selective catalytic or noncatalytic reduction (SCR or SNCR). The addition of chelators such as ethylenediaminetetraacetic acid (EDTA), methyliminodiacetic acid (MIDA), nitrilotriacetic acid (NTA), and dimercaptopropanesulfonic acid (DMPS) also improves NO removal (Aslam and Tahira, 2016). The decision to use any of the methods depends on cost, the level of removal, and the resource availability of the industry.

12.3.4 Particulates

Particulates in the flue gas may increase medium turbidity and become toxic when suspended in the microalgae cultivation system. Currently, a wide range of technologies is available to effectively control dust and particulate matter in the flue gas. These technologies are utilized by industries as part of the mandatory requirements to control air pollution from their stacks. As such, particulate matter is significantly reduced to below the allowable level set by the regulations after passing through the air pollution control units. The main principles of the technologies are based on the separation of particulates from the flue gas using filtration and precipitation techniques (Hosseini et al., 2018). The particulate removal methods adopted by most industries include electrostatic precipitation, filter bag dust collectors, and wet-dry cyclones. Electrostatic precipitators are commonly installed to remove fine particulates and particulates containing heavy metals.

12.4 Microalgae strain tolerant to flue gas

Recently, the development of microalgae strains tolerant to flue gas has become an attractive option to improve microalgae cultivation systems. Numerous studies focus on microalgae tolerance to high concentrations of CO_2, as well as tolerance to the levels of SO_2 and NOx commonly present in flue gas (Cheng et al., 2019; Kumar et al., 2011; Thomas et al., 2016). Common methods for screening microalgae strains tolerant to high CO_2 include the introduction of diluted CO_2 from gas cylinders or the addition of $NaHCO_3$ to the medium. Likewise, the use of sulfite and bisulfite salts to induce the formation of SO_x compounds, and the use of nitrite instead of nitrate salts in the medium are the common methods to obtain strains tolerant to high SO_2 and NO_x. Based on these screening methods, it was reported that a strain of *Chlorella* can grow under 40% CO_2 (Li et al., 2020), while *S. obliquus* can grow under CO_2 concentrations as high as 15% (Ma et al., 2019). In addition to having high tolerance characteristics, it is expected that tolerant strains also possess high efficiency in capturing CO_2 from the flue gas and a rapid growth rate to obtain high biomass yield.

Recent studies indicate that the tolerance to SO_2 and NO_x is critical when the native flue gas is used as the carbon source for microalgae growth. Tolerance to SO_2 and NO_x varies widely among microalgae species and within a microalgae species. For example, the growth of *Chlorella* strain LAMB 31 under the flue gas containing 30 and 129 ppm SO_2 and NOx, respectively, produced 0.39 g/L biomass on day 11 (Li et al., 2024). Similarly, *S. obliquus* PF3 showed tolerance to 100 ppm SO2 and NOx when supplied with 10% CO_2 (Ma et al., 2019). Additionally, *Nannochloropsis oculata* showed tolerance to 50 and 120 ppm SO_2 and NOx, respectively, when supplied with flue gas from coal-fired power plant, achieving a biomass productivity of 14.1 $g/m^2/d$ (Cheng et al., 2018).

Flue gas, after passing through the air pollution control units, may contain a very low amount of metals, which may serve as the source of trace elements for microalgae cultivation. However, when the metal concentration in the flue gas has a significant negative effect on microalgae growth, the use of microalgal species tolerant to heavy metals may become a better option.

12.5 Technology development of microalgae biomass production utilizing flue gas

The technology development of microalgae cultivation utilizing flue gas has been an active research and innovation area during the last three decades. In addition to obtaining the economic scalability of the technology, current global issues regarding resource sustainability, low carbon initiatives, circular economy, and clean environment are the driving forces behind the development of this technology. As a result, several technological advancements have been made to improve the process's efficiency, scalability, and economic viability. Some fundamental technology developments in microalgae biomass production utilizing flue gas as carbon source are as follows.

12.5.1 Cultivation systems

Different microalgae cultivation systems have been developed to optimize flue gas utilization. Open ponds, closed PBRs, and hybrid systems are commonly used cultivation setups. Open ponds have the advantage of lower operational costs and simpler control systems compared with PBRs, but they have drawbacks of high contamination rates, lower productivity, and large land requirements. Therefore, hybrid systems are developed to minimize the disadvantages of both systems. Cultivation system design and engineering advances have focused on improving light distribution, nutrient supply, mixing, CO_2 delivery to microalgae, and productivity. Integration of monitoring and control systems has also been explored to optimize growth conditions.

The most common open pond system operating in recent days for microalgae cultivation is raceway ponds. The maximum depth of raceway ponds allowing light penetration to the bottom of the pond is limited to 60 cm. Since flue gas is injected from the bottom of the pond, the retention time of CO_2 in the raceway pond system is very short before unfixed CO_2 is released into the air. Therefore, flue gas is commonly injected from many points spread across the bottom of the pond to ensure optimum supply of CO_2 to the cultivation system. In addition, mixing the culture medium increases the solubility and residence time of the CO_2 gas in the medium. Mixing the culture media is done by using horizontal paddle wheels installed at certain points in the raceway pond (Cheng et al., 2018; Zheng et al., 2018).

Several types of PBR designs have been developed recently (Tripathi et al., 2023). The most common type of PBR is tubular PBR, which can be installed in a horizontal or vertical position. The closed PBR is designed to improve control of microalgae growth under environmental conditions. In addition to nutrients and CO_2 supply, other factors critical to microalgae growth in PBR systems are light intensity and mixing (Xu et al., 2023; Yadav and Sen, 2017). In a closed PBR system, flue gas can be injected at the bottom or at certain points of the tubes, and then mixed using circulating pumps to maintain long retention time of the diluted CO_2 in the medium. Mixing the medium also enables microalgae cells to evenly receive light and temperature within the tubes. In general, design improvements of PBRs for flue gas utilization are directed toward improving mass transfer efficiency. This is associated with the distribution of microalgal cells, nutrients, CO_2, light, and temperature along the column of PBRs while maintaining low energy consumption. It is expected that high mass efficiency produces a high CO_2 conversion rate.

12.5.2 Strain selection and genetic engineering

A wide range of microalgae species exist in various environments, but only few of them are suitable for flue gas utilization. Flue gas tolerance and biomass production performance are known to be species dependent. Therefore, strain selection is critical in maximizing CO_2 utilization and biomass productivity. Given the complexity among determining factors, selections are made on strains having one or two key traits with high productivity in the presence of flue gas. Hence, research efforts have focused on screening and selecting microalgae strains that have a high affinity for CO_2, fast growth rates, and tolerance to flue gas contaminants (Cheng et al., 2019), including tolerance to low pH (acidophilic) and high temperature (thermophilic). Strain selection is performed by growing microalgae at various CO_2, SO_2, or NOx concentrations, from either native or simulated flue gas. For example, *Chlorella pyrenoidosa* Chick was selected based on robust growth at various pH ranges and under simulated flue gas consisting of 10% CO_2, 18 ppm SO_2, and 149 ppm NOx (Nad et al., 2023). In another study, two out of 13 microalgae strains, *Micractinium pusillum* KMC8 and *Scenedesmus acutus* NCIM5584, have been obtained from strain screening that have tolerance to dissolved flue gas (Chauhan et al., 2022).

In addition to screening under flue gas environments, another strategy to obtain strains tolerant to flue gas is by isolating strains from environments surrounding the industrial plants. For example, *C. fusca* LEB 111, isolated from the stabilization pond of a coal-fired power plant in Brazil, showed comparable CO_2 biofixation rates when grown using flue gas and CO_2 (da Silva Vaz et al., 2016). It is suggested that the strain has developed the adaptation mechanisms in the presence of pollutants over a period of time.

Recently, genetic engineering techniques are also being explored to enhance the efficiency of CO_2 fixation and biomass production in microalgae (Barati et al., 2021). The techniques include random mutagenesis (Beacham et al., 2015) and targeted genetic modification (Ng et al., 2017). Genetic engineering of microalgae is still limited to laboratory experiments in confined environments, since there is a potential risk of negative impacts when the engineered microalgae are released into the natural environment.

12.5.3 Integration with power plants and industrial facilities

Research has focused on developing integrated systems that directly capture and utilize flue gas within power plants or industrial facilities for microalgae cultivation. This involves designing efficient CO_2 capture units and microalgae cultivation systems within or adjacent to the existing infrastructure. Integration not only reduces transport costs but also enhances the process's overall energy and resource efficiency. Under this scenario, microalgae biomass is converted to lipids or other valuable products. A pilot-scale study to implement this concept was carried out in a steel industry in Taiwan (Kao et al., 2014). PBR units utilizing flue gas were installed next to the stacks of a hot stove, coke oven, and power plant of the industry. *Chlorella* sp. MTF-15 showed the highest specific growth rate and lipid production when aerated with the flue gas from the power plant and the hot stove, respectively, that is, 0.73/day and 0.96 g/L. Although promising results were obtained from this study (Kao et al., 2014), scaling up the project remains a challenge, particularly related to controlling environmental conditions for maximum yield and developing optimum configurations of new cultivation facilities within the industry.

12.5.4 Wastewater treatment and nutrient recycling

Flue gas combined with wastewater treatment has attracted a lot of attention as a promising technology that implements low carbon initiatives. Under this scenario, pollutant substances, mainly nitrogen and phosphorus, in wastewater are utilized by microalgae as nutrient sources for their growth, thus reducing the presence of those pollutants in wastewater. Therefore, agroindustrial wastewater containing high nutrient levels of nitrogen and phosphorus is the most preferred type of application under this concept. In addition, when biomass production and its conversion are the main objectives of an industrial activity, nutrient recycling from waste streams could lessen the dependency on external fertilizer sources. This strategy improves sustainability and lowers running expenses.

Studies on the utilization of microalgae to treat flue gas and municipal wastewater were reported recently (Kumar et al., 2018; Tu et al., 2019; Yadav et al., 2019). Using *Chlorella vulgaris* in batch mode, chemical oxygen demand (COD) and CO_2 removal of 78% and 64%, respectively, have been achieved at the laboratory scale (Kumar et al., 2018). Moreover, SOx and NOx removal were found to be 62% and 63%, respectively, under the same culturing conditions. Likewise, Yadav et al. (2019) reported that COD and nitrate removal of *Chlorella* sp. were 92% and 100%, respectively, when supplied with 5% CO_2 from the flue gas. Another study showed that biomass production of *C. pyrenoides* increased by 83% in the presence of flue gas when cultured outdoors in columnar PBRs, using municipal wastewater as the growth medium (Tu et al., 2019).

Despite these advantages, implementation of flue gas coupled with wastewater treatment faces some challenges for large-scale application (Joun et al., 2023). Wastewater may contain compounds other than nitrogen and phosphorus with concentration levels that are harmful to microalgae growth. Wastewater also contains other microorganisms that may interfere with microalgae growth. As such, pretreatment of wastewater, such as sterilization or precipitation, is required, which may add additional operational costs and provide further technical challenges. In some cases, pretreatment includes dilution of wastewater before it is utilized for microalgae cultivation. Therefore, future studies and technology development are needed to address these challenges.

Overall, the technology development in flue gas utilization for microalgae biomass production aims to improve efficiency, reduce costs, and advance the process's commercial viability. Ongoing research and innovation in this field focus on addressing key challenges, optimizing system design, and exploring novel approaches to further enhance flue gas utilization for sustainable microalgae biomass production.

12.6 Current application of flue gas utilization for microalgae cultivation

The application of flue gas for microalgae production is gaining more attention in various industries and research institutions. While commercial-scale implementations are still elusive, several pilot projects and research initiatives are underway to explore the feasibility and potential of this technology. Nevertheless, flue gas from several industries has been used for microalgae cultivation at a small scale, for example, in power plant industries, iron and steel industries, and waste-to-energy treatment plants.

12.6.1 Power plants

Flue gas from power plants, especially those fueled by coal or natural gas, is a significant source of CO_2 emissions. Several power plants have initiated pilot projects to capture and utilize flue gas for microalgae cultivation. The captured CO_2 is redirected to PBRs or open ponds, where microalgae use it as a carbon source for biomass production. These initiatives aim to reduce CO_2 emissions, utilize wastewater as nutrient source for microalgal growth, and explore the possibility of generating valuable coproducts from microalgae biomass. Examples of these activities were previously reported by several research groups (Cheng et al., 2018; Kao et al., 2014; Mohler et al., 2019; Yadav et al., 2019) and extensively discussed elsewhere (Cheng et al., 2019). In particular, a PBR system has been installed with total capacity of 1200 L next to the site of the power plant stacks, and it demonstrated that *Chlorella* culture was capable of utilizing CO_2, SO_2, and NO_x from the flue gas for biomass production (Kao et al., 2014). Likewise, Mohler et al. (2019) developed a cyclic flow PBR system using vertical tubes with total volume of 1200 L in the power plant station, and obtained a biomass productivity of *S. acutus* of 0.1–0.2 g/L per day.

12.6.2 Iron and steel industries

The iron and steel industries are significant emitters of CO_2 due to their energy-intensive processes. Some iron and steel plants are exploring flue gas utilization for microalgae biomass production to mitigate their carbon footprint. Flue gas containing CO_2 is captured and supplied to microalgae cultivation systems, enabling the production of biomass while reducing carbon emissions. This approach aligns with their sustainability goals and supports the circular economy by converting waste streams into valuable resources. An example of recent application of flue gas utilization for microalgae cultivation is reported from a steel industry in China (Yang et al., 2022). In an effort to reduce carbon emissions from its production process, this industry conducted a pilot test from 2014 to 2016 by integrating the utilization of flue gas as CO_2 source in cultivating microalgae with the treatment of wastewater by using microalgae. Although results of this study showed that the microalgae could tolerate 10% CO_2 of the flue gas, improvement of fixation rate and efficiency by microalgae remains challenge for a larger-scale applications.

12.6.3 Waste-to-energy facilities

Waste-to-energy facilities, such as anaerobic digestion or landfill gas plants, generate flue gas containing CO_2 during the combustion of organic waste materials. Some facilities have integrated microalgae cultivation systems to capture and utilize this CO_2 for biomass production. By incorporating microalgae into their processes, these facilities enhance CO_2 utilization efficiency, reduce emissions, and potentially produce bioenergy or other valuable products from microalgae biomass. In addition, microalgae biomass, after its conversion to valuable products, can be used as feed in a waste-to-energy plant. A pilot study on the prospect of flue gas utilization from a waste-to-energy plant for microalgae cultivation has been reported in Czech (Nad et al., 2023). They reported that the biomass productivity and CO_2 fixation rate of *C. pyrenoidosa* Chick, when supplied with the flue gas containing 0–20% CO_2, 0–3000 mg SO_2/m^3, and 0–1250 mg NO/m^3, were 0.13 and 0.25 g/L/day, respectively. Although these results were lower than results of the same study at laboratory scale, this experiment provides the possibility of flue gas utilization from the waste-to-energy plant.

12.7 Challenges and strategies for the implementation of flue gas utilization in microalgae production

Despite many small-scale studies have proved the concept of flue gas utilization for microalgae production, a number of challenges remain to be overcome before moving to large-scale operations that are economically sound and feasible. This includes technical, economical, and regulatory issues. Some challenges and strategies are discussed as follows.

1. Integrating flue gas utilization systems with existing industrial infrastructure, such as power plants or manufacturing facilities, is potentially an intricate task for some corporations. One of the key reasons is that microalgae cultivation is alien to the core business of the existing facilities, thus generating efforts to provide additional manpower and experts, space allocation, and associated facilities. Key challenges are ensuring seamless integration, optimizing process compatibility, and minimizing disruption to existing operations.
2. When the utilization of flue gas for microalgae growth has been decided, the economic viability of flue gas utilization is a significant challenge. The costs associated with operation of carbon capture, cultivation infrastructure, downstream processing, and product extraction must be balanced against the value of the produced biomass or coproducts. As such, expanding the range of valuable products derived from microalgae biomass can enhance the economic viability of flue gas utilization. Therefore, research efforts are also focused on developing technologies to extract and utilize different components of microalgae

biomass, such as lipids for biofuel production, proteins for food or feed applications, and high-value compounds for various industries.

3. Although it is still in its infancy, the regulatory environment and policy frameworks are crucial for promoting flue gas utilization and offering incentives for its implementation. Developing supportive policies, such as carbon pricing or tax incentives, can facilitate the adoption and commercialization of flue gas utilization technologies.
4. Assessing the overall environmental impact and sustainability of flue gas utilization is important. Comprehensive life cycle assessments, considering factors such as energy consumption, water use, and land footprint, can provide insights into the environmental benefits and potential trade-offs associated with flue gas utilization for microalgae production.
5. Flue gas from different industries varies in composition and quantity, some of which contain contaminants that can inhibit microalgae growth. Moreover, it is common for concentration fluctuations of the flue gas components to occur on daily basis during plant operation. As such, addressing the challenges associated with variable flue gas composition and potential pollutants is crucial to ensure optimal microalgae growth and biomass production. Direct use of flue gas may reduce operational costs, but it may also increase the risk of toxic compounds in the gas.
6. Improving the efficiency of carbon capture by microalgae to increase flue gas utilization remains a technical challenge. Carbon capture efficiency depends on several key parameters, including efficient CO_2 biofixation rate, rapid growth, and tolerance to high stress due to unfavorable conditions of the flue gas (Zheng et al., 2018). Therefore, research and development efforts focus on developing cost-effective and energy-efficient methods for capturing and concentrating CO_2 from flue gas streams. Strategies and critical components for the improvement of CO_2 fixation by microalgae are briefly discussed.

As mentioned earlier, strain selection is one of the key components to improve the efficiency of carbon capture by microalgae. Screening microalgae strains with high affinity for CO_2 and fast growth rates can significantly increase flue gas utilization. Genetic engineering of microalgae can also be employed to improve their CO_2 uptake and utilization capabilities. This approach involves modifying the metabolic pathways of microalgae to enhance their carbon fixation efficiency and biomass productivity (Priyadharsini et al., 2022).

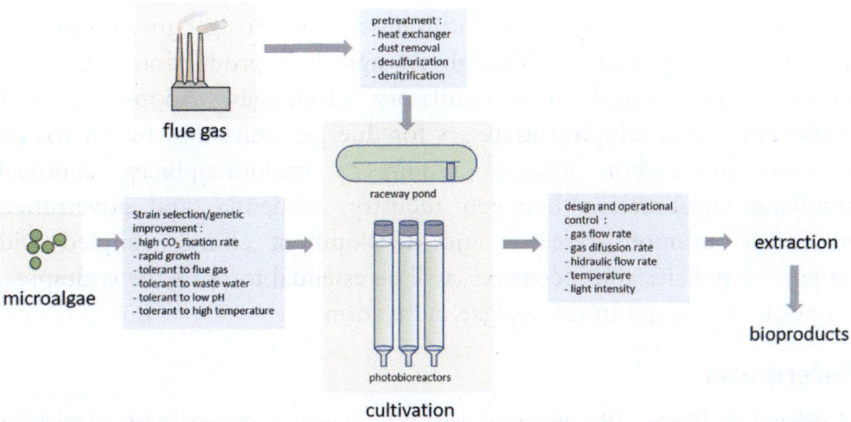

Figure 12.1 ***Strategies and challenges for CO_2 fixation from the flue gas for low carbon initiative.*** This figure shows possible challenges and strategies for CO_2 fixation from the flue gas for low carbon initiative, including strain selection or genetic improvement of the microalgae for high CO_2 fixation capacity, developing a suitable and effective system for microalgae cultivation adapted to local environment and situations, such as raceway ponds or PBRs, and pretreatment of the flue gas according to its characteristics before delivering it into the cultivation system.

Optimizing nutrient availability based on species-dependent manner is also essential for maximizing microalgae growth and CO_2 utilization. Understanding the nutrient requirements of microalgae and developing strategies to provide adequate and balanced nutrition can enhance their growth and biomass productivity. Coupling flue gas utilization with wastewater as a nutrient source provides an alternative for reducing operational costs.

Improving the design and operation of microalgae cultivation systems can enhance flue gas utilization. Exploring different cultivation configurations, such as closed PBRs or hybrid systems, can optimize light distribution, nutrient supply, and gas exchange, thereby improving CO_2 utilization efficiency. Fig. 12.1 shows the summary of strategies and challenges to improve flue gas utilization by microalgae.

12.8 Conclusion

Utilization of flue gas for microalgae biomass production has been an interesting topic during the last decade since it offers opportunity to develop a technologically sound low carbon initiative in industrial sector.

The proof of concept on flue gas utilization for microalgal growth has been shown by many studies, although its upscaling production faces some technical, economical, and regulatory challenges. Addressing these challenges and developing strategies for flue gas utilization by microalgae to meet low-carbon scenario requires a multidisciplinary approach involving collaboration between industry, academia, and government entities. Continuous research and development efforts, coupled with supportive policies and incentives, will be essential to drive the widespread adoption of flue gas in microalgae cultivation.

References

M., Ahmad, S., Gersen, 2014. Water solubility in CO2 mixtures: experimental and modelling investigation. Energy Procedia 63, 2402–2411. Available at https://doi.org/10.1016/j.egypro.2014.11.263, http://www.sciencedirect.com/science/journal/18766102.

Aouini, I., Ledoux, A., Estel, L., Mary, S., 2014. Pilot plant studies for CO2 capture from waste incinerator flue gas using MEA based solvent. Oil & Gas Science and Technology – Revue d'IFP Energies Nouvelles 69 (6), 1091–1104. Available at https://doi.org/10.2516/ogst/2013205.

Aslam, A., Tahira, A., Mughal, 2016. A review on microalgae to achieve maximal carbon dioxide (CO2) mitigation from industrial flue gases. International Journal of Research in Advent Technology 4 (9), 12–29. https://www.ijrat.org/archives/VOLUME-4-ISSUE-9. Available at https://www.ijrat.org/archives/VOLUME-4-ISSUE-9.

Barati, B., Zeng, K., Baeyens, J., Wang, S., Addy, M., Gan, S.-Y., et al., 2021. Recent progress in genetically modified microalgae for enhanced carbon dioxide sequestration. Biomass and Bioenergy 145, 105927. Available at https://doi.org/10.1016/j.biombioe.2020.105927.

I.A. Bashmakov, L.J. Nilsson, A. Acquaye, C. Bataille, J.M. Cullen, S. de la R. du Can, et al., 2022. Climate Change 2022: Mitigation of Climate Change. Contribution of Working Group III to the Sixth Assessment Report of the Intergovernmental Panel on Climate Change, Chapter 11. In: P.R. Shukla, J. Skea, R. Slade, Al. Khourdajie, R. van Diemen, D. McCollum, M. Pathak, S. Some, P. Vyas, R. Fradera, M. Belkacemi, A. Hasija, G. Lisboa, S. Luz, J. Malley, Cambridge University Press, Cambridge UK. doi:https://doi.org/10.1017/9781009157926.013.

Beacham, T.A., Macia, V.M., Rooks, P., White, D.A., Ali, S.T., 2015. Altered lipid accumulation in Nannochloropsis salina CCAP849/3 following EMS and UV induced mutagenesis. Biotechnology Reports 7, 87–94. Available at https://doi.org/10.1016/j.btre.2015.05.007. http://www.journals.elsevier.com/biotechnology-reports/.

Chauhan, D.S., Sahoo, L., Mohanty, K., 2022. Maximize microalgal carbon dioxide utilization and lipid productivity by using toxic flue gas compounds as nutrient source. Bioresource Technology 348. Available at https://doi.org/10.1016/j.biortech.2022.126784. https://www.sciencedirect.com/science/article/abs/pii/S0960852422001134.

Cheah, W.Y., Show, P.L., Chang, J.S., Ling, T.C., Juan, J.C., 2015. Biosequestration of atmospheric CO2 and flue gas-containing CO2 by microalgae. Bioresource Technology 184, 190–201. Available at https://doi.org/10.1016/j.biortech.2014.11.026, http://www.elsevier.com/locate/biortech.

Cheng, J., Yang, Z., Zhou, J., Cen, K., 2018. Improving the CO2 fixation rate by increasing flow rate of the flue gas from microalgae in a raceway pond. Korean Journal of Chemical Engineering 5 (2), 498–502. Available at https://doi.org/10.1007/s11814-017-0300-1. http://www.springerlink.com/content/120599/.

Cheng, J., Zhu, Y., Zhang, Z., Yang, W., 2019. Modification and improvement of microalgae strains for strengthening CO2 fixation from coal-fired flue gas in power plants. Bioresource Technology 291, 121850. Available at https://doi.org/10.1016/j.biortech.2019.121850.

Doucha, J., Straka, F., Lívanský, K., 2005. Utilization of flue gas for cultivation of microalgae (Chlorella sp.) in an outdoor open thin-layer photobioreactor. Journal of Applied Phycology 17 (5), 403–412. Available at https://doi.org/10.1007/s10811-005-8701-7.

Duarte, J.H., Fanka, L.S., Costa, J.A.V., 2016. Utilization of simulated flue gas containing CO2, SO2, NO and ash for Chlorella fusca cultivation. Bioresource Technology 214, 159–165. Available at https://doi.org/10.1016/j.biortech.2016.04.078, http://www.elsevier.com/locate/biortech.

Du, K., Wen, X., Wang, Z., Liang, F., Luo, L., Peng, X., et al., 2019. Integrated lipid production, CO2 fixation, and removal of SO2 and NO from simulated flue gas by oleaginous Chlorella pyrenoidosa. Environmental Science and Pollution Research 26 (16), 16195–16209. Available at https://doi.org/10.1007/s11356-019-04983-9.

de Godos, I., Mendoza, J.L., Acién, F.G., Molina, E., Banks, C.J., Heaven, S., et al., 2014. Evaluation of carbon dioxide mass transfer in raceway reactors for microalgae culture using flue gases. Bioresource Technology 153, 307–314. Available at https://doi.org/10.1016/j.biortech.2013.11.087.

Van Den Hende, S., Vervaeren, H., Boon, N., 2012. Flue gas compounds and microalgae: (Bio-)chemical interactions leading to biotechnological opportunities. Biotechnology Advances 30 (6), 1405–1424. Available at https://doi.org/10.1016/j.biotechadv.2012.02.015.

Hosseini, N.S., Shang, H., Scott, J.A., 2018. Biosequestration of industrial off-gas CO2 for enhanced lipid productivity in open microalgae cultivation systems. Renewable and Sustainable Energy Reviews 92, 458–469. Available at https://doi.org/10.1016/j.rser.2018.04.086. https://www.sciencedirect.com/science/article/pii/S136403211830279X.

Hunt, A.J., Sin, E.H.K., Marriott, R., Clark, J.H., 2010. Generation, capture, and utilization of industrial carbon cioxide. ChemSusChem: Chemistry & Sustainability Energy & Materials 3 (3), 306–322. Available at https://doi.org/10.1002/cssc.200900169. https://chemistry-europe.onlinelibrary.wiley.com/doi/full/10.1002/cssc.200900169.

Ighalo, J.O., Dulta, K., Kurniawan, S.B., Omoarukhe, F.O., Ewuzie, U., Eshiemogie, S.O., et al., 2022. Progress in microalgae application for CO2 sequestration. Cleaner Chemical Engineering 3, 100044. Available at https://doi.org/10.1016/j.clce.2022.100044.

Joun, J., Sirohi, R., Yu, B.S., Yang, H.E., Sim, S.J., 2023. Utilizing wastewater and flue gas for enhancing biomass and high-value products from microalgae. Phytochemistry Reviews 22 (4), 861–877. Available at https://doi.org/10.1007/s11101-023-09876-x. https://www.springer.com/journal/11101.

Kao, C.Y., Chen, T.Y., Chang, Y.B., Chiu, T.W., Lin, H.Y., Chen, C.D., et al., 2014. Utilization of carbon dioxide in industrial flue gases for the cultivation of microalga Chlorella sp. Bioresource Technology 166, 485–493. Available at https://doi.org/10.1016/j.biortech.2014.05.094, http://www.elsevier.com/locate/biortech.

Kumar, K., Dasgupta, C.N., Nayak, B., Lindblad, P., Das, D., 2011. Development of suitable photobioreactors for CO2 sequestration addressing global warming using green algae and cyanobacteria. Bioresource Technology 102 (8), 4945–4953. Available at https://doi.org/10.1016/j.biortech.2011.01.054.

Kumar, P.K., Vijaya Krishna, S., Verma, K., Pooja, K., Bhagawan, D., Himabindu, V., 2018. Phycoremediation of sewage wastewater and industrial flue gases for biomass generation from microalgae. African Journal of Chemical Engineering 25, 133–146. Available at https://doi.org/10.1016/j.sajce.2018.04.006. https://www.journals.elsevier.com/south-african-journal-of-chemical-engineering/.

Larki, I., Zahedi, A., Asadi, M., Forootan, M.M., Farajollahi, M., Ahmadi, R., et al., 2023. Mitigation approaches and techniques for combustion power plants flue gas emissions: a comprehensive review. Science of the Total Environment 903. Available at https://doi.org/10.1016/j.scitotenv.2023.166108, http://www.elsevier.com/locate/scitotenv.

Lee, J.S., Kim, D.K., Lee, J.P., Park, S.C., Koh, J.H., Cho, H.S., et al., 2002. Effects of SO2 and NO on growth of Chlorella sp. KR-1. Bioresource Technology 82 (1), 1–4. Available at https://doi.org/10.1016/S0960-8524(01)00158-4.

Li, J., Tang, X., Pan, K., Zhu, B., Li, Y., Ma, X., et al., 2020. The regulating mechanisms of CO2 fixation and carbon allocations of two Chlorella sp. strains in response to high CO2 levels. Chemosphere 247, 125814. Available at https://doi.org/10.1016/j.chemosphere.2020.125814.

Li, J., Tang, X., Pan, K., Zhu, B., Wang, Z., Zhao, Y., 2024. Application study for the high CO2 tolerant Chlorella strain by flue gas culture: evaluation of growth performance and adaptive mechanisms. Chemical Engineering Journal 479, 147700. Available at https://doi.org/10.1016/j.cej.2023.147700.

Ma, S., Li, D., Yu, Y., Li, D., Yadav, R.S., Feng, Y., 2019. Application of a microalga, Scenedesmus obliquus PF3, for the biological removal of nitric oxide (NO) and carbon dioxide. Environmental Pollution 252, 344–351. Available at https://doi.org/10.1016/j.envpol.2019.05.084. https://www.journals.elsevier.com/environmental-pollution.

Mohler, D., Wilson, M.H., Kesner, S., Schambach, J.Y., Vaughan, D., Frazar, M., et al., 2019. Beneficial re-use of industrial CO2 emissions using microalgae: demonstration assessment and biomass characterization. Bioresource Technology 293. Available at https://doi.org/10.1016/j.biortech.2019.122014, http://www.elsevier.com/locate/biortech.

Nad, M., Brummer, V., Lošák, P., Máša, V., Sukačová, K., Tatarová, D., et al., 2023. Waste-to-energy plants flue gas CO2 mitigation using a novel tubular photobioreactor while producing Chlorella algae. Journal of Cleaner Production 385. Available at https://doi.org/10.1016/j.jclepro.2022.135721. https://www.sciencedirect.com/science/article/abs/pii/S0959652622052957.

Nagappan, S., Tsai, P.C., Devendran, S., Alagarsamy, V., Ponnusamy, V.K., 2020. Enhancement of biofuel production by microalgae using cement flue gas as substrate. Environmental Science and Pollution Research 27 (15), 17571–17586. Available at https://doi.org/10.1007/s11356-019-06425-y. https://link.springer.com/journal/11356.

Ng, I.S., Tan, S.I., Kao, P.H., Chang, Y.K., Chang, J.S., 2017. Recent developments on genetic engineering of microalgae for biofuels and bio-based chemicals. Biotechnology journal 12 (10). Available at https://doi.org/10.1002/biot.201600644. http://onlinelibrary.wiley.com/journal/10.1002/(ISSN)1860-7314.

Pikula, K.S., Chernyshev, V.V., Zakharenko, A.M., Chaika, V.V., Waissi, G., Hai, L.H., et al., 2019. Toxicity assessment of particulate matter emitted from different types of vehicles on marine microalgae. Federation Environmental Research 179. Available at https://doi.org/10.1016/j.envres.2019.108785. http://www.elsevier.com/inca/publications/store/6/2/2/8/2/1/index.htt.

Priyadharsini, P., Nirmala, N., Dawn, S.S., Baskaran, A., SundarRajan, P., Gopinath, K.P., 2022. J, Genetic improvement of microalgae for enhanced carbon dioxide sequestration and enriched biomass productivity: review on CO2 bio-fixation pathways modifications. Algal Research 66. Available at https://doi.org/10.1016/j.algal.2022.102810. https://www.sciencedirect.com/science/article/pii/S2211926422001813.

Ronda, S.R., Kethineni, C., Parupudi, L.C.P., Thunuguntla, V.B.S.C., Vemula, S., Settaluri, V.S., et al., 2014. A growth inhibitory model with SOx influenced effective growth rate for estimation of algal biomass concentration under flue gas atmosphere. Bioresource Technology 152, 283–291. Available from. Available at https://doi.org/10.1016/j.biortech.2013.10.091, http://www.elsevier.com/locate/biortech.

da Silva Vaz, B., Costa, J.A.V., de Morais, M.G., 2016. CO2 biofixation by the Cyanobacterium spirulina sp. LEB 18 and the green alga Chlorella fusca LEB 111 grown using gas effluents and solid residues of thermoelectric origin. Applied Biochemistry and Biotechnology 178 (2), 418–429. Available at https://doi.org/10.1007/s12010-015-1876-8. http://www.springer.com/humana+press/journal/12010.

S. Solomon, D. Qin, M. Manning, Z. Chen, M. Marquis, K. Averyt, et al., 2007. Climate Change 2007: The Physical Science Basis. Contribution of Working Group I to the Fourth Assessment Report of the Intergovernmental Panel on Climate Change. Cambridge University Press, Cambridge, United Kingdom and New York, NY, USA. In: Susan. Solomon, Dahe. Qin, Martin. Manning, Zhenlin. Chen, Melinda. Marquis, Kristen. Averyt, Melinda M.B. Tignor, Henry LeRoy. Miller, Cambridge University Press, Cambridge, UK and New York, NY, USA.

Song, D., Jiang, T., Rao, C., 2022. Review of policy framework for the development of carbon capture, utilization and storage in China. International journal of environmental research and public health 19 (24), 16853. Available at https://doi.org/10.3390/ijerph192416853. https://www.mdpi.com/1660-4601/19/24/16853.

Speight, J.G., 2018. Natural Gas: A Basic Handbook, 2nd Edition. Gulf Professional Publishing, Cambridge.

Thomas, D.M., Mechery, J., Paulose, S.V., 2016. Carbon dioxide capture strategies from flue gas using microalgae: a review. Environmental Science and Pollution Research 23 (17), 16926–16940. Available at https://doi.org/10.1007/s11356-016-7158-3. http://www.springerlink.com/content/0944-1344.

Tripathi, S., Choudhary, S., Meena, A., Poluri, K.M., 2023. Carbon capture, storage, and usage with microalgae: a review. Environmental Chemistry Letters 21. Available at https://doi.org/10.1007/s10311-023-01609-y. https://link.springer.com/article/10.1007/s10311-023-01609-y.

Tu, R., Jin, W., Han, S.-fang, Zhou, X., Wang, J., Wang, Q., et al., 2019. Enhancement of microalgal lipid production in municipal wastewater: fixation of CO2 from the power plant tail gas. Biomass and Bioenergy 131, 105400. Available at https://doi.org/10.1016/j.biombioe.2019.105400.

UNFCCC, K. 1998. Protocol to the United Nations Framework Convention on Climate Change. UNFCCC, New York.

UNFCCC, 2015. The Paris Agreement. UNFCCC, New York.

Varshney, P., Mikulic, P., Vonshak, A., Beardall, J., Wangikar, P.P., 2015. Extremophilic micro-algae and their potential contribution in biotechnology. Bioresource Technology 184, 363–372. Available at https://doi.org/10.1016/j.biortech.2014.11.040, http://www.elsevier.com/locate/biortech.

Wang, B., Xu, Y.-F., Sun, Z.-L., 2022. Mass transfer characteristics and effect of flue gas used in microalgae culture. Applied Microbiology and Biotechnology 106. Available at https://doi.org/10.1007/s00253-022-12206-4. https://link.springer.com/article/10.1007/s00253-022-12206-4.

Wen, Z., Di, J., Liu, S., Han, J., Lee, J.C.K., 2018. Evaluation of flue-gas treatment technologies for municipal waste incineration: a case study in Changzhou, China. Journal of Cleaner Production 184, 912–920. Available at https://doi.org/10.1016/j.jclepro.2018.02.282. https://www.sciencedirect.com/science/article/abs/pii/S0959652618306140.

Xu, P., Li, J., Qian, J., Wang, B., Liu, J., Xu, R., et al., 2023. Recent advances in CO2 fixation by microalgae and its potential contribution to carbon neutrality. Chemosphere 319. Available at https://doi.org/10.1016/j.chemosphere.2023.137987. https://www.sciencedirect.com/science/article/abs/pii/S0045653523002540.

Yadav, G., Dash, S.K., Sen, R., 2019. A biorefinery for valorization of industrial waste-water and flue gas by microalgae for waste mitigation, carbon-dioxide sequestration and algal biomass production. Science of the Total Environment 688, 129–135. Available at https://doi.org/10.1016/j.scitotenv.2019.06.024, http://www.elsevier.com/locate/scitotenv.

Yadav, G., Sen, R., 2017. Microalgal green refinery concept for biosequestration of carbon-dioxide vis-à-vis wastewater remediation and bioenergy production: recent technological advances in climate research. Journal of CO2 Utilization 17, 188–206. Available at https://doi.org/10.1016/j.jcou.2016.12.006. http://www.journals.elsevier.com/journal-of-co2-utilization/.

Yanagi, M., Watanabe, Y., Saiki, H., 1995. CO2 fixation by Chlorella sp. HA-1 and its utilization. Energy Conversion and Management 36 (6-9), 713–716. Available at https://doi.org/10.1016/0196-8904(95)00104-L.

Yang, Y., Xu, W., Wang, Y., Shen, J., Wang, Y., Geng, Z., et al., 2022. Progress of CCUS technology in the iron and steel industry and the suggestion of the integrated application schemes for China. Chemical Engineering Journal 450. Available at https://doi.org/10.1016/j.cej.2022.138438. http://www.elsevier.com/inca/publications/store/6/0/1/2/7/3/index.htt.

Zheng, Q., Xu, X., Martin, G.J.O., Kentish, S.E., 2018. Critical review of strategies for CO2 delivery to large-scale microalgae cultures. Chinese Journal of Chemical Engineering 26 (11), 2219–2228. Available at https://doi.org/10.1016/j.cjche.2018.07.013.

Zhu, B., Sun, F., Yang, M., Lu, L., Yang, G., Pan, K., 2014. Large-scale biodiesel production using flue gas from coal-fired power plants with Nannochloropsis microalgal biomass in open raceway ponds. Bioresource Technology 174, 53–59. Available at https://doi.org/10.1016/j.biortech.2014.09.116, http://www.elsevier.com/locate/biortech.

CHAPTER 13

Life cycle assessment on sustainable biorefinery of microalgae

Nur Azalina Suzianti Feisal[1,2], Noor Haziqah Kamaludin[3,4], Tengku Nilam Baizura Tengku Ibrahim[4,5], Mohd. Azwan Ahmad[1,2], Najwa Mat Yasin[6], Zoe Tan[2], Taishi Yazawa[7], and Mia Yang Ang[8,9]

[1]MSU Centre for Climate Resilience and Strategy (m-CREST), Management and Science University, Shah Alam, Selangor, Malaysia
[2]Department of Diagnostic and Allied Health Science, Faculty of Health and Life Sciences, Management and Science University, Shah Alam, Selangor, Malaysia
[3]Center for Environmental Health and Safety Study, Faculty of Health Sciences, Universiti Teknologi MARA, Puncak Alam, Selangor, Malaysia
[4]Occupational Health and Safety Risk Management (OHSeRM) Research Initiative Group, Faculty of Health Sciences, Universiti Teknologi MARA, Puncak Alam, Selangor, Malaysia
[5]Department of Environmental Health, Faculty of Health Sciences, Universiti Teknologi MARA Cawangan Pulau Pinang, Bertam, Pulau Pinang, Malaysia
[6]Faculty of Ocean Engineering Technology (FTKK), Universiti Malaysia Terengganu, Kuala Terengganu, Terengganu, Malaysia
[7]Institute of Industrial Science, The University of Tokyo, Tokyo, Japan
[8]Department of Biomedical Sciences, Jeffrey Cheah Sunway Medical School, Faculty of Medical and Life Sciences, Sunway University, Petaling Jaya, Selangor, Malaysia
[9]Sunway Microbiome Centre, Faculty of Medical and Life Sciences, Sunway University, Petaling Jaya, Selangor, Malaysia

13.1 Introduction

Biorefineries promote the conversion of biomass to a diverse range of products and energy utilization. These products encompass, along with intermediates and end products, spanning categories including applications in food, feed, materials, and chemicals (Takkellapati et al., 2018). SDG 13, Climate Action, is particularly interconnected with biorefinery production methods, as safeguarding and preserving the environment is a fundamental element within the realm of sustainability (Degnachew et al., 2021). The integration of environmental sustainability has become increasingly prevalent in various strategies and development initiatives. This entails considering the environmental and natural resource impacts to ensure the well-being of future generations. The concept of sustainability is reflected in the "Sustainable Development Goals (SDGs)." The goals of these initiatives are to tackle poverty, improve the standard of living, decrease the pollution and depletion of natural resources, and promote the development

Applied Phycology
DOI: https://doi.org/10.1016/B978-0-443-24758-3.00007-1

of a society that is more equitable, progressive, inclusive, and equal (Asadikia et al., 2021). Biomass is available from a variety of sources, including food scraps, agricultural residues, animal waste, industrial waste, natural waste, and so on. However, biomass has been classified into five major types: garbage, wood waste, alcohol fuels, crop waste, and landfill gas. Garbage consists of domestic waste such as food, paper, cotton, and wool products, as well as organic components of household and commercial or industrial waste. Meanwhile, wood waste can come from furniture, land clearing, and agriculture, and it is currently the largest biomass energy resource. Crop waste includes grassy and woody plants, agricultural or forestry residues, oil-rich algae, and damaged plants or fruit. Recycled biomass energy can even be obtained from landfill fumes, which are a source of methane, the primary component of natural gas. Fuels, electricity, and other products that would typically be made from fossil fuels can be produced using biomass.

13.2 Biorefinery overview

Biorefineries are complex systems that comprehensively process biomass to produce value-added products and alternative energy sources, incorporating modern scientific advancements, innovations in technology, and sustainable strategies (Arias et al., 2023). Biorefineries have been highlighted as prospective options due to their ability to efficiently upgrade all biomass components to fulfill the demands of diverse production industries. This means that various biomass components may be managed and converted into useful products, such as biofuels and biochemicals, offering a flexible and sustainable approach to resource utilization. Thus, to facilitate the deployment of biorefineries, research must be conducted on novel processing lines, methods for expanding biomass upgrading at the industrial level, and avenues for raising the technological readiness level (TRL) (Singh et al., 2022a). Although there are numerous and ever-changing opportunities provided by biorefining, integrating a sustainable biorefinery should take priority over obstacles related to technology, finances, the environment, and society (Barragán-Ocaña et al., 2023). Biorefining relies on cooperation between various stakeholders, including the academic and industrial sectors, as well as funding for research, development, and innovation, to produce products with added value.

13.2.1 Definition and concept

Biorefinery (Fig. 13.1) can be defined as a distinct concept based on a sophisticated system and evolution. The National Renewable Energy

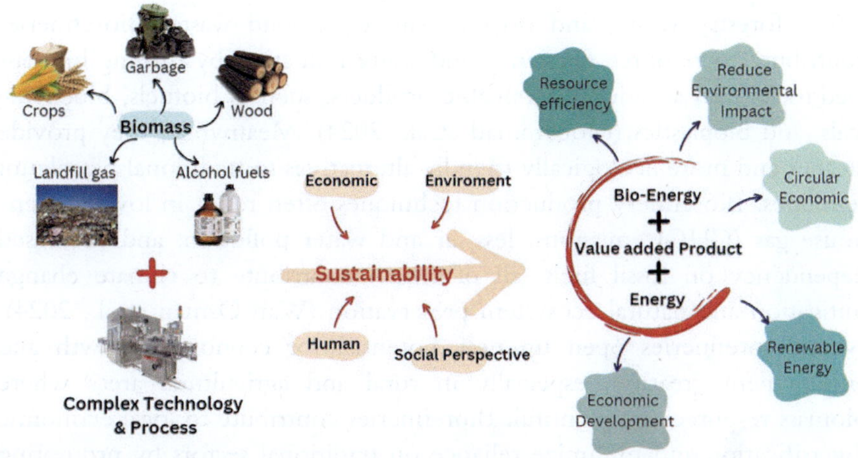

Figure 13.1 Fundamental mechanisms of biorefinery.

Laboratory defines a biorefinery as a biomass conversion process that produces numerous fuels, power, and chemicals, all of which contribute to the development of domestic bio-based industries. This concept has been applied in the petroleum refinery process (Palmeros Parada et al., 2017). Processed biomass can be transformed into a variety of commercial products and energy through a sustainable process (Cherubini et al., 2009). A biorefinery is an extensive concept modeled (Fig. 13.1) on the petrochemical refinery that produces a multitude of valuable products (Biomass Cofiring: A Renewable Alternative for Utilities, 2000). Thus, the primary concept of a biorefinery is to ensure the sustainability of economic, environmental, human, and social perspectives as integral components in transforming biomass into energy, bioenergy, and value-added products. This sustainable biorefinery concept incorporates the reduction of raw material utilization, energy consumption, and biomass dumping. In the global context of the energy crisis, biorefinement offers an alternative for generating biofuels and other processes to produce value-added products (Kumar et al., 2020).

Biorefineries play an important role in fostering sustainability through numerous fundamental mechanisms, such as resource efficiency, reduced environmental impact, economic development, renewable energy production, and the circular economy. In terms of resource efficiency, biorefineries enable the effective use of biomass resources, such as agricultural waste, forestry waste, and organic municipal solid waste. Biorefineries contribute to resource efficiency and waste reduction by turning biomass feedstocks into a variety of valuable products, such as biofuels, biochemicals, and bioplastics (Mohammad et al., 2024). Meanwhile, they provide cleaner and more ecologically friendly alternatives to traditional petroleum refineries. Biorefinery production techniques often result in lower greenhouse gas (GHG) emissions, less air and water pollution, and decreased dependency on fossil fuels, all of which contribute to climate change mitigation and natural ecosystem preservation (Wan Osman et al., 2024). Next, biorefineries open up new potential for economic growth and employment creation, especially in rural and agricultural areas where biomass resources are plentiful. Biorefineries contribute to local economic diversification and minimize reliance on traditional sectors by promoting the development of a bio-based economy (Erdiwansyah et al., 2024). Biorefineries generate renewable energy, such as biofuels, biogas, and bioelectricity. These energy sources can replace fossil fuels in transportation, heating, and power generation, lowering GHG emissions and improving energy security.

The purpose of a sustainable biorefinery is to tackle the need to gradually replace fossil fuels due to their impending depletion and to help reduce global warming caused by energy demands (Mat Aron et al., 2020). Anaerobic digestion is still the best option for producing renewable energy in certain situations, such as processing manure, but it has several drawbacks that make it necessary to develop biorefining technology and implement thorough management plans, particularly in developing nations (Khoshnevisan et al., 2021). Biorefining provides an extensive variety of options for using biomass to produce chemical products, along with additional materials that can be utilized to generate energy (Modahl et al., 2015). Biorefineries are fundamental to the circular economy concept, which involves using resources in a closed-loop system to reduce waste and increase resource efficiency. Biorefineries help to accelerate the transition to a more sustainable and circular resource management paradigm by vaporizing biomass feedstocks and creating high-value products. Biorefineries have an important and innovative function in the circular economy, offering simultaneous economic and environmental advantages (Barragán-Ocaña et al., 2023). A circular economy can benefit greatly from the use of biomass in biorefineries to produce bioenergy and biomaterials, which can lessen GHG emissions and have a positive environmental impact (Leong et al., 2021). The biorefining of apple pomace to create valuable products, decrease waste, and promote sustainability is an example of how biorefineries are essential to the advancement of the economy because they give value to waste and enable circularity (Awasthi et al., 2021).

13.3 Environmental considerations in biorefinery operations

The uncontrolled generation of waste resulting from consumption patterns is becoming a notable concern in environmental management (Abubakar et al., 2022). Simultaneously, this issue is intertwined with the environmental dimension as it helps mitigate emissions associated with the mentioned activities. Moreover, from a societal standpoint, the establishment of small-scale local biorefineries holds promise for fostering job creation and encouraging farmers to adopt more sustainable agricultural practices. These practices include enhancing crop yields, reducing the reliance on fertilizers and pesticides, and promoting more sustainable use of water sources (Dhanaraju et al., 2022). Carrying out an environmental impact assessment of biorefineries is a vital step in thoroughly assessing their sustainability within a broader context (Perez-Almada et al., 2023). In this regard, future biorefineries should be designed with a comprehensive approach to biomass feedstock utilization, aligning with the principles of the circular economy concerning energy consumption, water usage, raw material sourcing, and emission levels (Vural Gursel et al., 2022). At the forefront of contemporary sustainability and circular economy objectives lies the concept of biorefining. It entails the production of various marketable bioproducts from diverse biomass feedstock materials, employing a range of conversion technologies, including physical, chemical, biochemical, and thermochemical processes (Clauser et al., 2021). The large-scale progression of biorefineries continues to face significant challenges and barriers, including difficulties in mobilizing biomass resources, economic viability concerns, and uncertainties surrounding their environmental impacts (Perez-Almada et al., 2023). Recognizing the correlation and trade-offs between economic and environmental performance is crucial for the development and expansion of sustainable biorefinery value chains.

13.3.1 Energy consumption

One essential factor in meeting society's energy and material sustainability needs is the adoption of an economy that relies on renewable resources derived from biomass for the generation of both materials and energy (Popp et al., 2021). With the growing population, the need for an adequate energy supply has become imperative, necessitating diverse forms of energy at every stage. Factors such as economic expansion, technological advancements, improved living standards, urbanization, industrial development, transportation needs, climate control requirements, population behaviors, and cultural influences collectively contribute to the rising energy demand, resulting in its continuous growth (Khan et al., 2021b; Kalak, 2023). The energy demand is not solely attributed to population growth, (Fig. 13.2) (Holechek et al., 2022).

In the 19th century, the agricultural sector witnessed the introduction of novel machinery, notably steam engines. The widespread adoption of these agricultural machines, coupled with the increased use of chemicals by farmers, marked the onset of Agricultural Era 2.0. This era was characterized by enhanced efficiency and productivity in farming practices, bringing about the integration of advanced technologies and equipment on farms (Hemathilake and Gunathilake, 2021). In essence, the energy output relative to the energy inputs in the production of most biofuels is insufficient to ensure profitability (Janota et al., 2023). Wind, solar, and biofuels all pose significant adverse environmental impacts, particularly

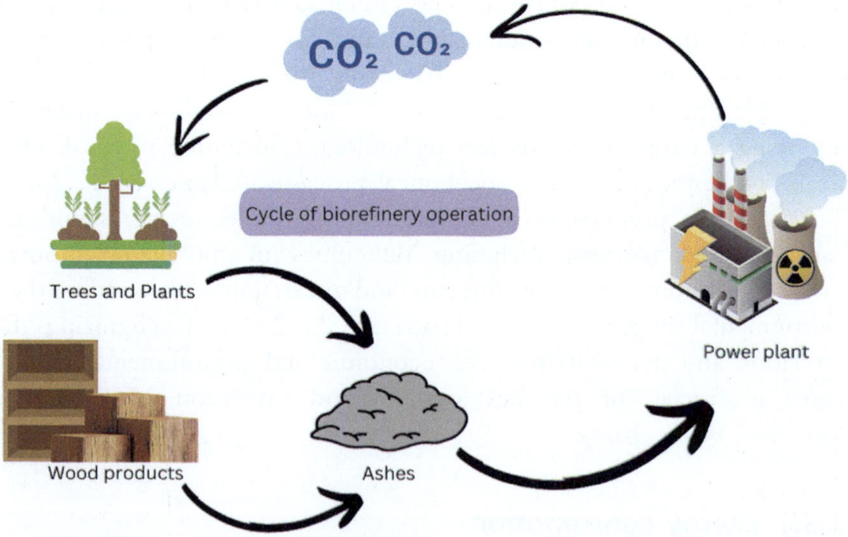

Figure 13.2 The cycle of biorefinery operation.

when expanded to meet most of the world's energy demands (Osman et al., 2023). Biomass can be decomposed naturally by microorganisms or incinerated in thermal facilities such as power plants, where chemical energy is converted into mechanical or electrical energy. During natural composting, biomass releases carbon in the form of methane or CO_2 into the atmosphere, thereby completing the closed cycle within the biorefinery energy system.

Approximately 58% of fossil fuels are utilized in the transportation of individuals and goods through road, rail, air, and marine travel. In 2016, the transportation sector alone accounted for 16% of the total global GHG emissions, underscoring the urgent requirement for eco-friendly alternatives to gasoline and diesel (Solaymani, 2022). Numerous nations have implemented policy measures aimed at boosting the uptake of electric vehicles and have declared electrification targets. As an example, the United Kingdom aims to cease the sale of new petrol and diesel cars and vans by 2030, mandating that all vehicles possess substantial zero-emission capabilities. When considering various renewable energy options, biorefinery is arguably a promising source of renewable energy within the transportation sector.

13.3.2 Raw material sourcing

The effective use of resources is intricately connected to raw material consumption; adopting a comprehensive approach to utilization leads to a decrease in the consumption of new materials. The implementation of circularity action plans, highlighting the integrated use of resources, coupled with the creation of biorefinery facilities dedicated to converting waste streams into raw material streams, is pivotal in averting situations where regions with the greatest demand for new materials are the least sustainable (Lizundia et al., 2022). Fig. 13.3 serves as the primary raw material in a biorefinery, encompassing a range of sources such as food crops, energy crops like switchgrass or prairie perennials, crop residues like corn stover, wood waste, by-products that include both mill residues and noncommercial biomass found in the woods, and animal manure (Hernández-Chaverri et al., 2021; Casau et al., 2022). In recent years, the definition of biomass has expanded to encompass a variety of sources, including algae, construction debris, municipal solid waste, yard waste, and food waste (Wang and Tester, 2023). Biorefinery systems can be categorized based on the diversity of raw materials they utilize, as

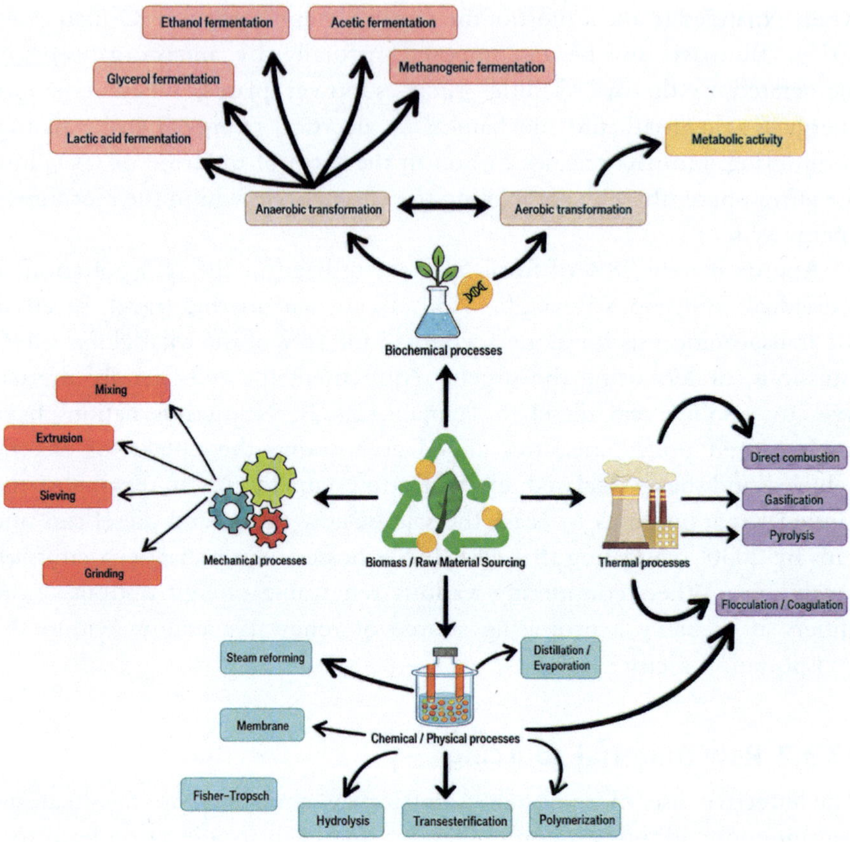

Figure 13.3 Overall framework of the biomass/raw material sourcing within biorefinery operations.

biorefineries process entire crops using the whole plant as the substrate, similar to an oil refinery where oilseeds such as rapeseed, sunflower, and soybeans are employed (Bartkowiak et al., 2022). Refineries adopting environmentally friendly practices utilize nonedible portions of plants or entire energy crops such as wet biomass for agriculture or the food industry (Saleem, 2022). Hence, there exist numerous concepts for biorefinery systems varying based on the utilized raw materials, the technological processes employed, and the resultant products (Takkellapati et al., 2018). From this standpoint, it is demonstrated that substantial portions of fossil raw material in the economy could be substituted with biomass by the years 2030 and 2050 (Sanders and Bos, 2013). Transforming biomass into value-added products via the "sugar platform" is a fundamental aspect of biorefinery concepts aimed at substituting fossil-based products (Chaturvedi et al., 2020). Glucose serves as a primary source for subsequent processing through various pathways in the development of industrial bioproducts, including biopolymers, biochemicals, and biofuels. Consideration of environmental aspects and impacts is a critical criterion for biorefinery concepts within the biomass industry.

13.4 Life cycle assessment in biorefinery

Biorefineries play a pivotal role in the transition toward a sustainable bioeconomy by converting biomass into valuable products such as biofuels, biochemicals, and biomaterials. However, to ensure their sustainability, it is essential to comprehensively evaluate their environmental impacts throughout their life cycle (Yao et al., 2024). Life cycle assessment (LCA) provides a systematic framework to quantify and assess the environmental burdens associated with biorefinery processes, from raw material extraction to end-of-life disposal (Kovacs et al., 2022). LCA is defined as a tool for evaluating the inputs, outputs, and potential environmental impacts of a product system throughout its life cycle, which can help understand the environmental impacts of a product from the acquisition of raw materials to final disposal. This systematic approach allows stakeholders to identify areas for improvement, prioritize resource-efficient strategies, and make informed decisions to minimize environmental footprints (Alaloul et al., 2021). Each stage in LCA presents opportunities for optimizing resource utilization, reducing emissions, and minimizing environmental burdens, as shown in Fig. 13.4. For example, evaluating different feedstock options based on their environmental performance can guide biorefinery operators in selecting sustainable biomass sources (Kumar and Verma, 2021). Additionally, assessing energy consumption, emissions of GHGs, and other pollutants during conversion processes aids in optimizing process efficiency and reducing environmental impacts.

Furthermore, LCA enables the comparison of different biorefinery pathways and technologies, helping to identify the most environmentally favorable options. By quantifying environmental indicators such as carbon footprint, water usage, and land occupation, stakeholders can prioritize investments in technologies with lower environmental footprints and higher resource efficiency (Shabir et al., 2023). Importantly, LCA also

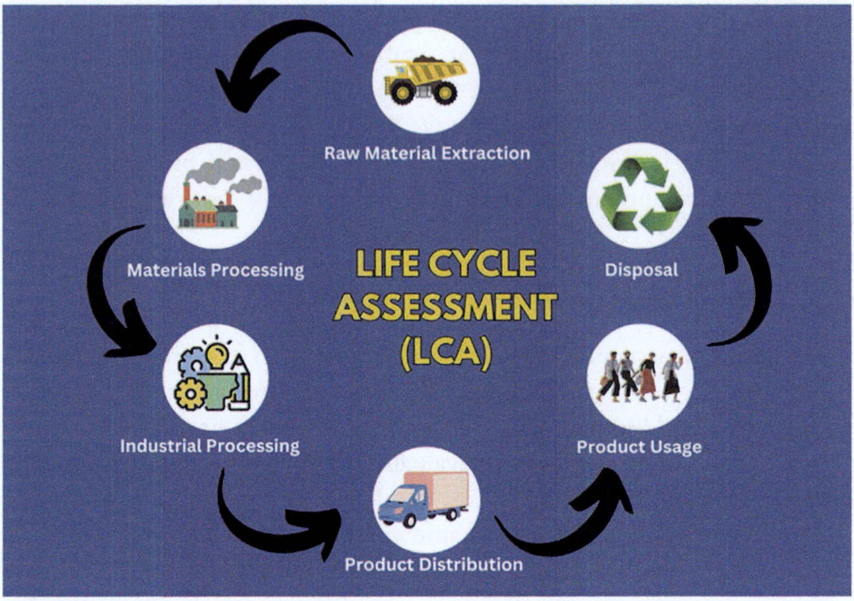

Figure 13.4 Life cycle assessment (LCA) overview.

facilitates the identification of potential trade-offs and unintended consequences across different life-cycle stages (Ravikumar et al., 2024). For instance, while certain biorefinery processes may reduce GHG emissions compared to fossil fuel-based alternatives, they could pose challenges in terms of land use change or water resource depletion (Khan et al., 2021a). Through comprehensive LCA studies, these trade-offs can be evaluated, and mitigation strategies can be developed to ensure overall environmental sustainability. Moreover, LCA can be utilized iteratively throughout the development and optimization of biorefinery processes (Khoo et al., 2019). By conducting LCA at different stages of technology development, researchers and practitioners can identify opportunities for improvement and innovation, leading to the design of more sustainable and efficient biorefinery systems. This iterative approach fosters continuous learning and improvement, driving the evolution of the economy toward greater sustainability over time. Furthermore, LCA can enhance transparency and communication among stakeholders involved in the biorefinery value chain (Petit et al., 2018). By sharing LCA results and methodologies, stakeholders can foster dialog, build trust, and collaborate toward common sustainability goals. Transparency in LCA studies also enables consumers to make informed choices by considering the environmental implications of the products they use, thereby driving market demand for sustainable alternatives.

13.4.1 Life cycle assessment methodology

The application of LCA in the context of biorefinery operations is instrumental in gauging the environmental footprint of biomass conversion processes. LCA offers a structured framework to meticulously evaluate and quantify the environmental impacts associated with various stages of biorefinery operations, ranging from the procurement of raw materials to the disposal of end products (Ali et al., 2023). Through a series of methodological steps, LCA enables stakeholders to gain comprehensive insights into the sustainability performance of biorefinery systems, facilitating informed decision-making and continual improvement initiatives. In essence, the LCA methodology enables stakeholders to dissect the environmental footprint of biorefinery operations into distinct stages, allowing for a granular examination of the processes involved (Barahmand and Eikeland, 2022). For instance, in the procurement phase, LCA can assess the environmental impact of different biomass sourcing strategies, such as the cultivation of dedicated energy crops versus the utilization of agricultural residues or forestry by-products (Martin et al., 2023). This analysis can help identify the most sustainable feedstock options based on factors like land use, water consumption, and carbon emissions. During the conversion processes within the biorefinery, LCA can quantify the environmental implications of various technologies and production methods (Yu et al., 2022). For example, comparing different biofuel production pathways, such as ethanol fermentation versus biodiesel transesterification, allows stakeholders to discern which processes have lower energy consumption, reduced GHG emissions, and minimized waste generation (Singh et al., 2022b). Research conducted by Osman et al. (2023) compared the environmental performance of different lignocellulosic biomass conversion pathways using LCA methodology, highlighting the importance of feedstock selection and process optimization in reducing environmental impacts. Similarly, a study by Kadiyala et al. (2016) evaluated the life-cycle GHG emissions of biofuels produced from various feedstocks and conversion technologies, underscoring the significance of LCA in guiding sustainable bioenergy development.

13.4.2 Goal and scope definition

At the onset of an LCA in the biorefinery sector, stakeholders define clear objectives and delineate the scope of the assessment. This involves specifying the functional unit under analysis, such as the quantity of biofuel produced, establishing the system boundaries to be considered (including all relevant processes from raw material acquisition to product disposal), and selecting pertinent impact categories for evaluation (Liu et al., 2021). These steps lay the foundation for a comprehensive analysis of the environmental impacts associated with biomass conversion processes, guiding stakeholders in making informed decisions to enhance sustainability (Talwar and Holden, 2022). Given the diverse range of products and processes within biorefineries, careful consideration is given to the selection of impact categories, such as GHG emissions, energy consumption, water usage, land occupation, and biodiversity impacts. Furthermore, this step also involves considering temporal and geographical aspects to ensure the relevance and applicability of the assessment (Rashid et al., 2023). The integration may specify the time frame over which the assessment will be conducted and delineate the geographic scope to account for regional variations in environmental factors and regulatory frameworks. Studies by Sadaghiani et al. (2023) emphasized the need for a clear goal and scope definition to ensure the relevance and reliability of LCA results in assessing the environmental sustainability of bioenergy systems. Similarly, a study by Khoshnevisan et al. (2020) showed the significance of specifying system boundaries and impact categories in LCA studies to facilitate consistent and transparent assessments of biorefinery operations. By laying the groundwork for a systematic analysis of environmental impacts, goal and scope definition ensure that LCA results are relevant, reliable, and actionable for enhancing sustainability in biorefinery operations.

13.4.3 Inventory analysis

The inventory analysis in the LCA phase entails the systematic collection, compilation, and quantification of input and output data across all stages of the biorefinery life cycle (Islam et al., 2016). This encompasses activities such as raw material extraction, transportation, biomass conversion processes, product distribution, and end-of-life scenarios. For biorefineries, this involves characterizing resource inputs such as biomass feedstocks, water, and energy and quantifying emissions and waste outputs generated throughout the production chain (Dunn, 2019). For example, in assessing the environmental performance of biofuel production, inventory analysis may involve quantifying the amount of biomass feedstock, fossil fuels, and electricity consumed during the conversion process (Quek and Balasubramanian, 2014). This data collection process may draw upon a range of sources, including process documentation, industry databases, scientific literature, and empirical measurements. Furthermore, inventory analysis may entail assessing the environmental consequences of transportation activities associated with biorefinery operations (Moutik et al., 2023). This includes quantifying the energy consumption and emissions associated with transporting raw materials to the biorefinery, as well as distributing finished products to end users (Benavides et al., 2017). By considering the entire supply chain, inventory analysis provides a holistic perspective on the environmental impacts of biorefinery operations. Research conducted by Lu et al. (2017) showed a comprehensive inventory analysis to quantify the life-cycle GHG emissions of biofuels produced from various feedstocks and conversion technologies, highlighting the significance of data collection and quantification in assessing environmental impacts. Similarly, a study by Zoppi et al. (2023) utilized inventory analysis to assess the energy and environmental performance of different lignocellulosic ethanol production pathways, emphasizing the importance of considering resource inputs and emissions throughout the entire biorefinery life cycle (Moutik et al., 2023). Through systematically collecting and quantifying data, inventory analysis enables informed decision-making and the identification of opportunities for improving the environmental sustainability of biorefinery operations.

13.4.4 Impact assessment

Following the inventory analysis in LCA, the collected data are subjected to an impact assessment process to elucidate the environmental implications of biorefinery operations. Impact assessment involves translating the inventory data into quantitative indicators of environmental impact across predefined impact categories (Esnouf et al., 2019). This step encompasses the utilization of impact assessment models, environmental fate and exposure factors, and characterization factors to quantify the potential effects on ecosystems, human health, and natural resources (Belyanovskaya et al., 2020). In the biorefinery context, impact categories may encompass carbon footprint, water footprint, land use change, eutrophication potential, acidification potential, and toxicity impacts (Bishop et al., 2021). One key aspect of impact assessment is the categorization and characterization of environmental impacts according to predefined impact categories. These categories encompass a wide range of environmental indicators, including GHG emissions, energy consumption, water usage, land occupation, eutrophication potential, acidification potential, and toxicity impacts (Woods et al., 2016). Currently, in assessing the environmental performance of biofuel production, stakeholders may evaluate the carbon footprint, water footprint, and potential impacts on biodiversity associated with different feedstocks and conversion technologies. For example, researchers may utilize life-cycle impact assessment models such as ReCiPe or IMPACT methods that come from SimaPro (LCA software) to assess the potential effects of biorefinery operations on human health, ecosystems, and natural resources (Prioux et al., 2022). Moreover, impact assessment facilitates the identification of environmental hotspots, trade-offs, and sensitivities within the biorefinery life cycle. The studies by Cai et al. (2018) conducted on the impact assessment of different biorefinery pathways for producing bio-based chemicals highlighted the importance of considering multiple environmental indicators to assess overall sustainability. Similarly, a study by Homagain et al. (2015) utilized impact assessment to evaluate the environmental impacts of various bioenergy systems, emphasizing the need for robust characterization of environmental impacts to inform policy and investment decisions. By quantifying environmental impacts across various impact categories, impact assessment enables informed decision-making and facilitates the identification of opportunities for improving the sustainability performance of biorefinery operations.

13.4.5 Interpretation

Interpretation is a crucial phase in the LCA process within biorefinery operations, as it involves synthesizing and contextualizing the results of impact assessment to derive meaningful insights and recommendations (Andrade et al., 2022). This phase enables stakeholders to understand the implications of LCA findings, identify areas for improvement, and make informed decisions to enhance the environmental sustainability of biomass conversion processes (Sabet et al., 2023). One key aspect of interpretation is analyzing the environmental hotspots and trade-offs identified through impact assessment. By examining the results of LCA, stakeholders can pinpoint processes or inputs within the biorefinery life cycle that have the most significant environmental impacts. For example, if the impact assessment reveals that a particular feedstock option or conversion technology contributes disproportionately to GHG emissions or water usage, stakeholders can prioritize efforts to mitigate these impacts through process optimization or alternative sourcing strategies (Pfister et al., 2017). For instance, if LCA results indicate that a certain biofuel production pathway offers lower environmental impacts compared to fossil-fuel alternatives, stakeholders may use this information to advocate for policy support or investment incentives to promote the adoption of sustainable biorefinery technologies (Jeswani et al., 2020). Research by KC et al. (2020) emphasized the need for critical interpretation of LCA results to ensure that findings are relevant, reliable, and actionable for stakeholders. Similarly, a study by Saranya and Ramachandra (2020) utilized interpretation to identify opportunities for improving the environmental performance of biofuel production processes, highlighting the importance of considering trade-offs and unintended consequences in decision-making. By sharing LCA findings with relevant stakeholders, including policymakers, investors, consumers, and industry partners, stakeholders can build consensus around sustainability goals and drive collective action toward more sustainable biorefinery practices. The interpretation is a critical phase in the LCA process within biorefinery operations, enabling stakeholders to derive meaningful insights from LCA findings and make informed decisions to enhance the environmental sustainability of biomass conversion processes.

13.5 Case studies and applications

The application of LCA plays a crucial role in understanding and mitigating the environmental impacts of sustainable biorefinery operations (Saavedra del Oso et al., 2023). These assessments provide real-world examples of how different approaches and technologies can influence the overall sustainability of biorefinery processes. By examining specific cases, researchers and stakeholders gain valuable insights into the challenges and opportunities associated with sustainable biorefinery practices. Such cases involve the comparison of various feedstock options for biofuel production (Soleymani Angili et al., 2021). Researchers may analyze the environmental impacts of using different biomass sources, such as agricultural residues, energy crops, or algae, to produce biofuels (Giraldi-Díaz et al., 2018). Through LCA, they can quantify factors like GHG emissions, energy consumption, and land use change associated with each feedstock

Table 13.1 The application of life cycle assessment (LCA) in environmental impacts on sustainable biorefinery.

No.	Case study types	LCA model	LCA input	LCA output	Application	References
1	Comparison of biofuel production pathways	SimaPro	Inventory data on biomass feedstocks, energy consumption, emissions	Environmental impact assessment (e.g., carbon footprint, water usage)	Identifying the most sustainable biofuel production methods	Cavalaglio et al. (2023), Spekreijse et al. (2019)
2	Assessment of lignocellulosic biomass conversion technologies	Gabi	Data on raw material inputs, energy consumption, emissions	Quantification of environmental burdens (e.g., greenhouse gas emissions, resource depletion)	Guiding technology selection and process optimization	Wang et al. (2020), Mohamed et al. (2021)
3	Evaluation of various biomass sourcing strategies	OpenLCA	Data on feedstock acquisition, transportation, land use	Analysis of environmental impacts throughout the supply chain	Optimizing feedstock procurement practices for reduced environmental footprint	Maier et al. (2019), Stančin et al. (2023)
4	LCA of biochemical production from agricultural residues	Umberto NXT LCA	Input-output data on resource use, emissions, waste generation	Identification of environmental hot spots and trade-offs	Informing decision-making in biochemical production process design	Bhakar and Singh (2018), Lodato et al. (2020)

5	Comparative LCA of bio-based chemical production pathways	SimaPro	Inventory data on feedstock utilization, energy inputs, emissions	Assessment of environmental performance indicators (e.g., acidification potential, eutrophication potential)	Supporting selection of sustainable chemical production routes	Forte et al. (2016), Forte et al. (2016), Mohd Yusof et al. (2019)
6	Analysis of bioethanol production from various feedstocks	GaBi	Data on biomass composition, energy requirements, waste generation	Evaluation of environmental impacts throughout the life cycle	Guiding feedstock selection and process optimization for enhanced sustainability	Keller et al. (2020), Ott et al. (2023)
7	LCA of biodiesel production from algae	OpenLCA	Input-output data on algae cultivation, processing, and transportation	Quantification of environmental impacts (e.g., energy consumption, water usage)	Informing sustainable algae-based biodiesel production strategies	Kiemel et al. (2022)
8	Comparative assessment of biorefinery coproducts utilization	Umberto LCA+	Data on coproduct generation, utilization pathways, emissions	Analysis of environmental benefits and trade-offs associated with coproduct utilization	Supporting decision-making for maximizing value and minimizing waste in biorefinery operations	Abubakar Jumare et al. (2019), Mazur and Olenchuk (2023)

(*Continued*)

Table 13.1 (Continued)

No.	Case study types	LCA model	LCA input	LCA output	Application	References
9	Environmental impact assessment of bioplastics production	SimaPro	Input-output data on raw material extraction, processing, disposal	Quantification of environmental burdens (e.g., carbon footprint, toxicity impacts)	Guiding the development of sustainable bioplastic production processes	Bałdowska-Witos et al. (2020), Kamalakkannan et al. (2022)
10	Life-cycle analysis of integrated biorefinery systems	GaBi	Data on biomass feedstock sourcing, conversion processes, product distribution	Evaluation of overall environmental performance indicators (e.g., cumulative energy demand, land occupation)	Supporting the design and optimization of integrated biorefinery concepts	Pacana et al. (2023), Slotyuk et al. (2024)

option. This analysis helps stakeholders identify the most environmentally sustainable feedstock choices for biofuel production, guiding decision-making in biorefinery operations (Raschke et al., 2021). Table 13.1 offers a comprehensive overview of various case study types, LCA models utilized, inputs, outputs, and applications concerning the evaluation of environmental impacts on sustainable biorefinery operations. Each case study type delves into a specific facet of biorefinery sustainability assessment, spanning from the comparison of diverse biofuel production pathways to scrutinizing the environmental repercussions of integrated biorefinery systems.

LCA models such as SimaPro, GaBi, OpenLCA, and Umberto NXT LCA stand as prevalent tools used in conducting LCAs within biorefinery studies. These models serve as robust frameworks for quantifying environmental impacts throughout the entire life cycle of biorefinery processes. They empower stakeholders to meticulously assess various factors, including GHG emissions, energy consumption, water usage, and land occupation, thereby facilitating a comprehensive understanding of the environmental implications of biorefinery operations (Vega et al., 2019). Inputs for LCA studies in biorefineries encompass a wide array of data, including details on feedstock sourcing, conversion processes, energy inputs, emissions, and waste generation. These inputs form the basis for the subsequent analysis and provide crucial insights into the resource utilization and environmental burdens associated with biorefinery operations (Yin et al., 2024).

On the other hand, outputs derived from LCA analysis consist of environmental impact indicators, meticulously quantified to provide insights into the sustainability performance of biorefinery operations. These indicators serve as valuable metrics for gauging the environmental footprint of biorefinery processes and assist stakeholders in identifying areas for improvement and optimization (Valdez-Valdes et al., 2024). The applications of these case studies are multifaceted and encompass various dimensions of sustainable biorefinery operations. They play pivotal roles in informing technology selection, optimizing process efficiency, guiding feedstock sourcing strategies, and supporting decision-making in biorefinery design and operation. By leveraging the insights gleaned from these case studies, stakeholders can make informed decisions aimed at enhancing the sustainability and environmental responsibility of biorefinery operations, thereby contributing to the advancement of a more sustainable bioeconomy. The application of LCA in evaluating environmental impacts on sustainable biorefinery operations plays a crucial role in promoting sustainability and driving innovation within the biorefinery sector. By providing a systematic and holistic approach to assessing environmental footprints, LCA enables stakeholders to make informed decisions, optimize processes, and advance toward a more environmentally responsible bioeconomy.

13.6 Environmental impact categories

In the context of LCA applied to biorefinery operations, assessing environmental impacts encompasses a broad range of considerations aimed at understanding the consequences of biomass conversion processes on the surrounding environment (Ogunleye et al., 2024). Biorefineries play a pivotal role in the transition toward sustainable bio-based economies by converting biomass into various valuable products, including biofuels, biochemicals, and biomaterials. However, the sustainability of these processes hinges on the careful evaluation and mitigation of their environmental footprints. LCA in biorefineries involves a holistic examination of the inputs, outputs, and associated consequences throughout the entire life cycle of biorefinery operations. Through LCA assessment, the environmental impact can be categorized into three variables: GHG emissions, energy consumption, and land occupation.

13.6.1 Greenhouse gas emissions

This category includes the quantification of emissions such as carbon dioxide (CO_2), methane (CH_4), and nitrous oxide (N_2O) throughout the biorefinery life cycle, from feedstock production to product distribution (Zhang et al., 2020). Evaluating GHG emissions is critical for understanding the impact of biorefinery operations on climate change. One significant source of GHG emissions in biorefineries is the combustion of fossil fuels for energy during various stages of the process, such as feedstock harvesting, transportation, and conversion (Mesfun et al., 2023). Additionally, emissions may arise from indirect sources, such as land use change associated with feedstock cultivation and methane release from the anaerobic decomposition of organic matter. Xu et al. (2022) conducted a comprehensive LCA of biofuels, highlighting the significance of considering GHG emissions throughout the entire life cycle of biorefinery processes. Similarly, a study by Johannes and Xuan (2024) assessed the carbon footprint of lignocellulosic ethanol production, emphasizing the need to minimize emissions from energy-intensive processes, such as feedstock pretreatment and fermentation. For instance, integrating carbon capture and storage (CCS) technologies can help capture and store carbon dioxide emissions from biorefinery processes, thereby mitigating their impact on the atmosphere. Additionally, optimizing energy efficiency and utilizing renewable energy sources can reduce reliance on fossil fuels and minimize GHG emissions associated with energy consumption.

13.6.2 Energy consumption

This category involves evaluating the energy inputs required for various processes within the biorefinery, including feedstock cultivation, conversion, and product transportation (Ahorsu et al., 2018). Examining energy use will assist in discovering possibilities to improve energy efficiency and reduce reliance on fossil fuels. Energy consumption includes the energy inputs needed for the various stages of the biorefinery life cycle. For example, in the production of biofuels such as biodiesel or bioethanol, significant energy is often consumed during the conversion process, where biomass undergoes biochemical or thermochemical treatments to extract and refine biofuels (Grande et al., 2021). Additionally, energy is required for the cultivation of feedstock crops, such as irrigation, fertilization, and mechanized harvesting. Furthermore, energy is consumed during transportation activities, including the transportation of feedstock from farms to biorefineries and the distribution of bioproducts to end users. Several studies have emphasized the importance of assessing energy consumption in biorefinery operations. Research by Morales-Vera et al. (2022) conducted an LCA of lignocellulosic ethanol production, highlighting the significance of energy inputs in determining the overall environmental performance of biorefinery processes. Similarly, a study by Wang et al. (2024) evaluated the energy balance of biofuel production pathways, emphasizing the need to minimize energy consumption, which enables stakeholders to identify energy-intensive processes and explore opportunities for energy optimization and renewable energy integration within biorefinery operations. Additionally, utilizing renewable energy sources, such as solar, wind, or biomass-derived energy, can further reduce the carbon footprint of biorefinery operations and enhance their sustainability.

13.6.3 Land occupation

This category involves assessing the land area required for feedstock cultivation, biorefinery infrastructure, and other activities associated with biorefinery operations (Makepa and Chihobo, 2024). Evaluating land occupation helps to understand the potential impacts on biodiversity, soil quality, and land use change, especially in regions with limited available land. Assessing land occupation within biorefinery operations is vital for understanding the spatial footprint of biomass conversion processes and their potential impacts on land use and biodiversity (Burland and von Cossel, 2023). Land occupation encompasses the land area required for various activities associated with biorefinery operations, including feedstock cultivation, biorefinery infrastructure, and ancillary facilities. For example, in the cultivation of energy crops such as sugarcane or oil palm, extensive land areas are often dedicated to feedstock production, leading to land use change and potential impacts on ecosystems and biodiversity (Acosta et al., 2016). Additionally, land is occupied by biorefinery facilities, including processing plants, storage facilities, and transportation infrastructure, further contributing to land occupation. Several studies have emphasized the importance of assessing land occupation in biorefinery operations. Research by Borole (2015) conducted a comprehensive analysis of land use change associated with biofuel production, highlighting the significance of land occupation in determining the overall sustainability of bioenergy pathways. Similarly, a study by Koponen et al. (2018) evaluated the land use impacts of bioenergy sustainability assessments. Yet, adopting agroforestry or mixed-cropping systems can enhance land productivity and biodiversity while reducing the land area required for feedstock cultivation. Additionally, implementing sustainable land management practices, such as no-till agriculture or integrated pest management, can improve soil quality and reduce the environmental footprint of biorefinery operations.

13.7 Greenhouse gas emissions

Interest in renewable resources has grown as worries about the availability and expense of exploiting fossil-fuel resources have increased. However, lowering GHG emissions is being suggested more and more as a crucial element in accomplishing the world's sustainable development objectives. As a result, there is now more interest in using renewable biomass resources to produce a wide range of essential compounds. Because renewable biomass resources can be processed into chemicals, they may eventually take the place of fossil crude oil as a source of carbon (Kajaste, 2014). The extensive use of fossil fuels has resulted in many environmental issues, including pollution, resource depletion, and carbon dioxide (CO_2)

emissions. These issues emphasize the drawbacks of an imbalanced ratio of resource demands related to this nonrenewability. A significant source of raw materials for a variety of goods, including biofuels, is biomass obtained from agricultural products, as seen by the increased interest in alternative renewable resources such as CO_2-neutralizing materials (Jeswani et al., 2020).

Carbon dioxide (CO_2), methane (CH_4), nitrous oxide (N_2O), perfluorocarbons (PFCs), hydrofluorocarbons (HFCs), and sulfur hexafluoride (SF_6) are the principal GHGs. Since the study of climate change is still in its infancy, scientists are continuously learning about the intricate connection between human activity and climate change (Dervash et al., 2023). Nonetheless, there is compelling evidence that the climate is being significantly impacted by the rise in GHG emissions from human sources over the last 100 years. Human lives and development are already being negatively impacted by climate change, especially global warming (Mikhaylov et al., 2020). Sea levels are rising, and extreme weather events are occurring more often and intensely (Upadhyay, 2020), where 2016 was the hottest year on record for the entire planet. There have been two more consecutive years of heat records being broken after this. Given the continued increase in global GHG emissions, these patterns are probably here to stay (A Planet in Peril, 2020). Globally, governments monitor and report GHG emissions originating from human activity. The consequences of climate change may be lessened, and our world can be preserved for future generations by adopting measures to limit GHG emissions.

Recognizing the dire implications of these rising GHG emissions and their established role in exacerbating climate change and its effects on our planet, it is crucial to delve deeper into practical measures, particularly in the transportation sector, where significant reductions in emissions can be achieved through informed fuel choices and sustainable practices. The data on GHG emissions per kilometer traveled provides valuable insights into the environmental impact of different transportation modes (Table 13.2). When considering fuel for transportation, various options have different GHG emission levels per kilometer traveled. Fischer–Tropsch diesel from biomass emits 15–55 gCO_2-eq./km, bioethanol from lignocellulose emits 25–50 gCO_2-eq./km, biogas emits 25–100 gCO_2-eq./km, bioethanol from sugar cane emits 50–75 gCO_2-eq./km, biodiesel (rapeseed, soy, and sunflower) emits 80–140 gCO_2-eq./km, and bioethanol from other crops (corn, sugar beet, and wheat) emits 100–195 gCO_2-eq./km. To reduce these emissions, strategies such as promoting the use of lower-emission

Table 13.2 Greenhouse gas (GHG) emissions per kilometer traveled with renewable fuels versus fossil fuels.

Fuel for transportation	GHG emission (gCO_2-eq./km)
Fischer–Tropsch diesel from biomass	15–55
Bioethanol from lignocellulose	25–50
Biogas	25–100
Bioethanol from sugarcane	50–75
Biodiesel (rapeseed, soy, sunflower)	80–140
Bioethanol from other crops (corn, sugar beets, wheat)	100–195
Natural gas	155–185
Diesel	185–220
Gasoline	210–220

fuels like Fischer–Tropsch diesel from biomass and bioethanol from lignocellulose, increasing the adoption of biogas, and encouraging sustainable practices in agriculture for biofuel production can play a significant role in mitigating the environmental impact of transportation.

Countries around the world have undergone diverse development cycles since 1990, with notable shifts in economic structures and population dynamics. The EU27 + United Kingdom and the United States of America (USA), for instance, have experienced significant growth in their service sectors, surpassing traditional manufacturing and heavy industries as primary contributors to national gross domestic product (GDP) (Andrejiová et al., 2020). In contrast, developing economies like China, India, Brazil, and several Southeast Asian nations have witnessed unprecedented economic expansion driven by industrialization and accompanied by substantial population growth. These socioeconomic changes have coincided with global-scale technological advancements spurred by climate change mitigation policies and market forces, leading to an increased adoption of renewable energy sources worldwide (Foster et al., 2023).

In parallel with economic transformations, there has been a notable shift in energy consumption patterns across different regions. Developed economies, like those in Europe and other high-income nations, have witnessed a decline in coal usage alongside an uptick in renewable energy adoption and initial increases in natural gas consumption, followed by a decline in recent years (Oreggioni et al., 2021). These shifts have been facilitated by advancements in energy production technologies, such as natural gas-fired combined cycles, fuel economy standards for vehicles, and improved heating and cooling systems in buildings. Conversely, fast-growing industrializing economies have met their rising energy demands primarily through increased solid fuel usage while also scaling up renewable and natural gas utilization (Khaleel et al., 2022; Sikder et al., 2022). The top ten GHG-emitting countries, including China, the United States, the European Union, India, and others, collectively account for a significant portion of global emissions. Efforts to address climate change were underscored by the Paris Climate Agreement, ratified by over 189 nations responsible for most global emissions. Despite progress in curbing GHG emissions from the energy sector since the 1990s, continued efforts across all industries are necessary to achieve the ambitious goal of reducing emissions to zero by 2050 and mitigating the worst impacts of climate change (Bacchetta, 2023; Yadav et al., 2023; Cordonier Segger, 2016).

13.7.1 Carbon footprint

The term "carbon footprint" is often used by meteorologists and the public in discussions about taking responsibility and actions to address global warming as well as climate change. The usage of the term has been steadily increasing for years, driven by its relevance and urgency. Currently, the phrase is used in many media as well as in the realms of business and government. The definition of "carbon footprint" has become a prominent issue. The phrase "carbon footprint" lacks a universally accepted scholarly meaning despite its widespread use. There is uncertainty and potential misunderstanding about the definition, evaluation, and specific unit to be used for the phrase (Durojaye et al., 2020; Pandey et al., 2011). The term "carbon footprint" is widely used in the Ecological Footprinting field to refer to a specific amount of GHG emissions linked to human activities related to production and consumption that are considered impactful on climate change. However, this is the only aspect where there is agreement, since there is no consensus on how to assess or quantify carbon footprint. One definition is based on direct CO_2 emissions compared to total life-cycle GHG emissions; however, the specific units of measurement are ambiguous (Sarkodie, 2021; Scrucca et al., 2021). The biodiesel life cycle emits about 40%–65% less GHG emissions compared to the conventional diesel life cycle. Bioethanol technologies significantly impact GHG emissions based on the technology used. The total life-cycle emissions of corn-derived bioethanol may be around 80%–90% lower than those of competing fossil fuels. Bioethanol derived from sugar cane can reduce fossil-fuel emissions by 75%–80% (Hill et al., 2006; Merritt and Barragán-Ocaña, 2023). The ultimate outcomes are influenced significantly by the quantities and kinds of fossil fuels that are used as energy carriers throughout the life cycle for the production, transportation, and processing of the feedstock. Additionally, non-CO_2 emissions like N_2O produced during the cultivation period must be considered. Furthermore, the efficiency of the conversion process, the utilization of biomass for energy, and the yield of feedstock during cultivation are crucial factors. The energy and mass balances are affected by the bioenergy plant's capacity and the project's size. Large-scale projects may lead to significant land use changes that may impact carbon reserves in the soil (Pan et al., 2022).

Several studies have focused on understanding and addressing carbon dioxide (CO_2) emissions and their impact on climate change through various lenses. For instance, Yang et al. (2020) mapped CO_2 emissions in China, driven by global consumption patterns, highlighting key manufacturing hubs as carbon footprint hotspots and emphasizing the need for collaborative mitigation efforts. Yuliani et al. (2023) underscored the significant contribution of LPG gas consumption in small industries to CO_2 emissions, advocating for sustainable practices and energy-efficient measures. Shigetomi et al. (2021) quantified household carbon footprints associated with lifestyle choices and socioeconomics, providing insights into potential reductions in CO_2 emissions based on specific actions. Additionally, Hu et al. (2021) proposed strategies for controlling energy consumption and CO_2 emissions in Guangdong Province, emphasizing the importance of integrated approaches and data-supported decision-making to effectively address urban industry emissions (Yang et al., 2020; Hu et al., 2021; Shigetomi et al., 2021; Yuliani et al., 2023).

Studies have provided valuable insights into the environmental impacts and sustainability of biorefinery operations. Ebrahimian and Mohammadi (2023) compared the environmental impacts of bio-based 2,3-butanediol (BDO) with its fossil-based counterpart, showing significant improvements in human health, climate change, and resource scarcity by utilizing wood residues as sustainable feedstocks for BDO production. Berger et al. (2022) focused on integrating waste and emission valorization technologies with biorefinery production, highlighting the technical feasibility and economic value of CO_2 recycling and electrolysis for syngas production in bio-IBN integrated systems. Giwa et al. (2022) evaluated the energy consumption and GHG emissions of an integrated multiproduct biorefinery, emphasizing the importance of process optimization for emission reduction. Bhatt et al. (2022) studied the air pollutant emissions from an advanced biorefinery producing renewable diesel blend stock (RDB) from lignocellulosic biomass, underscoring the need for mitigating air pollutant emissions to promote sustainable practices. Additionally, Wang et al. (2023) explored the role of microalgae in biorefineries for carbon dioxide fixation and bioresource production, assessing economic and environmental impacts. Arora et al. (2020) characterized the GHG emissions from CO_2 supply for algal biorefineries, while Van Roijen and Miller (2024) discussed production pathways for GHG-negative bio-based plastics from various feedstocks. These studies collectively contribute to understanding and promoting sustainability in biorefinery operations.

13.7.2 Methane and nitrous oxide emissions

Methane (CH_4), along with nitrous oxide (N_2O), is one of the two significant GHGs (GHG) that are increasingly recognized for their role in global change. Methane has a relative global warming tendency 25 times greater than CO_2 over 100 years. Its levels have more than doubled since 1800, and it accounts for around 20% of global radiative forcing. Natural wetlands and human activities, including biomass burning, are the main sources of CH_4, whereas upland soils are the primary CH_4 sink (Kabange et al., 2023). N_2O has a global warming potential 298 times greater than CO_2 over 100 years and accounts for around 7% of radiative forcing. The atmospheric concentration of N_2O has risen by 18% from preindustrial levels, with a consistent annual increase of 0.26% over the last several decades (Griffis et al., 2017; Ussiri and Lal, 2012). The rise in atmospheric N_2O levels is mainly linked to reactive nitrogen inputs via synthetic nitrogen fertilizer and animal manure applications, farmland expansion, as well as activities related to fossil-fuel combustion alongside biomass burning (Aryal et al., 2022). N_2O generation and consumption in soils occur via both living organisms and nonliving processes. Worldwide N_2O emissions are very uncertain, in the range of 6.7–36.6 TgN/yr, in comparison to worldwide CH_4 emission estimates. Various environmental conditions impact the generation and use of CH_4 as well as N_2O (Saikawa et al., 2014).

Global warming might lead to an increase in net CH_4 absorption in upland ecosystems by stimulating microbial CH_4 oxidation and enhancing CH_4 diffusivity under drier soil conditions (Guo et al., 2023). Additionally, global warming generally leads to higher N_2O emissions in terrestrial ecosystems by boosting nitrifier activity as well as nitrogen supply via mineralization. Yet, it may also lower N_2O emissions by causing soil dryness and promoting plant growth and nitrogen absorption (Joos et al., 2020). Increased atmospheric CO_2 levels may have varying effects on CH_4 intake in upland habitats and can enhance CH_4 emissions within water-saturated environments by enhancing photosynthesis and carbon input into the soil, which in turn promotes the development of methanogenic bacteria (D'Imperio et al., 2023). Under high nitrogen supply, elevated atmospheric CO_2 can greatly increase N_2O emissions by raising soil moisture and soil labile carbon levels. However, in nitrogen-limited ecosystems, the impact may be minimal or even negative because enhanced plant growth could reduce nitrogen availability for nitrifiers and denitrifiers. Nitrogen fertilization may significantly reduce methane consumption in grassland and forest ecosystems and may either decrease or increase methane emissions in rice fields, depending on the nitrogen fertilizer quantity, while also boosting nitrous oxide emissions. Furthermore, the interactions between various components are equally significant (Butterbach-Bahl et al., 2013). Tropospheric ozone pollution may decrease photosynthesis, leading to a reduction in CH_4 emissions, and either stimulate or reduce N_2O emissions by affecting litter amount and quality. Nitrogen deposition and increased atmospheric CO_2 were shown to collaboratively decrease CH_4 emissions from wetlands and boost CH_4 absorption in upland soil. Another study determined that temperature and elevated atmospheric CO_2 jointly altered the seasonal fluctuation of CH_4 emissions. The interaction between atmospheric CO_2 and nitrogen deposition may synergistically enhance soil-accessible nitrogen and labile carbon, leading to a significant rise in soil N_2O emissions (Tian et al., 2015).

Research on 15N-nitrate fed to silage and released as nitrogen oxides (NO_X) or N_2O shows losses ranging from 0.9% to 2% as N_2O. There are concerns that these measurements may not fully capture the actual nitrate conversion owing to the methods used (Butterbach-Bahl et al., 2013). Recent research found high levels of NO_X near silos during feed-out, although the overall projected nitrogen loss was less than 0.0001% of the total biomass nitrogen. This indicates that net N_2O emissions during end-of-storage activities may be minimal. Regrettably, most data in forage and silage literature lack information on either the total gas produced or the total biomass from which gases were released, or both (Ribaudo et al., 2011; Wang et al., 2021). It is challenging to accurately determine the exact amount of GHG released during storage. Research on GHG emissions from the decomposition of various biomass sources has reached comparable findings. Manure-derived N_2O emissions may provide from 0.1% to 2.6% of the total nitrogen in biomass. Research on GHG emissions from composting mixed hay found that 0.5% of the original nitrogen was lost as N_2O and 1.7% of the initial carbon was lost as CH_4. The extent of decomposition in manure and hay composting may vary compared to biofuel feedstock storage, but the number of emissions from lost biomass might be equal (Geronimo et al., 2022; Nordahl et al., 2023; Wang et al., 2021).

To effectively reach the goal of Net Zero emissions by the year 2050, addressing both CH_4 and N_2O emissions is crucial. These strategies align with the United Nations Framework Convention on Climate Change (UNFCCC) Paris Agreement, aiming to keep the global temperature rise below 1.5°C (Sun et al., 2021). Methane, with a shorter atmospheric lifespan of approximately 9–12.4 years, offers a rapid and effective means for short-term climate warming mitigation. The United Nations Environment Programme (UNEP) and the Climate and Clean Air Coalition (CCAC) report highlights the cost-effectiveness of methane mitigation, especially in the context of eliminating fossil fuel-related climate impacts (Vellinga and Groenestein, 2022). However, progress in reducing methane emissions, particularly from gas fields, remains insufficient, underscoring the need for improved monitoring and leak prevention as outlined in the UK's Sixth Carbon Budget. Conversely, nitrous oxide, a less discussed yet significant GHG, has contributed about 0.2 W/m^2 of radiative forcing since the year 1750 (Sun et al., 2021). Its atmospheric concentration has risen sharply, primarily due to agricultural practices and the use of nitrogen fertilizers. CH_4 and N_2O require targeted mitigation strategies, although these can be complex and varied, from improving wastewater treatment processes to enhancing agricultural fertilizer use efficiency. Both gases underscore the intertwined challenges of climate change, agriculture, and land use, necessitating a nuanced approach that considers the sociocultural aspects of mitigation (Hassan et al., 2022; Hristov et al., 2013).

Methane and N_2O mitigation present particular opportunities in developing countries, which are vulnerable to climate change and face air pollution challenges. Simple, cost-effective strategies can significantly impact, emphasizing the need for technological solutions and capacity-building (Nisbet et al., 2021). Moreover, reducing methane emissions also yields public health benefits by decreasing ozone formation and associated respiratory issues. While methane's potent warming potential and its impact on air quality make it a critical target for immediate action, N_2O's long atmospheric lifetime and rising emissions highlight a pressing, yet often overlooked, area of concern (Forabosco et al., 2017). Both require global cooperation, effective policies, and a commitment to transitioning away from fossil-fuel dependence and toward sustainable agricultural practices. The synergistic reduction of CO_2, methane, and N_2O emissions is essential for a comprehensive approach to mitigating global warming and achieving the ambitious Net Zero emissions goal by 2050 (Tanneberger et al., 2021).

13.8 Conclusion

Biorefineries serve a significant function in mitigating the environmental repercussions stemming from the overreliance on fossil fuels. These facilities undergo continuous evolution due to ongoing research into novel pathways for biomass upgrading. Carbon dioxide storage and utilization, valorization of stillage, utilization of multiple feedstocks, and proposing new products based on emerging trends form the foundation for enhancing biorefinery sustainability. This chapter provides insights into key challenges and opportunities aimed at enhancing the sustainability of biorefineries. Specifically, it underscores the importance of minimizing waste streams, exploring novel pathways for biomass upgrading, and integrating multiple feedstocks to optimize biorefinery design. Meanwhile, this chapter offers valuable insights and information for guiding the development of biorefinery initiatives and promoting environmental conservation efforts.

AI disclosure

During the preparation of this work, the author(s) used Grammarly in order to correct any grammar mistakes. After using this tool/service, the author(s) reviewed and edited the content as needed and take(s) full responsibility for the content of the publication.

References

Abubakar Jumare, I., Bhandari, R., Zerga, A., 2019. Environmental life cycle assessment of grid-integrated hybrid renewable energy systems in Northern Nigeria. Sustainability 11 (21), 5889. Available at https://doi.org/10.3390/su11215889.

Abubakar, I.R., Maniruzzaman, K.M., Dano, U.L., AlShihri, F.S., AlShammari, M.S., Ahmed, S.M.S., et al., 2022. Environmental sustainability impacts of solid waste management practices in the Global South. International Journal of Environmental Research and Public Health 19 (19). Available at https://doi.org/10.3390/ijerph191912717, http://www.mdpi.com/journal/ijerph.

Acosta, L., Magcale-Macandog, D., Kumar, K., Cui, X., Eugenio, E., Macandog, P., et al., 2016. The role of bioenergy in enhancing energy, food and ecosystem sustainability based on societal perceptions and preferences in Asia. Agriculture 6 (2), 19. Available at https://doi.org/10.3390/agriculture6020019.

Ahorsu, R., Medina, F., Constantí, M., 2018. Significance and challenges of biomass as a suitable feedstock for bioenergy and biochemical production: a review. Energies 11 (12), 3366. Available at https://doi.org/10.3390/en11123366.

Alaloul, W.S., Altaf, M., Musarat, M.A., Faisal Javed, M., Mosavi, A., 2021. Systematic review of life cycle assessment and life cycle cost analysis for pavement and a case study. Sustainability 13 (8), 4377. Available at https://doi.org/10.3390/su13084377.

Ali, S.S., Abdelkarim, E.A., Elsamahy, T., Al-Tohamy, R., Li, F., Kornaros, M., et al., 2023. Bioplastic production in terms of life cycle assessment: a state-of-the-art review. Environmental Science and Ecotechnology 15. Available at https://doi.org/10.1016/j.ese.2023.100254, www.journals.elsevier.com/environmental-science-and-ecotechnology.

Andrade, L.R.S., Felisardo, R.J.A., Cruz, I.A., Bilal, M., Iqbal, H.M.N., Mulla, S.I., et al., 2022. Integrated biorefinery and life cycle assessment of cassava processing residue–from production to sustainable evaluation. Plants 11 (24). Available at https://doi.org/10.3390/plants11243577, www.mdpi.com/journal/plants.

Andrejiová, M., Grincova, A., Marasová, D., 2020. Study of the percentage of greenhouse gas emissions from aviation in the EU-27 countries by applying multiple-criteria statistical methods. International Journal of Environmental Research and Public Health 17 (11), 3759. Available at https://doi.org/10.3390/ijerph17113759.

Arias, A., Feijoo, G., Moreira, M.T., 2023. Biorefineries as a driver for sustainability: key aspects, actual development and future prospects. Journal of Cleaner Production 418, 137925. Available at https://doi.org/10.1016/j.jclepro.2023.137925.

Arora, P., Chance, R., Hendrix, H., Realff, M.J., Thomas, V.M., Yuan, Y., 2020. Life cycle greenhouse gas emissions of different CO2supply options for an algal biorefinery. Journal of CO2 Utilization 40. Available at https://doi.org/10.1016/j.jcou.2020.101213, http://www.journals.elsevier.com/journal-of-co2-utilization/.

Aryal, B., Gurung, R., Camargo, A.F., Fongaro, G., Treichel, H., Mainali, B., et al., 2022. Nitrous oxide emission in altered nitrogen cycle and implications for climate change. Environmental Pollution 314. Available at https://doi.org/10.1016/j.envpol.2022.120272, https://www.journals.elsevier.com/environmental-pollution.

Asadikia, A., Rajabifard, A., Kalantari, M., 2021. Systematic prioritisation of SDGs: machine learning approach. World Development 140, 105269. Available at https://doi.org/10.1016/j.worlddev.2020.105269.

Awasthi, M.K., Ferreira, J.A., Sirohi, R., Sarsaiya, S., Khoshnevisan, B., Baladi, S., et al., 2021. A critical review on the development stage of biorefinery systems towards the management of apple processing-derived waste. Renewable and Sustainable Energy Reviews 143. Available at https://doi.org/10.1016/j.rser.2021.110972, https://www.journals.elsevier.com/renewable-and-sustainable-energy-reviews.

Bacchetta, S., 2023. From Rio to Paris: International Climate Change Treaties Between Consensus and Efficacy. Global Climate Justice. Available at https://www.e-ir.info/2023/02/07/from-rio-to-paris-international-climate-change-treaties-between-consensus-and-efficacy/.

Bałdowska-Witos, P., Kruszelnicka, W., Kasner, R., Tomporowski, A., Flizikowski, J., Kłos, Z., et al., 2020. Application of LCA method for assessment of environmental impacts of a polylactide (PLA) bottle shaping. Polymers 12 (2), 388. Available at https://doi.org/10.3390/polym12020388.

Barahmand, Z., Eikeland, M.S., 2022. Life cycle assessment under uncertainty: a scoping review. WORLD (Oakland, Calif.: 1993) 3 (3), 692–717. Available at https://doi.org/10.3390/world3030039.

Barragán-Ocaña, A., Merritt, H., Sánchez-Estrada, O.E., Méndez-Becerril, J.L., Longar-Blanco, M.D.P., Koller, M., 2023. Biorefinery and sustainability for the production of biofuels and value-added products: a trends analysis based on network and patent analysis. PLoS One 18 (1), e0279659. Available at https://doi.org/10.1371/journal.pone.0279659.

Bartkowiak, A., Bartkowiak, P., Kinelski, G., 2022. Efficiency of shaping the value chain in the area of the use of raw materials in agro-biorefinery in sustainable development. Energies 15 (17), 6260. Available at https://doi.org/10.3390/en15176260.

Belyanovskaya, A.I., Laratte, B., Rajput, V.D., Perry, N., Baranovskaya, N.V., 2020. The innovation of the characterisation factor estimation for LCA in the USETOX model. Journal of Cleaner Production 270. Available at https://doi.org/10.1016/j.jclepro.2020.122432, https://www.journals.elsevier.com/journal-of-cleaner-production.

Benavides, P.T., Cronauer, D.C., Adom, F., Wang, Z., Dunn, J.B., 2017. The influence of catalysts on biofuel life cycle analysis (LCA). Sustainable Materials and Technologies 11, 53–59. Available at https://doi.org/10.1016/j.susmat.2017.01.002.

Berger, N.J., Lindorfer, J., Fazeni, K., Pfeifer, C., 2022. The techno-economic feasibility and carbon footprint of recycling and electrolysing CO2 emissions into ethanol and syngas in an isobutene biorefinery. Sustainable Production and Consumption 32, 619–637. Available at https://doi.org/10.1016/j.spc.2022.05.014, http://www.journals.elsevier.com/sustainable-production-and-consumption/.

Bhakar, P., Singh, A.P., 2018. Life cycle assessment of groundwater supply system in a hyper-arid region of India. Procedia CIRP 69, 603–608. Available at https://doi.org/10.1016/j.procir.2017.11.050, http://www.sciencedirect.com/science/journal/22128271.

Bhatt, A.H., Zhang, Y., Davis, R., Heath, G., Ravi, V., 2022. Biorefinery upgrading of herbaceous biomass to renewable hydrocarbon fuels, Part 2: air pollutant emissions and permitting implications. Journal of Cleaner Production 362, 132409. Available at https://doi.org/10.1016/j.jclepro.2022.132409.

Biomass Cofiring: A Renewable Alternative for Utilities. U.S. Department of Energy, U.S. Department of Energy, 2000.

Bishop, G., Styles, D., Lens, P.N.L., 2021. Environmental performance comparison of bioplastics and petrochemical plastics: a review of life cycle assessment (LCA) methodological decisions. Resources, Conservation and Recycling 168, 105451. Available at https://doi.org/10.1016/j.resconrec.2021.105451.

Borole, A., 2015. Sustainable and efficient pathways for bioenergy recovery from low-value process streams via bioelectrochemical systems in biorefineries. Sustainability 7 (9), 11713–11726. Available at https://doi.org/10.3390/su70911713.

Burland, A., von Cossel, M., 2023. Towards managing biodiversity of European marginal agricultural land for biodiversity-friendly biomass production. Agronomy 13 (6), 1651. Available at https://doi.org/10.3390/agronomy13061651.

Bush, M.J., 2020. A planet in peril. Climate Change and Renewable Energy: How to End the Climate Crisis. Palgrave Macmillan, pp. 1–57. Available at https://doi.org/10.1007/978-3-030-15424-0_1.

Butterbach-Bahl, K., Baggs, E.M., Dannenmann, M., Kiese, R., Zechmeister-Boltenstern, S., 2013. Nitrous oxide emissions from soils: how well do we understand the processes and their controls? Philosophical Transactions of the Royal Society B: Biological Sciences 368 (1621). Available at https://doi.org/10.1098/rstb.2013.0122 20130122.

Cai, H., Han, J., Wang, M., Davis, R., Biddy, M., Tan, E., 2018. Life-cycle analysis of integrated biorefineries with co-production of biofuels and bio-based chemicals: co-product handling methods and implications. Biofuels, Bioproducts and Biorefining 12 (5), 815–833. Available at https://doi.org/10.1002/bbb.1893.

Casau, M., Dias, M.F., Matias, J.C.O., Nunes, L.J.R., 2022. Residual biomass: a comprehensive review on the importance, uses and potential in a circular bioeconomy approach. Resources 11 (4). Available at https://doi.org/10.3390/resources11040035, https://www.mdpi.com/2079-9276/11/4/35/pdf.

Cavalaglio, G., Mecca, I., Iodice, P., Giannoni, T., Gelosia, M., Nicolini, A., et al., 2023. Life cycle assessment of polyol production from lignin via organosolv and liquefaction treatments. Sustainability 15 (22), 15905. Available at https://doi.org/10.3390/su152215905.

Chaturvedi, T., Torres, A.I., Stephanopoulos, G., Thomsen, M.H., Schmidt, J.E., 2020. Developing process designs for biorefineries-definitions, categories, and unit operations. Energies 13 (6). Available at https://doi.org/10.3390/en13061493, https://www.mdpi.com/1996-1073/13/6/1493.

Cherubini, F., Jungmeier, G., Wellisch, M., Willke, T., Skiadas, I., Van Ree, R., et al., 2009. Toward a common classification approach for biorefinery systems. Biofuels, Bioproducts and Biorefining 3 (5), 534–546. Available at https://doi.org/10.1002/bbb.172.

Clauser, N.M., González, G., Mendieta, C.M., Kruyeniski, J., Area, M.C., Vallejos, M.E., 2021. Biomass waste as sustainable raw material for energy and fuels. Sustainability 13 (2), 794. Available at https://doi.org/10.3390/su13020794.

Cordonier Segger, M.-C., 2016. Advancing the Paris agreement on climate change for sustainable development. Cambridge International Law Journal 5 (2), 202–237. Available at https://doi.org/10.4337/cilj.2016.02.03.

D'Imperio, L., Li, B.B., Tiedje, J.M., Oh, Y., Christiansen, J.R., Kepfer-Rojas, S., et al., 2023. Spatial controls of methane uptake in upland soils across climatic and geological regions in Greenland. Communications Earth and Environment 4 (1). Available at https://doi.org/10.1038/s43247-023-01143-3, https://www.nature.com/commsenv/.

Degnachew, A., Hod, A., Soest, H.V., Vuuren, D.V., 2021. Mapping synergies and trades off to guide multi-level decision-making. Climate Change Measures and Sustainable Development Goals Book. PBL Publication.

Dervash, M.A., Yousuf, A., Ozturk, M., Bhat, R.A., 2023. Sources and Monitoring Tools of Atmospheric Carbon Dioxide. Springer Science and Business Media LLC, pp. 19–26. Available at http://doi.org/10.1007/978-3-031-26921-9_3.

Dhanaraju, M., Chenniappan, P., Ramalingam, K., Pazhanivelan, S., Kaliaperumal, R., 2022. Smart farming: internet of things (IoT)-based sustainable agriculture. Agriculture 12 (10), 1745. Available at https://doi.org/10.3390/agriculture12101745.

Dunn, J.B., 2019. Biofuel and bioproduct environmental sustainability analysis. Current Opinion in Biotechnology 57, 88–93. Available at https://doi.org/10.1016/j.copbio.2019.02.008, http://www.elsevier.com/locate/copbio.

Durojaye, O., Laseinde, T., Oluwafemi, I., 2020. A descriptive review of carbon footprint. Advances in Intelligent Systems and Computing 1026, 960–968. Available at https://doi.org/10.1007/978-3-030-27928-8_144, http://www.springer.com/series/11156.

Ebrahimian, F., Mohammadi, A., 2023. Assessing the environmental footprints and material flow of 2,3-butanediol production in a wood-based biorefinery. Bioresource Technology 387, 129642. Available at https://doi.org/10.1016/j.biortech.2023.129642.

Erdiwansyah, Gani, A., Mamat, R., Bahagia, Nizar, M., Yana, S., et al., 2024. Prospects for renewable energy sources from biomass waste in Indonesia. Case Studies in Chemical and Environmental Engineering 10, 100880. Available at https://doi.org/10.1016/j.cscee.2024.100880.

Esnouf, A., Heijungs, R., Coste, G., Latrille, É., Steyer, J.-P., Hélias, A., 2019. A tool to guide the selection of impact categories for LCA studies by using the representativeness index. Science of The Total Environment 658, 768–776. Available at https://doi.org/10.1016/j.scitotenv.2018.12.194.

Forabosco, F., Chitchyan, Z., Mantovani, R., 2017. Methane, nitrous oxide emissions and mitigation strategies for livestock in developing countries: a review. South African Journal of Animal Science 47 (3), 268–280. Available at https://doi.org/10.4314/sajas.v47i3.3, http://www.sasas.co.za/sites/sasas.co.za/files/Forabosco47Issue.pdf.

Forte, A., Zucaro, A., Basosi, R., Fierro, A., 2016. LCA of 1,4-butanediol produced via direct fermentation of sugars from wheat straw feedstock within a territorial biorefinery. Materials 9 (7), 563. Available at https://doi.org/10.3390/ma9070563.

Foster, V., Dim, J.U., Vollmer, S., Zhang, F., 2023. Understanding the challenge of decoupling transport-related CO2 emissions from economic growth in developing countries. World Development Sustainability 3, 100111. Available at https://doi.org/10.1016/j.wds.2023.100111.

Geronimo, C., Vergara, S.E., Chamberlin, C., Fingerman, K., 2022. Overlooked emissions: influence of environmental variables on greenhouse gas generation from woody biomass storage. Fuel 319, 123839. Available at https://doi.org/10.1016/j.fuel.2022.123839.

Giraldi-Díaz, M.R., De Medina-Salas, L., Castillo-González, E., De la Cruz-Benavides, M., 2018. Environmental impact associated with the supply chain and production of biodiesel from Jatropha curcas L. through life cycle analysis. Sustainability 10 (5), 1451. Available at https://doi.org/10.3390/su10051451.

Giwa, T., Akbari, M., Kumar, A., 2022. Assessment of the greenhouse gas emission footprint of a biorefinery over its life cycle. Energy Conversion and Management 271, 116327. Available at https://doi.org/10.1016/j.enconman.2022.116327.

Grande, L., Pedroarena, I., Korili, S.A., Gil, A., 2021. Hydrothermal liquefaction of biomass as one of the most promising alternatives for the synthesis of advanced liquid biofuels: a review. Materials 14 (18), 5286. Available at https://doi.org/10.3390/ma14185286.

Griffis, T.J., Chen, Z., Baker, J.M., Wood, J.D., Millet, D.B., Lee, X., et al., 2017. Nitrous oxide emissions are enhanced in a warmer and wetter world. Proceedings of the National Academy of Sciences of the United States of America 114 (45), 12081–12085. Available at https://doi.org/10.1073/pnas.1704552114, http://www.pnas.org/content/114/45/12081.full.pdf.

Guo, J., Feng, H., Peng, C., Chen, H., Xu, X., Ma, X., et al., 2023. Global climate change increases terrestrial soil CH4 emissions. Global Biogeochemical Cycles 37 (1). Available at https://doi.org/10.1029/2021GB007255, http://agupubs.onlinelibrary.wiley.com/hub/journal/10.1002/(ISSN)1944-9224/.

Hassan, M.U., Aamer, M., Mahmood, A., Awan, M.I., Barbanti, L., Seleiman, M.F., et al., 2022. Management strategies to mitigate N2O emissions in agriculture. Life (Chicago, Ill.: 1978) 12 (3). Available at https://doi.org/10.3390/life12030439, https://www.mdpi.com/2075-1729/12/3/439/pdf.

Hemathilake, D.M.K.S., Gunathilake, D.M.C.C., 2021. Agricultural productivity and food supply to meet increased demands. Future Foods: Global Trends, Opportunities, and Sustainability Challenges. Elsevier, Sri Lanka, pp. 539–553. Available at https://www.sciencedirect.com/book/9780323910019, http://doi.org/10.1016/B978-0-323-91001-9.00016-5.

Hernández-Chaverri, R.A., Buenrostro-Figueroa, J.J., Prado-Barragán, L.A., 2021. Biomass: biorefinery as a model to boost the bioeconomy in Costa Rica, a review. Agronomia Mesoamericana 32 (3), 1047–1070. Available at https://doi.org/10.15517/AM.V32I3.43736, https://revistas.ucr.ac.cr/index.php/agromeso/article/view/43736/47916.

Hill, J., Nelson, E., Tilman, D., Polasky, S., Tiffany, D., 2006. Environmental, economic, and energetic costs and benefits of biodiesel and ethanol biofuels. Proceedings of the National Academy of Sciences 103 (30), 11206–11210. Available at https://doi.org/10.1073/pnas.0604600103.

Holechek, J.L., Geli, H.M.E., Sawalhah, M.N., Valdez, R., 2022. A global assessment: can renewable energy replace fossil fuels by 2050? Sustainability 14 (8), 4792. Available at https://doi.org/10.3390/su14084792.

Homagain, K., Shahi, C., Luckai, N., Sharma, M., 2015. Life cycle environmental impact assessment of biochar-based bioenergy production and utilization in Northwestern

Ontario, Canada. Journal of Forestry Research 26 (4), 799–809. Available at https://doi.org/10.1007/s11676-015-0132-y.

Hristov, A.N., Oh, J., Firkins, J.L., Dijkstra, J., Kebreab, E., Waghorn, G., et al., 2013. SPECIAL TOPICS—Mitigation of methane and nitrous oxide emissions from animal operations: I. A review of enteric methane mitigation options1. Journal of Animal Science 91 (11), 5045–5069. Available at https://doi.org/10.2527/jas.2013-6583.

Hu, M., Chen, S., Wang, Y., Xia, B., Wang, S., Huang, G., 2021. Identifying the key sectors for regional energy, water and carbon footprints from production-, consumption- and network-based perspectives. Science of The Total Environment 764, 142821. Available at https://doi.org/10.1016/j.scitotenv.2020.142821.

Islam, S., Ponnambalam, S.G., Lam, H.L., 2016. Review on life cycle inventory: methods, examples and applications. Journal of Cleaner Production 136, 266–278. Available at https://doi.org/10.1016/j.jclepro.2016.05.144.

Janota, L., Vávrová, K., Bízková, R., 2023. Methodology for strengthening energy resilience with SMART solution approach of rural areas: local production of alternative biomass fuel within renewable energy community. Energy Reports 10, 1211–1227. Available at https://doi.org/10.1016/j.egyr.2023.07.057, http://www.journals.elsevier.com/energy-reports/.

Jeswani, H.K., Chilvers, A., Azapagic, A., 2020. Environmental sustainability of biofuels: a review. Proceedings of the Royal Society A: Mathematical, Physical and Engineering Sciences 476 (2243). Available at https://doi.org/10.1098/rspa.2020.0351.

Johannes, L.P., Xuan, T.D., 2024. Comparative analysis of acidic and alkaline pretreatment techniques for bioethanol production from perennial grasses. Energies 17 (5). Available at https://doi.org/10.3390/en17051048, http://www.mdpi.com/journal/energies/.

Joos, F., Spahni, R., Stocker, B.D., Lienert, S., Müller, J., Fischer, H., et al., 2020. N 2 O changes from the last glacial maximum to the preindustrial – part 2: terrestrial N 2 O emissions and carbon–nitrogen cycle interactions. Biogeosciences 17 (13), 3511–3543. Available at https://doi.org/10.5194/bg-17-3511-2020.

Kabange, N.R., Kwon, Y., Lee, S.M., Kang, J.W., Cha, J.K., Park, H., et al., 2023. Mitigating greenhouse gas emissions from crop production and management practices, and livestock: a review. Sustainability (Switzerland) 15 (22). Available at https://doi.org/10.3390/su152215889, http://www.mdpi.com/journal/sustainability/.

Kadiyala, A., Kommalapati, R., Huque, Z., 2016. Evaluation of the life cycle greenhouse gas emissions from different biomass feedstock electricity generation systems. Sustainability 8 (11), 1181. Available at https://doi.org/10.3390/su8111181.

Kajaste, R., 2014. Chemicals from biomass – managing greenhouse gas emissions in biorefinery production chains – a review. Journal of Cleaner Production 75, 1–10. Available at https://doi.org/10.1016/j.jclepro.2014.03.070.

Kalak, T., 2023. Potential use of industrial biomass waste as a sustainable energy source in the future. Energies 16 (4), 1783. Available at https://doi.org/10.3390/en16041783.

Kamalakkannan, S., Abeynayaka, A., Kulatunga, A.K., Singh, R.K., Tatsuno, M., Gamaralalage, P.J.D., 2022. Life cycle assessment of selected single-use plastic products towards evidence-based policy recommendations in Sri Lanka. Sustainability 14 (21), 14170. Available at https://doi.org/10.3390/su142114170.

KC, R., Aalto, M., Korpinen, O.-J., Ranta, T., Proskurina, S., 2020. Lifecycle assessment of biomass supply chain with the assistance of agent-based modelling. Sustainability 12 (5), 1964. Available at https://doi.org/10.3390/su12051964.

Keller, F., Lee, R.P., Meyer, B., 2020. Life cycle assessment of global warming potential, resource depletion and acidification potential of fossil, renewable and secondary feedstock for olefin production in Germany. Journal of Cleaner Production 250, 119484. Available at https://doi.org/10.1016/j.jclepro.2019.119484.

Khaleel, M.M., Alsharif, A., Khalefah Imbayah, I.I., 2022. Renewable energy technologies: recent advances and future predictions. African Journal of Advanced Pure and Applied Sciences 1 (3), 58–64. Available at https://aaasjournals.com/index.php/ajapas/article/view/52.

Khan, M.A.H., Bonifacio, S., Clowes, J., Foulds, A., Holland, R., Matthews, J.C., et al., 2021a. Investigation of biofuel as a potential renewable energy source. Atmosphere 12 (10). Available at https://doi.org/10.3390/atmos12101289, https://www.mdpi.com/2073-4433/12/10/1289/pdf.

Khan, N., Sudhakar, K., Mamat, R., 2021b. Role of biofuels in energy transition, green economy and carbon neutrality. Sustainability 13 (22), 12374. Available at https://doi.org/10.3390/su132212374.

Khoo, H.H., Eufrasio-Espinosa, R.M., Koh, L.S.C., Sharratt, P.N., Isoni, V., 2019. Sustainability assessment of biorefinery production chains: a combined LCA-supply chain approach. Journal of Cleaner Production 235, 1116–1137. Available at https://doi.org/10.1016/j.jclepro.2019.07.007, https://www.journals.elsevier.com/journal-of-cleaner-production.

Khoshnevisan, B., Tabatabaei, M., Tsapekos, P., Rafiee, S., Aghbashlo, M., Lindeneg, S., et al., 2020. Environmental life cycle assessment of different biorefinery platforms valorizing municipal solid waste to bioenergy, microbial protein, lactic and succinic acid. Renewable and Sustainable Energy Reviews 117, 109493. Available at https://doi.org/10.1016/j.rser.2019.109493.

Khoshnevisan, B., Duan, N., Tsapekos, P., Awasthi, M.K., Liu, Z., Mohammadi, A., et al., 2021. A critical review on livestock manure biorefinery technologies: sustainability, challenges, and future perspectives. Renewable and Sustainable Energy Reviews 135, 110033. Available at https://doi.org/10.1016/j.rser.2020.110033.

Kiemel, S., Rietdorf, C., Schutzbach, M., Miehe, R., 2022. How to simplify life cycle assessment for industrial applications—a comprehensive review. Sustainability 14 (23), 15704. Available at https://doi.org/10.3390/su142315704.

Koponen, K., Soimakallio, S., Kline, K.L., Cowie, A., Brandão, M., 2018. Quantifying the climate effects of bioenergy – Choice of reference system. Renewable and Sustainable Energy Reviews 81, 2271–2280. Available at https://doi.org/10.1016/j.rser.2017.05.292, https://www.journals.elsevier.com/renewable-and-sustainable-energy-reviews.

Kovacs, E., Hoaghia, M.-A., Senila, L., Scurtu, D.A., Varaticeanu, C., Roman, C., et al., 2022. Life cycle assessment of biofuels production processes in viticulture in the context of circular economy. agronomy 12 (6), 1320. Available at https://doi.org/10.3390/agronomy12061320.

Kumar, B., Bhardwaj, N., Agrawal, K., Chaturvedi, V., Verma, P., 2020. Current perspective on pretreatment technologies using lignocellulosic biomass: an emerging biorefinery concept. Fuel Processing Technology 199, 106244.https://doi.org/10.1016/j.fuproc.2019.106244.

Kumar, B., Verma, P., 2021. Life cycle assessment: blazing a trail for bioresources management. Energy Conversion and Management: X 10, 100063. Available at https://doi.org/10.1016/j.ecmx.2020.100063.

Leong, H.Y., Chang, C.K., Khoo, K.S., Chew, K.W., Chia, S.R., Lim, J.W., et al., 2021. Waste biorefinery towards a sustainable circular bioeconomy: a solution to global issues. Biotechnology for Biofuels 14 (1). Available at https://doi.org/10.1186/s13068-021-01939-5, http://www.biotechnologyforbiofuels.com/.

Liu, F., Dong, X., Zhao, X., Wang, L., 2021. Life cycle assessment of organosolv biorefinery designs with the complete use of biomass. Energy Conversion and Management 246, 114653. Available at https://doi.org/10.1016/j.enconman.2021.114653.

Lizundia, E., Luzi, F., Puglia, D., 2022. Organic waste valorisation towards circular and sustainable biocomposites. Green Chemistry 24 (14), 5429–5459. Available at https://doi.org/10.1039/d2gc01668k.

Lodato, C., Tonini, D., Damgaard, A., Fruergaard Astrup, T., 2020. A process-oriented life-cycle assessment (LCA) model for environmental and resource-related technologies (EASETECH). International Journal of Life Cycle Assessment 25 (1), 73–88. Available at https://doi.org/10.1007/s11367-019-01665-z, http://www.springerlink.com/content/0948-3349.

Lu, Y.-T., Lee, Y.-M., Hong, C.-Y., 2017. Inventory analysis and social life cycle assessment of greenhouse gas emissions from waste-to-energy incineration in Taiwan. Sustainability 9 (11), 1959. Available at https://doi.org/10.3390/su9111959.

Maier, J.M., Sowlati, T., Salazar, J., 2019. Life cycle assessment of forest-based biomass for bioenergy: a case study in British Columbia, Canada. Resources, Conservation and Recycling 146, 598–609. Available at https://doi.org/10.1016/j.resconrec.2019.02.035, www.elsevier.com/locate/resconrec.

Makepa, D.C., Chihobo, C.H., 2024. Barriers to commercial deployment of biorefineries: a multi-faceted review of obstacles across the innovation chain. Heliyon 10 (12). Available at https://doi.org/10.1016/j.heliyon.2024.e32649, https://www.sciencedirect.com/science/journal/24058440.

Martin, M., Elnour, M., Siñol, A.C., 2023. Environmental life cycle assessment of a large-scale commercial vertical farm. Sustainable Production and Consumption 40, 182–193. Available at https://doi.org/10.1016/j.spc.2023.06.020, http://www.journals.elsevier.com/sustainable-production-and-consumption/.

Mat Aron, N.S., Khoo, K.S., Chew, K.W., Show, P.L., Chen, W.H., Nguyen, T.H.P., 2020. Sustainability of the four generations of biofuels – a review. International Journal of Energy Research 44 (12), 9266–9282. Available at https://doi.org/10.1002/er.5557, http://onlinelibrary.wiley.com/journal/10.1002/(ISSN)1099-114X.

Mazur, Ł., Olenchuk, A., 2023. Life cycle assessment and building information modeling integrated approach: carbon footprint of masonry and timber-frame constructions in single-family houses. Sustainability 15 (21), 15486. Available at https://doi.org/10.3390/su152115486.

Merritt, H., Barragán-Ocaña, A., 2023. The impact of market factors on the development of eco-friendly energy technologies: the case of bioethanol. Clean Technologies and Environmental Policy 25 (2), 313–321. Available at https://doi.org/10.1007/s10098-021-02225-6.

Mesfun, S., Gustafsson, G., Larsson, A., Samavati, M., Furusjö, E., 2023. Electrification of biorefinery concepts for improved productivity—yield, economic and GHG performances. Energies 16 (21), 7436. Available at https://doi.org/10.3390/en16217436.

Mikhaylov, A., Moiseev, N., Aleshin, K., Burkhardt, T., 2020. Global climate change and greenhouse effect. Entrepreneurship and Sustainability Center, Russian Federation Entrepreneurship and Sustainability Issues 7 (4), 2897–2913. Available at https://doi.org/10.9770/jesi.2020.7.4(21), http://jssidoi.org/jesi/article/download/560.

Modahl, I.S., Brekke, A., Valente, C., 2015. Environmental assessment of chemical products from a Norwegian biorefinery. Journal of Cleaner Production 94, 247–259. Available at https://doi.org/10.1016/j.jclepro.2015.01.054, https://www.journals.elsevier.com/journal-of-cleaner-production.

Mohamed, U., Zhao, Y.-jie, Yi, Q., Shi, L.-juan, Wei, G.-qing, Nimmo, W., 2021. Evaluation of life cycle energy, economy and CO2 emissions for biomass chemical looping gasification to power generation. Renewable Energy 176, 366–387. Available at https://doi.org/10.1016/j.renene.2021.05.067.

Mohammad, R.E.A., Abdullahi, S.S.A., Muhammed, H.A., Musa, H., Habibu, S., Jagaba, A.H., et al., 2024. Recent technical and non-technical biorefinery development barriers and potential solutions for a sustainable environment: a mini review. Case Studies in Chemical and Environmental Engineering 9. Available at https://doi.org/10.1016/j.cscee.2023.100586, https://www.sciencedirect.com/science/journal/26660164.

Mohd Yusof, S.J.H., Roslan, A.M., Ibrahim, K.N., Abdullah, S.S.S., Zakaria, M.R., Hassan, M.A., et al., 2019. Life cycle assessment for bioethanol production from oil palm frond juice in an oil palm based biorefinery. Sustainability 11 (24), 6928. Available at https://doi.org/10.3390/su11246928.

Morales-Vera, R., Vásquez-Ibarra, L., Scott, F., Puettmann, M., Gustafson, R., 2022. Life cycle assessment of bioethanol production: a case study from poplar biomass growth in the U.S. Pacific Northwest. Fermentation 8 (12), 734. Available at https://doi.org/10.3390/fermentation8120734.

Moutik, B., Summerscales, J., Graham-Jones, J., Pemberton, R., 2023. Life cycle assessment research trends and implications: a bibliometric analysis. Sustainability 15 (18), 13408. Available at https://doi.org/10.3390/su151813408.

Nisbet, E.G., Dlugokencky, E.J., Fisher, R.E., France, J.L., Lowry, D., Manning, M.R., et al., 2021. Atmospheric methane and nitrous oxide: challenges alongthe path to Net Zero. Philosophical Transactions of the Royal Society A: Mathematical, Physical and Engineering Sciences 379 (2210). Available at https://doi.org/10.1098/rsta.2020.0457, http://rsta.royalsocietypublishing.org/.

Nordahl, S.L., Preble, C.V., Kirchstetter, T.W., Scown, C.D., 2023. Greenhouse gas and air pollutant emissions from composting. Environmental Science and Technology 57 (6), 2235–2247. Available at https://doi.org/10.1021/acs.est.2c05846, http://pubs.acs.org/journal/esthag.

Ogunleye, A., Flora, J., Berge, N., 2024. Life cycle assessment of hydrothermal carbonization: a review of product valorization pathways. Agronomy 14 (2), 243. Available at https://doi.org/10.3390/agronomy14020243.

Oreggioni, G.D., Monforti Ferraio, F., Crippa, M., Muntean, M., Schaaf, E., Guizzardi, D., et al., 2021. Climate change in a changing world: socio-economic and technological transitions, regulatory frameworks and trends on global greenhouse gas emissions from EDGAR v.5.0. Global Environmental Change 70, 102350. Available at https://doi.org/10.1016/j.gloenvcha.2021.102350.

Osman, A.I., Chen, L., Yang, M., Msigwa, G., Farghali, M., Fawzy, S., et al., 2023. Cost, environmental impact, and resilience of renewable energy under a changing climate: a review. Environmental Chemistry Letters 21 (2), 741–764. Available at https://doi.org/10.1007/s10311-022-01532-8, https://www.springer.com/journal/10311.

Ott, D., Goyal, S., Reuss, R., Gutzeit, H.O., Liebscher, J., Dautz, J., et al., 2023. LCA as decision support tool in the food and feed sector: evidence from R&D case studies. Environment Systems and Decisions 43 (1), 129–141. Available at https://doi.org/10.1007/s10669-022-09874-y.

Pacana, A., Siwiec, D., Bednárová, L., Petrovský, J., 2023. Improving the process of product design in a phase of life cycle assessment (LCA). Processes 11 (9), 2579. Available at https://doi.org/10.3390/pr11092579.

Palmeros Parada, M., Osseweijer, P., Posada Duque, J.A., 2017. Sustainable biorefineries, an analysis of practices for incorporating sustainability in biorefinery design. Industrial Crops and Products 106, 105–123. Available at https://doi.org/10.1016/j.indcrop.2016.08.052, www.elsevier.com/inca/publications/store/5/2/2/8/2/5.

Pan, S.Y., He, K.H., Lin, K.T., Fan, C., Chang, C.T., 2022. Addressing nitrogenous gases from croplands toward low-emission agriculture. npj Climate and Atmospheric Science 5 (1). Available at https://doi.org/10.1038/s41612-022-00265-3, https://www.nature.com/npjclimatsci/.

Pandey, D., Agrawal, M., Pandey, J.S., 2011. Carbon footprint: current methods of estimation. Environmental Monitoring and Assessment 178 (1-4), 135–160. Available at https://doi.org/10.1007/s10661-010-1678-y.

Perez-Almada, D., Galan-Martin, A., Contreras, M.D.M., Castri, E., 2023. Integrated techno-economic and environmental assessment of biorefineries: review and future research directions. Sustainable Energy Fuels 7. Available at https://doi.org/10.1039/D3E00405H.

Petit, G., Sablayrolles, C., Yannou-Le Bris, G., 2018. Combining eco-social and environmental indicators to assess the sustainability performance of a food value chain: a case study. Journal of Cleaner Production 191, 135–143. Available at https://doi.org/10.1016/j.jclepro.2018.04.156, https://www.journals.elsevier.com/journal-of-cleaner-production.

Pfister, S., Boulay, A.M., Berger, M., Hadjikakou, M., Motoshita, M., Hess, T., et al., 2017. Understanding the LCA and ISO water footprint: a response to Hoekstra (2016) "A critique on the water-scarcity weighted water footprint in LCA" Ecological Indicators 72, 352–359. Available at https://doi.org/10.1016/j.ecolind.2016.07.051, http://www.elsevier.com/locate/ecolind.

Popp, J., Kovács, S., Oláh, J., Divéki, Z., Balázs, E., 2021. Bioeconomy: biomass and biomass-based energy supply and demand. New Biotechnology 60, 76–84. Available at https://doi.org/10.1016/j.nbt.2020.10.004, http://www.elsevier.com.

Prioux, N., Ouaret, R., Belaud, J.P., 2022. Machine learning based framework for biorefinery environmental assessment. Italian Association of Chemical Engineering - AIDIC, France Chemical Engineering Transactions 96, 517–522. Available at https://doi.org/10.3303/CET2296087, https://www.aidic.it/cet/22/96/087.pdf.

Quek, A., Balasubramanian, R., 2014. Life cycle assessment of energy and energy carriers from waste matter - a review. Journal of Cleaner Production 79, 18–31. Available at https://doi.org/10.1016/j.jclepro.2014.05.082, https://www.journals.elsevier.com/journal-of-cleaner-production.

Raschke, A., Hernandez-Suarez, J.S., Nejadhashemi, A.P., Deb, K., 2021. Multidimensional aspects of sustainable biofuel feedstock production. Sustainability 13 (3), 1424. Available at https://doi.org/10.3390/su13031424.

Rashid, S.S., Harun, S.N., Hanafiah, M.M., Razman, K.K., Liu, Y.Q., Tholibon, D.A., 2023. Life cycle assessment and its application in wastewater treatment: a brief overview. Processes 11 (1). Available at https://doi.org/10.3390/pr11010208, http://www.mdpi.com/journal/processes.

Ravikumar, D., Keoleian, G.A., Walzberg, J., Heath, G., Heller, M.C., 2024. Advancing environmental assessment of the circular economy: challenges and opportunities. Resources, Conservation and Recycling Advances 21. Available at https://doi.org/10.1016/j.rcradv.2024.200203, https://www.sciencedirect.com/science/journal/26673789.

Ribaudo, M., Hansen, L.R., Livingston, M.J., Mosheim, R., Williamson, J., Delgado, J., 2011. Nitrogen in agricultural systems: implications for conservation policy. SSRN Electronic Journal. Available at https://doi.org/10.2139/ssrn.2115532.

Saavedra del Oso, M., Mauricio-Iglesias, M., Hospido, A., Steubing, B., 2023. Prospective LCA to provide environmental guidance for developing waste-to-PHA biorefineries. Journal of Cleaner Production 383, 135331. Available at https://doi.org/10.1016/j.jclepro.2022.135331.

Sabet, H., Moghaddam, S.S., Ehteshami, M., 2023. A comparative life cycle assessment (LCA) analysis of innovative methods employing cutting-edge technology to improve sludge reduction directly in wastewater handling units. Journal of Water Process Engineering 51, 103354. Available at https://doi.org/10.1016/j.jwpe.2022.103354.

Sadaghiani, S., Mafakheri, F., Chen, Z., 2023. Life cycle assessment of bioenergy production using wood pellets: a case study of remote communities in Canada. Energies 16 (15), 5697. Available at https://doi.org/10.3390/en16155697.

Saikawa, E., Prinn, R.G., Dlugokencky, E., Ishijima, K., Dutton, G.S., Hall, B.D., et al., 2014. Global and regional emissions estimates for N 2 O. Atmospheric Chemistry and Physics 14 (9), 4617–4641. Available at https://doi.org/10.5194/acp-14-4617-2014.

Saleem, M., 2022. Possibility of utilizing agriculture biomass as a renewable and sustainable future energy source. Heliyon 8 (2), e08905. Available at https://doi.org/10.1016/j.heliyon.2022.e08905.

Sanders, J.P.M., Bos, H.L., 2013. Raw material demand and sourcing options for the development of a bio-based chemical industry in Europe.: Part 2: sourcing options. Biofuels, Bioproducts and Biorefining 7 (3), 260–272. Available at https://doi.org/10.1002/bbb.1392.

Saranya, G., Ramachandra, T.V., 2020. Life cycle assessment of biodiesel from estuarine microalgae. Energy Conversion and Management: X 8, 100065. Available at https://doi.org/10.1016/j.ecmx.2020.100065.

Sarkodie, S.A., 2021. Environmental performance, biocapacity, carbon & ecological footprint of nations: drivers, trends and mitigation options. Science of The Total Environment 751, 141912. Available at https://doi.org/10.1016/j.scitotenv.2020.141912.

Scrucca, F., Barberio, G., Fantin, V., Porta, P.L., Barbanera, M., 2021. Carbon footprint: concept, methodology and calculation. Environmental Footprints and Eco-Design of Products and Processes. Springer, Italy, pp. 1–31. Available at springer.com/series/13340, http://doi.org/10.1007/978-981-15-9577-6_1.

Shabir, I., Dash, K.K., Dar, A.H., Pandey, V.K., Fayaz, U., Srivastava, S., Nisha, R., 2023. Carbon footprints evaluation for sustainable food processing system development: A comprehensive review. Future Foods 7, 100215. Available at https://doi.org/10.1016/j.fufo.2023.100215.

Shigetomi, Y., Kanemoto, K., Yamamoto, Y., Kondo, Y., 2021. Quantifying the carbon footprint reduction potential of lifestyle choices in Japan. Environmental Research Letters 16 (6), 064022. Available at https://doi.org/10.1088/1748-9326/abfc07.

Sikder, M., Wang, C., Yao, X., Huai, X., Wu, L., KwameYeboah, F., et al., 2022. The integrated impact of GDP growth, industrialization, energy use, and urbanization on CO2 emissions in developing countries: evidence from the panel ARDL approach. Science of The Total Environment 837, 155795. Available at https://doi.org/10.1016/j.scitotenv.2022.155795.

Singh, N., Singhania, R.R., Nigam, P.S., Dong, C.D., Patel, A.K., Puri, M., 2022a. Global status of lignocellulosic biorefinery: challenges and perspectives. Bioresource Technology 344. Available at https://doi.org/10.1016/j.biortech.2021.126415, www.elsevier.com/locate/biortech.

Singh, R., Langyan, S., Rohtagi, B., Darjee, S., Khandelwal, A., Shrivastava, M., et al., 2022b. Production of biofuels options by contribution of effective and suitable enzymes: technological developments and challenges. Materials Science for Energy Technologies 5, 294–310. Available at https://doi.org/10.1016/j.mset.2022.05.001.

Slotyuk, L., Part, F., Schlegel, M.-C., Akkerman, F., 2024. Life cycle assessment of the domestic micro heat and power generation proton exchange membrane fuel cell in comparison with the gas condensing boiler plus electricity from the grid. Sustainability 16 (6), 2348. Available at https://doi.org/10.3390/su16062348.

Solaymani, S., 2022. CO2 emissions and the transport sector in Malaysia. Frontiers in Environmental Science 9. Available at https://doi.org/10.3389/fenvs.2021.774164.

Soleymani Angili, T., Grzesik, K., Rödl, A., Kaltschmitt, M., 2021. Life cycle assessment of bioethanol production: a review of feedstock, technology and methodology. Energies 14 (10), 2939. Available at https://doi.org/10.3390/en14102939.

Spekreijse, J., Weide, T., Hageman, S.P.W., Reumerman, P., Wetter, C., Brügging, E., et al., 2019. Life cycle assessment on a biorefinery approach to pyrolysis oil for wood modification treatment. Applied Sciences 9 (20), 4233. Available at https://doi.org/10.3390/app9204233.

Stančin, H., Strezov, V., Mikulčić, H., 2023. Life cycle assessment of alternative fuel production by co-pyrolysis of waste biomass and plastics. Journal of Cleaner Production 414, 137676. Available at https://doi.org/10.1016/j.jclepro.2023.137676.

Sun, T., Ocko, I.B., Sturcken, E., Hamburg, S.P., 2021. Path to net zero is critical to climate outcome. Scientific Reports 11 (1). Available at https://doi.org/10.1038/s41598-021-01639-y, www.nature.com/srep/index.html.

Takkellapati, S., Li, T., Gonzalez, M.A., 2018. An overview of biorefinery-derived platform chemicals from a cellulose and hemicellulose biorefinery. Clean Technologies and Environmental Policy 20 (7), 1615–1630. Available at https://doi.org/10.1007/s10098-018-1568-5, http://springerlink.metapress.com/app/home/journal.asp?wasp=46wa6aej7g0jwmac279x&referrer=parent&backto=browsepublicationsresults,91,542.

Talwar, N., Holden, N.M., 2022. The limitations of bioeconomy LCA studies for understanding the transition to sustainable bioeconomy. International Journal of Life Cycle Assessment 27 (5), 680–703. Available at https://doi.org/10.1007/s11367-022-02053-w, http://www.springerlink.com/content/0948-3349.

Tanneberger, F., Abel, S., Couwenberg, J., Dahms, T., Gaudig, G., Günther, A., et al., 2021. Towards net zero co2 in 2050: an emission reduction pathway for organic soils in germany. Mires and Peat 27, 1–17. Available at https://doi.org/10.19189/MaP.2020.SNPG.StA.1951, http://mires-and-peat.net/pages/volumes/map27/map2705.php.

Tian, H., Chen, G., Lu, C., Xu, X., Ren, W., Zhang, B., et al., 2015. Global methane and nitrous oxide emissions from terrestrial ecosystems due to multiple environmental changes. Ecosystem Health and Sustainability 1 (1), 1–20. Available at https://doi.org/10.1890/ehs14-0015.1.

Upadhyay, R.K., 2020. Markers for global climate change and its impact on social, biological and ecological systems: a review. American Journal of Climate Change 09 (03), 159–203. Available at https://doi.org/10.4236/ajcc.2020.93012.

Ussiri, D., Lal, R., 2012. Soil Emission of Nitrous Oxide and its Mitigation. Springer Netherlands, United States, pp. 1–378. Available at http://doi.org/10.1007/978-94-007-5364-8, http://doi.org/10.1007/978-94-007-5364-8_1.

Valdez-Valdes, S.A., Tejeda-Benitez, L.P., González-Delgado, Á.D., 2024. Assessing the environmental impacts of the valorization of creole-antillean avocado via an extractive-based biorefinery in the Montes de María Region. Sustainability 16 (24), 11057. Available at https://doi.org/10.3390/su162411057.

Van Roijen, E., Miller, S.A., 2024. Towards the production of net-negative greenhouse gas emission bio-based plastics from 2nd and 3rd generation feedstocks. Journal of Cleaner Production 445, 141203. Available at https://doi.org/10.1016/j.jclepro.2024.141203.

Vega, G.C., Sohn, J., Bruun, S., Olsen, S.I., Birkved, M., 2019. Maximizing environmental impact savings potential through innovative biorefinery alternatives: an application of the TM-LCA framework for regional scale impact assessment. Sustainability 11 (14), 3836. Available at https://doi.org/10.3390/su11143836.

Vellinga T., K. 2022. Groenestein, 10 Questions and Answers about Methane, a Short-Lived Greenhouse Gas.

Vural Gursel, I., Elbersen, B., Meesters, K.P.H., van Leeuwen, M., 2022. Defining circular economy principles for biobased products. Sustainability 14 (19), 12780. Available at https://doi.org/10.3390/su141912780.

Wan Osman, W.N.A., Rosli, M.H., Mazli, W.N.A., Samsuri, S., 2024. Comparative review of biodiesel production and purification. Carbon Capture Science and Technology 13. Available at https://doi.org/10.1016/j.ccst.2024.100264, https://www.sciencedirect.com/science/journal/27726568.

Wang, J., You, S., Lu, Z., Chen, R., Xu, F., 2020. Life cycle assessment of bio-based levoglucosan production from cotton straw through fast pyrolysis. Bioresource Technology 307, 123179. Available at https://doi.org/10.1016/j.biortech.2020.123179.

Wang, C., Amon, B., Schulz, K., Mehdi, B., 2021. Factors that influence nitrous oxide emissions from agricultural soils as well as their representation in simulation models: a review. Agronomy 11 (4), 770. Available at https://doi.org/10.3390/agronomy11040770.

Wang, K., Tester, J.W., 2023. Sustainable management of unavoidable biomass wastes. Green Energy and Resources 1 (1), 100005. Available at https://doi.org/10.1016/j.gerr.2023.100005.

Wang, Y., Yang, S., Liu, J., Wang, J., Xiao, M., Liang, Q., et al., 2023. Realization process of microalgal biorefinery: the optional approach toward carbon net-zero emission. Science of The Total Environment 901, 165546. Available at https://doi.org/10.1016/j.scitotenv.2023.165546.

Wang, M., Ye, X., Bi, H., Shen, Z., 2024. Microalgae biofuels: illuminating the path to a sustainable future amidst challenges and opportunities. Biotechnology for Biofuels and Bioproducts 17 (1). Available at https://doi.org/10.1186/s13068-024-02461-0.

Woods, J.S., Veltman, K., Huijbregts, M.A.J., Verones, F., Hertwich, E.G., 2016. Towards a meaningful assessment of marine ecological impacts in life cycle assessment (LCA). Environment International 89-90, 48–61. Available at https://doi.org/10.1016/j.envint.2015.12.033, www.elsevier.com/locate/envint.

Xu, H., Ou, L., Li, Y., Hawkins, T.R., Wang, M., 2022. Life cycle greenhouse gas emissions of biodiesel and renewable diesel production in the United States. Environmental Science and Technology 56 (12), 7512–7521. Available at https://doi.org/10.1021/acs.est.2c00289, http://pubs.acs.org/journal/esthag.

Yadav, S., Samadhiya, A., Kumar, A., Majumdar, A., Garza-Reyes, J.A., Luthra, S., 2023. Achieving the sustainable development goals through net zero emissions: innovation-driven strategies for transitioning from incremental to radical lean, green and digital technologies. Resources, Conservation and Recycling 197, 107094. Available at https://doi.org/10.1016/j.resconrec.2023.107094.

Yang, Y., Qu, S., Cai, B., Liang, S., Wang, Z., Wang, J., et al., 2020. Mapping global carbon footprint in China. Nature Communications 11 (1). Available at https://doi.org/10.1038/s41467-020-15883-9.

Yao, Y., Xu, P., Li, J., Hu, H., Qi, Q., 2024. Advancements and applications of life cycle assessment in slope treatment: a comprehensive review. Sustainability 16 (1), 398. Available at https://doi.org/10.3390/su16010398.

Yin, T., Huhe, T., Li, X., Wang, Q., Lei, T., Zhou, Z., 2024. Research on life cycle assessment and performance comparison of bioethanol production from various biomass feedstocks. Sustainability 16 (5), 1788. Available at https://doi.org/10.3390/su16051788.

Yu, Z., Ma, H., Liu, X., Wang, M., Wang, J., 2022. Review in life cycle assessment of biomass conversion through pyrolysis-issues and recommendations. Green Chemical Engineering 3 (4), 304–312. Available at https://doi.org/10.1016/j.gce.2022.08.002.

Yuliani, A.R., Sari, M.M., Suryawan, I.W.K., 2023. Carbon footprint assessment of LPG gas usage in small industries: a case study of Sami Laris Swalayan Shopping Center. Journal of Earth and Marine Technology (JEMT) 3 (2), 98–103. Available at https://doi.org/10.31284/j.jemt.2023.v3i2.4509.

Zhang, Q., Xiao, J., Xue, J., Zhang, L., 2020. Quantifying the effects of biochar application on greenhouse gas emissions from agricultural soils: a global meta-analysis. Sustainability 12 (8), 3436. Available at https://doi.org/10.3390/su12083436.

Zoppi, G., Tito, E., Bianco, I., Pipitone, G., Pirone, R., Bensaid, S., 2023. Life cycle assessment of the biofuel production from lignocellulosic biomass in a hydrothermal liquefaction – aqueous phase reforming integrated biorefinery. Renewable Energy 206, 375–385. Available at https://doi.org/10.1016/j.renene.2023.02.011.

Index

Note: Page numbers followed by "*f*" and "*t*" refer to figures and tables, respectively.

C

T

9780443247583